Secondary metabolism
in plants and animals

Secondary metabolism

in plants and animals

Dr. rer. nat. habil. MARTIN LUCKNER

Professor für Pharmazeutische Biologie an der Sektion Pharmazie
der Martin-Luther-Universität Halle-Wittenburg

Translated by Mr. T. N. Vasudevan
with the assistance of Dr. J. L. Wray

ACADEMIC PRESS
NEW YORK   1972

First published in Germany as
DER SEKUNDÄRSTOFFWECHSEL IN PFLANZE UND TIER
in 1969 by VEB Gustav Fischer Verlag
Copyright 1969 by VEB Gustav Fischer Verlag, Jena

Published in the United States by
ACADEMIC PRESS, INC.,
111 *Fifth Avenue, New York, N.Y.* 10003
Library of Congress Catalog Card Number 76–182558
For Academic Press, Inc.: ISBN 0–12–459050
Printed in Great Britain

Dedicated to my teacher
Professor Dr. h.c. mult. Kurt Mothes

Foreword

Professor L. Fowden F.R.S.

Books in English concerned with secondary metabolic products are few in number: they are either specialist texts devoted to particular biogenetic families of compounds or, if broader in scope, are edited versions of symposium contributions that lack the continuity associated with a single-author book. I was then delighted when Professor Martin Luckner informed me that an English translation of his excellent book *The Secondary Metabolism of Plants and Animals* – first published in German in 1969 – had been commissioned and I have great pleasure in introducing Professor Luckner and his book to an enlarged readership.

For the past 15 years, the ancient city of Halle in Germany has housed a remarkable group of scientists, who have dedicated their energies to unravelling the many and varied problems encountered in the biogenesis of natural products. Led and inspired by their doyen, Professor Emeritus Kurt Mothes, these young scientists achieved a notable inter-disciplinary co-operation that involved the Institutes of Botany and Pharmacy of the Martin-Luther University and the German Academy of Sciences Institute of Plant Biochemistry. Luckner joined the Martin-Luther University in 1954 as a student of pharmacy and later came under the direct influence of Professor Mothes as he began research work on the biosynthesis of alkaloids in 1959. This has remained his principal interest, but now he is recognized also as an authority on methods of drug assay and standardization: indeed, his first venture in textbook authorship was his successful *Prüfung von Drogen*, published in 1966 by VEB Gustav Fischer Verlag of Jena, Last year, at the age of 35, he became Professor of Pharmaceutical Biology in his University, an appointment appropriately marking his scientific distinction and scholarship.

When preparing the present book, Professor Luckner derived both encouragement and advice from many colleagues in Halle, able to augment and complement his own expertise. The final product is a most useful text describing the structural chemistry and metabolic reactions of all major classes of secondary product. However, most books give a little emphasis to certain special interests of the author and this one is no exception; the reader will notice that secondary nitrogenous compounds feature predominantly in the second half of the book. I believe the emphasis is both correct and timely, for this category of secondary products, apart from the alkaloids, has previously received little attention in textbooks.

The book assumes that the reader has a sound knowledge of organic chemistry and of the concepts of basic intermediary metabolism. It presents informative discussions of many types of natural product that are important to advanced students and research workers in the fields of chemistry, biochemistry, agriculture, pharmacognosy and medicine. As the reader becomes acquainted with this book, he will surely come to respect the scholarship of Professor Luckner.

Contents

Preface to the English edition   *page xv*

Preface   *xvii*

List of Abbreviations   *xix*

PART A. Secondary metabolites in the life of plants and animals   1

1.   The position of secondary products in metabolism   3
2.   Ecological significance of secondary natural products   7
3.   Location of synthesis of secondary products in the cell and regulation
            of secondary metabolism   9

PART B. Methods of investigation of secondary metabolism   13

1.   Methods of working with isotopically labelled compounds   15
2.   Investigations with mutants   20
3.   Use of homogenates, enzymes and enzyme systems   22

PART C. Enzymes of general significance in secondary metabolism   25

1.   Activating enzymes   27
1.1   Transphosphorylases   27
1.1.1   Phosphorylated compounds   27
1.1.2   Transphosphorylation   28
1.2   Thiokinases and thiophorases   30

2.   Oxidoreductases and oxygenases   32
2.1   Dehydrogenases containing pyridine nucleotides   32
2.1.1   Mechanism of reaction   32
2.1.2   The oxidative deamination of amino acids   33
2.2   Flavin enzymes   34
2.2.1   Mechanism of reaction   34
2.2.2   Xanthine dehydrogenases and xanthine oxidases   35
2.3   Phenol oxidases   36
2.4   Peroxidases   36
2.4.1   Dehydrogenations and hydroxylations catalysed by peroxidases   36
2.4.2   Halogenation of organic molecules   37
2.5   Dioxygenases   38
2.6   Mixed function oxygenases   39

2.6.1	Structure and mechanism of reaction	39
2.6.2	Oxygenation of amines and thio compounds	40
2.6.3	Amino acid oxygenases	41
2.6.4	Hydroxylation of tetragonal carbon atoms and oxidative demethylation	41
2.6.5	Hydroxylation of unsaturated or aromatic compounds and the NIH shift	41
3.	Enzymes of 'one carbon' metabolism	45
3.1	Carboxylases and transcarboxylases	45
3.2	Transformylases, hydroxymethyltransferases and methyltransferases	46
3.3	Methyl transferases	48
3.4	α-Keto acid decarboxylases	50
4.	Transaminases and amino acid decarboxylases	52
PART D.	**Biosynthesis and metabolism of secondary plant and animal products**	**55**
1.	Secondary natural products originating from sugars	57
1.1	Formation of glycosides	59
1.1.1	*De novo* formation of the glycosidic bond in the biosynthesis of aldose-1-phosphates and the transfer of glycosyl groups to nucleoside phosphates	60
1.1.2	Formation of holosides and heterosides	62
1.2	Biosynthesis of 'abnormal' sugars	67
1.2.1	Formation of amino sugars	67
1.2.2	Formation of deoxy sugars	68
1.2.3	Formation of sugars with branched carbon chains	69
1.3	Oxidation of sugars	72
1.3.1	Formation and decarboxylation of uronic acids	72
1.3.2	The biosynthesis of onic acids and sugar dicarboxylic acids	73
1.3.3	The formation of ascorbic acid	74
1.4	Secondary natural products originating from sugars by reduction	75
1.4.1	Formation of aliphatic sugar alcohols	75
1.4.2	Biosynthesis of cyclitols	76
1.4.3	Biosynthesis of streptidine	78
2.	Formation of secondary natural products from acetate	81
2.1	Biosynthesis and properties of 'activated acetic acid'	81
2.2	Biosynthesis, transformations and degradation of fatty acids	83
2.2.1	The formation of saturated fatty acids by the fatty-acid-synthetase complex	83
2.2.2	Elongation of the carbon chain of fatty acids by the fatty-acid-elongation complex	85
2.2.3	Methylation of fatty acids	86
2.2.4	The nucleophilic substitution of fatty acids by fatty acids at the α-carbon atom	87

2.2.5   Biosynthesis of unsaturated and hydroxylated fatty acids and the
         formation of prostaglandins   87
2.2.6   Degradation of fatty acids   90
2.2.7   Formation of fatty acid esters   92
2.2.8   Biosynthesis of sphingolipids and plasmalogens   95
2.2.9   Formation of n-alkanes from fatty acids   97
2.2.10   Biosynthesis of acetylenic derivatives   98
2.3   Formation of polyketides   103
2.3.1   Formation of anthracene derivatives and secoanthraquinones   104
2.3.2   Biosynthesis of tetracyclines   106
2.3.3   Formation of griseofulvin   108
2.3.4   The biosynthesis of polyketide phenol carboxylic acids and phenol
         carboxylic acid derivatives in fungi and lichens   109
2.3.5   Formation of hemlock alkaloids, cycloheximide and other piperidine
         derivatives   112

3.   Secondary natural products originating from propionic acid   114
3.1   Biosynthesis of methyl fatty acids and macrolide antibiotics   115

4.   Biosynthesis of secondary products from the acids of the tricarboxylic
         acid and glyoxylic acid cycles   118
4.1   Formation of substituted citric and derivatives from oxaloacetic acid 118
4.2   Conversion of glyoxylic acid to oxalic acid   120

5.   The biosynthesis of secondary natural products from 'activated
         isoprene'   122
5.1   Formation of isopentenyl pyrophosphate ('activated isoprene') from
         acetyl CoA   123
5.2   Polymerization of isopentenyl pyrophosphate   124
5.3   The head-to-head condensation of farnesyl pyrophosphate and
         geranylgeranyl pyrophosphate with the formation of phytoene and
         squalene   127
5.4   Formation of monoterpenes, sesquiterpenes and diterpenes from
         geranyl-, farnesyl- and geranylgeranyl pyrophosphates   129
5.5   Biosynthesis of iridoid compounds   136
5.6   Formation of carotenoid compounds and vitamin A   140
5.7   Biosynthesis of tetracyclic and pentacyclic triterpenes   145
5.7.1   Formation of cyclic triterpene ring systems from squalene   147
5.7.2   Biosynthesis of various groups of steroids   153
5.8   Formation of vitamin D_2 and D_3 163

6.   Secondary products originating from dehydroquinic acid, shikimic
         acid, chorismic acid and anthranilic acid   166
6.1   The formation of quinic acid and chlorogenic acid   167
6.2   Shikimic acid as precursor of secondary natural products   168
6.2.1   The biosynthesis of naphthoquinone and anthraquinone derivatives   168
6.2.2   The conversion of shikimic acid to protocatechuic and gallic acids 170
6.3   Secondary natural products formed from chorismic acid   171

6.3.1   The conversion of chorismic acid to anthranilic acid, *p*-aminobenzoic acid, salicylic acid and *p*-hydroxybenzoic acid   171

6.3.2   Formation of ubiquinones from *p*-hydroxybenzoic acid   172

6.4   Formation of secondary natural products from anthranilic acid   174

6.4.1   Formation of 3-hydroxyanthranilic acid and its derivatives   174

6.4.2   Biosynthesis of quinoline alkaloids   176

6.4.3   Formation of quinazoline alkaloids   181

6.4.4   Synthesis of phenazines   182

7.   General pathways for the formation of secondary natural products from L-amino acids   184

7.1   Formation of D-amino acids   184

7.2   Biosynthesis of acylated amino acids   185

7.3   Synthesis of amines   186

7.4   Formation of methylated amino acids and betaines   188

7.5   Biosynthesis and degradation of cyanogenic glycosides   190

7.6   Formation of glucosinolates and mustard oils   194

7.7   General reactions in the formation of alkaloids   198

8.   Formation of secondary natural products from glycine   202

8.1   Formation of porphyrins, bile pigments and cobalamin derivatives   202

8.2   Formation of purine derivatives and compounds derived from purines   208

8.2.1   Biosynthesis of compounds with the purine ring system   208

8.2.2   Biosynthesis of purine alkaloids and methylated aminopurines   212

8.2.3   Formation and metabolism of pteridines   213

8.2.4   Formation of isoalloxazines (benzopteridines)   216

8.2.5   Conversion of purines to pyrrolopyrimidines   218

8.2.6   Degradation of purine ring system   219

8.3   Formation of glycine conjugates in animals   222

9.   Biosynthesis of secondary products from cysteine   224

9.1   Oxidation and decarboxylation of cysteine   225

9.2   Formation and metabolism of *S*-alkylcysteine derivatives and sulphoxides   228

9.3   Luciferin and luciferase   229

9.4   Biosynthesis of biotin   230

9.5   Formation of premercapturic acids and mercapturic acids   231

10.   Formation of secondary products from methionine   233

10.1   Sulphonium compounds as secondary natural products   234

10.2   Formation of ethylene   236

10.3   Formation and reactions of thiamine   237

11.   Valine as precursor of secondary natural products   240

12.   Leucine as precursor of secondary natural products   243

12.1   Formation of isopentenyl pyrophosphate from leucine   244

13.	Formation of secondary products from isoleucine	246
13.1	Formation of tiglic acid and α-methylbutyric acid	247
13.2	Biosynthesis of tenuazonic acid	247
14.	Biosynthesis of secondary products from aspartic acid	249
14.1	Biosynthesis and degradation of orotic acid and orotic acid derivatives	249
14.2	Formation of nicotinic acid and the alkaloids derived from nicotinic acid	252
14.3	Formation of 2,6-dipicolinic acid and fusaric acid	257
15.	Compounds of the glutamic acid-proline-ornithine-group as precursors of secondary natural products	259
15.1	Biosynthesis of tropane alkaloids	261
15.2	Formation of pyrrolizidine alkaloids	264
15.3	Formation of pyrrolidine ring of nicotine	266
15.4	Formation of ornithine and glutamine conjugates in animals	267
16.	Synthesis of secondary metabolites from lysine	269
16.1	Biosynthesis of pipecolic acid and pipecolic acid derivatives	272
16.2	Formation of punica-, sedum-, and lobelia alkaloids	273
16.3	Biosynthesis of quinolizidine alkaloids	275
16.4	Formation of the piperidine ring of anabasine	276
16.5	Biosynthesis of mimosine and desmosine	278
17.	Arginine as precursor of secondary natural products	279
17.1	Formation of secondary guanidine compounds	280
17.2	Conversion of homoarginine to lathyrine	281
18.	Biosynthesis of secondary natural products from histidine	282
19.	Formation of secondary natural products from tryptophan	289
19.1	Biosynthesis of indole alkylamines	292
19.2	Biosynthesis of ergoline alkaloids	293
19.3	Formation of β-carboline alkaloids and related compounds	295
19.4	Secondary products originating via indolenine derivatives	300
19.4.1	Formation of Calycanthus alkaloids	300
19.4.2	Biosynthesis of Cinchona alkaloids	301
19.4.3	Biosynthesis of pyrrolnitrins	304
19.5	Secondary products originating from the intermediates of tryptophan degradation by the pyrrolase pathway	305
19.5.1	Formation and degradation of kynurenic acid and kynurenic acid derivatives	306
19.5.2	Biosynthesis of ommochromes	309
19.5.3	Biosynthesis of pyridine carboxylic acids	310
19.5.4	Formation of quinazolines in *Pseudomonas aeruginosa*	312
19.6	Secondary products which originate from tryptophan by shortening or elimination of the side chain	313

20.	Formation of secondary natural products derived from phenylalanine and tyrosine	317
20.1	Formation of secondary products with the retention of amino nitrogen	318
20.1.1	Biosynthesis of phenylalkylamines	318
20.1.2	Formation of isoquinoline alkaloids by Mannich condensation	321
20.1.3	Biosynthesis of Erythrina and Amaryllidaceae alkaloids	328
20.1.4	Biosynthesis of colchicine	332
20.1.5	Formation of aristolochic acid	334
20.1.6	Formation of indole derivatives from dihydroxyphenylalanine and m-tyrosine	336
20.1.7	Formation of novobiocin and anisomycin	340
20.2	Formation of plastoquinones, tocopherol quinones and tocopherols from p-hydroxyphenylpyruvic acid	341
20.3	Formation of tropic acid, atropic acid and phenylglyceric acid	343
20.4	Cinnamic acid and cinnamic acid derivatives as secondary natural products	344
20.4.1	Formation of cinnamic acid and cinnamic acid derivatives	344
20.4.2	Biosynthesis of coumarins	345
20.4.3	Formation of cinnamic alcohol glucosides from cinnamic acids	348
20.4.4	Formation of lignin	349
20.4.5	Formation of lignans	353
20.5	Formation of secondary natural products from cinnamic acid derivatives and acetate	354
20.5.1	Biosynthesis of stilbene derivatives	355
20.5.2	Biosynthesis of compounds with flavan structure	357
20.5.3	Formation of aurones, isoflavones, 3-phenylcoumarins and rotenones	362
20.5.4	Biosynthesis of neoflavonoids (4-phenylcoumarins)	364
20.6	Shortening of the side chain of cinnamic acid and phenylethylamine derivatives with the formation of C_6C_2-, C_6C_1- and C_6-bodies	365
21.	Peptides, peptide derivatives and proteins possessing the character of secondary products	371
21.1	Formation of diketopiperazines and compounds derived from them	371
21.2	Formation of hydroxamic acids	373
21.3	Biosynthesis of penicillins and cephalosporins	375
21.4	Formation of the framework of the bacterial cell wall	378
21.5	Biosynthesis of cyclic polypeptides	380
21.6	Formation of sclerotins in insects	381
	Index	384

Preface to the English edition

I am extremely glad to hear that an English translation of my book on secondary metabolism in plants and animals is coming out almost immediately after the Japanese translation of the same. For this purpose the manuscript was revised and enlarged by an additional chapter on enzymes of secondary metabolism in order to emphasize the biochemical point of view on which the material under discussion is based.

I would like to thank Mr. Vasudevan and Dr. J. L. Wray very much for the translation and preparation of the English manuscript. I am especially grateful to Professor L. Fowden, Professor of Plant Chemistry in the Department of Botany and Microbiology at University College, London, not only for his kind Foreword, but also for his help in ensuring the advent of this English edition.

Halle, *August, 1971*   Martin Luckner

Preface

As well as the vitally essential compounds of primary metabolism which are very similar in almost all organisms, plants, and to a lesser extent, animals, produce an infinite variety of substances which frequently are revealed by their physiological action, colour, smell or taste. A great number of these substances have found entry into the materia medica, e.g. many alkaloids and antibiotics. Others serve the chemical industry as starting materials for the production of compounds used in therapeutics, e.g. certain steroid sapogenins and alkaloids from which steroid hormones are produced. A few others, like rubber, tannins and the fibres of the animal and plant kingdoms, possess technologically valuable properties.

The multitude of substances, which in general are of no direct significance to the organisms producing them, are now grouped together under the term secondary natural products. Their origin, conversions and degradations in living organisms are the subject matter of this book, which has an aim similar to that of K. Paech in *Biochemie und Physiologie der sekundären Pflanzenstoffe* (Springer Verlag, 1950). Since the appearance of this milestone, our knowledge of secondary product metabolism has increased so enormously that a renewed and extensive treatment of the subject including the secondary products of the animal kingdom appears to be justified. The growing interest in secondary natural products responsible for this development is not dependent only on their practical utility. They also present fascinating problems to the chemists and biologists. This research has not only enriched the chemistry of natural products, but has also furthered theoretical organic chemistry and given biology a more complete understanding of living phenomena.

Only the most important groups of natural products could be considered in the choice of material for this book. The formation and metabolism of primary natural products are discussed at the level which appeared necessary for the understanding of secondary product metabolism. Proof of the illustrated metabolic pathways and discussion of the reliability of particular results are omitted, apart from a few exceptions. The localization of the enzymes catalysing the reactions and the mechanism of regulation applicable to the formation of secondary natural products could only be touched on due to shortage of space. However, the original references and review articles cited at the end of individual sections offer a deeper penetration into the literature, and enable further study.

The book is intended as a textbook for students of the final semesters and therefore previous knowledge, especially of chemistry, biology and the biochemistry of primary metabolism, is taken for granted. In addition it is hoped that it will offer information on the biochemistry of secondary natural products to trained and professional biologists, pharmacists, chemists, agricultural chemists and physicians.

I wish to express my sincere thanks to Professor Kurt Mothes for his valuable suggestions. My thanks are also due to all those who helped in the formation of this book, especially to Professor Otto Beßler, Director of the Department of Pharmaceutical Biology (Pharmacognosy) in the Section of Pharmacy, Halle- and to my wife who offered valuable help in the preparation of the manuscript and other technical matters.

M. Luckner

List of abbreviations

⤳	Shift of one electron
⤳	Shift of two electrons
~	High-energy bond (cf. C.1.1)
ADP	Adenosine diphosphate
AMP	Adenosine monophosphate
ATP	Adenosine triphosphate
c	*cis*
CDP	Cytidine diphosphate
CMP	Cytidine monophosphate
CoA	Coenzyme A
CTP	Cytidine triphosphate
dADP	Deoxyadenosine diphosphate
DNA	Deoxyribonucleic acid
Dopa	Dihydroxyphenylalanine
dUDP	Deoxyuridine diphosphate
e^-	Electron
E	Enzyme
FAD	Flavin adenine dinucleotide, oxidized
$FADH_2$	Flavin adenine dinucleotide, reduced
FMN	Flavin mononucleotide, oxidized
$FMNH_2$	Flavin mononucleotide, reduced
GDP	Guanosine diphosphate
GTP	Guanosine triphosphate
NAD^+	Nicotinamide adenine dinucleotide, oxidized
NADH	Nicotinamide adenine dinucleotide, reduced
$NADP^+$	Nicotinamide adenine dinucleotide phosphate, oxidized
NADPH	Nicotinamide adenine dinucleotide phosphate, reduced
P	Phosphate (cf. C.1.1.2.)
PP	Pyrophosphate (cf. C.1.1.2.)
PPP	Triphosphate (cf. C.1.1.2.)
R	An undefined residue
RNA	Ribonucleic acid
S	Substrate
t	*trans*
TDP	Thymidine diphosphate
THF	Tetrahydrofolic acid
tRNA	Transfer ribonucleic acid
UDP	Uridine diphosphate

A. Secondary metabolites in the life of plants and animals

1. The position of secondary products in metabolism

The formation and metabolism of nucleic acids, most of the carbohydrates, amino acids and proteins, certain carboxylic acids etc. is, in principle, similar in all living organisms (or at least in the large groups of organisms). The vital reactions involved in these processes may all be termed reactions of basic or primary metabolism.

However, in addition to the reactions of primary metabolism, reactions frequently occur which are not necessarily vital, often differ from species to species, and may be considered as an expression of the chemical individuality of the organism. These reactions are grouped under the term secondary metabolism and the products formed are called secondary metabolites.

The extensive biochemical investigations of the last three decades have proved that secondary metabolism is very closely connected with primary metabolism. A sharp boundary cannot always be drawn between these two areas, however, and therefore at many places in the following discussion, the metabolism of compounds of primary metabolism is also discussed if this reveals a similarity to that of secondary metabolites as, for example, in the case of purine and pyrimidine nucleotides.

An important characteristic of secondary products is that they may be eliminated from primary metabolism. They may be considered as excretions which are of no importance to the living organisms producing them, either as a source of energy, or as a specific storage product. That does not imply that they are no more reactive. The substances deposited outside the protoplasm, e.g. in the cell wall, in dead cells or in special excretory tissues (lignin, rubber, pigments on the feathers of birds and wings of insects and seed coats, etc.), remain largely unchanged, unlike the water-soluble substances remaining in the reactive area of the cell. Change of metabolic conditions, e.g. due to ageing, may bring about further transformations, or for that matter the degradation of such excretions.

Thus, in the case of some *Nicotiana* species, nicotine disappears during ripening of seeds (cf. D.14.2). In other plants the alkaloid stored in the endosperm or the embryo is destroyed during germination. That such a degradation is of vital importance has not been proved, however. The idea of considering such nitrogen compounds as a metabolic reserve should be completely rejected, since they form only a fraction of the total nitrogen. The same thing holds in the case of sugars formed by degradation of secondary glycosides (e.g. the anthocyanins). These are of no economic importance to the plant.

If, however, secondary products have no importance as reserve materials, it is probable that their biosynthetic mechanisms are of physiological significance. The 'detoxication theory' must be referred to at this juncture. It has been shown to be quite probable that in a few cases the processes taking place during the formation of secondary metabolites lead to stabilization and detoxication of substances, which if accumulated in an unchanged form might be detrimental to the organism. Such processes are understood in more detail in animals. Here, the toxicity of substances introduced from outside, or originating during metabolism, e.g. of phenols and indoles, is reduced in the liver, and also partly in the kidneys, by hydroxylation or by methylation or 'conjugation' with amino acids, glucuronic acid or sulphuric acid.

An exceedingly interesting problem connected with this is the unequal distribution of secondary metabolites in living organisms. Although, as mentioned previously, animals are in no way free from secondary metabolites, there are few of them. We cannot say that alkaloids are solely plant bases because there are several animals known which can produce alkaloids. Thus, millipedes are known which produce quinazolines by a mechanism similar to that occurring in plants (cf. D.6.4.3). There are however, only 20 to 30 animal alkaloids in comparison to more than 5000 plant alkaloids. More extreme still is the situation obtaining in the case of terpenoids, phenolic glycosides and steroids. It would appear that the unequal distribution of secondary metabolites is an expression of the difference in excretory metabolism. The animal eliminates all unwanted accumulated substances of primary metabolism mainly through the kidneys. Thus, the necessity for the extensive synthesis of secondary metabolites (viz. the accumulation of metabolically active waste products of primary metabolism) is not required. The waste products of plant metabolism, on the other hand, are accumulated in the vacuoles, the cell walls, and, if lipophilic in character, in special excretory cells or spaces (volatile oil cells, resin ducts etc.). Vacuoles are well separated from the cytoplasm by means of tonoplasts. The substances accumulated here are by no means always harmless to the cytoplasm which has produced them. The vacuolar fluid, which is non-toxic in the cell since its contents cannot penetrate into the cytoplasm, is often extremely toxic if introduced into the cells from outside.

These relationships are of special importance in allowing cells to tolerate secondary metabolites. It is not enough that organisms can produce enzymes which are essential for the synthesis of their secondary metabolites. They must also develop mechanisms which can protect the cytoplasm from the products synthesized in it.

There are a few species of *Solanaceae* containing small amounts of nicotine in all tissues. If scions of these plants are grafted to tobacco roots, which normally produce nicotine in greater amounts and transport it to the shoot, the grafted scion suffers from nicotine poisoning, and the characteristic disruption of chlorophyll metabolism sets in. This may also occur if the leaves of the grafted plant are sprayed with a solution of nicotine. In both these cases the nicotine concentration in the cells exceeds a tolerable level.

The storage of water-soluble substances in the vacuoles is a little-investigated phenomenon. Often there are quite specific cells or tissues that are capable of such storage. This suggests that a special chemical system is necessary. Storage is a process of concentration. The transport is against the concentration gradient and proceeds by an active mechanism requiring energy. It has only been investigated in a few cases.

Of greater importance is the fact that the location of synthesis and of accumulation of secondary metabolites need not be identical. Between these two locations there is frequently a distance which has to be covered by transport. This distance may be quite short: e.g. formation in the cytoplasm near the cell walls, and accumulation in the cell walls or vacuoles; formation in the epithelial cells and accumulation in the neighbouring excretory spaces. On the other hand the distance may be great: e.g. formation of nicotine and atropine in the young portion of the roots and accumulation in the leaf cells. Such long-distance transport of substances is in many cases combined with further chemical conversions.

A comparison of the secondary metabolism of plants of different classes (taxa) shows that lower organisms are in no way incapable of producing and storing secondary metabolites.

Hydrophilic substances are somewhat rare in lower aquatic plants (algae) and a few lipophilic types of substances are more common. It cannot, however, be said that secondary metabolism is a property of the organism aquired as a result of a higher stage of development. Since micro-organisms have been little investigated, no final verdict regarding their capacity in this respect can be given at present. The number of antibiotics of different structures occurring in Actinomycetes, the number of pigments in moulds, colouring matters of lichens etc. are not yet assessable. The accumulation of such substances by typical soil microbes is remarkable, and may perhaps be an indication of an inhibition of excretion into the media.

The secondary products occurring in animals are not always produced by the organism itself but originate occasionally as a result of its nutritional pattern (e.g. the alkaloid castoramin, found in beavers, which resembles to a great extent the alkaloid desoxynupharidin present in the rhizomes of *Nuphar luteum*; the cardiac glycosides calactin and calotropin present in the excretions of the locust *Poekilocerus bufonius*, which ingests them when feeding on plants of *Asclepiadaceae*, or the cardiac glycosides in the adult butterfly *Danaus plexippus* and its larvae, which feeds on different species of *Asclepiadaceae*).

Secondary products stored in animals are found usually in specific tissues, e.g. the alkaloids in salamander and the cardioactive glycosides of toads are found in certain glands located in the skin; the defensive secretions of insects frequently in special defensive glands. Secondary products (e.g. cantharidin in insects) may, however, also occur disolved in the haemolymph, or they may be stored in the hair, skin or other tissues.

References for further reading

See A.3.

2. Ecological significance of secondary natural products

The usefulness of secondary products to the producer organism may be based on ecological grounds.

Certain secondary products have a mechanical function. Frey-Wyssling, who was the first to review excretion in higher plants in his book *Die Stoffausscheidung der höheren Pflanze*, considers cellulose as an excretory product without which the stability of the plant cells and their tissues would be inconceivable. This also applies to lignin. The development of higher forms of plants is linked with its 'invention' by nature.

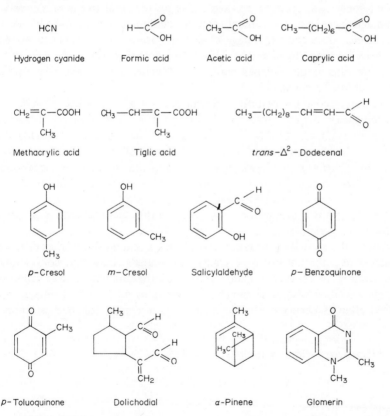

Fig. 1. Physiologically active compounds from the defensive secretions of arthropods.

Furthermore, secondary products are of great importance to the sexual and reproductive life of plants and animals. Of special significance are the aromatic substances of insects which are used to attract a partner and, in the case of higher plants, floral pigments and aroma, to attract insects for pollination. Other secondary products produced by animals, e.g. salamander alkaloids, the cardio-active glycosides of toads (cf. D.5.7.2), as well as a great number of substances from insects (fig. 1), and snake venoms, play a role in the attack on prey or in defence against bacteria, pathogenic fungi and other enemies.

The role of 'defence secretions' of plants is questionable since the resistance of individual types of organisms to 'poisonous' substances is different. Thus, for example, belladonna (*Atropa belladonna*) is highly toxic to human beings, owing to its tropane alkaloid content. It can, however, be consumed without any dangerous consequences by certain types of rabbits and pigs etc. since they produce an enzyme capable of degrading these alkaloids (cf. D.15.1).

All plants, whether they produce alkaloids or not, have particular plant or animal foes. If this were not so, the 'toxic' plants would in a short time overgrow the 'non-toxic' ones. Even if one supposes that 20% of the total number of plant species do produce alkaloids, it is obvious that this is only a very low percentage of the total. Only under extreme conditions, such as high altitude, or in desert areas, does the percentage of alkaloid-bearing plants present appear to have been favoured by selection. This is also probably true in the case of those plants growing in the Steppes and in the Mediterranean area, which produce unpleasant-tasting substances.

However, secondary natural products are not a very effective phylogenetic selection factor. Anyone who has been involved in the cultivation of spices and medicinal plants can confirm this. Apparently the content of secondary products is rather frequently limited by a diminution of vitality. Artificial infection of rye by *Claviceps purpurea*, ergot, indicates that those races possessing an average alkaloid content of 0·50% are high yielding (assessed as alkaloid production per acre), whereas those races having a higher alkaloid content possess a much lower growth rate, with the consequence that the total amount of alkaloid produced is lower.

Some of the secondary products formed act in the metabolism as effectors or regulators, e.g. the hormones of plants and animals. However, considering the great number of secondary products formed the amount of used compounds is small and their biological significance seems to be a random process which evolved after its occurrence rather than being a 'reason' for their selection.

References for further reading

See A.3.

3. Location of synthesis of secondary products in the cell and regulation of secondary metabolism

Little is known about the site of synthesis of secondary products in the cell. Results obtained during the study of primary metabolism, however, indicate that compartmentation of the cell by substructures plays an important part.

These structures may be quite simple. Thus, for example, in the multi enzyme complex (cf. C.2.6.1 and C.2.3.1) the different enzymes taking part in the reaction chain are spatially adjacent to one another in stoichiometric ratio. The whole system behaves like a single enzyme and yet the individual components of the complex may be separated. Such a system increases the efficiency and speed of the overall reaction since diffusion of intermediates from one locus of reaction to another is largely overcome.

In a similar manner the coupling of enzymes to a membrane system creates special reaction conditions. Extensive compartmentation of the cell is achieved by means of the organelles surrounded by semipermeable membranes. These include the mitochondria, different plastids, golgi bodies, endoplasmic reticulum and others. The division of the cell space into different areas of reaction is an important prerequisite for the sequence of metabolism and its regulation. This becomes quite obvious, when several distinct pools of the same substance are present. For example, it has been shown that malic acid formed in the mitochondria through the tricarboxylic acid cycle (cf. D.4) is soon metabolized further, while that formed in the cytoplasm by carboxylation of phosphoenolpyruvate behaves like a secondary metabolite and does not undergo any other appreciable reactions.

It is an important characteristic of secondary metabolic reactions that they are dependent to a great extent on the stage of development of the particular organism (phase dependence). Secondary metabolism is thus a part of the process of cell differentiation.

To understand the mechanisms by which the regulation of secondary metabolism proceeds it seems necessary to introduce and to discuss the terms 'primary' and 'secondary' differentiation. By primary differentiation stages of differential gene activity are understood within a cyclic process corresponding with the cycles of cell division. This phenomenon is clearly demonstrated by the varying enzyme pattern of embryogenic, meristematic or other self reduplicating cells at different stages of the division cycle.

By internal or external stimuli some cells escape from the cycle of primary differentiation. Genetic material unused in primary differentiation is activated, thus starting a process called secondary differentiation. Secondary differentiation covers all processes which are commonly included under the term 'cell differentiation', e.g. the synthesis of contractile protein filaments by myogenic cells as well as the formation of cells producing secondary products from the meristematic cells of higher plants or the formation of conidiospores from the hyphae of moulds. Due to differential gene activity in the course of primary differentiation, it is only at certain stages of development that meristematic cells are competent to respond to the stimuli causing secondary differentiation. This has been shown by synchronized mammalian cells and by cells of algae, bacteria and higher plants.

The formation of secondary compounds as a consequence of secondary differentiation is a comparatively simple process. The chain of chemical reactions and enzymes involved can be followed easily while often not causing marked changes in the viability of the experimental cells. Thus it seems that secondary metabolism is also a convenient model for investigating basic aspects of other fields of secondary differentiation.

Recently some compounds were found which participate in the regulation of secondary metabolism. Using micro-organisms and higher plants it was shown that compounds of the primary metabolism from which special secondary products originate are involved in the induction of those enzymes, which transfer these metabolites to secondary compounds. Similar results are obtained with some secondary metabolites themselves. Ferulic acid, e.g., induces the enzymes which transform this compound to the anthocyanin paeonidin and 6-methyl salicylic acid those which transfer itself into patulin. However, these inductions are possible only if the cells are competent to react with the effectors, the competence depending on the developmental stage and the determination of the cells.

In short, secondary metabolism is evidently regulated by the concerted action of the following two effects:

(a) The ability of the cell to respond to certain internal or external stimuli (Competence) and

(b) the actual presence of the suitable effectors at the correct time.

References for further reading

SECONDARY METABOLISM, GENERAL ASPECTS

Czapek, F. *Biochemie der Pflanzen* (Fischer Verlag, Jena, 1920/21).

Paech, K. *Biochemie und Physiologie der sekundären Pflanzenstoffe* (Springer Verlag, Berlin, 1950).

Schwarze, P. 'Der Stoffwechsel sekundärer Pflanzenstoffe, Einführung', *Handbuch der Pflanzenphysiologie*, Vol. X, ed. W. Ruhland (Springer Verlag, Berlin, 1958), pp. 1-23.

Zenk, M. H. Biochemie und Physiologie sekundärer Pflanzenstoffe, *Ber. dtsch. Bot. Ges.* **80** (1967), pp. 573–91.

Luckner, M. Was ist Sekundärstoffwechsel?, *Pharmazie* **26** (1971), pp. 717–24.

OCCURRENCE AND STRUCTURE OF SECONDARY PRODUCTS

Miller, M. W. *The Pfizer Handbook of Microbial Metabolites* (McGraw–Hill Book Company, New York, 1961).

Hegnauer, R. *Chemotaxonomie der pflanzen*, Vols. I–V (Birkhäuser Verlag, Basel, 1962–69).

Karrer, W. *Konstitution und Vorkommen der organischen Pflanzenstoffe, exkl. Alkaloide* (Birkhäuser Verlag, Basel, 1958).

Karrer, W. and Eugster, C. H. *Ergänzungsband* to the preceding book (Birkhäuser Verlag, Basel, 1967).

Florkin, M. and Scheer, B. J. (Eds.). *Chemical Zoology*, Vols. I–VII (Academic Press, New York, 1967–71).

Campbell, J. W. (Ed.). *Comparative Biochemistry of Nitrogen Metabolism*, Vols. I and II (Academic Press, New York, 1970).

Raffauf, R. F. *A Handbook of Alkaloids and Alkaloid Containing Plants* (Wiley Interscience, New York, 1970).

Turner, W. B. *Fungal Metabolites* (Academic Press, London, 1971).

THE RELATIONS OF SECONDARY METABOLISM TO GENERAL METABOLISM

Bu'Lock, J. D. 'Aspects of Secondary Metabolism in Fungi', *Biogenesis of Antibiotic Substances*, ed. Z. Vaněk and Z. Hošťálek (Academic Press, New York, 1965), pp. 61–71.

Rimpler, H. Zur taxonomischen Wertigkeit chemischer Merkmale: Exkretionsmechamismen und Exkrete, *Planta medica* 13 (1965), pp. 412–17.

Demain, A. L. Regulatory Mechanisms and the Industrial Production of Microbial Metabolites, *Lloydia* 31 (1968), pp. 395–418.

Hess, D. *Biochemische Genetik* (Springer Verlag, Berlin, 1968).

Böhm, H. 'Genetik des Alkaloidmerkmals', *Biosynthese der Alkaloide*, ed. K. Mothes and H. R. Schütte (VEB Deutscher Verlag der Wissenschaften, Berlin, 1969), pp. 21–39.

Mothes, K. 'Biologie der Alkaloide', *Biosynthese der Alkaloide*, ed. K. Mothes and H. R. Schütte (VEB Deutscher Verlag der Wissenschaften, Berlin, 1969), pp. 1–20.

Weinberg, E. D. 'Biosynthesis of Secondary Metabolites', *Advances in Microbial Physiology,* ed. A. H. Rose and J. F. Wilkinson (Academic Press, London, 1970), pp. 1–44.

THE SIGNIFICANCE OF SECONDARY METABOLISM

Detoxication

Mothes, K. Zur Problematik der metabolischen Exkretion bei Pflanzen, *Naturwiss.* 53 (1966), pp. 317–23.

Williams, R. T. 'The Biogenesis of Conjugation and Detoxication Products', *Biogenesis of Natural Compounds,* ed. P. Bernfeld (Pergamon Press, Oxford, 1967), pp. 589–639.

Schnepf, E. 'Sekretion und Exkretion bei Pflanzen', *Protoplasmatologia* VIII/8 (Springer Verlag, Vienna, 1969).

Frey-Wyssling, A. Betrachtungen über die pflanzliche Stoffelimination, *Ber. Schweizer. Bot. Ges.* 80 (1970), pp. 454–66.

Ecological significance

Reichstein, T. Cardenolide (herzwirksame Glykoside) als Abwehrstoffe bei Insekten, *Naturwiss. Rundschau* **20** (1967), pp. 499–511.

Habermehl, G. Chemie und Biochemie von Amphibiengiften, *Naturwiss.* **56** (1969), pp. 615–22.

Schildknecht, H. Die Wehrchemie von Land- und Wasserkäfern, *Angew. Chem.* **82** (1970), pp. 17–25.

Brunner, R. and Machek, G. (Eds.). *Die Antibiotica*, Vols. I–III (Verlag Hans Carl, Nürnberg, 1962–70).

Gottlieb, D. and Shaw, P. D. (Eds.). *Antibiotics*, Vols. I and II (Springer Verlag, Berlin, 1967).

Sondheimer, E. and Simeone, J. B. (Eds.). *Chemical Ecology* (Academic Press, New York, 1970).

Synge, L. M. Proteine und Gifte in Pflanzen, *Nuturwiss. Rundschau* **24** (1971), pp. 56–61.

Whittaker, R. H. and Feeny, P. P. Allelochemics: Chemical Interactions between Species, *Science* **171** (1971), pp. 757–70.

SECONDARY PRODUCTS AS REGULATORS OF THE METABOLISM

Pastan, I. Biochemistry of Nitrogen Containing Hormones, *Ann. Rev. Biochem.* **35** (1966), pp. 369–404.

Heftmann, E. *Steroid Biochemistry* (Academic Press, New York, 1970).

Burgus R. and Guillemin, R. Hypothalamic Releasing Factors, *Ann. Rev. Biochem.* **39** (1970), pp. 499–526.

Behrens, O. K. and Grinnan E. L. Polypeptide Hormones, *Ann. Rev. Biochem.* **38** (1969), pp. 83–112.

Wightman, F. and Setterfield, G. (Eds,), *Biochemistry and Physiology of Plant Growth Substances* (The Runge Press, Ottawa, 1968).

Wilkins, M. B. (ed.). *The Physiology of Plant Growth and Development* (McGraw–Hill, London, 1969).

Miller, C. O. 'Plant Hormones', *Biochemical Actions of Hormones,* Vol. I, ed. G. Litwack (Academic Press, New York, 1970), pp. 503–18.

Skoog, F. and Armstrong, D. J. Cytokinins, *Ann. Rev. Plant Physiology* **21** (1970), pp. 359–84.

B. Methods of investigation of secondary metabolism

1. Methods of working with isotopically labelled compounds

The fate of a compound in metabolism could only be followed with great difficulty until a short time ago. However, in the last few years our understanding of the biosynthesis and metabolism of natural products has been greatly facilitated by administering ('feeding') isotopically labelled compounds. Our knowledge in this field has therefore increased almost explosively since about the end of the Second World War.

The isotopes which are most important in biosynthetic studies are ^{14}C, ^{2}H, ^{3}H, ^{15}N, ^{18}O, ^{35}S and ^{32}P. The label may be introduced into a compound either during its synthesis, or by isotope exchange after synthesis. The radioactive isotopes ^{14}C, ^{3}H, ^{35}S and ^{32}P can be detected with the help of special counters. The stable isotopes ^{2}H and ^{18}O may be determined with the help of mass spectroscopy. An optical spectroscopic determination is also possible in the case of ^{15}N.

The feeding of isotopically labelled compounds enables one to determine into which substance the labelled compound, or more exactly the labelled portion of the compound, is incorporated (i.e. for which substances they are precursors) and to what extent it happens (i.e. the value of the so-called incorporation rate). For the calculation of the incorporation rate the radioactivity [given in Curie $(C) = 3\cdot7.10^{10}$ disintegrations per second, in Milicurie (mC) or in Microcurie (μC)] or the deviation from the normal isotopic composition (given in atom-%) in the compound fed and in the product formed are compared with one another.

The results are usually expressed as follows:

(a) In the case of the administration of radioactive labelled compounds, the ratio

$$\frac{\text{mmol of end product formed . specific activity of end product}}{\text{mmol of precursor fed . specific activity of precursor}}$$

is known as the *absolute incorporation rate* and the specific activity is the radioactivity of a particular substance usually given in mC/mmol.

In the case of administration of compounds which are labelled with stable isotopes, the absolute incorporation rate is the ratio

$$\frac{\text{mmol of end product formed . atom-\% excess in end product}}{\text{mmol of precursor fed . atom-\% excess in the precursor}}$$

The absolute incorporation rate is usually expressed as a percentage in both cases.

(b) The ratio

$$\frac{\text{Specific activity (or atom-\% excess) in end product}}{\text{Specific activity (or atom-\% excess) in precursor}}$$

is known as the *specific incorporation rate* (or 'specific radiochemical yield').

The specific incorporation rate is also usually given in percentage.

(c) The ratio of the specific radioactivity, or atom-% excess, in precursor to the specific activity, or atom-% excess, in the end product is known as the *dilution*. The activity or atom-% excess in precursor is fixed as 1.

$$\text{dilution} = 1 : \frac{\text{specific activity (atom-\% excess) in precursor}}{\text{specific activity (or atom-\% excess) in end product}}$$

The absolute incorporation rate in which the total quantity of the precursor administered is correlated to the portion that is incorporated into the particular end product under investigation is usually less suited for the assessment of biochemical investigations than is the specific incorporation rate or the dilution of the precursor. It depends, to a great extent, on the quantity of the substance under investigation which is synthesized during the experimental time and is therefore usually subject to great variations. In addition, the exact determination of the total quantity of the end product is a prerequisite for the calculation of the absolute incorporation rate, a condition which cannot usually be satisfied.

The specific incorporation rate (or the dilution of the precursor) permits the calculation of the portion of the compound formed which originates from the precursor fed and that which originates from the pool of the endogenous precursor present in the organism. Thus, in the case of a specific incorporation rate of 0.1% (or a dilution of $1:1000$), every thousandth molecule of the compound under investigation is formed from the isotopically labelled precursor which is introduced from outside. The specific incorporation rate is therefore dependent on the ratio of the endogenous and exogenous precursor and is influenced by the absolute rate of synthesis during the interval of the experiment only as long as this ratio is altered. The determination of the specific activity of the end product necessary for the calculation is more simple than the absolute quantity, since only a small portion of the end product has to be purified to constant specific activity.

The specific incorporation rate (or the dilution of the precursor) permits conclusions regarding the closeness of relationship between the end product formed and the precursor fed, since it is usually higher if the added precursor can be changed to the end product in one, or at least a few, steps. The height of the incorporation rate must, however, not be overrated, since the size of the endogenous precursor pool, limits of permeability for the precursor entering the cell from outside etc., play a considerable role.

The specific incorporation rates in experiments with higher plants are most frequently smaller than 0·01%, but may amount to more than 90% in experiments with micro-organisms. The reason for this difference lies, above all, in the fact that in micro-organisms the compounds fed reach the actual site of synthesis comparatively easily, while in the case of organisms of complicated structure they must pass through a number of tissues which do not synthesize the required end product, and during transport are utilized in other reactions to a greater extent. In micro-organisms the experimental conditions can be handled in such a manner that the amount of the substance under investigation which is present at the end of the experiment is exclusively synthesized during the time of experiment, and the incorporation rates obtained are not small since the amount of the unlabelled substance under investigation which was present at the beginning of the experiment is included in the calculation.

The experimental difficulties faced in biochemical experiments on higher plants may be reduced in a few cases by using tissue cultures (callus cultures or cell suspensions). Tissue cultures can be fed like cultures of micro-organisms (the incorporation rates may be considerably higher than those obtained by using whole plants or parts of plants). Their cells are mostly uniform and because of their delicate cell walls they are also suitable for the preparation of homogenates (cf. B.3).

They are, however, not universally applicable for the investigation of secondary metabolism. In agreement with the fact that secondary products are synthesized only by certain tissues in higher plants and are the result of differentiation of the cells, the more or less undifferentiated cells of the tissue culture usually do not have the capacity to produce secondary products in large quantities. The experience obtained by molecular biologists in recent years suggests that before a general application of tissue culture may be possible, redifferentiation of the cells and induction of the enzymes taking part in the reactions of secondary metabolism must first be obtained.

In order to establish a direct biogenetical relationship between the isotopically labelled precursor fed and the end product isolated, it is necessary to degrade the isolated compound with the help of chemical methods and to determine individually the isotopic content of the particular atoms under investigation.

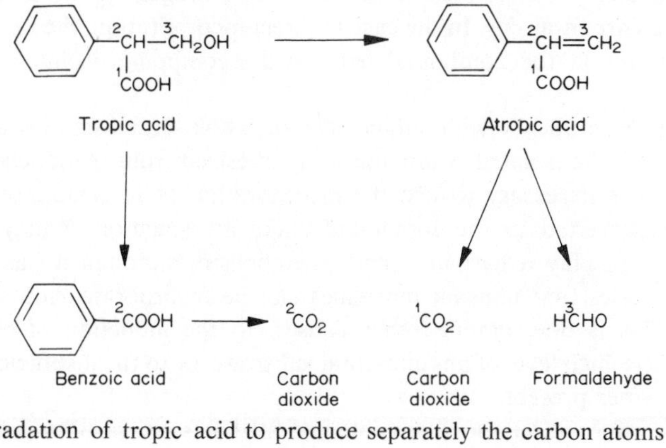

Fig. 2. Degradation of tropic acid to produce separately the carbon atoms of the side chain.

The degradation of tropic acid is briefly discussed as an example. Tropic acid originates from phenylalanine by an intramolecular shift of the carboxyl group of the amino acid (cf. D.20.3). In order to determine the isotope concentration at the individual carbon atoms of the side chain separately (fig. 2), tropic acid is first oxidatively converted to benzoic acid and this compound is subsequently decarboxylated. Carbon atom 2 is separated from all other carbon atoms and obtained as CO_2. By conversion of tropic acid to atropic acid and decarboxylation of this substance carbon atom 1 is obtained as CO_2. The methylene group can, in addition, be degraded by a periodate oxidation so that carbon atom 3 is removed as formaldehyde. By means of these reactions it is possible to determine the isotope concentration in all three carbon atoms of the side chain separately.

If there is a direct biogenetical relation between the substance fed and the isolated substance, i.e. if the substance fed is a direct precursor of the isolated compound, then only those atoms directly corresponding to those labelled in the substance fed will be labelled. If, however, the substance fed first enters the pathways of general metabolism, and the isolated product is synthesized from its conversion products, then most of the atoms will be labelled. Thus a smearing (randomization) of labelling occurs which will be greater the longer the time between feeding and extracting the material. Biosynthetic experiments with isotopically labelled compounds in which the pertinent isotope is not localized do not usually stand up to rigorous scientific criticism.

Since only the labelled portion of the molecule fed can be determined, it is necessary to label all the important atoms. This is achieved in most cases by separate experiments, in order to enable one to follow the behaviour of a particular atom in metabolism. For certain investigations, however, multiple labelling of the precursor compound is necessary. Either different atoms of the same element, or different elements participating in the structure of a compound, could be labelled with isotope. For example, if it is to be determined whether the whole molecule of an amino acid including the amino group is incorporated into an alkaloid, it is necessary to label the nitrogen of the amino group as well as the carbon atoms. In the case of direct incorporation, the ratio of both the labelled atoms in the compound fed and the compound isolated must remain the same.

Feeding experiments with intact cells or whole organisms do not show the way in which the isolated compound is synthesized from the labelled precursor. It is therefore necessary to test the incorporation of possible intermediates. If these are converted to the compound under investigation, it may be assumed that they also play a part in normal metabolism, although it has been shown in certain cases that unusual molecules can be incorporated into certain compounds. This is due, among other causes, to the induction of enzymes after exposure to a high level of an 'unnatural substrate' or to the insufficient specificity of the enzymes present.

Further information about possible intermediates can be obtained by so-called competition experiments. Here the specific incorporation rates of a labelled compound are determined with and without, the addition of greater amounts of another unlabelled compound, which is assumed to be an intermediate in the metabolic pathway by which the labelled compound is converted to the end product. If this is the case, the unlabelled compound will be incorporated into the end product, and the specific incorporation rate of the active compound will be reduced by the addition of the inactive one. If the unlabelled compound is not an intermediate then the specific incorporation rate will remain unchanged.

By such experiments it is possible to study hypothetical intermediates even when they are not available in the radioactive form. Wrong interpretations can, however, be made, due to little or no uptake of the unlabelled compound.

The material used for the experiments should not contain any other organisms, thus excluding the possibility that the compounds fed are changed, for example by micro-organisms, before they are taken up by the organism under investigation. In the case of non-sterile experiments, it is not possible to determine whether the compound fed itself, or a product originating from it is incorporated into the compound isolated. In the case of micro-organisms, cultures which are free from other organisms do not present any difficulty. Also in recent years, methods have been developed which enable one to grow higher plants (e.g. seedlings) or parts of higher plants (e.g. roots) as well as of animals, under sterile conditions; however, the experimental work involved is so great that this possibility is seldom exploited.

Since the capacity to produce certain compounds may change in the course of development of an organism, or since in a few cases only special tissues are capable of synthesizing the substance under investigation (cf. A.3), the physiology of formation must always be studied before biosynthetic investigations are undertaken. Whether the formation of the substance under investigation actually takes place during the experimental period can easily be detected in control experiments by feeding glucose-U-^{14}C, ^{14}CO$_2$ or similar substances which enter primary metabolism easily. If there is synthesis, then the isolated secondary product will be labelled after a sufficiently long interval, since it is ultimately formed from the products of primary metabolism. If no labelled secondary product is detected after feeding these unspecific precursors for some time, then the feeding of potential specific precursors would be of little use.

References for further reading

Swain, T. 'Methods Used in the Study of Biosynthesis', *Biosynthetic Pathways in Higher Plants,* ed. J. B. Pridham and T. Swain (Academic Press, London, 1965), pp. 9–36.

Schütte, H. R. *Radioactive Isotope in der organischen Chemie and Biochemie* (VEB Deutscher Verlag der Wissenschaften, Berlin, 1966).

Bubner, M. and Schmidt, L. *Die Synthese Kohlenstott-14-markierter orgamische Verbindungen* (VEB George Thieme Verlag, Leipzig, 1966).

Simon, H. and Floss, H. G. *Bestimmung der Isotopen-verteilung in markierten Verbindungen* (Springer Verlag, Berlin, 1967).

2.   Investigations with mutants

Most of the reactions which take place in living organisms are catalysed by ferments (enzymes). The formation of the enzyme system of a particular cell is regulated by the genetic material present in it and the information stored in it ultimately determines which enzymes can be synthesized in a cell. By changes in the genetic material (so-called mutations), the formation of certain enzymes is prevented, or at least the enzymes formed are so changed that they are no longer catalytically active. Organisms with changed genetic material, in contrast to the wild strain, are known as mutants. If certain enzymes are no longer detectable then the organisms are known as defective mutants.

If the enzyme in question, whether changed or no longer synthesized, is vital for the particular organism then metabolism and growth come to a halt. However, the vital metabolite which is no longer synthesized in the mutant may be introduced into the cells from outside, after which metabolism and growth proceed normally. In cases where the enzyme is missing the substrate of the missing enzyme accumulates and is usually excreted into the surrounding medium as such, or in a slightly modified form.

Since in most metabolic pathways the starting materials are converted to the end products through many intermediary stages, the reaction chain may be blocked at different places by mutation. By feeding compounds which result in formation of the end product and by isolation of the substance accumulated by a particular mutant, it is possible to elucidate the biochemical reaction chain and to determine the sequence of the individual reaction steps.

Biosynthetic research with the help of mutants has attained great importance only in the micro-organisms. Since they are small, the comparatively simple constitution of their genetic material (especially in the case of bacteria), as well as their saprophytic nutrition and fast multiplication, allows large numbers of mutants to be obtained easily by artificial means, e.g. by treatment with X-rays or chemical agents.

Studies with mutants in higher plants and animals are rendered more difficult because of the fact that these organisms are usually diploid, triploid or polyploid, i.e. every cell contains two, three or several sets of chromosomes. Mutations in which usually one chromosome is changed are therefore noticeable only when the mutant gene is dominant, i.e. asserts itself in action compared to the non-mutant genes. If the gene is recessive, the mutation can be recognized if, by breeding and crossing, cells are obtained in which all the corresponding chromosomes are formed from the mutant. Moreover, plants and animals are multicellular organisms, in which case at least a larger amount of tissue, or the whole organ, must be subjected to the action of mutagenic agents in order to obtain mutants experimentally. If mutation takes place only in a single cell, it would only be noticeable when that particular cell is still in the embryonic stage, i.e. is capable of division, and a large number of cells with changed genetic material originate from it.

Mutations which affect vital metabolic pathways are usually detectable when the defective mutants exhibit metabolic activity and growth on addition of the end product which is not synthesized. Detection of the required mutant is, however, difficult when the mutation does not affect the vital areas, e.g. most of the reactions of secondary metabolism. The identification of a suitable mutant is especially difficult in the absence of a simple test (as is possible, for example, in the case of the production of coloured compounds, by the visual comparison of the wild strain and the mutant). In addition, the enzymes taking part in secondary metabolism are often of low specificity and frequently further transform the intermediates accumulated after mutation, by pathways which are not observed in the wild strain (cf. C). In these cases it is not the intermediates of the original synthetic pathway that are accumulated in the mutants, but compounds with new structures. The interpretation of the experimental results is thus made more complicated.

References for further reading

Swain, T. 'Methods Used in the Study of Biosynthesis', *Biosynthetic Pathways in Higher Plants,* ed. J. B. Pridham and T. Swain (Academic Press, London, 1965), pp. 9–36.

Böhm, H. 'Genetik des Alkaloidmerkmals', *Biosynthese der Alkaloide,* ed. K. Mothes and H. R. Schütte (VEB Deutscher Verlag der Wissenschaften, Berlin, 1969), pp. 21–39.

Bergmann, F. Mutationschimären, Rohmaterial züchterischer Weiterbearbeitung, *Umschau* **67** (1967), pp. 791–7.

Hopwood, D. A. 'The Isolation of Mutants', *Methods of Microbiology,* ed. J. R. Norris and D. W. Ribbons (Academic Press, London, 1970), pp. 363–433.

3. Use of homogenates, enzymes and enzyme systems

Biosynthetic investigations are still frequently carried out with whole organisms, parts of organisms (e.g. certain organs), tissues or intact cells. The compounds to be fed are usually introduced into animals and plants by injection, and in the case of higher plants, also via the roots (in hydroculture) or by means of a cotton thread passing through the shoot. Cut shoots which absorb the solution of the compound to be fed by transpiration, and pieces of whole organs which are shaken together with a solution of the precursor, or tissue cultures, are often used as well.

Before the compound fed by any of these methods reaches the actual site of biosynthesis it must pass through many barriers, of which the most important appears to be the outer layer of the cytoplasm of the cell. Transport through this membrane occurs exclusively with the help of very specific transport mechanisms so that little, or none, of the compounds fed may be absorbed.

Moreover, the compounds taken up into the cell may not reach the site of biosynthesis due to compartmentation (cf. A.3), but may be utilized in another part of the cell, or stored in the case of plant cells in the vacuoles. It has been shown in a few instances that substances entering the cells from outside have a different metabolic fate to those which originate in the cell itself. This may be the case, for example, if the endogenous intermediate is strongly bound to an enzyme complex while the exogenous precursor exists in the free form in the cytoplasm.

Most of these problems caused by cell structure may be avoided if, instead of whole cells, homogenates, certain fractions of homogenates, or enzyme preparations are used. The induction and *de novo* formation of enzymes during the course of the experiment will also be ruled out. In working with homogenates, the structure of the cells and tissues is first destroyed by chemical or physical methods (e.g. the action of enzymes, ultrasonic waves, grinding or freezing) and a preparation (homogenate) is obtained which still includes all the constituents of the starting material. It is possible to divide this homogenate into individual fractions and ultimately to obtain a more or less pure enzyme preparation from them by centrifugation, precipitation and chromatography. Even when certain enzymes are structurally bound, concentration of the organelles containing the enzyme is often possible by differential centrifugation, gradient centrifugation and similar methods.

The disappearance of the precursors introduced (substrates) as well as the formation of certain metabolic products can be detected by direct or indirect methods with the help of enzyme preparations. In a few cases isotopic exchange, catalysed by enzymes, is suitable for the identification of reactions, that is if the substances taking part in the catalytic reaction remain in a measurable equilibrium with each other (the appearance of the end product formed or the disappearance of the substrate need not be proved in this case).

Thus, for example, the first step of activation of carboxylic acids (cf. C.1.2) in the following reaction is usually shown by the exchange of radioactivity between ^{32}P-pyrophosphate and ATP. In this example the radioactivity in ATP incorporated during the reverse reaction is determined.

$$R-COOH + ATP \xrightleftharpoons{enzyme} R-CO-AMP + PP$$

Working with enzymes of secondary metabolism is, however, made difficult by the fact that the quantities of the enzymes involved are extraordinarily small. In addition, technical difficulties are encountered in working with metabolically active differentiated cells of higher plants (cf. A.3), which are due both to the structure of these cells and their storage capacity for secondary natural products (cf. A.1). During homogenization, pieces of the tough cell wall break the organelles, and the mixing of the cell sap, often made highly toxic with phenolic compounds, with the cytoplasm can inhibit or destroy enzyme systems. For this reason the enzymology of secondary products in higher plants has been elucidated only to a small extent.

The finding that isolated latex of *Hevea brasiliensis* was capable of synthesizing rubber gave rise to the hope that it might be used to obtain biochemically active preparations from higher plants. Synthesis of triterpenes in *Euphorbia* latex and of alkaloids in *Papaver* latex has also been observed.

The latex which is present in special cells, the latex vessels, and which is usually obtained by cutting the organs carrying latex, alters with the physiological condition of the plant. The results obtained, therefore, are reproducible only with difficulty.

Young latex vessels are filled with a fluid in which a large number of small vacuoles are present. They unite in the course of development with the central vacuole which is gradually being formed as parts of the cytoplasm are degraded and the decomposed material is fused with it. The sap of the central vacuole no longer contains organelles, but contains enzymes which may be considered to be all that is left of the cytoplasm.

While fluid, together with vacuoles, flows out of young latex vessels, only vacuolar sap flows out of older ones. The thin cytoplasmic lining still existing along the wall and which may be rich in endoplasmic reticulum, ribosomes, mitochondria and golgi bodies, all indicative of specialized biochemical functions, remains, on the contrary, in the cell. The latex of young vessels, therefore, shows a different biochemical capacity to that of older cells.

References for further reading

Swain, T. 'Methods Used in the Study of Biosynthesis', *Biosynthetic Pathways in Higher Plants*, ed. J. B. Pridham and T. Swain (Academic Press, London, 1965), pp. 9–36.

Mothes, K and Meissner, L. Über Milchsaft, *Forschungen und Fortschritte* **38** (1964), pp. 328–31.

C.   Enzymes of general significance in secondary metabolism

Enzymology of secondary metabolism is in its infancy at the present time. Characterization of the enzymes taking part in the metabolism of secondary natural products may be regarded as one of the major aims in investigations of secondary metabolism, which may be achieved in the near future. In recent years it has been possible, especially in the case of micro-organisms, to identify a few enzymes of secondary metabolism *in vitro*, and to purify and characterize them in detail. The results obtained from such studies indicate that three types of reactions occur during the biosynthesis and degradation of secondary natural products:

(a) Reactions which are catalysed by the enzymes of secondary metabolism

These reactions constitute the largest and the most important group. Obvious differences between the enzymes of secondary and primary metabolism have not yet been found.

The following criteria help to determine whether a particular enzyme belongs to secondary metabolism:

The enzyme is formed when the particular metabolic pathway is detectable (in the case of micro-organisms, for example, at the beginning of the idiophase, cf. A.3). Though an accurate detection of *de novo* synthesis of enzymes is only possible with difficulty, certain indications are comparatively easy to obtain.

The enzyme must possess a sufficient amount of substrate specificity to rule out the possibility that it also reacts with the compounds of primary metabolism.

(b) Reactions which are catalysed by the enzymes of primary metabolism

Enzymes of primary metabolism which possess little substrate specificity (e.g. certain alcohol dehydrogenases, cf. C.2.1, and thiokinases, cf. C.1.2) may also react with the products of secondary metabolism. However, the importance of such reactions for secondary metabolism is probably less than previously supposed.

(c) Reactions which proceed spontaneously

Besides enzymatically regulated reactions, certain spontaneous reactions are important in secondary metabolism. Examples of such reactions are the coupling of radicals (cf. D.20.1.2) and the synthesis of Schiff bases (cf. D.15 and D.15.3). Considering the whole field of secondary metabolism, however, the part played by spontaneous reactions is very limited.

The general properties of a few enzyme groups whose individual members frequently participate in the reactions of secondary metabolism are discussed in the following sections.

1. Activating enzymes

1.1 Transphosphorylases

1.1.1 Phosphorylated compounds

Phosphorylated compounds (esters, amides and anhydrides of phosphoric acid) participate in a number of metabolic reactions. During the hydrolysis of the bonds through which the phosphate group is linked to the acceptor molecule, a large amount of energy may be set free (more than about 6 Kcal) in the case of a number of substances (high-energy phosphates). In the case of other substances, on the contrary, the energy set free is less (approx. 3 Kcal, low-energy phosphates) (cf. table 1).

Table 1. Energy liberated during the hydrolysis of a few compounds of biochemical interest under standard conditions.

			cal/mol
Phosphoenol pyruvate → Pyruvate	+ Phosphate		13000
Acetylphosphate → Acetate	+ Phosphate		10500
Creatine phosphate → Creatine	+ Phosphate		9000
ATP → AMP	+ Pyrophosphate		7600
ATP → ADP	+ Phosphate		7400
Pyrophosphate → Phosphate	+ Phosphate		6500
Aldose-1-phosphate → Aldose	+ Phosphate		5000
Phosphate ester → Alcohol	+ Phosphate		3000

Mesomeric structures of phosphate anion → (A simple form of writing) The dotted lines represent the fractional bonds participating in the resonance →

The energy content of the phosphate bond is dependent, to a considerable extent, on the degree of disturbance of the resonance of the phosphorylated compound. In the formation of phosphates with low energy content, the resonance of the phosphate anion, which is dependent on the easy convertibility of the mesomeric structures I–IV into one another, is diminished. One of the four mesomeric formulae is excluded from resonance in the bonding of a phosphate group with an acceptor.

In the case of energy-rich phosphates, resonance of the acceptor molecule is also diminished. In acyl phosphates, for example, the resonance of the carboxyl growth is disturbed, in guanidine phosphates that of the guanidine group, and in the case of the di- and triphosphates (e.g. ADP and ATP), the resonance of the phosphate group is disturbed (fig. 3).

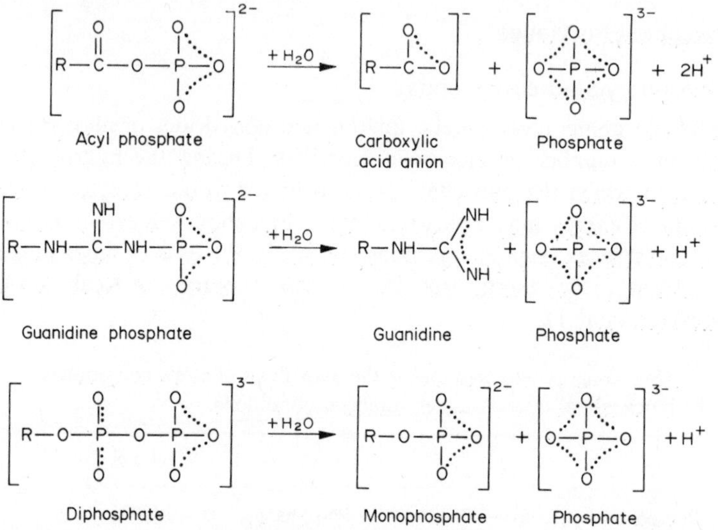

| Acyl phosphate | Carboxylic acid anion | Phosphate |

| Guanidine phosphate | Guanidine | Phosphate |

| Diphosphate | Monophosphate | Phosphate |

Fig. 3. New possibilities of resonance after the decomposition of high-energy phosphates. (The dotted lines indicate the existence of resonance.)

This diminution of resonance causes polarization of the bonds in the phosphorylated compounds. The stronger the polarization, the easier the electrophilic and nucleophilic substitution of a substance or its capacity to substitute. It is therefore reactive and is termed 'activated'. Activated phosphates are, for example, the sugar phosphates (cf. D.1.1.1) and isopentenyl pyrophosphate (cf. D.5.1).

References for further reading   Cf. C.1.1.2.

1.1.2 Transphosphorylation

The enzymes termed transphosphorylases catalyse the transfer of a phosphate group to other compounds according to the following equation:

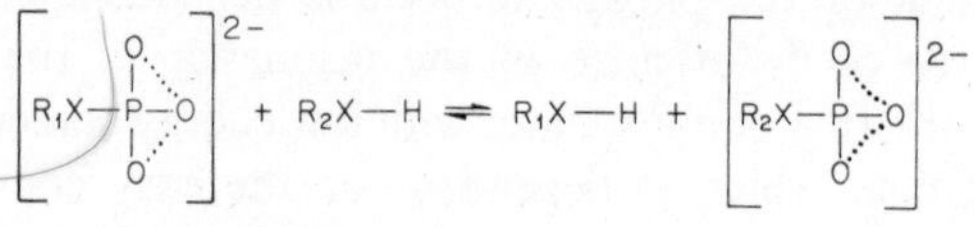

$X = O$ or N

A still shorter form of writing is usually used in which the group

$$\left[\begin{array}{c} O \\ \| \\ -P\!\!-\!\!O \\ | \\ O \end{array}\right]^{2-}$$

is represented as P:

$$R_1X\!\!-\!\!P \;+\; R_2X\!\!-\!\!H \rightleftharpoons R_1X\!\!-\!\!H \;+\; R_2X\!\!-\!\!P$$

One should, however, be careful since P standing alone represents the phosphate anion

$$\left[\begin{array}{c} O \\ \| \\ O\!\!-\!\!P\!\!-\!\!O \\ | \\ O \end{array}\right]^{3-}$$

Whenever a high-energy phosphate is meant this attribute is indicated by the symbol '~' between X and P.

If a greater amount of energy is set free during transphosphorylation then the reaction is practically irreversible. However, if the phosphate bond that is formed has approximately the same energy content as the one which was cleaved, then both phosphorylated compounds remain in equilibrium with each other. Therefore considerations of the energy content of the numerous phosphorylated compounds present in metabolism are of great importance in the understanding of possible reactions, their directions and rates.

In all living organisms a phosphate cycle exists. During oxidative reactions of primary metabolism (e.g. oxidative phosphorylations within the respiratory chain) free phosphoric acid is first incorporated into high-energy phosphates (especially ATP). The high-energy phosphates are then decomposed to phosphates with low energy content by means of transphosphorylation reactions. These are then decomposed hydrolytically, whereby free phosphoric acid is again formed which may be used further for the formation of energy rich phosphates.

Thus the energy-utilizing processes taking place in an organism are directly or indirectly linked with the decomposition of high-energy phosphates. By means of transphosphorylases, phosphorylated intermediates are formed whose further transformation creates sufficient free energy to allow favourable equilibrium conditions.

References for further reading

Boyer, P. D., Lardy, H. and Myrbäck, K. (Eds.). *The Enzymes*, Vol. 5 (Academic Press, New York, 1961).

Crane, R. K. 'Transfer of Phosphate Groups, Section a: Phosphokinases', *Comprehensive Biochemistry*, Vol. 15 ed. M. Florkin and E. H. Stotz (Elsevier Publishing Company, Amsterdam, 1964), pp. 200–11.

Cori, C. F. and Brown, D. H. 'Transfer of Phosphate Groups, Section b: Phosphomutases', *Comprehensive Biochemistry*, Vol. 15, ed. M. Florkin and E. H. Stotz (Elsevier Publishing Company, Amsterdam, 1964), pp. 212–29.

Racker, E. *Mechanisms in Bioenergetics* (Academic Press, New York, 1965).

Wang, J. H. Oxidative and Photosynthetic Phosphorylation Mechanisms, *Science* **167** (1970), pp. 25–30.

1.2 Thiokinases and thiophorases

Thioesters of coenzyme A (cf. D.9.1) are important intermediates in carboxylic acid metabolism. Two different reaction mechanisms are involved in the *de novo* synthesis of thioester grouping by thiokinases. Acylphosphate is formed in the first case (*a*) and in the second (*b*) an acyl-AMP ester occurs as the intermediate.

(a)
$$R-C\overset{O}{\underset{OH}{}} + ATP \longrightarrow R-C\overset{O}{\underset{O\sim P}{}} + ADP$$

$$R-C\overset{O}{\underset{O\sim P}{}} + CoA \longrightarrow R-C\overset{O}{\underset{CoA}{}} + P$$

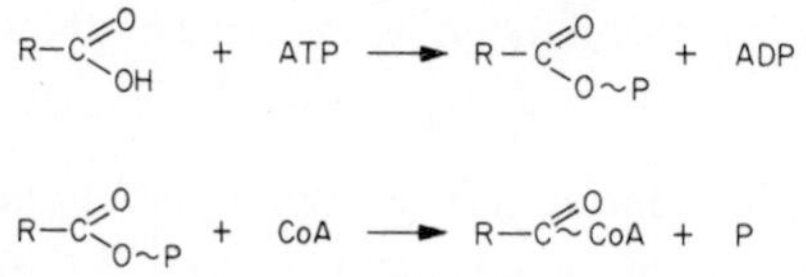

(b)
$$R-C\overset{O}{\underset{OH}{}} + ATP \longrightarrow R-C\overset{O}{\underset{O\sim AMP}{}} + PP$$

$$R-C\overset{O}{\underset{O\sim AMP}{}} + CoA \longrightarrow R-C\overset{O}{\underset{CoA}{}} + AMP$$

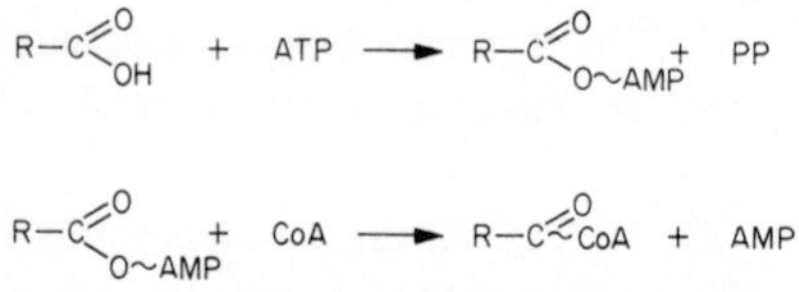

Thiophorases catalyse an exchange of the acyl group. Enzymes of this group thus catalyse the following reaction:

$$R_1-COOH + R_2-CO\sim CoA \rightleftharpoons R_1-CO\sim CoA + R_2-COOH$$

The resonance of the carboxyl group is disturbed by the thioester grouping and the CO-grouping has a considerable carbonyl character (as in the case of acyl phosphates, cf. C.1.1). Because of this the oxygen atom carries a negative, and the carbon atom a positive, fractional charge. The positive charge induces a

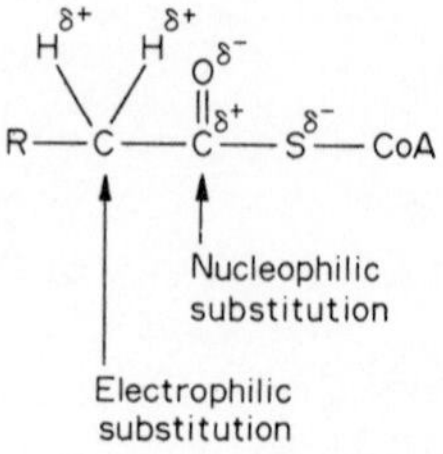

Fig. 4. Reactivity of CoA esters.

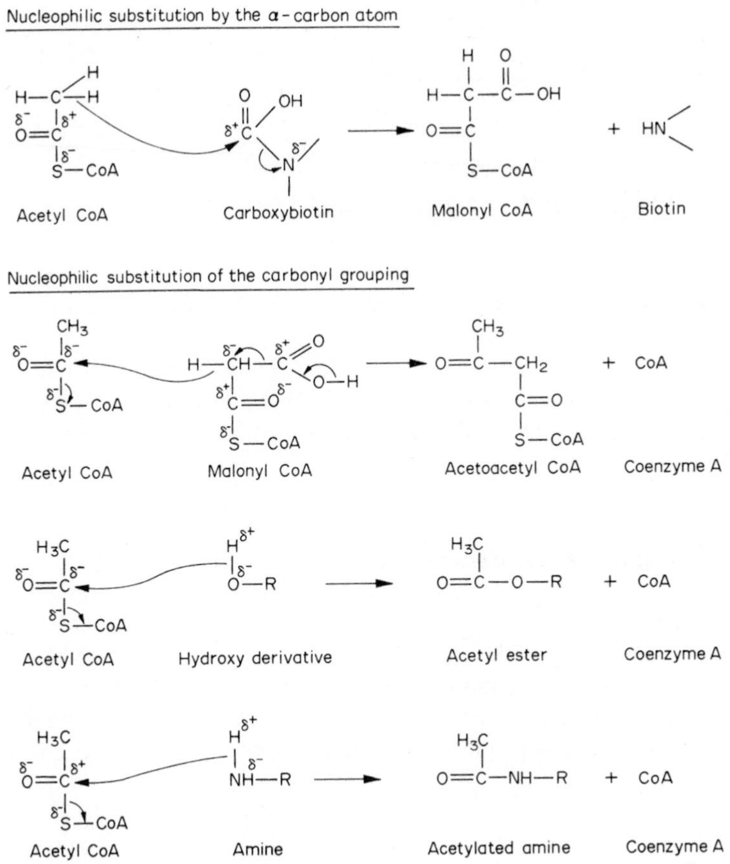

Nucleophilic substitution by the α-carbon atom

Acetyl CoA   Carboxybiotin   →   Malonyl CoA   +   Biotin

Nucleophilic substitution of the carbonyl grouping

Acetyl CoA   Malonyl CoA   →   Acetoacetyl CoA   +   CoA   Coenzyme A

Acetyl CoA   Hydroxy derivative   →   Acetyl ester   +   CoA   Coenzyme A

Acetyl CoA   Amine   →   Acetylated amine   +   CoA   Coenzyme A

Fig. 5 Possible reactions of acetyl CoA.

fractional negative charge at the α-carbon atom, causing a weakening of the C–H bonds. Thioesters can, therefore, substitute nucleophilic substances at the α-carbon atom and undergo themselves nucleophilic substitution at the carbon atom of the carbonyl group (figs. 4 and 5). They are thus 'activated' acid derivatives and can be easily transformed further in metabolism (cf. D.2 and D.3).

References for further reading

Jaenicke, L. and Lynen, F. 'Coenzyme A', *The Enzymes*, Vol. 3, ed. P. D. Boyer, H. Lardy and K. Myrbäck (Academic Press, New York, 1960).

Goldman, P. and Vagelos, Roy P. 'Acyl-Transfer Reactions (CoA-Structure, Function)', *Comprehensive Biochemistry*, Vol. 15, ed. M. Florkin and E. H. Stotz (Elsevier Publishing Company, Amsterdam, 1964, pp. 71–92.

2.   Oxidoreductases and oxygenases

While oxidoreductases (dehydrogenases or oxidases) catalyse the introduction or the removal of hydrogen or electrons, the oxygenases bring about the incorporation of oxygen which originates from molecular oxygen. Pyridine nucleotides, as well as flavin and porphyrin derivatives, are of special significance in hydrogenation and dehydrogenation reactions occurring in secondary metabolism.

2.1 Dehydrogenases containing pyridine nucleotides

2.1.1 Mechanism of reaction

In a large number of dehydrogenations the hydrogen of the substrate is transferred to the pyridine nucleotides NAD$^+$ and NADP$^+$ (cf. D.14.2). The general formulation given in fig. 6 may be written for the hydrogenations and dehydrogenations taking place with the participation of pyridine nucleotide coenzymes.

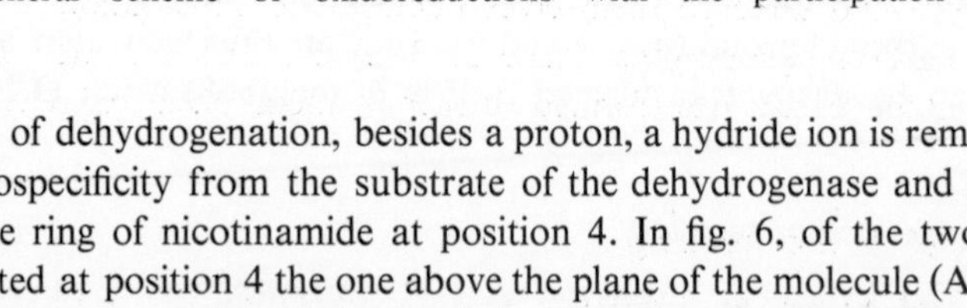

Fig. 6. General scheme of oxidoreductions with the participation of pyridine nucleotides.

In the case of dehydrogenation, besides a proton, a hydride ion is removed under strict stereospecificity from the substrate of the dehydrogenase and is linked to the pyridine ring of nicotinamide at position 4. In fig. 6, of the two hydrogen atoms located at position 4 the one above the plane of the molecule (A-side of the molecule) is known as H$_A$ and the one below the plane (B-side of the molecule) as H$_B$. H$_A$ and H$_B$ are not equivalent in relation to their transferability to substrates by dehydrogenases. According to the stereospecificity of the particular apoenzyme present, one or other hydrogen atom participates in the oxidation-reduction reactions. A few dehydrogenases are illustrated in table 2 with their stereospecificity in relation to the pyridine nucleotides.

Table 2. Stereospecificity of a few dehydrogenases with respect to H_A and H_B of their pyridine nucleotide coenzymes

H_A by	Transfer of hydrogen from H_B by
Alcohol dehydrogenases (cf. D.20.6)	D-Glucose-6-phosphate dehydrogenase (cf. D.1.3.1)
Dihydrofolic acid reductase (cf. D.8.2.3)	L-Glutamic acid dehydrogenase (cf. C.2.1.1)
Dihydroorotic acid dehydrogenase (cf. D.14.1)	Dihydrolipoic acid amide dehydrogenase (cf. D.2.1)
Mevaldic acid reductase (cf. D.5.1)	UDP-Glucose dehydrogenase (cf. D.1.3.1)

The pyridine nucleotides are only loosely bound to the enzyme protein. Usually only the apoenzyme, and not the pyridinoprotein, is obtained on isolation of the enzyme.

References for further reading

See C.2.1.2.

2.1.2 The oxidative deamination of amino acids

A reaction in which the NH_2-group of α-amino acids is liberated as ammonia, and an α-keto acid is formed is known as an oxidative deamination. Glutamic acid dehydrogenase, which transfers the hydrogen to the NAD^+ coenzyme in a reversible reaction, is the most important enzyme of this group. α-Ketoglutaric acid and ammonia are formed from glutamic acid. An imino acid which decomposes spontaneously is formed as an intermediate.

(a)
$$\underset{\text{Glutamic acid}}{\overset{\text{CH}_2-\text{CH}_2-\text{CH}-\text{COOH}}{\underset{\text{COOH}\quad\text{NH}_2}{|}}} \quad + \quad NAD \rightleftharpoons \underset{\text{Imino acid}}{\overset{\text{CH}_2-\text{CH}_2-\text{C}-\text{COOH}}{\underset{\text{COOH}\quad\text{NH}}{|}}} + NADH + H^+$$

(b)
$$\underset{\text{Imino acid}}{\overset{\text{CH}_2-\text{CH}_2-\text{C}-\text{COOH}}{\underset{\text{COOH}\quad\text{NH}}{|}}} \quad + \quad H_2O \rightleftharpoons \underset{\alpha-\text{Ketoglutaric acid}}{\overset{\text{CH}_2-\text{CH}_2-\text{C}-\text{COOH}}{\underset{\text{COOH}\quad\text{O}}{|}}} + NH_3$$

The NH_2-group of amines and other amino acids may be transferred by transamination (cf. C.4) to α-ketoglutaric acid, thus forming glutamic acid again. Glutamic acid dehydrogenase is specific for glutamic acid, and the major portion of the ammonia liberated by oxidative deamination is formed through this reaction.

Besides glutamic acid dehydrogenase, two other amino acid oxidases have been isolated from animals and micro-organisms, the substrate specificity of which is comparatively small. One type oxidatively deaminates D-amino acids and the other

L-amino acids. The enzymes contain FAD as coenzyme and the first step in the reaction is reduction of FAD to $FADH_2$, with the intermediate formation of an imino acid. The reduced coenzyme is oxidized by means of oxygen (cf. C.2.2.1).

Oxidative deamination is important in the degradation of amino acids in primary metabolism. The carbonyl derivatives (ketoacids) thus formed are precursors of a number of alkaloids (cf. D.7.6).

References for further reading

Kaplan, N. O. 'The Pyridine Coenzymes', *The Enzymes*, Vol. 3, ed. P. D. Boyer, H. Lardy and K. Myrbäck (Academic Press, New York, 1960), pp. 105–69.

Boyer, P. D., Lardy, H. and Myrbäck, K. (Eds.). *The Enzymes*, Vol. 7 (Academic Press, New York, 1963).

Colowick, S. P., van Eys, J. and Park, J. H. 'Dehydrogenation', *Comprehensive Biochemistry*, Vol. 14, ed. M. Florkin and E. H. Stotz (Elsevier Publishing Company, Amsterdam, 1966), pp. 1–98.

Sund, H. 'The Pyridine Nucleotide Coenzymes', *Biological Oxidations*, ed. T. P. Singer (Interscience Publishers, New York, 1968), pp. 603–39.

Sund, H. 'The Pyridine Nucleotide-Dependent Dehydrogenases', *Biological Oxidations*, ed. T. P. Singer (Interscience Publishers, New York, 1968), pp. 641–705.

2.2 Flavin enzymes

2.2.1 Mechanism of reaction

The flavin enzymes contain the riboflavin derivatives flavin mononucleotide (FMN) and flavin adenine dinucleotide (FAD) (cf. D.8.2.4) as the prosthetic group. They act as the carrier of hydrogen or electrons and participate in a large number of oxidoreductions. Flavin enzymes can be oxidized or reduced in a one-step or two-step reaction (fig. 7). In the case of the one-step mechanism the addition and elimination of a hydride ion plays an important role and in the two-step transformation a proton and an electron are involved. A flavin radical appears as the intermediate. This is further oxidized, reduced or disproportionated according to the following equation:

$$2 \text{ Flavin-H} \rightleftharpoons \text{Flavin-H}_2 + \text{Flavin}$$

An enzyme which catalyses the one-step reaction is, for example, glucose oxidase (cf. D.1.3.2). An enzyme which only eliminates an electron takes part in the formation of ethylene from methionine (cf. D. 10.2). Hydrogen peroxide is formed when flavin enzymes react with oxygen (cf. D.1.3.2 and D.10.2).

References for further reading

See C.2.2.2.

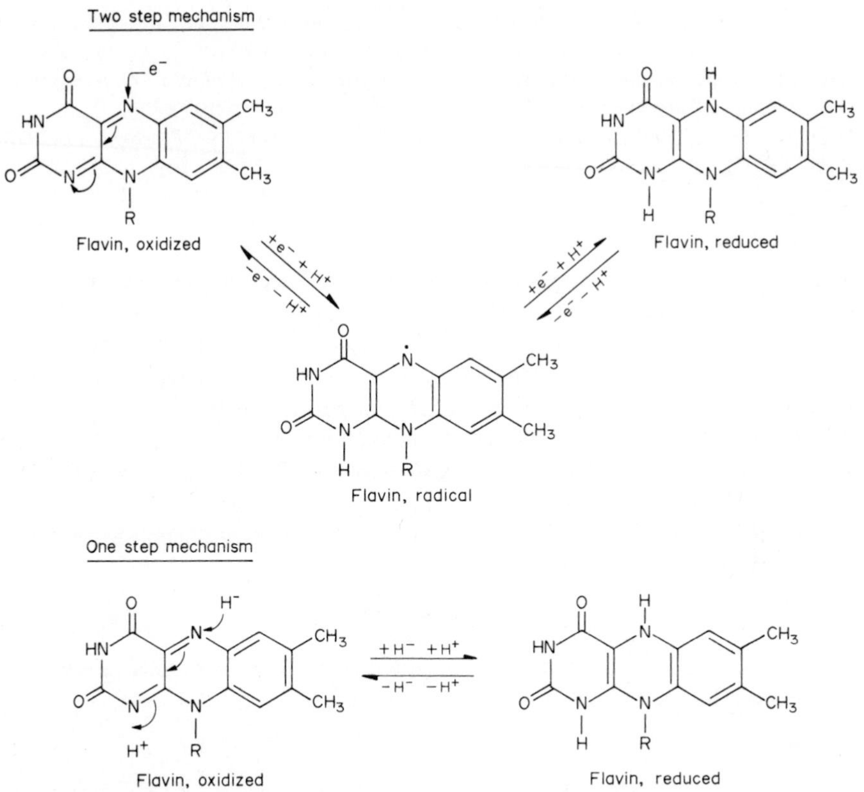

Two step mechanism

Flavin, oxidized

Flavin, reduced

Flavin, radical

One step mechanism

Flavin, oxidized

Flavin, reduced

Fig. 7. Oxidoreduction of flavins.

2.2.2. Xanthine dehydrogenases and xanthine oxidases

Xanthine dehydrogenases (usually termed xanthine oxidases if they are capable of reacting with molecular oxygen) are widespread in nature. Milk and the liver of mammals and birds are important sources for the preparation of these enzymes.

Xanthine oxidases are metalloflavoproteins which contain iron and molybdenum as well as FAD. They have little substrate specificity. Besides xanthine and hypoxanthine (cf. D.8.2.6), pteridines (cf. D.8.2.6), imidazole derivatives (cf. D.18), NADH and aldehydes are also dehydrogenated. The eliminated hydrogen may be transferred to NAD^+, oxygen, quinones and cytochrome C.

References for further reading

Boyer, P. D., Lardy, H. and Myrbäck, K. (Eds.). *The Enzymes*, Vol. 7 (Academic Press, New York, 1963).

Ehrenberg, A. and Hemmerich, P. 'Flavoenzymes: Chemistry and Molecular Biology', *Biological Oxidations*, ed. T. P. Singer (Interscience Publishers, New York, 1968), pp. 239–62.

Palmer, G. and Massey, V. 'Mechanism of Flavoprotein Catalysis', *Biological Oxidations,* ed. T. P. Singer (Interscience Publishers, New York, 1968), pp. 263–300.

Rajagopalan, K. V. and Handler, P. 'Metalloflavoproteins', *Biological Oxidations,* ed. T. P. Singer (Interscience Publishers, New York, 1968), pp. 301–37.

Neims, A. H. and Hellerman, L. 'Flavoenzyme Catalysis', *Ann. Rev. Biochem.* **39** (1970), pp. 867–88.

2.3 Phenol oxidases

Phenol oxidases are enzymes containing copper. Their catalytic action depends (as is the case with many iron containing enzymes, cf. C.2.5 and C.2.6) on change of valency of the metal, which is reduced by the substrate and oxidized by oxygen. The most important enzymes of this group belong to the laccases (such an enzyme preparation was first obtained from the Japanese lac tree *Rhus vernicifera*). Others, for example certain mixed function oxygenases of the tyrosinase type (cf. D.20.1.6), and the peroxidases (cf. C.2.4), also have properties similar to those of the phenol oxidases.

Phenol oxidases are capable of producing radicals from phenols according to the following equation:

$$E-Cu^{2+} + S-OH \longrightarrow E-Cu^{+} + S-O^{.} + H^{+}$$

The monovalent copper of the prosthetic group of the enzymes is subsequently regenerated by the action of molecular oxygen:

$$2\,E-Cu^{+} + \tfrac{1}{2}O_2 + 2\,H^{+} \longrightarrow 2\,E-Cu^{2+} + H_2O$$

Kinetic measurements have shown that the substrate radicals formed react further by reactions which are not enzymically mediated (cf. C). Several subsequent reactions may occur at the same time and a large number of different compounds may be formed (cf. D.20.4.4).

References for further reading

Malmström, B. G. and Rydén, L. 'The Copper-Containing Oxidases', *Biological Oxidations,* ed. T. P. Singer (Interscience Publishers, New York, 1968), pp. 415–38.

2.4 Peroxidases

2.4.1 Dehydrogenations and hydroxylations catalysed by peroxidases

Peroxidases catalyse the oxidation of their substrates utilizing hydrogen peroxide. Most of the plant peroxidases (e.g. the well-investigated peroxidase of horseradish) contain ferriprotoporphyrin IX (haemin) as the coenzyme. The prosthetic group of the animal peroxidases is similar, but its exact structure is not yet known. At the present time only one enzyme of this group is known whose prosthetic group possesses a fundamentally different structure (a flavin enzyme from *Streptococcus faecalis*).

Peroxidases usually transfer two hydrogen atoms of the substrate to hydrogen peroxide, forming two molecules of water. The reaction (e.g. the oxidation of phenols) may, however, also proceed in such a way that, at any one time, only one atom of hydrogen is eliminated from the substrate and the substrate radicals so formed may react further, spontaneously (cf. C.2.3). The peroxidases also introduce hydroxyl groups into certain substrates.

It is now more or less certain that the peroxidases first break down hydrogen peroxide to form hydroxyl radicals. They then cleave either two hydrogen radicals from the substrate (equation a) or one hydroxyl radical replaces a hydrogen radical of the substrate which then reacts with the second hydroxyl radical to form water (equation b).

(a)

$$SH_2 + OH^{\cdot} \longrightarrow SH^{\cdot} + H_2O$$

$$SH^{\cdot} + OH^{\cdot} \longrightarrow S + H_2O$$

$$SH_2 + 2OH^{\cdot} \longrightarrow S + 2H_2O$$

(b)

$$SH_2 + OH^{\cdot} \longrightarrow SHOH + H^{\cdot}$$

$$H^{\cdot} + OH^{\cdot} \longrightarrow H_2O$$

$$SH_2 + 2OH^{\cdot} \longrightarrow SHOH + H_2O$$

Peroxidases have little substrate specificity.

References for further reading

Paul, K. G. 'Peroxidases', *The Enzymes*, Vol. 8, ed. P. D. Boyer, H. Lardy and K. Myrbäck (Academic Press, New York, 1963), pp. 227–74.

Brill, A. S. 'Peroxidases and Catalase', *Comprehensive Biochemistry*, ed. M. Florkin and E. H. Stotz (Elsevier Publishing Company, Amsterdam, 1966), pp. 447–79.

Gibson, Q. H. 'Cytochromes', *Biological Oxidations*, ed. T. P. Singer (Interscience Publishers, New York, 1968), pp. 379–413.

2.4.2 Halogenation of organic molecules

Halogenoperoxidases catalyse the substitution of hydrogen by a halogen atom (cf. D.2.4.2 and D.2.4.3) according to the following equation:

$$SH + Halogen^{-} + H_2O_2 \longrightarrow S\text{-}Halogen + H_2O + OH^{-}$$

Such an enzyme isolated from Caldariomyces is similar in many properties to horse-radish peroxidase and contains ferriprotoporphyrin IX as the prosthetic group. A similar peroxidase has been detected in the thyroid of rats.

The halogenoperoxidases also possess little substrate specificity. They are also capable of reacting with chloride as well as bromide and iodide ions. Probably these halogen anions are first oxidized by means of hydroxyl radicals to the level of free radical and then substitute the substrate:

$$OH^{\cdot} + Halogen^{-} \longrightarrow Halogen^{\cdot} + OH^{-}$$

$$SH + Halogen^{\cdot} \longrightarrow S\text{-}Halogen + H^{\cdot}$$

$$H^{\cdot} + OH^{-} \longrightarrow H_2O$$

Fluoride ions cannot be utilized by halogenoperoxidases. This finding is in agreement with the fact that H_2O_2 (or OH-radicals) are not capable of oxidizing fluorine.

References for further reading

Fowden, L. The Occurrence and Metabolism of Carbon-Halogen Compounds, *Proc. Roy. Soc.* **B 171** (1968), pp. 1–89.

2.5 Dioxygenases

The oxidation of organic compounds by means of molecular oxygen is catalysed by oxygenases. Dioxygenases introduce both the atoms of an oxygen molecule into the substrate.

All dioxygenases investigated up until now contain iron, which is either directly linked with the enzyme protein or exists as ferroprotoporphyrin IX (haem). The metal activates the oxygen linked to the enzyme, and this is associated with polarization of the charge of the complex:

$$E-Fe^{2+} + O_2 \rightleftharpoons E-Fe^{2+}\text{-}O\text{-}O \rightleftharpoons E-Fe^{3+}\text{-}O\text{-}O$$

The enzyme-oxygen complex adds to the substrate with the elimination of a proton. The oxygen molecule is thus transferred to the substrate and the $E-Fe^{2+}$ complex is again formed.

Frequently C–C double bonds are broken and two carbonyl groups formed (cf. D.19.5 and D.19.5.3) by means of dioxygenases. Possibly cyclic peroxides are intermediate products. The sequence of reactions outlined in fig. 8 may be formulated for the enzyme metapyrocatechase which breaks the aromatic ring of catechol adjacent to both the hydroxyl groups. In certain cases degradation of the carbon chain does not occur, e.g. in the formation of prostaglandins (cf. D.2.3.2).

References for further reading

Hayaishi, O. 'Oxygenases (Oxygen-Transfering Enzymes)', *Biological Oxidations*, ed. T. P. Singer (Interscience Publishers, New York, 1968), pp. 581–601.

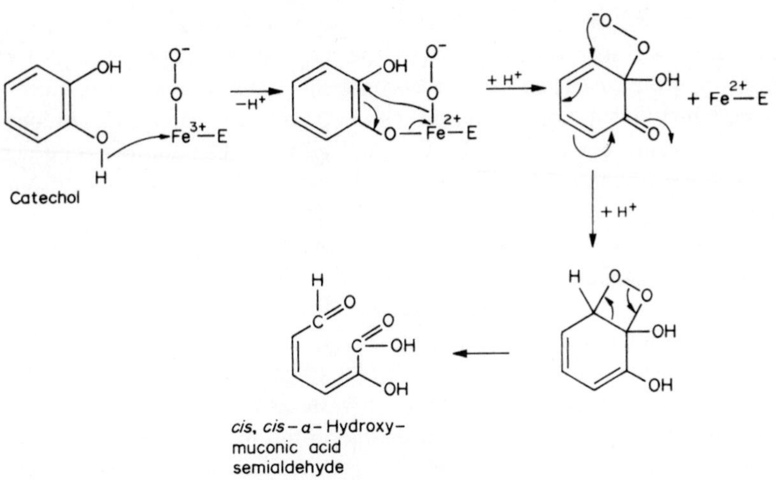

Fig. 8. Possible mechanism of reaction of metapyrocatechase-oxygen complex with catechol.

2.6 Mixed function oxygenases

2.6.1 Structure and mechanism of reaction

Mixed function oxygenases are enzymes which transfer only one oxygen atom of an oxygen molecule to the substrate while the second is, as a result, reduced to water.

$$S + O_2 + RH_2 \rightleftharpoons SO + H_2O + R$$

RH_2 = Reduced cofactor
R = Oxidized cofactor

A large number of mixed function oxygenases (e.g. the steroid hydroxylases) consist of a number of individual enzymes which are combined to form a multi-enzyme complex (fig. 9). The reaction between molecular oxygen and the

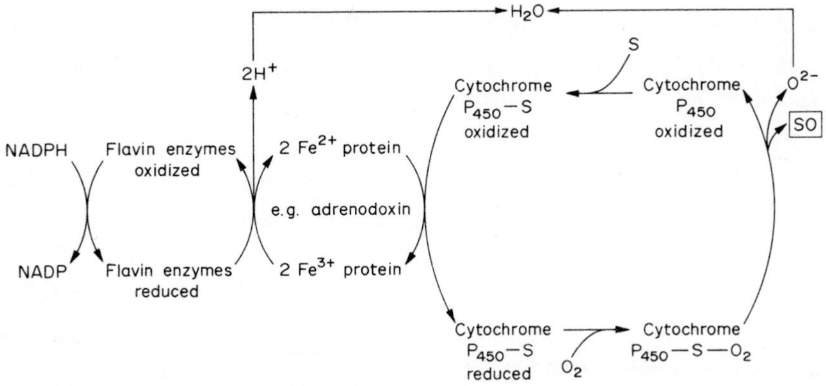

Fig. 9. Structure of a typical mixed function oxygenase system.

substrate (the actual oxygenation) is in the case of these enzyme complexes catalysed by a cytochrome (cf. D.8.1). The substrate is first linked to the oxidised iron-protoporphyrin. Then the cytochrome-substrate complex is reduced by an enzyme which has a 'non-haem-iron' as the prosthetic group. The reduced cytochrome-substrate complex reacts with molecular oxygen and breaks down with the liberation of the oxygenated substrate and the formation of oxidized cytochrome.

Pyridine nucleotides serve as the electron donor for most mixed function oxygenases. Although NADPH takes part most frequently, NADH or both compounds may also be used in certain cases. The hydrogen of the reduced pyridine nucleotides is accepted by flavin enzymes (cf. C.2.2). These are in turn oxidized by the non-haem-iron proteins (e.g. rubredoxin, adrenodoxin) mentioned above. In the case of a few enzymes other redox systems play a role, e.g. in the case of phenylalanine hydroxylase, dihydrobiopterin/tetrahydrobiopterin is involved. The whole process resembles electron transfer within the respiratory chain, an extremely important process in primary metabolism.

In a few, widespread, mixed function oxygenases (e.g. in enzyme systems which are capable of oxidizing the proline and lysine groups in the polypeptide precursors of collagen) ascorbic acid (cf. D.1.3.3) serves as the reducing equivalent in place of pyridine nucleotides. In a number of cases, e.g. amino acid oxygenases (cf. D.2.6.3) or tyrosinase (cf. D.20.1.6), the necessary hydrogen atoms are directly eliminated from the substrate during the mixed function oxidation (internal mixed function oxidation).

Several groups of mixed function oxidations can be distinguished on the basis of substrate specificity. The types of reactions catalysed by these enzymes which are most important for secondary metabolism are discussed in the following.

References for further reading   Cf. C.2.6.5.

2.6.2 Oxygenation of amines and thio compounds

A number of mixed function oxygenases are capable of oxygenating amines and thio compounds to hydroxy derivatives and oxides respectively. The following types of reactions are involved:

$$\geq\!NH \longrightarrow \geq\!N-OH \quad \text{(cf. D.21.2.)}$$

$$\geq\!N \longrightarrow \geq\!N\!\rightarrow\!O \quad \text{(cf. D.14.2.)}$$

$$-S-H \longrightarrow -S-OH \quad \text{(cf. D.9.1.)}$$

$$\geq\!S \longrightarrow \geq\!S\!\rightarrow\!O \quad \text{(cf. D.9.2.)}$$

References for further reading   Cf. C.2.6.5.

2.6.3 Amino acid oxygenases

Amino acid oxygenases are internal mixed function oxygenases. They catalyse the following reaction:

$$R-CH-COOH \;+\; O_2 \longrightarrow R-C=O \;+\; CO_2 \;+\; H_2O$$

Amino acid (C_n)   Acid amide (C_{n-1})

γ-Guanidobutyramide is formed from L-arginine, δ-aminovaleramide from L-lysine and indole-3-acetamide from L-tryptophan.

References for further reading   Cf. C.2.6.5

2.6.4 Hydroxylation of tetragonal carbon atoms and oxidative demethylation

During the hydroxylation of tetragonal carbon atoms by means of mixed function oxygenases a hydrogen atom is stereospecifically replaced by a hydroxyl group. The hydrogen atoms located in the neighbourhood do not undergo any change either with respect to their position or their configuration. The mixed function oxygenases directly attack the electrons of the C–H bond.

The oxidative elimination of O-methyl and N-methyl groups is probably initiated by hydroxylation of the CH_3-group as shown below. The hydroxy derivatives formed decompose spontaneously to amino or hydroxy compounds and formaldehyde.

$$>N-CH_3 \longrightarrow >N-CH_2-O-H \rightleftharpoons >N-H + HCHO$$
$$-O-CH_3 \longrightarrow -O-CH_2-O-H \rightleftharpoons -O-H + HCHO$$

Only in those few cases where they are immediately stabilized by glycosylation can the hydroxylated primary products be directly detected.

References for further reading   Cf. C.2.6.5.

2.6.5 Hydroxylation of unsaturated or aromatic compounds and the NIH shift.

An epoxide is formed as the first reaction product if mixed function oxygenases attack a C–C double bond. In the case of aromatic compounds the epoxides are very unstable. Epoxide derivatives may be further transformed by various reactions (fig. 10).

Pathway I leads through the addition of water to dihydrodiol, which can either be dehydrogenated to dihydroxy-derivatives (cf. D.19.5.1), or may be transformed to monohydroxylated compounds with the elimination of water.

Pathway II leads directly to the monohydroxylated compounds. The epoxide is opened by an electrophilic agent (e.g. a proton) and the cationic intermediate formed later stabilizes with the elimination of a proton (cf. D.5.7.1).

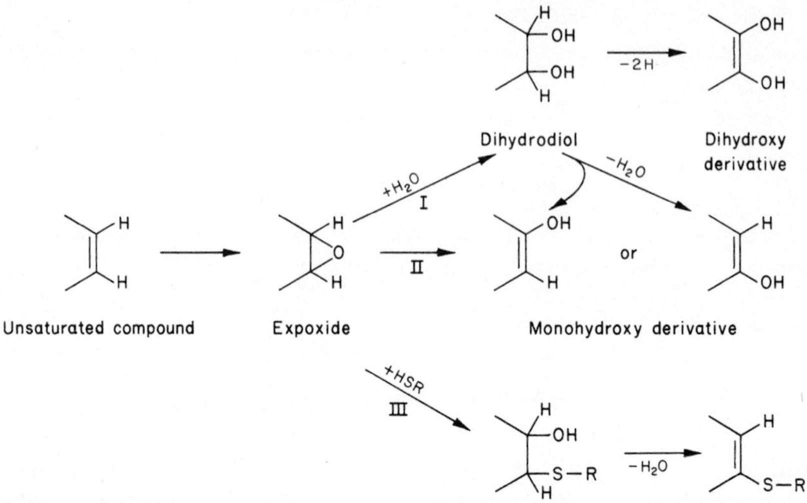

Fig. 10. Possibilities of reaction of unsaturated compounds after mixed function oxygenation.

Pathway III consists of the degradation of the epoxide by means of a compound containing an SH-group. Thioethers are formed from it by the elimination of water (cf. D.9.5).

A shift of substituents is observed in the case of pathway II. This shift was observed by workers at the National Institute of Health and is termed the 'NIH-shift' after the initial letters of this institute. Enzymes in which such a shift can be observed are phenylalanine hydroxylase (cf. D.20), tryptophan-5-hydroxylase (cf. D.19.1) and 4-hydroxyphenylpyruvate oxygenase (cf. D.20.2). In the formation of 5-hydroxytryptophan from tryptophan, a tritium atom at position 5 is shifted to position 4, and in the formation of tyrosine one at position 4 is shifted to positions 3 and 5 which are equivalent to one another.

This conversion is based on the 1,2 shift of a hydride ion or the anion of another substituent located at the hydroxylated C-atom (e.g. a chloride or a bromide anion or a negatively charged alkyl group). It takes place when the compound shown as I in the figs. 11 and 12 cannot stabilize itself by the shift of electrons (as, for example, in the formation of 2,5-dihydroxybenzoic acid, fig. 12). From product II that is formed the grouping that is most loosely bound is preferentially eliminated. In the presence of both tritium and a hydroxyl group (in the formation of 2,5-dihydroxybenzoic acid), the tritium atom is eliminated, while in the presence of tritium and hydrogen the hydrogen atom is removed since the C–T bond is considerably stronger than the C–H bond.

Shifts of hydride ions comparable to the NIH shift have also been found during the methylation of compounds with isolated double bonds (cf. D.2.2.3). The intermediate product originating by addition of the methyl cation (cf. fig. 43) corresponds to intermediate I in fig. 11 and stabilizes itself according to the rules

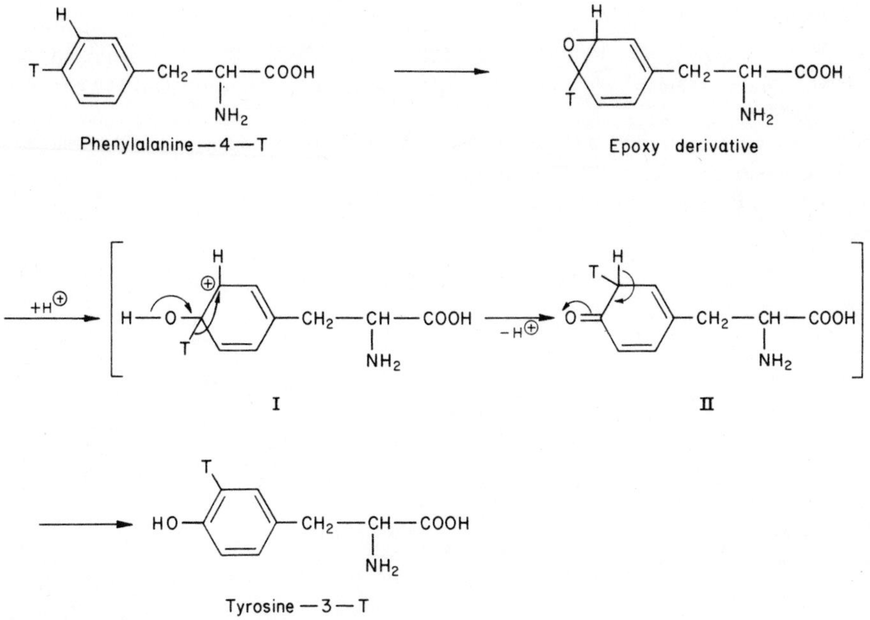

Fig. 11. NIH shift during the hydroxylation of phenylalanine.

mentioned above. Corresponding intermediary products, and also the shift of the hydride ion, are to be expected during the attack of an isopentenyl cation, but so far they have not been detected.

The formation of an epoxide as an initial reaction in mixed function oxidation provides a possible simple explanation for the formation of compounds from phenylalanine hydroxylated at the *m*-position (cf. D.6.4.2) and the formation of 4-hydroxytryptamine (cf. D.19.1). It is sufficient in these cases for the epoxide ring of the intermediate formed to be opened at the other, 'unusual', side.

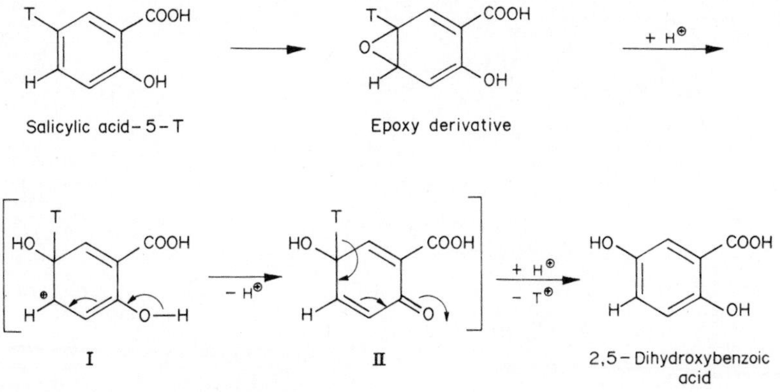

Fig. 12. Hydroxylation of salicylic acid.

References for further reading

Guroff, G., Daly, J. W., Jerina, D. M., Benson, J., Witkop, B., and Udenfriend, S. Hydroxylation Induced Migration: The NIH-Shift, *Science* (*Washington*) **157** (1967), pp. 1524–30.

Nover, L. Chemische und biologische Aspekte der Wirkungsweise mischfunktioneller Oxygenasen, *Pharmazie* **24** (1969), pp. 361–78.

Hayaishi, O. Enzymic Hydroxylation, *Ann. Rev. Biochem.* **38** (1969), pp. 21–44.

3. Enzymes of 'one carbon' metabolism

3.1 Carboxylases and transcarboxylases

It was shown for the first time in 1958 that biotin (cf. D.9.4) participates in carboxylation and decarboxylation reactions and that it is the prosthetic group of acetyl-CoA carboxylase (cf. D.2.2). Other biotin-containing enzymes have since been isolated (cf. D.3 and D.12.1).

These enzymes are either transcarboxylases, which catalyse the following reactions:

$$R_1\text{—COOH} + \text{Biotin—enzyme} \longrightarrow R_1 + CO_2\text{—Biotin—enzyme}$$

$$CO_2\text{—Biotin—enzyme} + R_2 \longrightarrow R_2 + \text{COOH} + \text{Biotin—enzyme}$$

$R_1 = $ Carboxy group donor
$R_2 = $ Carboxy group acceptor

or carboxylases which, utilizing ATP, can activate free carbon dioxide:

$$HCO_3^- + ATP + \text{Biotin—enzyme} \longrightarrow CO_2\text{—Biotin—enzyme} + ADP + P$$

The mechanism of the carboxylation reaction was first worked out in the case of α-methylcrotonyl carboxylase (cf. D.12.1) and these studies have since been extended to other carboxylases and transcarboxylases. These studies have shown that the carboxy group reacts with biotin to form N^1-carboxybiotin (fig. 13).

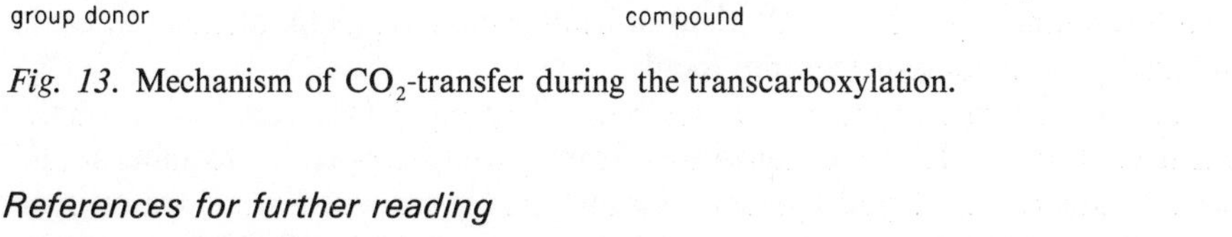

Carboxy   Biotin   Decarboxylated N^1— Carboxybiotin
group donor   compound

Fig. 13. Mechanism of CO_2-transfer during the transcarboxylation.

References for further reading

Ochoa, S. and Kaziro, Y. 'Carboxylases and the Role of Biotin', *Comprehensive Biochemistry*, Vol. 16, ed. M. Florkin and E. H Stotz (Elsevier Publishing Company, Amsterdam, 1965), pp. 210–49.

Knappe, J. Mechanism of Biotin Action, *Ann. Rev. Biochem.* **39** (1970), pp. 757–76.

3.2 Transformylases, hydroxymethyltransferases and methyltransferases

Tetrahydrofolic acid (THF) holds a central position in one-carbon metabolism (cf. D.8.2.3). Formyl groups, as well as hydroxymethyl and methyl groups, may be transferred to other compounds from the corresponding THF derivatives with the help of the enzymes mentioned in the above title.

Serine is the most important source of one-carbon units. In the presence of THF this amino acid can be degraded by the enzyme serine hydroxymethylase to form glycine and N^5,N^{10}-methylene THF (fig. 14). On hydrolysis of the latter compound THF and formaldehyde are formed. N^5,N^{10}-methylene THF therefore is also called 'activated formaldehyde'.

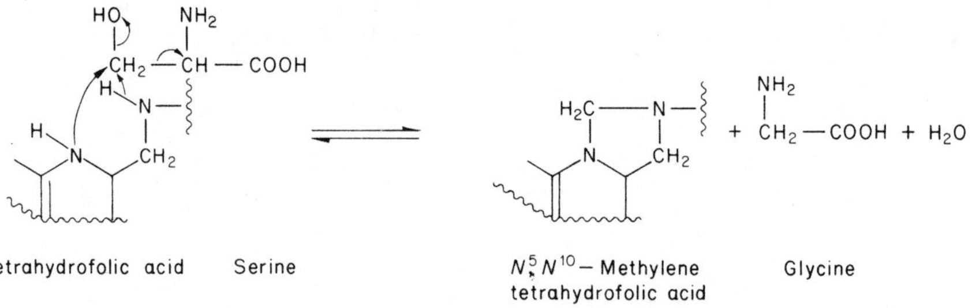

Tetrahydrofolic acid   Serine $N^5_*N^{10}$ – Methylene tetrahydrofolic acid   Glycine

Fig. 14. De novo formation of N^5,N^{10}-methylenetetrahydrofolic acid

N^5,N^{10}-Methylene THF can be converted to N^{10}-formyl-THF ('activated formic acid') by oxidation, and to N^5-methyl-THF ('activated methanol') by reduction (fig. 15). Both reactions are reversible and are catalysed by dehydrogenases (cf. C.2.1). It is not yet clear whether the formation of N^{10}-formyl-THF proceeds via N^5,N^{10}-methylenedihydrofolic acid or via anhydrocitrovorum factor. Anhydrocitrovorum factor is converted to N^{10}-formyl-THF by means of the enzyme, methenyl-THF cyclohydrolase.

N^{10}-formyl-THF may also be formed from formic acid, THF and ATP in a reaction catalysed by the enzyme formyl-tetrahydrofolic-acid synthetase (fig. 16). In rare cases the N^5-formyl derivative may be formed by an analagous reaction and converted to N^{10}-formyl-THF in a subsequent ATP-dependent reaction. N^{10}-formyl THF may be obtained via N^5-formimino-THF, which originates either from formimino glycine (cf. D.8.2.6) or from formimino glutamic acid (cf. D.18), and also via anhydrocitrovorum factor.

Anhydrocitrovorum factor as well as N^{10}-formyl-THF (and to a smaller extent N^5-formyl-THF) are active as formyl group carriers in metabolism (cf. D.8.2.1 and D.7.2). Hydroxymethyl groups are transferred in analogous reactions through the participation of N^5,N^{10}-methylene THF (cf. D.14.1), as in the formation of serine (fig. 14).

The transfer of methyl groups from N^5-methyl-THF to acceptor molecules has been shown to occur in the formation of thymine and 5-methylcytosine (cf.

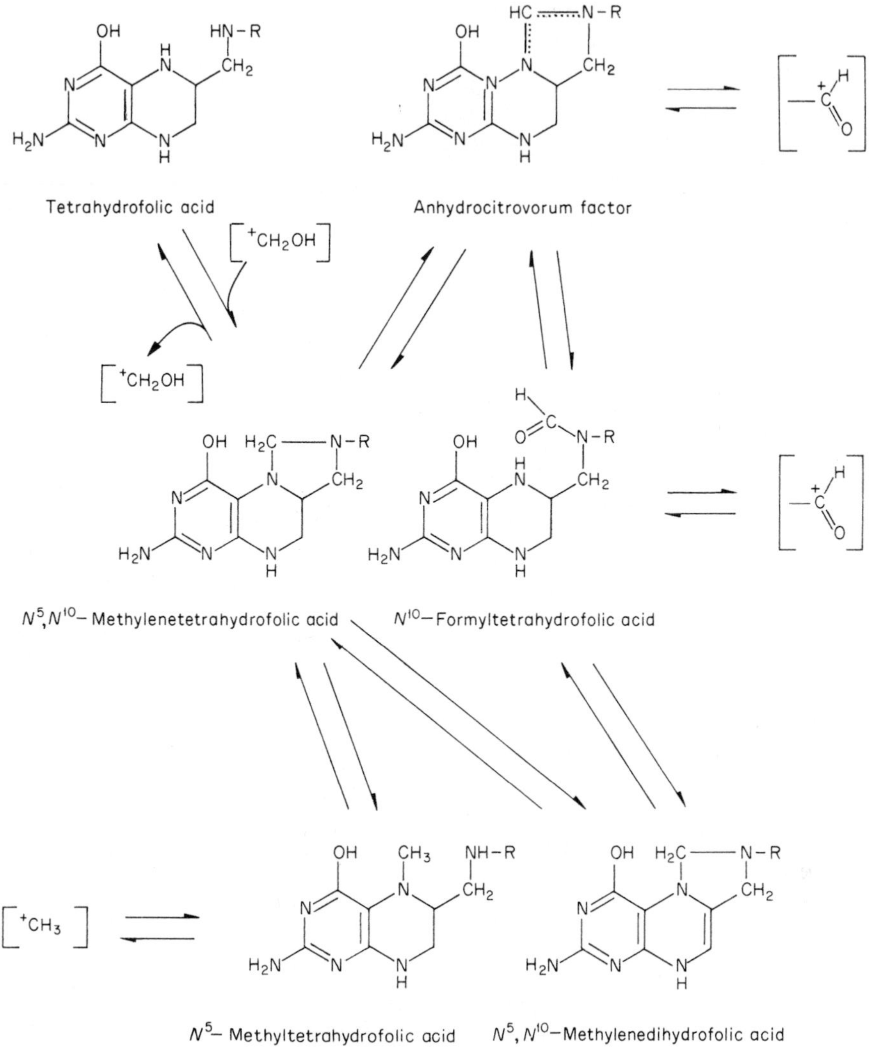

Fig. 15. Conversions of N^5, N^{10}-methylenetetrahydrofolic acid.

D.14.1) and in the biosynthesis of methionine (cf. D.10). The latter reaction is catalysed by the enzyme N^5-methyltetrahydrofolic acid homocysteine methyltransferase which contains cobalamin as the prosthetic group. Methylcobalamin is formed as the intermediate (fig. 141). Methionine is the most important donor of methyl groups in metabolism (cf. C.3.3).

References for further reading

See C.3.3.

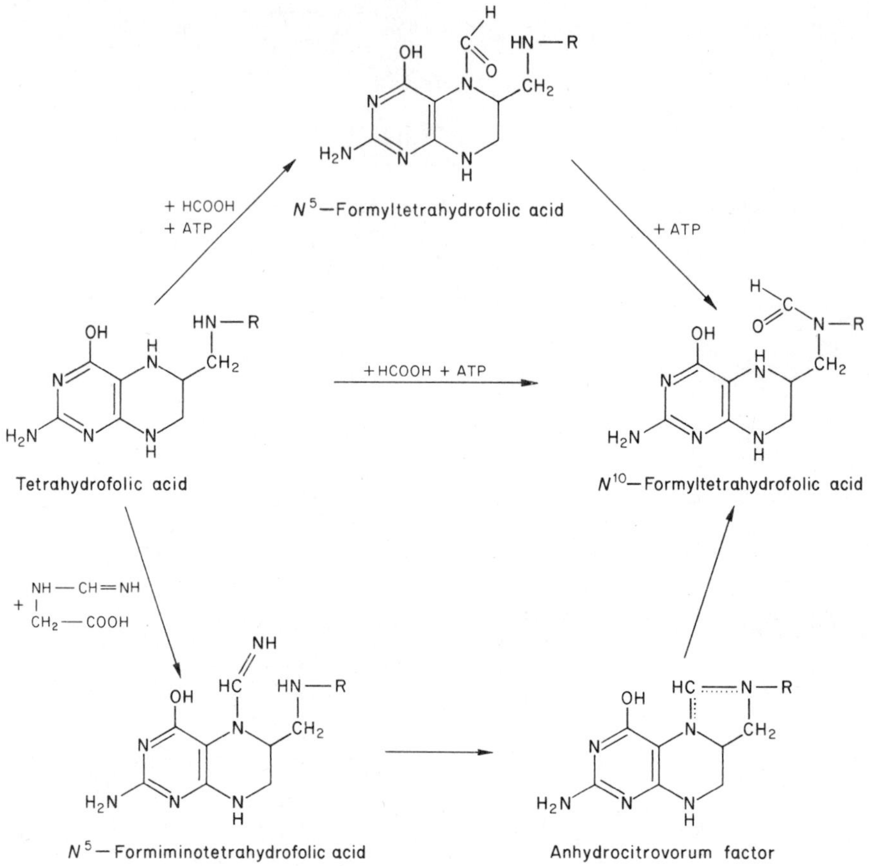

Fig. 16. Formation of N^5-formyl-, N^5-formimino-, N^{10}-formyltetrahydrofolic acid and anhydrocitrovorum factor.

3.3 Methyl transferases

The methyl groups present in many secondary natural products originate from the *S*-methyl group of methionine by means of a transmethylation reaction. The activation of methionine necessary for this transfer takes place by reaction with ATP, with the formation of the sulphonium compound, *S*-adenosyl methionine. The methyl group is easily eliminated from this compound and transferred to other compounds, forming *S*-adenosyl-L-homocysteine as a result (fig. 17). This latter compound may be hydrolysed to adenosine and homocysteine. A few compounds which are formed by methylation with *S*-adenosyl methionine are given in table 3.

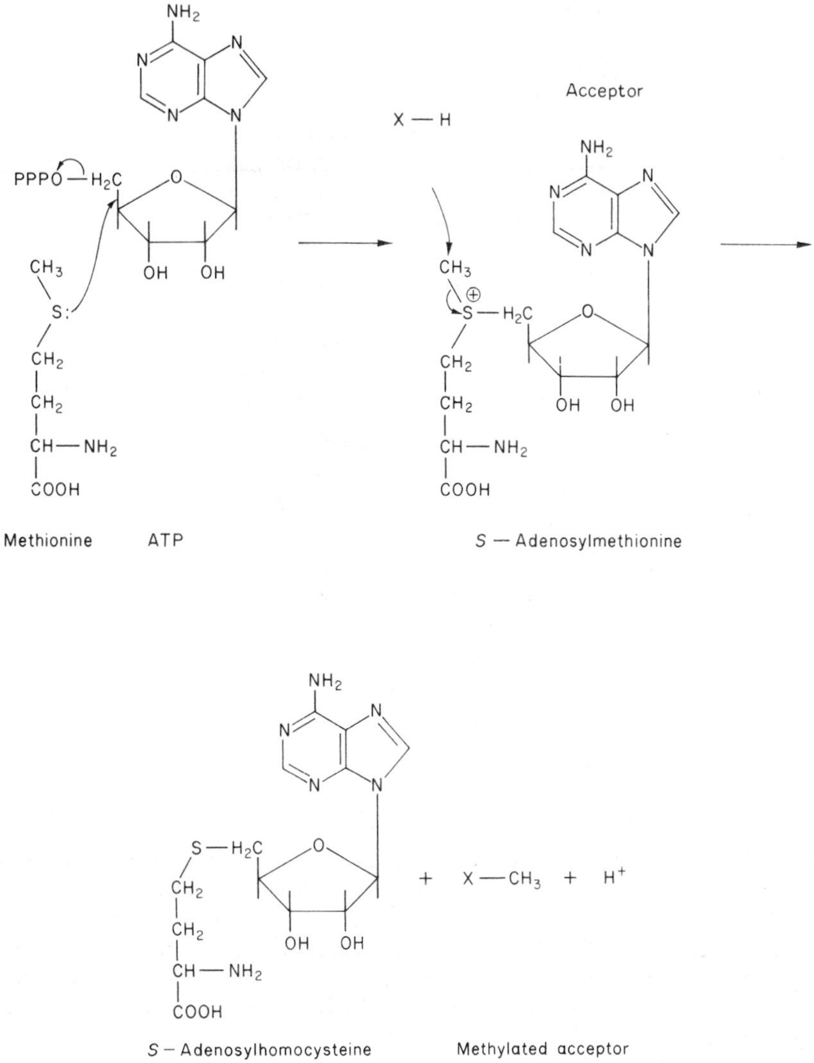

Fig. 17. Formation and degradation of *S*-adenosyl methionine.

During transmethylation all the hydrogen atoms of the methyl group are transferred. However, during the *C*-methylation of unsaturated compounds a hydrogen atom is eliminated in a secondary reaction (cf. D.2.3.1 and D.5.7.2). References have recently accumulated which indicate that preceded methylation reactions are of great importance in the flow of compounds in particular pathways of secondary metabolism (cf. D.20.1.2).

Table 3. Some compounds whose
methyl groups originate from
methionine

(*a*) Compounds with *O*-methyl groups

Pectin	(cf. D.1.1.2)
Colchicine	(cf. D.20.1.4)
Ferulic acid, sinapic acid	(cf. D.20.4.1)
Lignin	(cf. D.20.4.4)

(*b*) Compounds with *N*-methyl groups

Betaine	(cf. D.7.2)
N^1-methyladenine	(cf. D.8.2.2)
Nicotine	(cf. D.14.2)
Gramine	(cf. D.20.1.1)

(*c*) Compounds with *C*-methyl groups

Mycarose and noviose	(cf. D.1.2.3)
Tuberculostearic acid	(cf. D.2.3.1)
Ergosterol	(cf. D.5.7.2)
Vitamin B_{12}	(cf. D.8.1)

(*d*) Compounds with *S*-methyl groups

S-Methylcysteine	(cf. D.9.2)

(*e*) Other compounds

Trimethylarsine
Dimethylselenide

References for further reading

Mudd, H. and Cantoni, G. L. 'Biological Transmethylation, Methyl Group Neo-genesis and Other "One Carbon" Metabolic Reactions Dependent upon Tetra-hydrofolic Acid', *Comprehensive Biochemistry*, Vol. 15, ed. M. Florkin and E. H. Stotz (Elsevier Publishing Company, Amsterdam, 1964), pp. 1–47.

Luckner, M. 'Purine, Pteridine und Alloxazine', *Biosynthese der Alkaloide*, ed. K. Mothes and H. R. Schütte (VEB Deutscher Verlag der Wissenschaften, Berlin, 1969), pp. 568–92.

Schütte, H. R. 'Methylierung und Transmethylierung', *Biosynthese der Alkaloide*, ed. K. Mothes and H. R. Schütte (VEB Deutscher Verlag der Wissenschaften, Berlin, 1969), pp. 123–67.

Blakley, R. L. *The Biochemistry of Folic Acid and Related Pteridines* (North-Holland Publishing Company, Amsterdam, 1969).

3.4 α-Keto acid decarboxylases

Enzymes which degrade α-keto acids to CO_2 and an aldehyde shorter by one carbon atom are widespread in nature. They contain thiamine pyrophosphate (cf. D.10.3) as coenzyme. It is thought that the keto acid is added in a reversible reaction to carbon atom 2 of the thiazole ring of thiamine pyrophosphate forming an α-hydroxy acid derivative. This compound is then decarboxylated to form an aldehyde and thiamine pyrophosphate again. In the specific case of pyruvate decarboxylase, α-lactyl-2'-thiamine pyrophosphate ('activated pyruvate') is formed first from pyruvate and is then converted to acetaldehyde via α-hydroxy-ethyl thiamine pyrophosphate ('activated acetaldehyde') (fig. 17a) (cf. D.2.1 for the formation of acetyl-CoA from α-hydroxy-ethyl-2'-thiamine pyrophosphate).

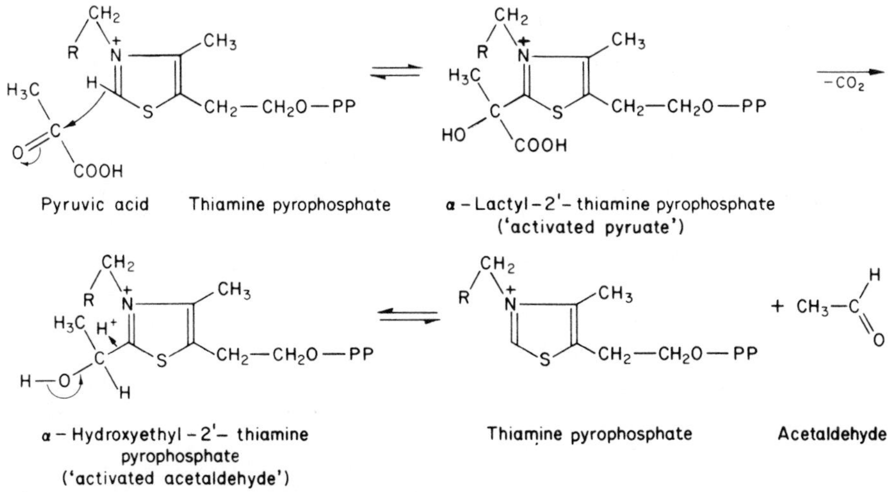

Fig. 17a. Formation of acetaldehyde from pyruvate.

References for further reading

Utter, M. F. 'Nonoxidative Carboxylation and Decarboxylation', *The Enzymes*, Vol. 5, ed. P. D. Boyer, H. Lardy and K. Myrbäck (Academic Press, New York, 1961), pp. 319–40.

4. Transaminases and amino acid decarboxylases

The coenzyme of both these groups of enzymes is pyridoxal-5′-phosphate, a compound which is capable of forming Schiff bases with both amino acids as well as amines with the elimination of water (fig. 18). The Schiff base provides a conjugated system of double bonds extending from the reaction site to the strongly electrophilic pyridine nitrogen atom, and aiding the displacement of a pair of electrons adjacent to the α-carbon of the amino acid. Depending on the type of enzyme, either a proton is eliminated, as in the case of amino acids and amines (amino acid transaminases, aldehyde transaminases, cf. D.10.2 and D.7.3) (fig. 18, pathway *a*), or the carboxyl group is lost as in the case of amino acids (amino acid decarboxylases, cf. D.7.3) (fig. 18, pathway *b*).

The intermediate II formed by pathway *a* is hydrolytically degraded to pyridoxamine-5′-phosphate and an α-keto acid (or an aldehyde), after which the amino group of the pyridoxamine may be transferred to another α-keto acid or an aldehyde (transamination).

The intermediate IV formed by pathway *b* (decarboxylation) forms an amine and pyridoxal-5′-phosphate by hydrolysis. The latter compound may enter the cycle again.

References for further reading

Guirard, B. M. and Snell, E. E. 'Vitamin B_6 Function in Transamination and Decarboxylation Reactions', *Comprehensive Biochemistry*, Vol. 15, ed. M. Florkin and E. H. Stotz (Elsevier Publishing Company, Amsterdam, 1964), pp. 138–99.

Meister, A. *Biochemistry of the Amino Acids*, Vol. I (Academic Press, New York, 1965).

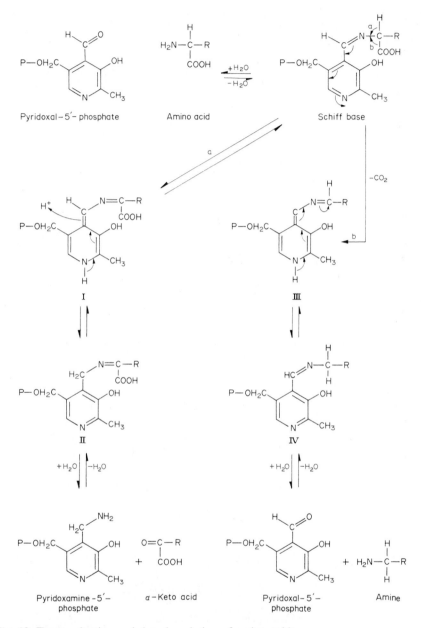

Fig. 18. Transamination and decarboxylation of amino acids.

D. Biosynthesis and metabolism of secondary plant and animal products

All secondary natural products are derived ultimately from compounds which originate during primary metabolism. Two groups of precursors may be differentiated.

Substances like acetyl-CoA and mevalonic acid, as well as certain sugars and a few amino acids which are produced and metabolized in great quantities during primary metabolism, belong to one group. Their conversion to secondary natural products may be considered as a deviation from a purely utilitarian purpose, the so-called 'luxury' of metabolism.

On the other hand there are substances like the aromatic amino acids phenylalanine, tyrosine and tryptophan, which can obviously only be completely degraded with difficulty, and are therefore converted by means of various reactions to a form capable of being stored or excreted.

In the following chapters the substances discussed are arranged according to the precursors from which they arise and, consequently, according to their relationship with primary metabolism. Biogenetically related substances are discussed together. Analogous reactions starting from other precursors and structurally similar substances which are synthesized in different ways are as far as possible frequently referred to by cross references.

1. Secondary natural products originating from sugars

Polyhydroxy aldehydes (aldoses) or polyhydroxy ketones (ketoses) are referred to as sugars, and according to the number of oxygen atoms contained in the molecule they may be further differentiated into trioses, tetroses, pentoses, hexoses etc. The most important naturally occurring simple sugars (monosaccharides) are shown in fig. 19. A consideration of their structure shows that most of them carry an oxygen grouping at each carbon atom. In the deoxy sugars, however (e.g. deoxyribose, rhamnose and fucose), certain carbon atoms do not possess an oxygen-containing group.

According to the above definition then L-rhamnose and L-fucose could be considered as methyl pentoses. However, the conventional nomenclature of 'deoxyhexoses' for sugars with six carbon atoms and five oxygen atoms, and of 'deoxypentoses' for those with five carbon atoms and four oxygen atoms will be used in the following discussion.

In the formation of cardioactive glycosides and pregnane derivatives occurring in higher plants (cf. D.5.7.2) sugars participate which differ in structure from those common in nature. They possess 'unusual' configurations or methylated hydroxy groups on certain carbon atoms (cf. fig. 20).

Sugars, and substances derived from them, occupy a central position in the metabolism of all organisms. They play a large part in the fixation of carbon dioxide by green plants, as well as in the storage of carbon and as substrates for respiration and fermentation.

The majority of sugars occurring in nature are synthesized by a few fundamental reaction mechanisms. The following are of special importance:

(a) Epimerizations, i.e. change of configuration at a certain carbon atom, e.g. the conversion of glucose to galactose (cf. D.1.1.1).

(b) Isomerizations, i.e. conversion of aldose to ketose, e.g. conversion of glucose to fructose and vice versa (cf. D.1.4.1).

(c) Aldol condensation reactions, i.e. the addition of an aldehyde to the α-C-atom of another carbonyl compound, e.g. the formation of fructose-1,6-diphosphate from dihydroxyacetone phosphate and glyceraldehyde-3-phosphate, or the addition of a dihydroxyacetone group to phosphoglyceraldehyde or erythrose-4-phosphate, with the formation of fructose-6-phosphate or sedoheptulose-7-phosphate respectively by means of the enzyme transaldolase.

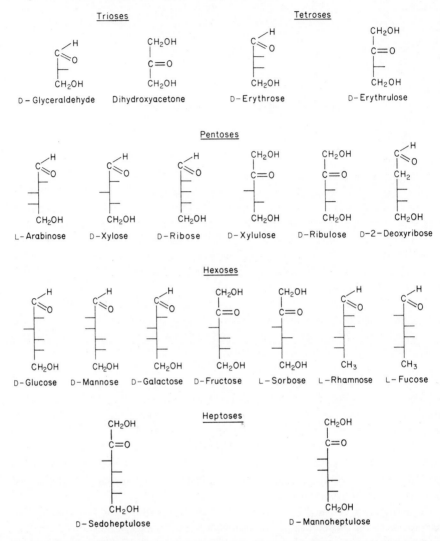

Fig. 19. The most important naturally occurring monosaccharides. The horizontal lines represent the number and position of secondary hydroxyl groups.

    (*d*) Transketolase reactions, i.e. the transfer of a two-carbon fragment from one aldose phosphate to another, e.g. the conversion of ribose-5-phosphate to sedoheptulose-7-phosphate.

    (*e*) Decarboxylations, e.g. of UDP-D-glucuronic acid with the formation of UDP-D-xylose (cf. D.1.3.1).

These reactions are dealt with in this book only to the level necessary for an understanding of the formation of carbohydrates involved in secondary metabolism.

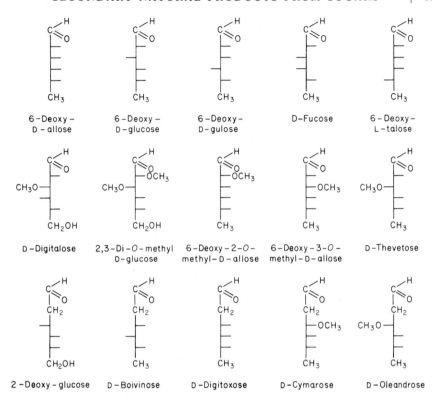

Fig. 20. Some abnormal sugars from cardioactive glycosides and pregnane derivatives. The horizontal lines represent the number and position of secondary hydroxyl groups.

References for further reading

Axelrod, B. 'Mono- and Oligosaccharides', *Plant Biochemistry*, ed. J. Bonner and J. E. Varner (Academic Press, New York, 1965), pp. 231–57.

Reichstein, T. Cardenolid- und Pregnanderivate, *Naturwiss.* **54** (1967), pp. 53–67.

Bernfeld, P. 'The Biogenesis of Carbohydrates', *Biogenesis of Natural Compounds,* ed. P. Bernfeld (Pergamon Press, Oxford, 1967), pp. 315–476.

1.1 Formation of glycosides

With the exception of D-glucose and D-fructose which occur in large amounts in the free form, most other sugars exist mainly as derivatives, or as polymers. Glycosides, i.e. compounds in which the glycosidic hydroxyl group of the sugar is replaced by another substituent, are also widespread. Depending on the atom through which the substitution takes place, glycosides may be classified as *O*-glycosides (e.g. alcohols, phenols or acids act as the substituent), *N*-glycosides and *S*-glycosides (amines and compounds with sulphhydryl groups are involved here). When the sugar is directly connected with a carbon atom (e.g. in case of vitexin, cf. D.20.5.2, or aloin) one speaks of *C*-glycosyl compounds (*C*-glycosides). In contrast to the real glycosides, these are not hydrolysable by acids, alkali or enzymes, and can only be degraded by oxidation (fig. 21). If glucose takes part in the formation of glycosides the compounds are called glucosides, if rhamnose, then rhamnosides, and if ribose, ribosides, etc.

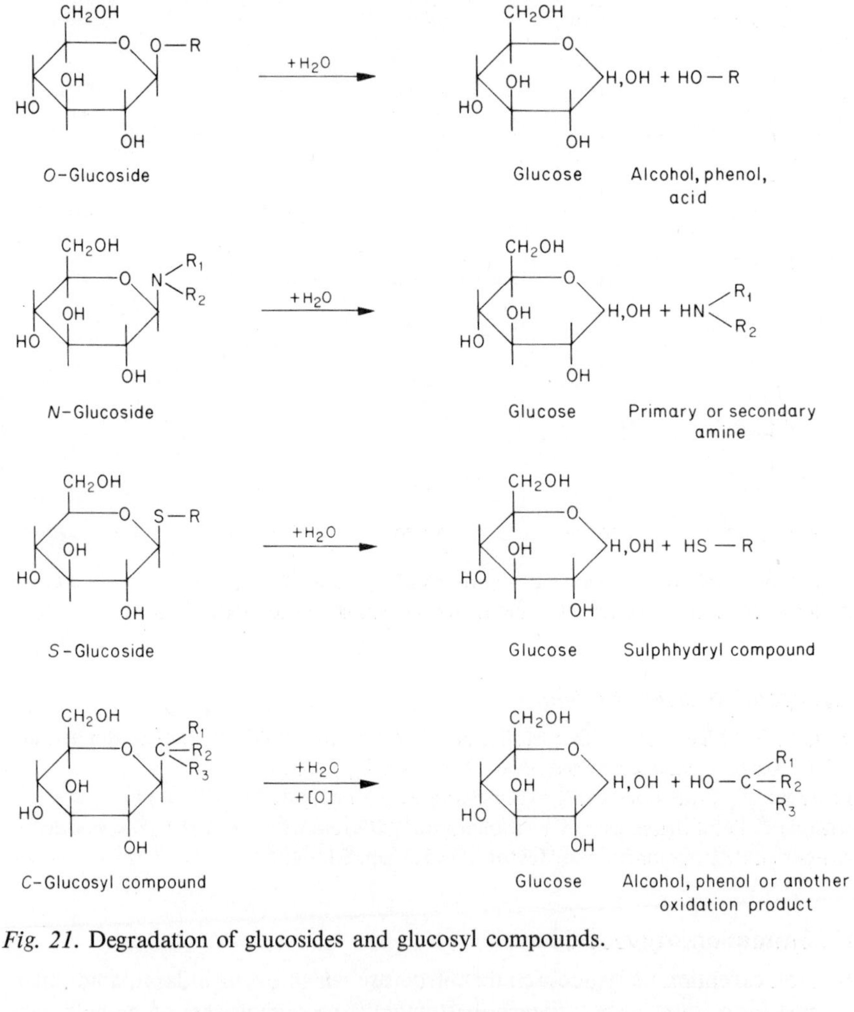

Fig. 21. Degradation of glucosides and glucosyl compounds.

1.1.1 *De novo* formation of the glycosidic bond in the biosynthesis of aldose-1-phosphates and the transfer of glycosyl groups to nucleoside phosphates

Phosphorylated derivatives are of prime importance in the metabolism of sugars. The most important reaction involved in the formation of sugar phosphates is the kinase-catalysed phosphorylation of free sugars by ATP. Glucose-6-phosphate and ADP are formed from glucose and ATP in this reaction.

Aldose-1-phosphates may be formed from 6-phosphates by the action of mutases. Since an aldose-1,6-diphosphate is formed as an intermediate, an intramolecular shift of the phosphate group is ruled out. Glucose-1-phosphate, for example, is formed from glucose-6-phosphate in this reaction, with glucose-1,6-diphosphate as an intermediate. The 1-phosphates of mannose, ribose, galactose and other sugars are synthesized in a similar manner.

Aldose-1-phosphates may also be formed through the action of kinases, e.g. in the synthesis of galactose-1-phosphate from galactose and L-arabinose-1-phosphate from L-arabinose.

Sugars phosphorylated at the glycosidic hydroxyl group are the glycosides primarily formed in metabolism. All other representatives of this class are

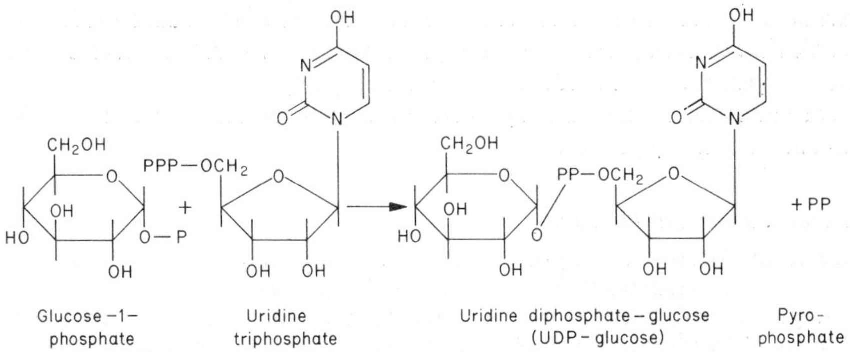

Fig. 22. Formation of uridine diphosphate glucose.

synthesized from them by transglycosylation reactions. Of special significance is the transfer of the glycosyl groups to nucleoside triphosphates. Nucleoside diphosphate sugars and pyrophosphate are formed as a result of such transfer. D-Glucose-1-phosphate, D-galactose-1-phosphate, D-xylose-1-phosphate, L-arabinose-1-phosphate, L-rhamnose-1-phosphate and D-glucuronic acid-1-phosphate are the sugar phosphates most often involved (fig. 22). The nucleotide most frequently taking part in this reaction is uridine triphosphate, although other nucleoside triphosphates are also active. A few of the sugar nucleotides which are found in higher plants are shown in table 4.

In addition to the *de novo* formation from nucleoside triphosphates, certain exchange reactions are also involved in the biosynthesis of some nucleoside diphosphate sugars. Galactose-1-phosphate, for example, reacts with uridine diphosphate glucose (UDP-glucose) to form glucose-1-phosphate and UDP-galactose. Epimerizations are also of importance. Epimerization, at carbon atom 4 especially, is of wide occurrence, e.g. the formation of UDP-galactose from

Table 4. Important sugar nucleotides from higher plants

Nucleotide	Sugar
UDP	D-Glucose, D-galactose, D-fructose, D-xylose, L-arabinose, D-glucuronic acid, D-galacturonic acid, L-rhamnose, D-acetyl glucosamine, D-acetyl galactosamine
GDP	D-Glucose, D-mannose, L-fucose
ADP	D-Glucose
TDP	D-Glucose, D-galactose, L-rhamnose
dUDP	D-Glucose, D-galactose

UDP-glucose. Such epimerization reactions also link the pairs UDP-glucuronic acid/UDP-galacturonic acid, UDP-arabinose/UDP-D-xylose and UDP-N-acetylglucosamine/UDP-N-acetylgalactosamine (cf. fig. 31).

The epimerases catalysing these reactions contain NAD^+, which, during the reaction is reduced to NADH. The reaction therefore is an oxidation-reduction in which the hydroxyl group of the sugar involved is first dehydrogenated to a carbonyl group and then reduced to a hydroxyl group.

The nucleoside diphosphate sugars are the most important intermediates in the formation of other glycosides.

References for further reading

Axelrod, B. 'Mono- and Oligosaccharides', *Plant Biochemistry,* ed. J. Bonner and J. E. Varner (Academic Press, New York, 1965), pp. 231–57.

Barber, G. A. 'Nucleotides and Carbohydrate Metabolism, A Summary', *Biosynthetic Pathways in Higher Plants,* ed. J. B. Pridham and T. Swain (Academic Press, London, 1965), pp. 117–21.

Hassid, W. Z. Transformation of Sugars in Plants, *Ann. Rev. Plant Physiol.* **18** (1967), pp. 253–80.

Bernfeld, P. 'The Biogenesis of Carbohydrates', *Biogenesis of Natural Compounds,* ed. P. Bernfeld (Pergamon Press, Oxford, 1967), pp. 315–476.

1.1.2 Formation of holosides and heterosides

Holosides are compounds which are built up of several sugars glycosidically linked to one another. According to the number of sugar residues they are further classified as oligosaccharides (di-, tri-, tetrasaccharides etc.) and the polysaccharides. Polysaccharides which are built up of glucose are called glucans and the corresponding compounds from galactose are called galactans.

The most widespread disaccharide is sucrose. In addition, lactose occurs in large amounts in the animal kingdom. Besides sucrose, the trisaccharide raffinose, and the tetrasaccharide stachyose derived from it, are encountered quite frequently in higher plants. In the Gentianaceae the trisaccharide gentianose serves a similar purpose as a storage carbohydrate. Besides the above-mentioned compounds, other higher molecular weight sugars occur.

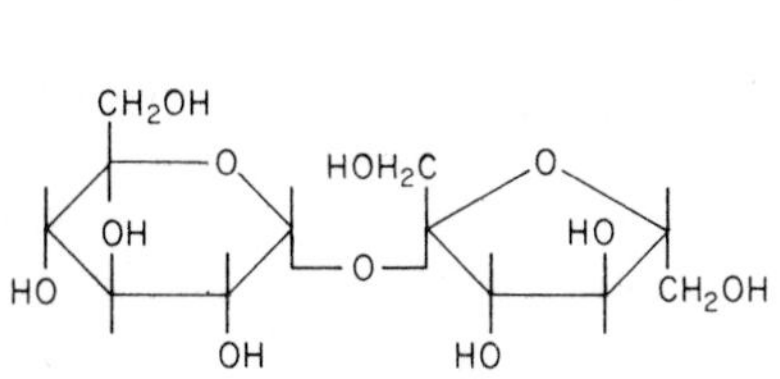

Sucrose

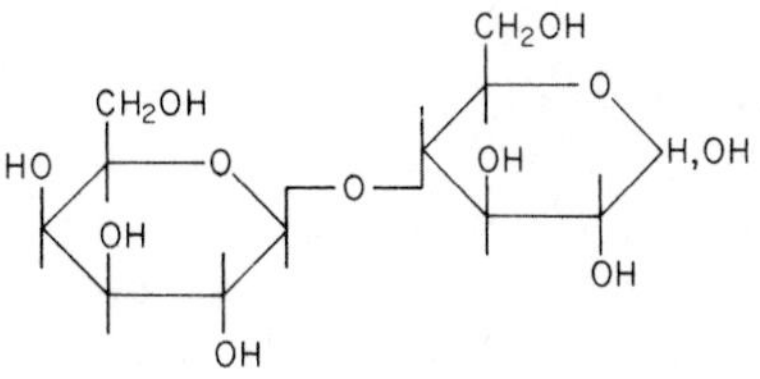

Lactose

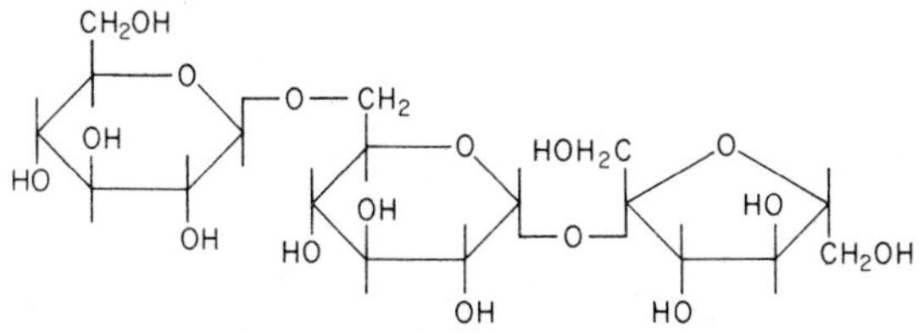

Gentianose

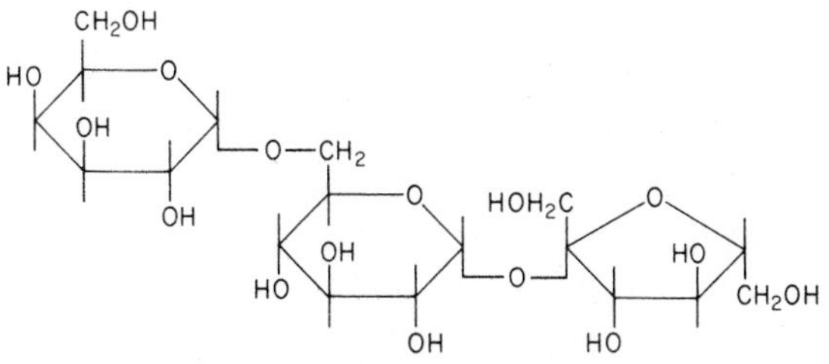

Raffinose

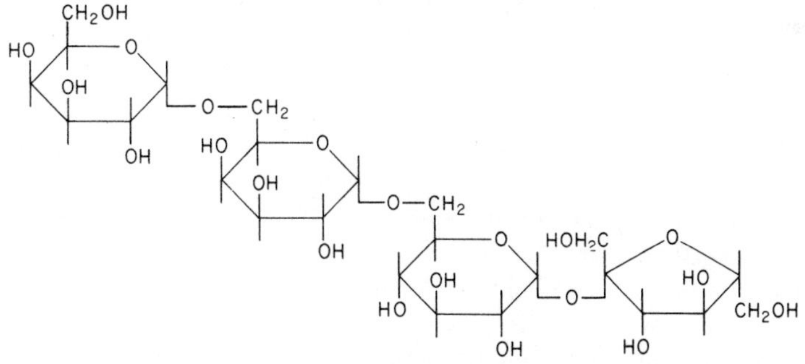

Stachyose

Fig. 23. A few widespread di-, tri- and tetrasaccharides.

Polysaccharides perform two main functions in living organisms. They serve on the one hand as structural elements (especially in plants and micro-organisms), and on the other as storage products.

In most plants, carbohydrates which can be mobilized in case of need are stored as reserve starch. This consists of 10–30% amylose, a linear glucan and about 70–90% amylopectin, a branched glucan. Glycogen, a carbohydrate similar to amylopectin, but more highly branched, serves as a reserve carbohydrate in animal cells (especially in liver and muscle cells).

In the case of higher plants, which have larger amounts of carbohydrate available than nitrogenous compounds, the structural polysaccharides of the cell wall are free of nitrogen. In the case of saprophytes, however, such as bacteria and fungi, to whom a larger nitrogen supply is available, the structural elements contain nitrogen (e.g. N-acetyl glucosamine, cf. D.21.4).

The most important carbohydrate of the plant cell is cellulose. About 10 billion tons of carbon are fixed per year by the plant kingdom in the form of cellulose. In young cells the cellulose fibrils are embedded in a matrix of other polysaccharides, of which the pectins (built partly from methylated galacturonic acid esters, cf. D.1.4.2, although other sugars are also used, e.g. 2-O-methyl-D-xylose and 2-O-methyl-L-fucose), D-xylans, L-arabans, glucomannans and mannans are of special significance. A large amount of lignin built up from phenylpropane units also occurs in the wall of certain cells (cf. D.20.4.4).

In heterosides, besides one or several sugars the molecule contains an aglycone such as an alcohol, steroid or nitrile. Since heterosides are in most cases more strongly hydrophilic than the aglycones, due to the sugar moiety, the solubility of the aglycone is improved by glycosylation in plants and animals. While in animal organisms 'conjugation' occurs mostly with uronic acids (cf. D.8.1), condensation with real sugars occurs more frequently in plants. One does, however, find in the animal kingdom, e.g. in insects, glycosides formed with the participation of sugars (cf. D.21.6), and in the plant kingdom, glucuronides such as quercetin-D-glucopyranoside uronic acid in *Phaseolus vulgaris*.

The biosynthesis of holo- and heterosides takes place by means of reversible transglycosylation reactions. Nucleoside diphosphate sugars are of special significance as glycosyl donors. In addition, aldose-1-phosphates, oligo- and polysaccharides as well as certain heterosides can transfer a glycosyl residue (tables 5 and 6). The great number of possible acceptors, and the frequently observed linkage of several sugars with one another, are the reason for the multiplicity of naturally occurring glycosides.

Growing carbohydrate chains are elongated in almost all cases by one carbohydrate unit at a time. The glycosyl groups are usually added at the non-reducing end of the chain. This is, naturally, the only possible mechanism in the case of heterosides, but has also been shown to be the case in the formation of most of the holosides, e.g. in the case of glycogen, starch, xylodextrin and chondroitin sulphate.

Certain bacterial polysaccharides, however, e.g. the O-antigen of the lipopolysaccharide from the cell wall of *Salmonella newington*, are formed by addition to the reducing end of the molecule. The sugars galactose, rhamnose and mannose which serve as building blocks are first converted to a lipid bound state, and this activated trisaccharide is then added to the reducing end of the growing polysaccharide chain. The mechanism corresponds to the elongation of fatty acid chains (cf. D.2.2.1) and of protein synthesis.

Branching of the sugar chains of glycogen and amylopectin is brought about by transglycosylases which can convert $a,1\rightarrow4$ bonds to $a,1\rightarrow6$ bonds. The enzyme active in glycogen formation is an amylo- $(a,1,4\rightarrow a,1,6)$ transglucosylase.

The degradation of holosides is carried out by hydrolysing or phosphorylating enzymes. Examples of enzymes which act hydrolytically are β-amylase (this enzyme splits off maltose from the non-reducing end of amylose and amylo-pectin), a-amylase (the enzyme forms oligosaccharides of the dextrin type from starch, which can be further degraded to maltose) and isoamylase (this enzyme splits $a,1\rightarrow6$ bonds in amylopectin). Phosphorylase, a phosphorylating enzyme, is active in glycogen degradation. In the presence of inorganic phosphate, phosphorylase splits off one glucose unit at a time as glucose 1-phosphate.

β-Glucosidases of the emulsin type are involved in the hydrolysis of most heterosides (cf. D.7.3).

Table 5. Formation of some low molecular glycosides

Glycosyl acceptor	Glycosyl donor	Glycoside formed
β-D-Fructose	a-D-Glucose-1-phosphate; UDP-D-Glucose[1]	Sucrose
β-D-Fructose-6-phosphate	UDP-D-Glucose	Sucrose-6-phosphate[2]
Purines, pyrimidines	a-D-Ribose-1-phosphate	Nucleosides
Aglycone	UDP-D-Glucose	Glucosides
Aglycone	UDP-D-Glucuronic acid	Glucuronides
Phosphate	Sucrose	Glucose-1-phosphate
Sucrose	UDP-D-Galactose	Raffinose
Raffinose	Galactinol[3]	Stachyose
Stachyose	Galactinol[3]	Verbascose

(1) The reverse of this reaction is important in the breakdown of sucrose. The UDP-glucose formed may be used in the synthesis of reserve starch (table 6).

(2) After hydrolysis of the sucrose-6-phosphate to sucrose, the reaction is no longer reversible. Thus the accumulation of large amounts of sucrose is possible, e.g in sugar cane and sugar beet.

(3) Galactinol (cf. D.1.4.2) has only been shown to be a glucosyl donor in higher plants. Similar glycosides possibly play a role as glycosyl donors in the metabolism of micro-organisms.

Table 6. Formation of some polysaccharides

Polymer	Repeating unit in the polymer	Glycosyl donor
Glycogen[1]	α-D-Glucose (1→4)	α-D-Glucose-1-phosphate and UDP-D-glucose
Starch[1]	α-D-Glucose (1→4)	UDP-D-Glucose, ADP-D-Glucose and dADP-D-Glucose
Dextran	α-D-Glucose (1→4)	Sucrose
Xylodextrin	β-D-Xylose (1→4)	UDP-D-Xylose
Cellulose	β-D-Glucose (1→4)	GDP-D-Glucose[2] and UDP-D-glucose
Hyaluronic acid	β-D-Glucuronic acid (1→3)– β-N-Acetyl glucosamine (1→4)	UDP-D-Glucuronic acid and UDP-N-Acetyl glucosamine
Chondroitin sulphate	β-D-Glucuronic acid (1→3)– β-N-Acetyl galactosamine (1→4)	UDP-D-Glucuronic acid and UDP-N-Acetyl galactosamine
Pectin	β-D-Galacturonic acid (1→4)[3]	UDP-D-Galacturonic acid
Callose	β-D-Glucose (1→3)	UDP-D-Glucose
Alginic acid	β-D-Mannuronic acid (80%)/ L-Guluronic acid (20%) (1→4)	GDP-D-Mannuronic acid and GDP-L-Guluronic acid[4]

(1) UDP- and ADP-glucose are of special importance in the synthesis of glycogen and starch. As stated before, phosphorylases are involved in degradation. However, the overall phosphorylase reaction is readily reversible and, in the presence of a primer molecule, starch may be synthesized from α-D-glucose-1-phosphate. In the case of potato phosphorylase, the smallest primer molecule is maltotriose, but it is not very active. The significance of the contribution to starch synthesis *in vitro* by phosphorylase is, however, open to question. ADP-Glucose, the most important glycosyl donor in the biosynthesis of starch in green plants, cannot be used for the biosynthesis of glycogen. dADP-Glucose can serve in the *in vitro* synthesis of starch, but does not occur in higher plants.

(2) In higher plants GDP-D-glucose is the most important glycosyl donor in the synthesis of cellulose.

(3) The carboxyl groups are partly methylated in the formation of pectin. The methyl ester of UDP-galacturonic acid cannot be polymerized.

(4) GDP-L-Guluronic acid is probably formed by epimerization from GDP-D-mannuronic acid.

References for further reading

Axelrod, B. 'Mono- and Oligosaccharides', *Plant Biochemistry*, ed. J. Bonner and J. E. Varner (Academic Press, New York, 1965), pp. 231–57.

Akazawa, T. 'Starch, Inulin and Other Reserve Polysaccharides', *Plant Biochemistry*, ed. J. Bonner and J. E. Varner (Academic Press, New York, 1965), pp. 258–97.

Nordin, J. H. and Kirkwood, S., Biochemical Aspects of Plant Polysaccharides, *Ann. Rev. Plant Physiol.* **16** (1965), pp. 393–414.

Marx-Figini, M. and Schulz, G. V. Zur Biosynthese der Cellulose, *Naturwiss.* **53** (1966), pp. 466–74.

Bernfeld, P. 'The Biogenesis of Carbohydrates', *Biogenesis of Natural Compounds*, ed. P. Bernfeld (Pergamon Press, Oxford, 1967), pp. 315–476.

Williams, R. T. 'The Biogenesis of Conjugation and Detoxication Products', *Biogenesis of Natural Compounds*, ed. P. Bernfeld (Pergamon Press, Oxford, 1967), pp. 589–639.

Hassid, W. Z. Biosynthesis of Oligosaccharides and Polysaccharides in Plants, *Science* **165** (1969), pp. 137–44.

Lamport, D. T. A. Cell Wall Metabolism, *Ann. Rev. Plant Physiol.* **21** (1970), pp. 235–70.

Heath, E. C. Complex Polysaccharides, *Ann. Rev. Biochem.* **40** (1971), pp. 29–56.

Leloir, F. Two Decades of Research on the Biosynthesis of Saccharides, *Science* 172 (1971), pp. 1299–1303.

1.2 Biosynthesis of 'abnormal' sugars

1.2.1 Formation of amino sugars

Amino sugars (also called deoxy amino sugars, since they contain amino groups in place of hydroxyl groups) occur on a large scale in the cell walls of certain bacteria. They are also constituents of certain antibiotics (cf. D.1.4.3 and D.3.1), of chitin, widely occurring in animals (e.g. in insects, cf. D.21.6) and fungi, as well as of heparin, chondroitin sulphate and hyaluronic acid.

The synthesis of 2-glucosamine has been studied most closely. 2-Glucosamine-6-phosphate is first formed from fructose-6-phosphate by the action of the corresponding enzymes and in the presence of glutamine as amino donor. The reaction includes a reduction step at carbon atom 2 and a simultaneous oxidation of the neighbouring CH_2OH-group (fig. 24). The enzyme catalysing this reaction can be detected in micro-organisms as well as in mammals. In certain cases glutamine can be replaced by ammonia as amino donor.

In certain organisms 2-glucosamine-6-phosphate may be converted in a subsequent reaction to *N*-acetyl glucosamine-6-phosphate. This reaction, catalysed by the enzyme glucosamine-6-phosphate-*N*-acylase and using acetyl CoA as the acetyl donor (fig. 24), occurs in animals as well as in micro-organisms.

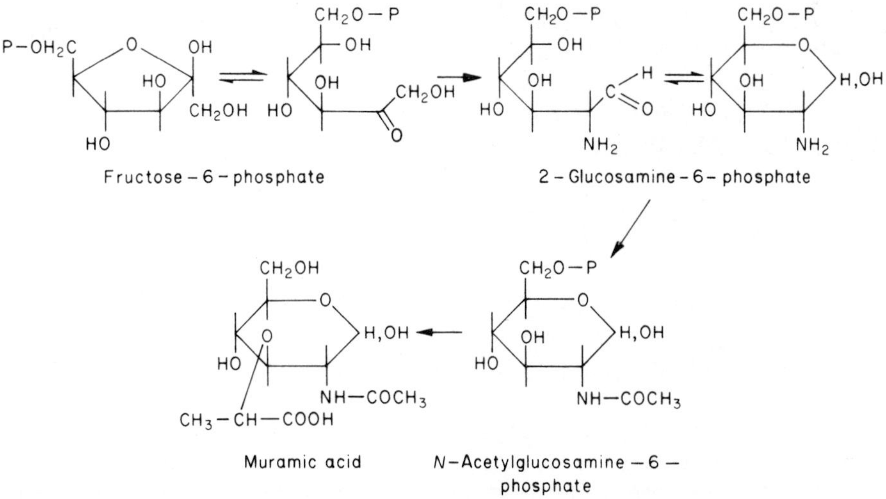

Fig. 24. Formation and conversion of amino sugars.

UDP-N-acetylglucosamine which is formed from N-acetylglucosamine-6-phosphate as described in D.1.1.1 can be polymerized to chitin with elimination of UDP. In addition, muramic acid (fig. 24), an important constituent of the cell wall of many bacteria (cf. D.21.4), is formed from N-acetylglucosamine by etherification with lactic acid.

The dimethylamino sugar, mycaminose, which occurs in macrolide antibiotics (cf. D.3.1) and is synthesized by a number of *Streptomyces* species, and 2-methylamino-L-glucose which is a constituent of the streptomycin molecule (cf. D.1.4.3), are derived from glucose. However, nothing is known about the exact mechanism of formation. The N-methyl groups of both the sugars arise from methionine (cf. C.3.3).

References for further reading

Mendicino, J. and Picken, J. M., 'Biosynthesis of Streptomycin', *Biosynthesis of Antibiotics*, ed. J. F. Snell (Academic Press, New York, 1966), pp. 121–40.

Corcoran, J. W. and Chick, M. 'Biochemistry of the Macrolide Antibiotics', *Biosynthesis of Antibiotics*, ed. J. F. Snell (Academic Press, New York, 1966), pp. 159–201.

Vanek, Z. and Majer, J. 'Macrolide Antibiotics', *Antibiotics*, Vol. II, ed. D. Gottlieb and P. D. Shaw (Springer Verlag, Berlin, 1967), pp. 154–88.

Rinehart, K. L. and Schimbor, R. F. 'Neomycins', *Antibiotics*, Vol. II, ed. D. Gottlieb and P. D. Shaw (Springer Verlag, Berlin, 1967), pp. 359–72.

Horner, W. H. 'Streptomycin', *Antibiotics*, Vol. II, ed. D. Gottlieb and P. D. Shaw (Springer Verlag, Berlin, 1967), pp. 373–99.

Bernfeld, P. 'The Biogenesis of Carbohydrates', *Biogenesis of Natural Compounds*, ed. P. Bernfeld (Pergamon Press, Oxford, 1967), pp. 315–476.

1.2.2 Formation of deoxy sugars

2-Deoxy sugars occur in all living organisms. Ribose and deoxyribose, which are present in nucleic acids, are of special significance in primary metabolism. Deoxyribose is formed from ribose at the level of nucleoside diphosphates (e.g. ADP, GDP, UDP and CDP), by means of a reaction catalysed by dehydrogenases (cf. C.2.1) with NADPH as hydrogen donor (fig. 25).

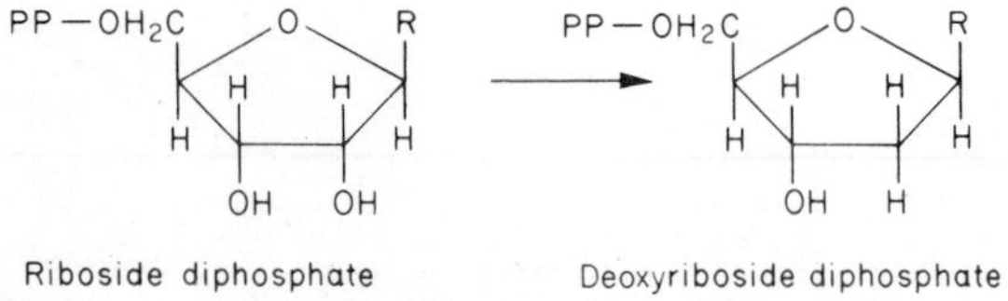

Riboside diphosphate   Deoxyriboside diphosphate

Fig. 25. Conversion of a nucleoside diphosphate to a deoxynucleoside diphosphate.

The biosynthesis of other naturally occurring 2-deoxy sugars, e.g. digitoxose, cymarose and oleandrose (cf. D.1), has not so far been investigated.

The 6-deoxyhexoses, L-rhamnose and L-fucose, originate from glucose and mannose respectively. In *Acetobacter aerogenes*, GDP-mannose is first converted,

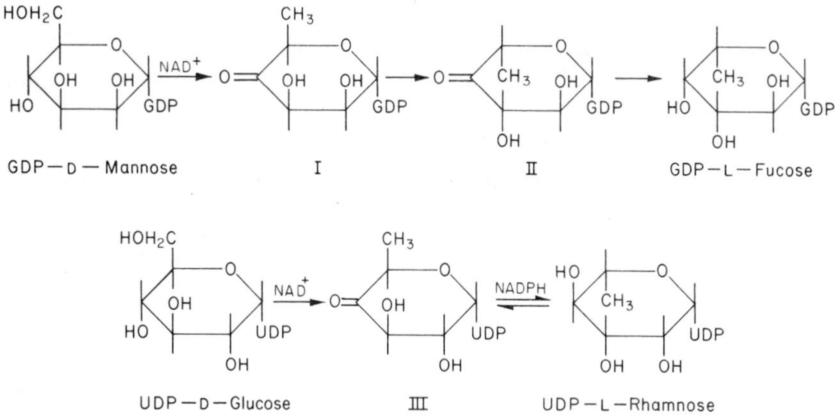

Fig. 26. Formation of 6-deoxyhexoses.

in the presence of NAD^+, to compound I which possesses a keto group at carbon atom 4 and a methyl group at carbon atom 6. Isomerization (probably via the enol form) occurs at carbon atoms 3 and 5 to give compound II, and GDP-L-fucose is formed from this (fig. 26) by reduction of the keto group at carbon atom 4.

In other micro-organisms, UDP-L-rhamnose is formed from UDP-D-glucose in a similar way through compound III (fig. 26).

Other methylpentoses outlined in fig. 20 are probably formed in an analogous manner. Experiments in this direction have, however, not yet been carried out.

References for further reading

Hassid, W. Z. Transformation of Sugars in Plants, *Ann. Rev. Plant Physiol.* **18** (1967), pp. 253–80.

1.2.3 Formation of sugars with branched carbon chains

Sugars with a branched carbon chain (cf. fig. 27) have only been found in higher plants and micro-organisms. Apiose, the constituent of the flavone glycoside, apiin, and hamamelose, the sugar component of hamamelis tannin, have been known for a long time. Recently many branched sugars have been detected as components of certain antibiotics (cf. D.1.4.3 and D.3.1).

Two reaction mechanisms are involved in the biosynthesis of these compounds:

(a) The branching of the sugar chains occurs by means of methylation of carbon atoms. This type of reaction occurs in the formation of the sugars mycarose and noviose, which are found in micro-organisms as constituents of the antibiotics erythromycin and novobiocin (cf. D.3.1). The methyl groups at carbon atom 3 and carbon atom 5 of mycarose and noviose

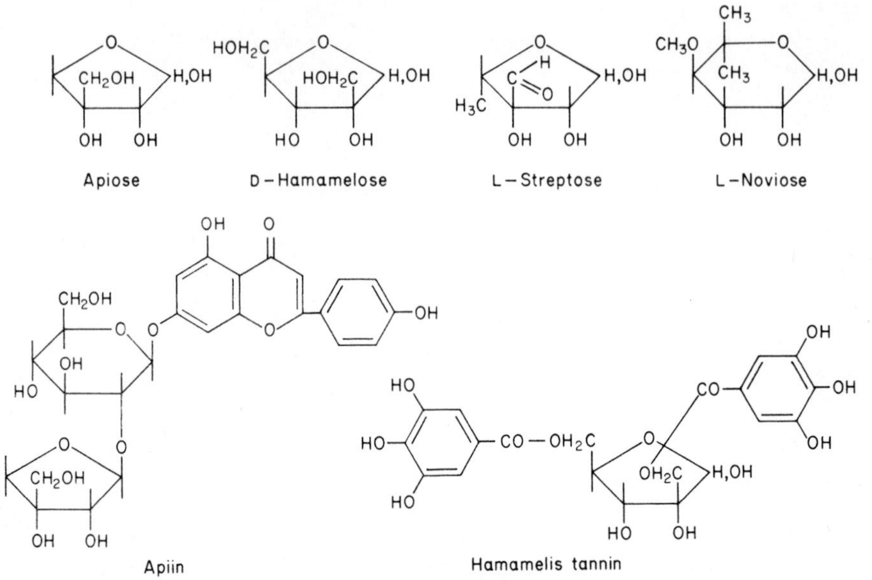

Fig. 27. Some sugars with branched carbon chain and compounds which contain such sugars.

respectively were labelled after feeding methionine-$^{14}CH_3$ and all the three hydrogen atoms of the methyl group are retained, in contrast to the situation obtained in the formation of tuberculostearic acid (cf. D.2.2.3) and ergosterol (cf. D.5.7.2). The rest of the molecule originates from glucose. Mycarose thus may be formed (fig. 28) in a way similar to that proposed for the formation of 6-deoxy sugars (cf. D.1.2.2).

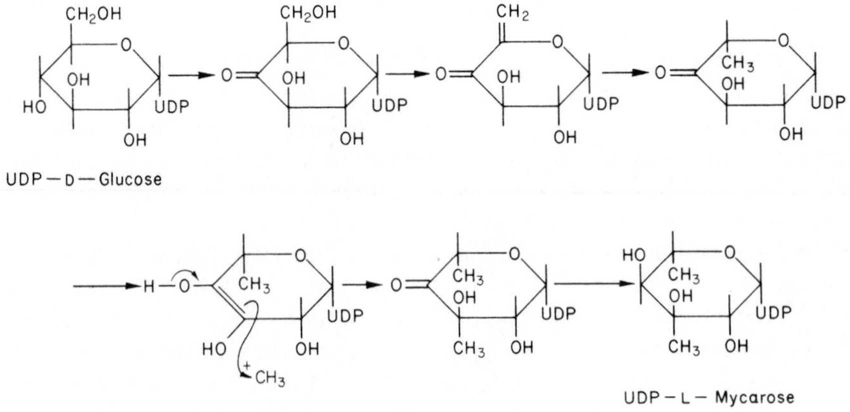

Fig. 28. Possible way of formation of L-mycarose from D-glucose.

In the case of noviose, one of the two methyl groups at carbon atom 5 originates from glucose, and the other from methionine.

(b) The branching is formed by the rearrangement of the sugar chain.

This biosynthetic pathway has been established in the case of apiose and streptose (cf. D.1.4.3). Both sugars originate from glucose and the 'additional' CH$_2$OH or CHO group is derived from carbon atom 3 of glucose.

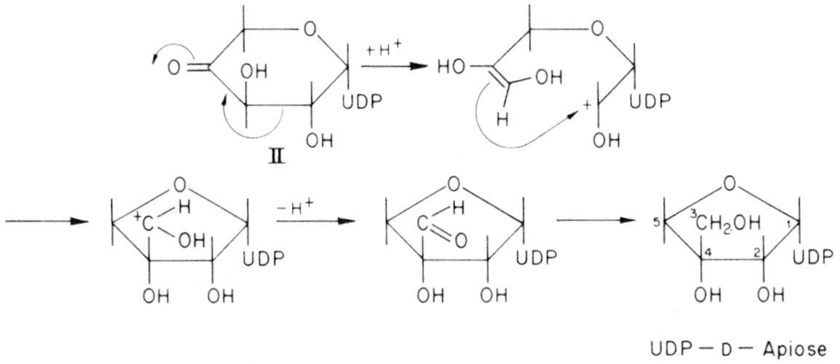

UDP — D — Apiose

Fig. 29. Possible way for the formation of apiose from glucuronic acid.

The biosynthesis of apiose (fig. 29) probably starts with the ketose II (cf. D.1.3.1) which is formed as an intermediate in the formation of D-xylose from D-glucuronic acid.

The biosynthesis of streptose possibly begins with intermediate III, which is produced during the formation of rhamnose (cf. D.1.2.2, fig. 26), and may then be converted to UDP-streptose via compounds IV, V and VI (fig. 30).

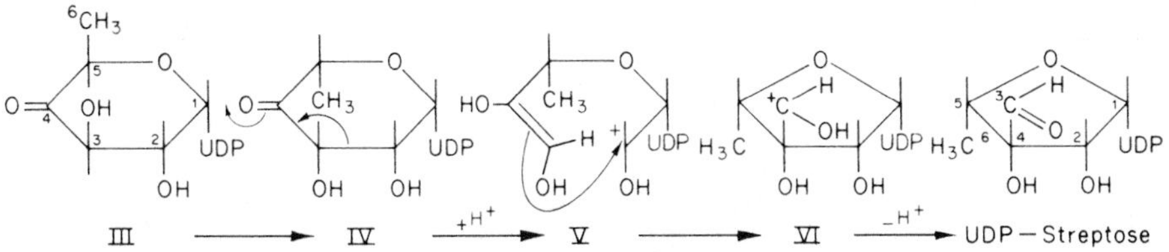

UDP — Streptose

Fig. 30. Possible way for the formation of UDP-streptose.

References for further reading

Mendicino, J. and Picken, J. M. 'Biosynthesis of Streptomycin', *Biosynthesis of Antibiotics*, ed. J. F. Snell (Academic Press, New York, 1966), pp. 121–40.

Grisebach, H. *Biosynthetic Patterns in Microorganisms and Higher Plants* (John Wiley, New York, 1967).

Kominek, L. A. 'Novobiocin', *Antibiotics*, Vol. II, ed. D. Gottlieb and P. D. Shaw (Springer Verlag, Berlin, 1967), pp. 231–9.

Horner, W. H. 'Streptomycin', *Antibiotics*, Vol. II, ed. D. Gottlieb and P. D. Shaw (Springer Verlag, Berlin, 1967), pp. 373–99, 447–8.

1.3 Oxidation of sugars

1.3.1 Formation and decarboxylation of uronic acids

UDP-glucose can be readily oxidized to UDP-glucuronic acid by UDP-glucose dehydrogenase with NAD^+ as coenzyme (cf. C.2.1.1). The intermediate aldehyde which might be expected to occur has not yet been found and probably does not exist in the free state. GDP-D-mannuronic acid and UDP-D-galacturonic acid are formed from GDP-D-mannose and UDP-D-galactose respectively, in a similar way (fig. 31).

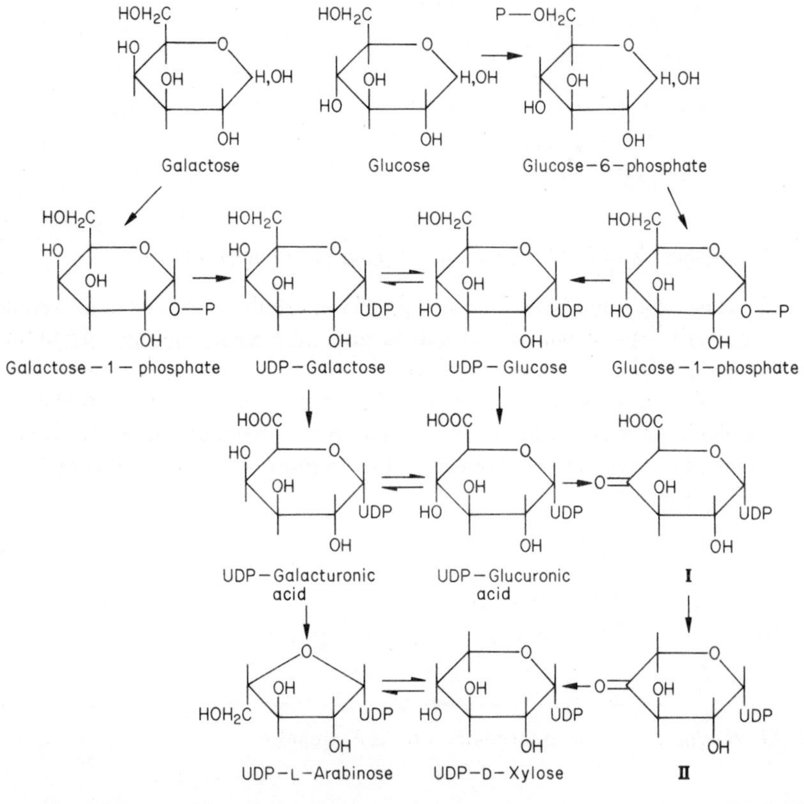

Fig. 31. Formation and decarboxylation of uronic acids.

UDP-D-galacturonic acid and UDP-D-glucuronic acid can be enzymatically decarboxylated to UDP-L-arabinose and UDP-D-xylose respectively. Since NAD^+ is necessary for this reaction, keto sugars probably occur as intermediates, which in the formation of UDP-D-xylose may possess the structures depicted as I and II (fig. 31).

The polymerization products of uronic acids and the above-mentioned pentoses are widespread as constituents of plant cell walls. They form the so-called hemicelluloses in which fibrils of cellulose are embedded (cf. D.1.1.2). The poly-uronides are also precursors of pectin and of a series of strongly swelling sub-stances namely hyaluronic acid and alginic acid (cf. table 6). In animals the formation of uronides serves as a process of hydrophilization of certain metabolic products before excretion (cf. A.2.4.1).

References for further reading

Hassid, W. Z. Transformation of Sugars in Plants, *Ann. Rev. Plant Physiol.* **18** (1967), pp. 253–80.

Bernfeld, P. 'The Biogenesis of Carbohydrates', *Biogenesis of Natural Compounds*, ed. P. Bernfeld (Pergamon Press, Oxford, 1967), pp. 315–476.

1.3.2 The biosynthesis of onic acids and sugar dicarboxylic acids

The onic acids, which occur comparatively rarely in nature, originate either from aldohexoses by oxidation, or from uronic acids by reduction (fig. 32). Preparations from higher plants and cell-free extracts of red algae convert D-glucose to

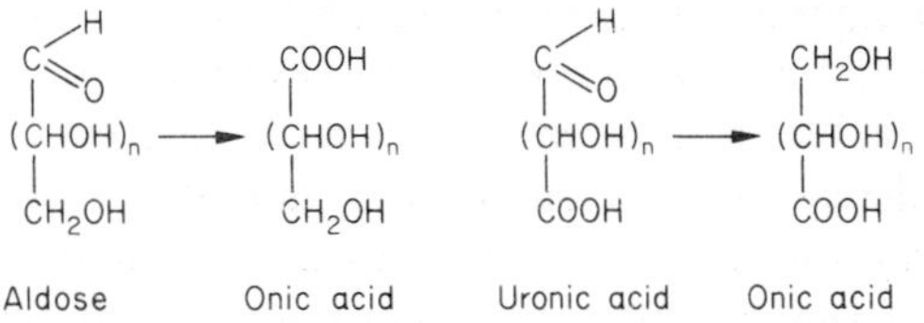

$$
\begin{array}{cccc}
\underset{|}{\overset{H}{C}}\!\!\diagdown\!\!O & \overset{COOH}{\underset{|}{}} & \underset{|}{\overset{H}{C}}\!\!\diagdown\!\!O & \overset{CH_2OH}{\underset{|}{}} \\
(CHOH)_n & (CHOH)_n & (CHOH)_n & (CHOH)_n \\
| & | & | & | \\
CH_2OH & CH_2OH & COOH & COOH
\end{array}
$$

Aldose Onic acid Uronic acid Onic acid

Fig. 32. Formation of onic acids from aldoses and uronic acids.

D-gluconic acid, D-galactose to D-galactonic acid and maltose, lactose and cello-biose to the corresponding aldobionic acids. The conversion of glucose to gluconic acid is widespread in micro-organisms.

In all cases investigated, oxygen serves as the oxidizing agent and hydrogen peroxide is formed. The enzymes catalysing these reactions are flavoenzymes (cf. C.2.2).

Aldoses can also be converted to the corresponding onic acid in animals, by dehydrogenases (cf. C.2.1), Conversion of D-glucuronic acid to L-gulonic acid is carried out by enzymes of this same group in animals and micro-organisms, and involves reduction of the aldehyde group of the uronic acid (cf. D.1.3.3).

Enzymes which oxidize uronic acids to the corresponding dicarboxylic acids have been found in *Phaseolus aureus*. D-Glucuronic acid and D-galacturonic acid are converted to D-glucaric acid and D-galactaric acid respectively (fig. 33).

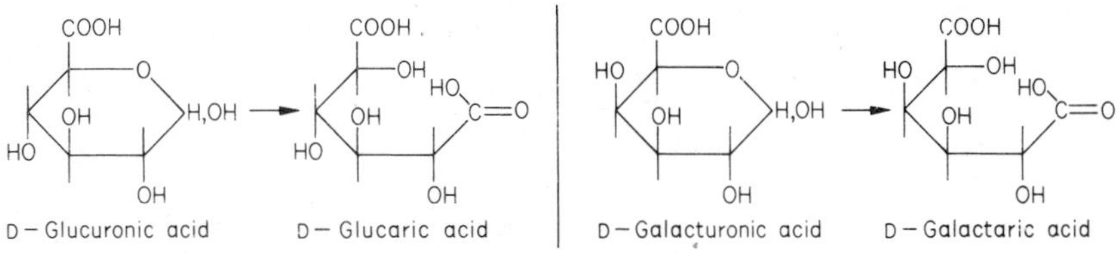

D – Glucuronic acid D – Glucaric acid   D – Galacturonic acid D – Galactaric acid

Fig. 33. Formation of D-glucaric and D-galactaric acid.

References for further reading

Axelrod, B. 'Mono- and Oligosaccharides', *Plant Biochemistry*, ed. J. Bonner and J. E. Varner (Academic Press, New York, 1965), pp. 231–57.

Hassid, W. Z. Transformations of Sugars in Plants, *Ann. Rev. Plant Physiol.* **18** (1967), pp. 253–80.

Bernfeld, P. 'The Biogenesis of Carbohydrates', *Biogenesis of Natural Compounds*, ed. P. Bernfeld (Pergamon Press, Oxford, 1967), pp. 315–476.

1.3.3 The formation of ascorbic acid

Ascorbic acid is needed in daily quantities of about 70 mg to sustain full stamina, and is an essential nutrient for human beings. In the case of insufficiency the symptoms of scurvy appear.

We have as yet no complete understanding of the mode of action of ascorbic acid. It is, however, known that this substance is involved in certain hydroxylation reactions which are catalysed by mixed function oxygenases (cf. C.2.6.1), in the reduction of folic acid to tetrahydrofolic acid (cf. D.8.2.3), as well as in the regulation of the redox equilibrium between Fe^{2+} and Fe^{3+} and Cu^{+} and Cu^{2+}.

In both plants and animals ascorbic acid is formed from D-glucuronic acid. UDP-glucuronic acid is first converted to D-glucuronic acid lactone via D-glucuronic acid-1-phosphate. This compound is then reduced at carbon atom 1 to form L-gulonic acid. (Since in the numbering of the carbon atoms of carbohydrates the most highly oxidized carbon atom is given the lowest possible number, the original carbon atom 6 of glucuronic acid becomes carbon atom 1 of gulonic acid.)

After the conversion of gulonic acid to the corresponding γ-lactone, the

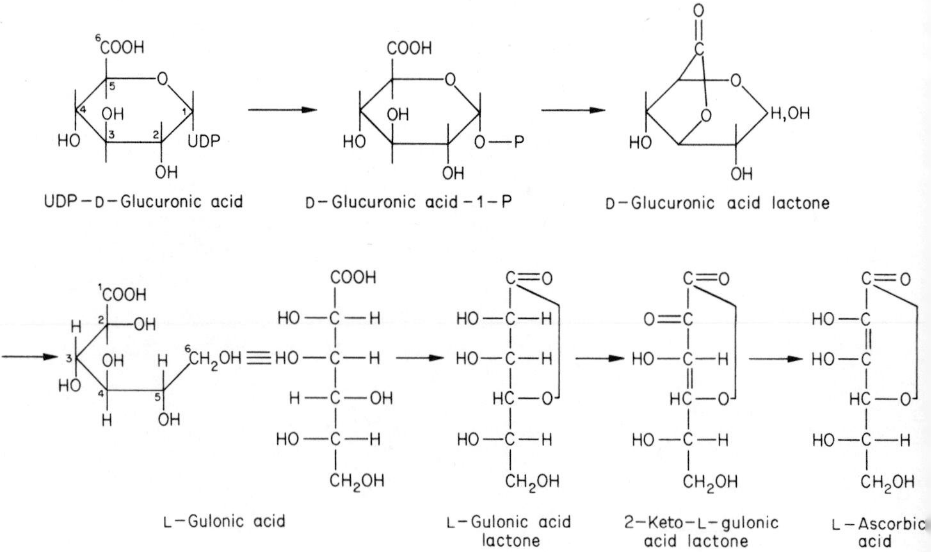

Fig. 34. The biosynthesis of L-ascorbic acid.

hydroxyl group at carbon atom 2 is oxidized to a keto group. The 2-keto-L-gulonic acid lactone formed is subsequently converted to L-ascorbic acid by enolization (fig. 34).

Direct conversion of D-glucuronic acid to L-gulonic acid by isomerization at carbon atom 5 has not yet been conclusively established.

Ascorbic acid is produced commercially from glucose by a semi-biochemical process. Glucose is first converted to sorbitol by catalytic hydrogenation; this alcohol is then converted to L-sorbose (cf. D.1.4.1) by cells of *Acetobacter sub-oxydans*. Sorbose, after conversion to the diacetonide to protect the hydroxyl groups in the positions 3, 4, 5 and 6, is chemically oxidized at carbon atom 1 to form 2-keto-L-gulonic acid, which can then be easily converted to L-ascorbic acid.

References for further reading

Bersin, T. *Biochemie der Vitamine* (Akademische Verlagsgesellschaft, Frankfurt/Main, 1966).

Cheldelin, V. H. and Baich, A. 'The Biosynthesis of the Water-Soluble Vitamins', *Biogenesis of Natural Compounds*, ed. P. Bernfeld (Pergamon Press, Oxford, 1967), pp. 679–742.

Sebrell, W. H. and Harris, R. S. *The Vitamins*, Vol. I (Academic Press, New York, 1967).

Loewus, F. Carbohydrate Interconversions, *Ann. Rev. Plant Physiol.* **22** (1971), pp. 337–64.

1.4 Secondary natural products originating from sugars by reduction

1.4.1 Formation of aliphatic sugar alcohols

Aliphatic sugar alcohols, i.e. polyalcohols without a carbonyl group, are extraordinarily widespread in nature. Examples are glycerol, erythritol, ribitol

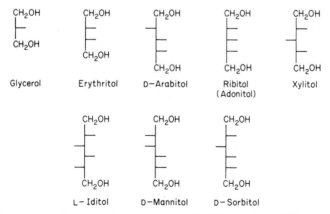

Fig. 35. Some naturally occurring sugar alcohols. The horizontal lines represent the number and position of the secondary hydroxyl groups.

(adonitol) and mannitol, which are found in animals as well as in plants and micro-organisms.

These compounds originate from free sugars by a reversible reaction catalysed by dehydrogenases with NADH or NADPH as coenzyme (cf. C.2.1). In this way aldoses and ketoses can be converted into one another through sugar alcohols, e.g. D-glucose and D-fructose remain in equilibrium with each other via sorbitol.

A few of the commonly occurring polyols and the sugars serving as their precursors are illustrated in fig. 35 and table 7.

Table 7. Sugars and the related sugar alcohols

Sugar alcohols	Sugars
Glycerol	Glyceraldehyde
Erythritol	Erythrose
Xylitol	D-Xylulose and L-xylulose
D-Arabitol	D-Xylulose and D-ribulose
Ribitol (adonitol)	L-Ribulose and D-ribulose
D-Sorbitol	D-Glucose, L-sorbose and D-fructose
L-Iditol	L-Sorbose
D-Mannitol	D-Fructose

References for further reading

Bernfeld, P. 'The Biogenesis of Carbohydrates', *Biogenesis of Natural Compounds*, ed. P. Bernfeld (Pergamon Press, Oxford, 1967), pp. 315–476.

1.4.2. Biosynthesis of cyclitols

Carbocyclic compounds with several hydroxyl groups are called cyclitols. The most important naturally occurring representatives of this class are the inositols, which possess a six-membered ring and six hydroxyl groups, and the quercitols, in which a hydroxyl group is replaced by a hydrogen atom.

Inositols are found in animals (skeletal muscle) as well as in higher and lower plants. *Meso*-inositol (myoinositol) is a growth-promoting substance in moulds. Its vitamin character in human beings is disputed.

Meso-inositol hexaphosphate (phytic acid) is of importance in the storage of phosphates in plant tissues, e.g. in seeds. In wheat endosperm the phosphate of phytic acid may be transferred to ADP by a phosphotransferase to form ATP.

1-0-α-D-Galactopyranosyl-*meso*-inositol (galactinol) is formed from *meso*-inositol and UDP-galactose. Galactinol is widespread in higher plants and, like the sugar nucleotides, can act as a glycosyl donor (cf. D.1.1.2).

In plants and animals the inositols are derived from glucose-6-phosphate. This is converted in animal organisms and micro-organisms to *meso*-inositol-1-

phosphate, probably via 5-keto-glucose-6-phosphate and 1-phospho-*meso*-inosose-2. This reaction is catalysed by the enzyme glucose-6-phosphate cyclase. *Meso*-inositol-1-phosphate is then dephosphorylated to *meso*-inositol by a specific phosphatase (fig. 36).

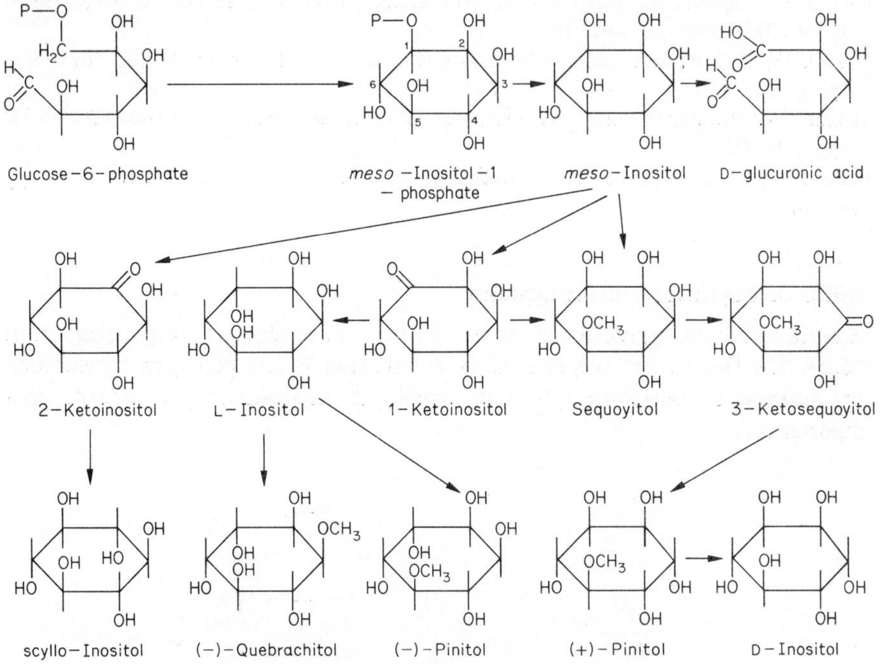

Glucose−6−phosphate   *meso* −Inositol−1 − phosphate   *meso*−Inositol   D−glucuronic acid

2−Ketoinositol   L−Inositol   1−Ketoinositol   Sequoyitol   3−Ketosequoyitol

scyllo−Inositol   (−)−Quebrachitol   (−)−Pinitol   (+)−Pinitol   D−Inositol

Fig. 36. Synthesis and degradation of inositols.

Meso-inositol is probably the precursor of all other naturally occurring inositols. Only phytic acid appears to originate directly from inositol-1-phosphate. The isomerization of individual carbon atoms proceeds via keto inositols (inososes). *S*-Adenosylmethionine serves as the methyl group donor (cf. C.3.3). Sequoyitol, scylloinositol, (−)-quebrachitol, (−)-inositol, (+)-inositol and D-inositol which are found in higher plants probably originate by the pathways outlined in fig. 36.

In both plants and animals the inositols are degraded to uronic acids. D-Glucuronic acid is formed from *meso*-inositol by cleavage of the bond between carbon atoms 1 and 6. There exists therefore, another way, besides the one described in D.1.3.1. by which glucose may be converted to glucuronic acid. Since, as mentioned in D.1.1.1, glucuronic acid is converted to galacturonic acid, which in turn is a precursor for pectins (cf. D.1.1.2) and for certain pentoses (cf. D.1.3.1), the inositols can serve as the precursors for these compounds. It is also thought that the methylated sugars which are found in the cell walls of a great many higher plants (cf. D.1.1.2) are formed from the corresponding substituted inositols.

References for further reading

Bersin, T. *Biochemie der Vitamine* (Akademische Verlagsgesellschaft, Frankfurt, 1966).

Anderson, L. and Wolter, K. E. Cyclitols in Plants: Biochemistry and Physiology, *Ann. Rev. Plant Physiol.* **17** (1966), pp. 203–22.

Kindl, H., Scholda, R. and Hoffmann-Ostenhof, O. Biosynthese der Cyclite, *Angew. Chem.* **73** (1966), pp. 198–206.

Hassid, W. Z. Transformation of Sugars in Plants, *Ann. Rev. Plant Physiol.* **18** (1967), pp. 253–80.

Tanner, W. Biochemie und Physiologie der Cyclite, *Ber. dtsch. Bot. Ges.* **80** (1967), pp. 592–607.

Loewus, F. Carbohydrate Interconversions, *Ann. Rev. Plant Physiol.* **22** (1971), pp. 337–64.

1.4.3 Biosynthesis of streptidine

Streptomyces sp. produce a large number of antibiotics originating from sugars. The best-known representative of this group is streptomycin. In addition, the antibiotics dihydrostreptomycin, neomycin and kanamycin are of great importance.

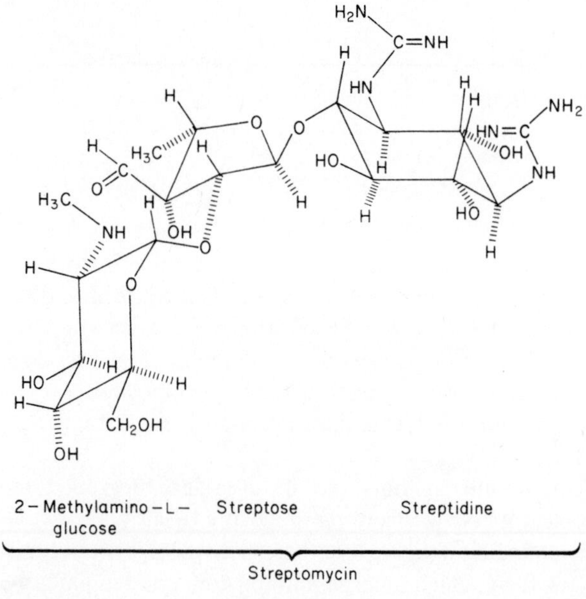

2 – Methylamino – L – glucose   Streptose   Streptidine

Streptomycin

Fig. 37. Streptomycin.

The streptomycin molecule consists of three parts glycosidically linked to one another (fig. 37). We shall only deal with the biosynthesis of the aglycone, streptidine, here (fig. 38). The formation of 2-methylamino-L-glucose and streptose has already been discussed in D.1.2.1 and D.1.2.3.

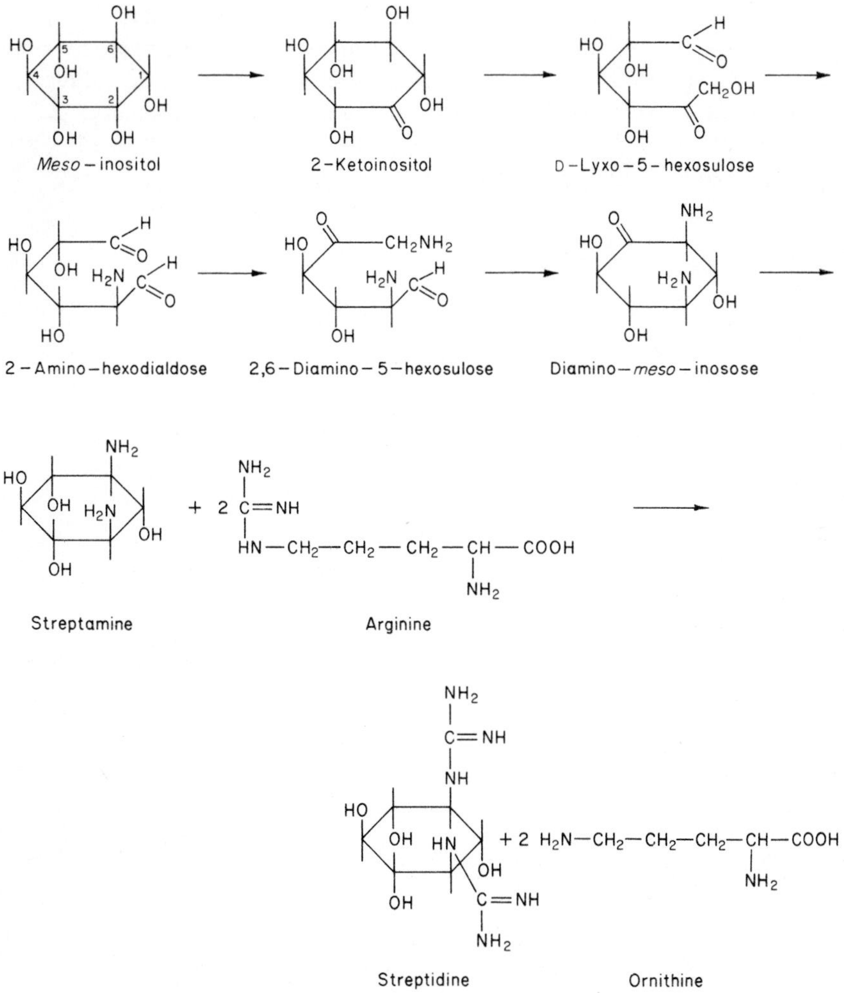

Meso – inositol

2-Ketoinositol

D – Lyxo – 5 – hexosulose

2 – Amino – hexodialdose

2,6 – Diamino – 5 – hexosulose

Diamino – meso – inosose

Streptamine

Arginine

Streptidine

Ornithine

Fig. 38. Possible pathway for formation of streptidine from *meso*-inositol. (Note the formulation of *meso*-inositol in contrast to that given in fig. 36).

It is now known that streptidine is formed from *meso*-inositol. It is thought that this compound is first oxidized to 2-keto inositol followed by splitting of the carbon ring to form D-lyxo-5-hexosulose. Streptamine may be formed from this compound by two amination steps which are thought to proceed by an intramolecular oxidation-reduction, as in the formation of 2-glucosamine-6-phosphate from fructose-6-phosphate and glutamine (cf. D.1.2.1), followed by cyclization by an aldol reaction and subsequent reduction. Streptamine is phosphorylated before conversion to streptidine. Then two amidine groups originating from arginine (cf. D.17.1) are transferred to streptamine phosphate by an amidinotransferase.

It is not known whether streptidine phosphate, free streptidine formed by elimination of the phosphate group, or a derivative of streptidine acts as the aglycone in the formation of streptomycin. The mechanism of glycosidization has also not been investigated. However, streptose and 2-methylamino-L-glucose are probably converted to nucleoside diphosphate sugars and then transferred by means of a transglycosidase (cf. D.1.1.2).

References for further reading

Mendicino, J. and Picken, J. M. 'Biosynthesis of Streptomycin', *Biosynthesis of Antibiotics*, ed. J. F. Snell (Academic Press, London, 1966), pp. 121–40.

Horner, W. H. 'Streptomycin', *Antibiotics*, Vol. II, ed. D. Gottlieb and P. D. Shaw (Springer Verlag, Berlin, 1967), pp. 373–99, 447–8.

2. Formation of secondary natural products from acetate

2.1 Biosynthesis and properties of 'activated acetic acid'

The thioester formed from acetic acid and coenzyme A (cf. D.9.1) is known as 'activated acetic acid' (acetyl CoA) (fig. 39).

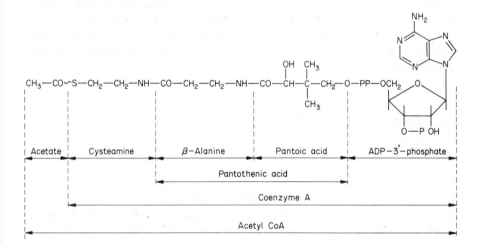

Fig. 39. Acetyl coenzyme A.

Acetyl CoA may be formed in several ways. The most important are:

(a) *Oxidative decarboxylation of pyruvate*

Pyruvate, which is formed in large quantities during the metabolism of amino acids and carbohydrates may be converted to acetyl CoA and CO_2 by a complex reaction involving the participation of thiamine pyrophosphate, lipoic acid and coenzyme A.

The step in the reaction which supplies the energy present in the thioster bond is the transfer of 'activated acetaldehyde' (cf. C.3.4) to lipoic acid (fig. 40).

The formation of acetyl lipoic acid is an oxidoreduction in which lipoic acid is reduced and 'acetaldehyde' oxidized. The acetyl group from acetyl lipoic acid

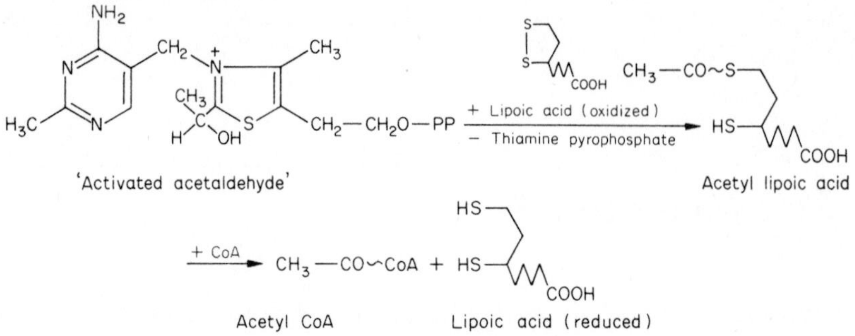

Fig. 40. Formation of acetyl CoA from activated acetaldehyde.

is then transferred to coenzyme A by ester exchange. The reduced lipoic acid is again oxidized to the disulphide form by means of a dehydrogenase (cf. C.2.1).

All the enzymes involved in this chain of reactions are united in a multienzyme complex. Lipoic acid is attached to the enzyme protein by means of an acid amide linkage.

(b) Thiolytic cleavage of fatty acids
Cf. D.2.2.6.

(c) Activation of acetic acid
Enzyme systems activating acetic acid have been detected in animals as well as in plants and micro-organisms. Activation proceeds in most bacteria via acetyl phosphate and in almost all other organisms via acetyl-AMP (cf. C.1.2).

The most important stage of primary metabolism requiring acetyl CoA is the tricarboxylic acid cycle in which the acetyl residue is finally degraded to hydrogen bound to coenzymes and CO_2. Acetyl CoA is also a precursor of a great number of secondary natural products.

Due to polarization of the molecule caused by the thioester bonding, the carboxyl carbon and the carbon atom of the methyl group can be easily substituted by nucleophilic and electrophilic reactions respectively (cf. C.1.2). The CoA residue is lost in a nucleophilic substitution.

An example of an electrophilic substitution is the condensation of acetyl CoA and CO_2 to form malonyl CoA catalysed by acetyl CoA carboxylase (cf. fig. 5). CO_2 is first bound to biotin (cf. C.3.1) and transferred from there to the acetyl residue. Acetyl CoA undergoes nucleophilic substitution by malonyl CoA in the biosynthesis of fatty acids and polyketides (cf. D.2.2.1 and D.2.3).

The formation of esters and acetylated amines is also possible by nucleophilic substitution with alcohols and phenols, or amines. These various reaction possibilities are the reason for the innumerable substances derived from acetyl CoA.

References for further reading

Stumpf, P. K. 'Lipid Metabolism', *Plant Biochemistry*, ed. J. Bonner and J. E. Varner (Academic Press, New York, 1965), pp. 322–45.

Mudd, J. B. Fat Metabolism in Plants, *Ann. Rev. Plant Physiol.* **18** (1967), pp. 229–52.

2.2 Biosynthesis, transformations and degradation of fatty acids

2.2.1 The formation of saturated fatty acids by the fatty-acid-synthetase complex

Saturated and unsaturated aliphatic monocarboxylic fatty acids occur in all living organisms. They are components of oils, waxes and other lipids. Since fatty acids with an even number of carbon atoms (in contrast to those having an odd number of carbon atoms) preponderate in nature, those with sixteen or eighteen carbon atoms being the most frequent, it was originally supposed that they were synthesized from a precursor with two carbon atoms.

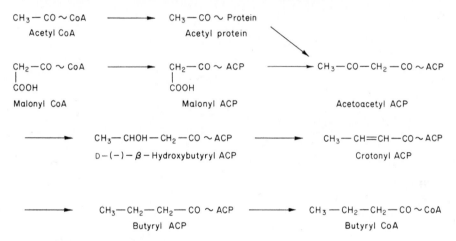

Fig. 41. Biosynthesis of fatty acids.

Today much is known about the formation of fatty acids. The saturated ones are formed by the pathway illustrated in fig. 41. The enzymes involved are united in a multi-enzyme complex, the 'fatty-acid synthetase', which in mammals and fungi is especially stable.

The synthesis of fatty acids begins with the linkage of a molecule of acetyl CoA to an SH-group of a subunit of the enzyme complex with elimination of the CoA grouping. Simultaneously a molecule of malonyl CoA (cf. D.2.1) is linked to another subunit, the acyl-carrier protein (ACP), again with the loss of CoA. The linkage of the malonyl residue to the ACP is through the SH-group of a 4-phospho-pantetheine molecule which in turn is attached to a serine residue in the peptide chain. As is the case with acetyl CoA, this linkage is also an ester interchange in which the coenzyme A is replaced by another sulphur-containing compound. The malonyl group then substitutes the acetyl residue at the carbonyl grouping by a nucleophilic reaction (cf. C.1.2), releasing it from the enzyme protein, and forming acetoacetyl-ACP.

Reduction of the acetoacetate bound to ACP forms D-(−)-β-hydroxybu-tyryl-ACP, from which crotonyl-ACP is formed by the elimination of water. Subsequent reduction forms butyryl-ACP. By means of a transferase also present in the enzyme system, the butyryl residue is transferred to CoA, setting free the SH-group of pantotheine which can again react with a molecule of malonyl CoA.

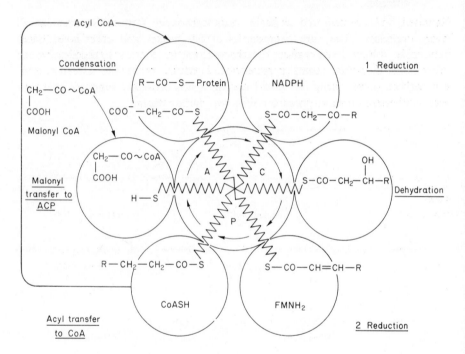

Fig. 42. Model for the structure of the multi-enzyme complex, the 'fatty-acid synthetase'.

Like acetyl CoA, butyryl CoA can be attached to the enzyme protein and enter the cycle again. In this manner a stepwise elongation of the carbon chain by two carbon atoms is possible. Chain elongation stops after the formation of acids with about sixteen to twenty carbon atoms owing to, as yet, unknown causes.

Though good electron micrographs are available, little is known of the structure of the fatty-acid synthetase. The properties of the multi-enzyme complex can, however, be explained by means of the model shown in fig. 42 in which the acyl-carrier protein with the 4-phospho-pantetheine residue, revolving around single bonds, stands at the centre. The pantetheine molecule is approximately 15 Å long. The SH-group might therefore be expected to reach the active centres of the enzyme proteins of the complex. This supposition helps to explain the regulation of individual steps of the reaction sequence simply (cf. fig. 42).

From the point of view of reaction kinetics, the linkage of the substrate with ACP is advantageous since 'accidental' diffusion of the intermediates to the reactive centres of the enzymes is prevented. Further, the action of other enzyme systems (e.g. those of fatty acid degradation) on the intermediates of the synthetic chain is impossible, and the regulation of the whole metabolic pathway is comparatively simple (cf. A.3).

Instead of acetyl CoA, other acyl CoA derivatives, (such as propionyl CoA, cf. D.3; isobutyryl CoA, cf. D.11; isovaleryl CoA, cf. D.12.1; and α-methylbutyryl CoA, cf. D.13.1) can act as starting molecules in some cases. Fatty acids with a branched carbon chain or with an odd number of carbon atoms would then be formed (cf. D.2.2.9).

References for further reading

Stumpf. P. K. 'Lipid Metabolism', *Plant Biochemistry*, ed. J. Bonner and J. E. Varner (Academic Press, New York, 1965), pp. 322–45.

Stoffel, W. Über Biosynthese und biologischen Abbau hochungesättigter Fettsäuren, *Naturwiss.* **53** (1965), pp. 621–30.

Lynen, F. Der Weg von der 'aktivierten Essigsäure' zu den Terpenen und Fettsäuren, *Angew. Chem.* **77** (1965), pp. 929–44.

Mudd, J. B. Fat Metabolism in Plants, *Ann. Rev. Plant Physiol.* **18** (1967), pp. 229–52.

Strickland, K. P. 'The Biogenesis of the Lipids', *Biogenesis of Natural Compounds*, ed. P. Bernfield (Pergamon Press, Oxford, 1967), pp. 103–205.

Green, D. E. and Allmann, D. W. 'Biosynthesis of Fatty Acids', *Metabolic Pathways*, Vol. II, ed. D. M. Greenberg (Academic Press, New York, 1968), pp. 37–67.

2.2.2 Elongation of the carbon chain of fatty acids by the fatty-acid-elongation complex

Enzyme systems for the elongation of fatty acid chains occur in both plants and animals (these are not identical to the fatty-acid-synthetase complex). Fatty acids with very long carbon chains synthesized by these enzyme systems are precursors for waxes (cf. D.2.2.7) and alkanes (cf. D.2.2.9).

The fatty-acid-elongation complex which is found in the outer membrane of mitochondria from heart and liver in rat has been comparatively well investigated. This enzyme system lengthens the carbon chain of previously synthesized fatty acids by two or four carbon atoms, using acetyl CoA (the fatty-acid synthetase would require malonyl CoA for this purpose, cf. D.2.2.1). Nothing is known at the present time about the intermediates of this reaction chain, but they are probably similar to those occurring during the *de novo* synthesis of fatty acids.

References for further reading

Colli, W., Hinkle, P. C. and Pullman, M. E. Characterization of the Fatty Acid Elongation System in Soluble Extracts and Membrane Preparations of Rat Liver Mitochondria, *J. biol. Chemistry* **214** (1969), pp. 6432–43.

2.2.3 Methylation of fatty acids

C-methylation of fatty acids has been detected in the case of tubercle bacteria. Thus tuberculostearic acid is obtained from oleic acid via 10-methylene stearic acid (fig. 43).

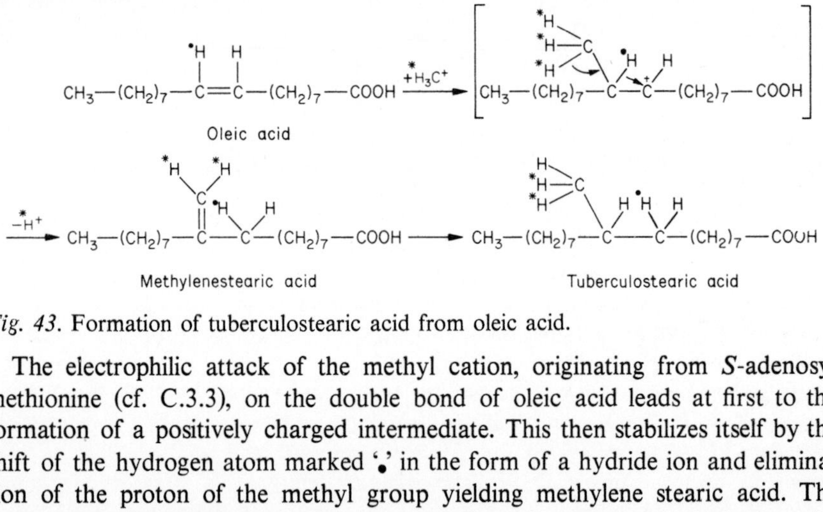

Fig. 43. Formation of tuberculostearic acid from oleic acid.

The electrophilic attack of the methyl cation, originating from *S*-adenosyl methionine (cf. C.3.3), on the double bond of oleic acid leads at first to the formation of a positively charged intermediate. This then stabilizes itself by the shift of the hydrogen atom marked '•' in the form of a hydride ion and elimination of the proton of the methyl group yielding methylene stearic acid. The mechanism of this shift is analogous to the NIH shift observed in the hydroxylation of aromatic systems (cf. C.2.6.5).

Methylene stearic acid is converted to tuberculostearic acid by reduction. Thus only two of the three original hydrogen atoms of the methyl group of methionine are present in the methyl group at carbon atom 10 of this compound.

References for further reading

See under D.2.2.1.

2.2.4 The nucleophilic substitution of fatty acids by fatty acids at the α-carbon atom

This type of reaction has been found in mycobacteria. By the condensation of two molecules of palmitic acid, which probably exist as the CoA-ester, corynomycolic acid and the ketone, palmiton, are formed by reduction and decarboxylation respectively via a β-keto acid (fig. 44.). After feeding palmitic acid-1-^{14}C the carbon atoms marked '*' were found to be radioactive.

The nucleophilic substitution of the α-carbon atom of one of the palmitic acid molecules by the activated carboxyl group of the second is similar to the reaction mechanism proceeding during the biosynthesis of fatty acids (cf. D.2.2.1 and D.2.2.2) and methyl fatty acids (cf. D.3.1).

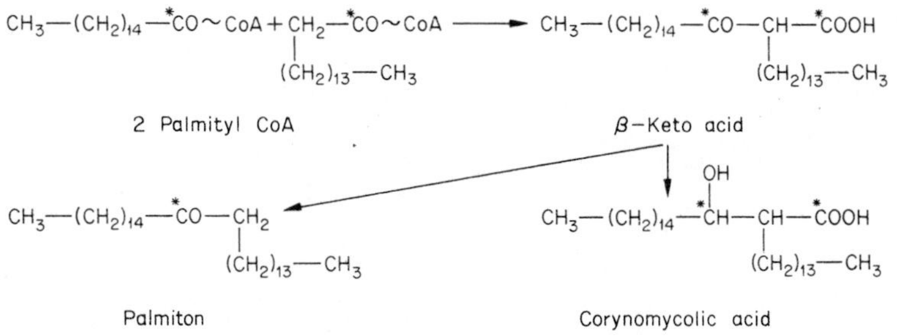

Fig. 44. Formation of corynomycolic acid and palmiton.

References for further reading

See under D.2.2.1.

2.2.5 Biosynthesis of unsaturated and hydroxylated fatty acids and the formation of prostaglandins

Unsaturated fatty acids arise through two independent pathways.

(a) In bacteria (and possibly, in certain cases, in higher plants) the last reduction step in the cycle of elongation of the carbon chain by fatty-acid synthetase does not take place (cf. D.2.2.1). The unsaturated fatty acid is then converted to the CoA-ester directly and may subsequently be further elongated without the loss of the double bond.

In animal cells also, an enzyme system is present in the membrane of the endoplasmic reticulum which lengthens the carbon chain of existing saturated and unsaturated fatty acids, using malonyl CoA.

(b) In animals, plants and fungi double bonds may be introduced into the saturated fatty-acid molecule at a later stage. Thus, for example, stearyl CoA is converted to linolyl CoA via oleyl CoA. Since, in addition to

NADH or NADPH, molecular oxygen is required in this reaction, hydroxy acids are probably formed as intermediates (cf. C.2.6). Though there is no conclusive proof for this, it has been shown in *Ricinus communis* that ricinoleic acid is formed from oleic acid by the action of a similar enzyme system (fig. 45). It is not clear whether ricinoleic acid can be converted to linoleic acid. Linoleic acid is not a precursor for oleic acid.

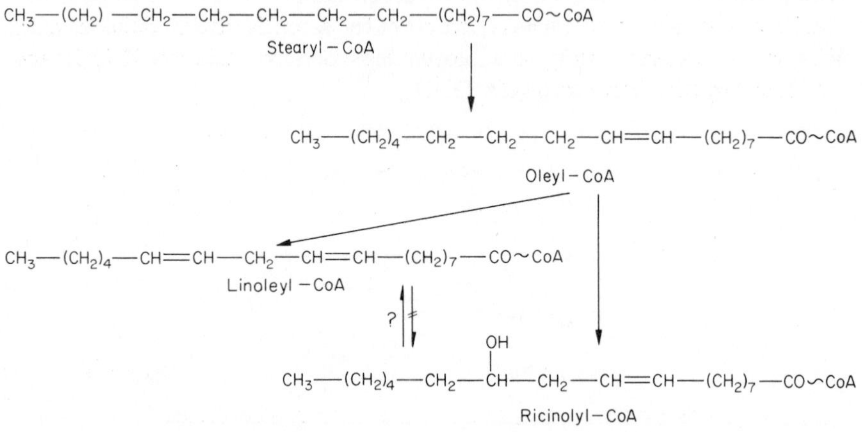

$$CH_3 — (CH_2) — CH_2 — CH_2 — CH_2 — CH_2 — CH_2 — (CH_2)_7 — CO \sim CoA$$

Stearyl – CoA

$$CH_3 — (CH_2)_4 — CH_2 — CH_2 — CH_2 — CH = CH — (CH_2)_7 — CO \sim CoA$$

Oleyl – CoA

$$CH_3 — (CH_2)_4 — CH = CH — CH_2 — CH = CH — (CH_2)_7 — CO \sim CoA$$

Linoleyl – CoA

?

$$CH_3 — (CH_2)_4 — CH_2 — \overset{\overset{OH}{|}}{CH} — CH_2 — CH = CH — (CH_2)_7 — CO \sim CoA$$

Ricinolyl – CoA

Fig. 45. Formation of unsaturated fatty acids and ricinoleic acid.

A few poly-unsaturated fatty acids with twenty carbon atoms which originate from linoleic acid and linolenic acid by elongation of the carbon chain and additional dehydrogenation are utilized in human beings and a number of animals during the formation of prostaglandins. Thus, an explanation of the vitamin nature of unsaturated fatty acids, which has been known for some time, has been found. Prostaglandins effect, among other things, the stimulation of the uterus, the widening of the genital tubes in males and inhibition of the mobilization of free fatty acids caused by noradrenalin.

Prostaglandin E_1 originates from bishomo-γ-linolenic acid ($\Delta^{8, 11, 14}$-all-*cis*-eicosatrienoic acid), prostaglandin E_2 from arachidonic acid ($\Delta^{5, 8, 11, 14}$-all-*cis*-eicosatetraenoic acid) and prostaglandin E_3 from $\Delta^{5, 8, 11, 14, 17}$-all-*cis*-eicosapentaenoic acid (fig. 46). All the oxygen atoms in the molecule of the prostaglandins arise from molecular oxygen, and both oxygen atoms in the ring are derived from a single oxygen molecule.

The synthesis of prostaglandin E_1 may be represented as outlined in fig. 47. It probably begins with the stereospecific addition of an oxygen molecule, activated by a dioxygenase (cf. C.2.5), at carbon atom 9 or at carbon atom 11 of bishomo-γ-linolenic acid (compound I). This is followed by a shift of electrons to form the cyclic peroxide II. Compound III may arise from this by an intramolecular oxidation-reduction. Attack by a mixed function oxygenase on the

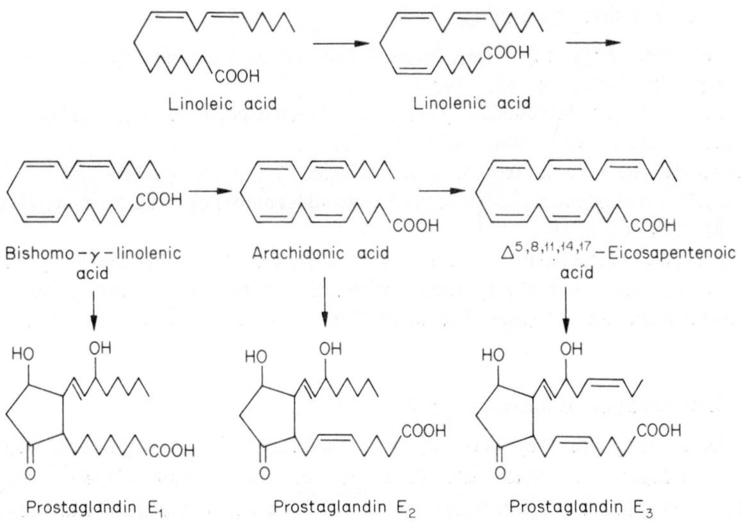

Fig. 46. Biosynthesis of prostaglandins.

double bond between carbon atom 14 and carbon atom 15 (cf. C.2.6.5) may lead via the epoxide IV to compound V, which is converted to prostaglandin E_1 by the elimination of a proton.

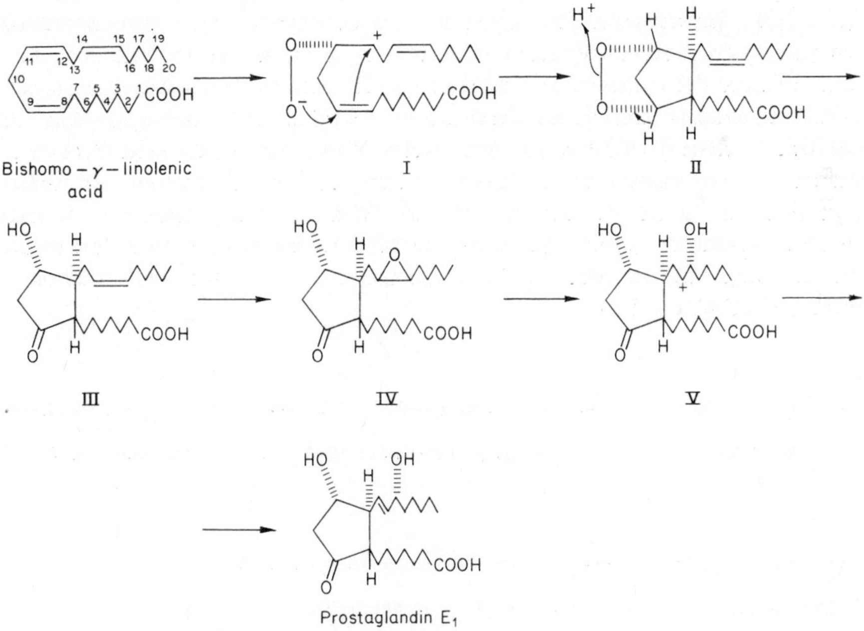

Fig. 47. Possible pathway for conversion of bishomo-γ-linolenic acid to prostaglandin E_1.

References for further reading

Stoffel, W. Über Biosynthese und biologischen Abbau hochungesättigter Fettsäuren, *Naturwiss.* **53** (1965), pp. 621–30.

Stumpf, P. K. 'Lipid Metabolism', *Plant Biochemistry*, ed. J. Bonner and J. E. Varner (Academic Press, New York, 1965), pp. 322–45.

Samuelsson, B. Die Prostaglandine, *Angew. Chem.* **77** (1965), pp. 445–52.

Bergström, S. Prostaglandins: Members of a New Hormonal System, *Science (Washington)* **157** (1967), pp. 382–91.

Mudd, J. B. Fat Metabolism in Plants, *Ann. Rev. Plant Physiol.* **18** (1967), pp. 229–52.

James, A. T. 'Fatty Acid Biosynthesis in Plants', *Perspectives in Phytochemistry*, ed. J. B. Harborne and T. Swain (Academic Press, London, 1969), pp. 91–106.

2.2.6 Degradation of fatty acids

Fatty acids may be degraded in three different and independent ways. Of these, β-oxidation, in which the bond between the α- and β-carbon atoms is cleaved, is of special importance, and is found in all organisms that have been investigated.

(a) β-Oxidation

In the case of saturated fatty acids degradation begins by activation with ATP. The acyl-AMP thus formed reacts with coenzyme A to give the fatty acid CoA (cf. C.1.2). The activated fatty acid is then dehydrogenated by a flavin-containing enzyme system, and converted to a β-keto acid by subsequent addition of water and a further dehydrogenation in which NAD$^+$ acts as the hydrogen acceptor. This compound is thiolytically degraded in the presence of coenzyme-A by the enzyme β-ketoacyl thiolase, to form acetyl CoA and a fatty-acid CoA ester shorter by two carbon atoms than the starting compound, which in turn undergoes the same series of reactions (fig. 48). While the fatty acids with an even number of carbon atoms are totally degraded to acetyl CoA the degradation of fatty acids with an odd number of carbon atoms ends with the formation of propionyl CoA (cf. D.3).

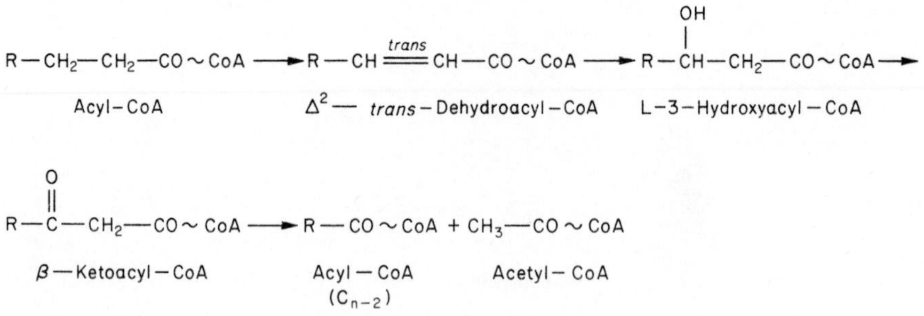

$$R-CH_2-CH_2-CO \sim CoA \longrightarrow R-CH \overset{trans}{=\!=\!=} CH-CO \sim CoA \longrightarrow R-\overset{OH}{\underset{|}{C}}H-CH_2-CO \sim CoA \longrightarrow$$

Acyl–CoA Δ^2- *trans*–Dehydroacyl–CoA L–3–Hydroxyacyl–CoA

$$R-\overset{O}{\overset{||}{C}}-CH_2-CO \sim CoA \longrightarrow R-CO \sim CoA + CH_3-CO \sim CoA$$

β—Ketoacyl–CoA Acyl–CoA (C_{n-2}) Acetyl–CoA

Fig. 48. β-Oxidation of fatty acids.

Unsaturated fatty acids must first be converted to compounds which can be attacked by the enzymes of β-oxidation. In the case of linoleic acid, three molecules of acetyl CoA are formed by β-oxidation in the normal way at the beginning of the degradation, giving rise to $\Delta^{3,6}$-*cis*, *cis*-dodecadienoyl CoA. The degradation of this compound, however, is only possible after rearrangement and shift of the *cis*-double bond from carbon atom 3 to carbon atom 2, with the formation of $\Delta^{2,6}$-*trans,cis*-dodecadienoyl CoA. This compound structurally resembles Δ^2-dehydroacyl CoA through which β-oxidation always proceeds (cf. fig. 48), and can therefore be further degraded. After elimination of two molecules of acetyl CoA, Δ^2-*cis*-octenoyl CoA is formed. This compound is first hydrated by means of a hydratase to D-3-hydroxycaproyl CoA, and is then converted to the L-compound by epimerase. This latter compound is then susceptible to the enzymes of β-oxidation, and may be degraded to four molecules of acetyl CoA (fig. 49).

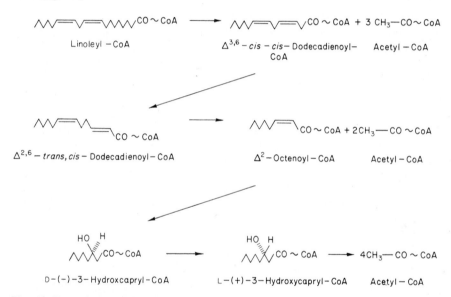

Fig. 49. Degradation of linoleic acid.

(b) α-Oxidation

In the case of α-oxidation the bond between the carboxyl group and the carbon atom at the α-position is broken.

The degradation of fatty acids by α-oxidation is not as important as β-oxidation, but has, however, been shown to exist in plants and animals. Thus, a peroxidase (cf. C.2.4) is found in the cotyledons of *Arachis hypogea*, the peanut, which oxidatively decarboxylates long-chain fatty acids to carbon dioxide, and an aldehyde shorter by one carbon atom. With enzyme preparations from the leaves of green plants and from animals α-oxidation is initiated by the formation of an L-α-hydroxy acid by means of a mixed function oxygenase (cf. C.2.6). The hydroxy acid is then oxidized to a keto acid and this is possibly oxidatively decarboxylated (fig. 50).

$$R-CH_2-COOH \longrightarrow R-CHOH-COOH \longrightarrow R-CO-COOH \longrightarrow R-COOH$$

Fatty acid (C_n)   L$-\alpha-$Hydroxy fatty acid (C_n) $\alpha-$keto acid (C_n)   Fatty acid (C_{n-1})

Fig. 50. Most important pathway for the α-oxidation of fatty acids.

The fatty acid shorter by one carbon atom than the original may then go through the same series of reactions, α-Oxidation also probably plays a role in the formation of alkanes (cf. D.2.2.9) and acetylenic compounds (cf. D.2.2.10).

(c) ω-Oxidation

In a few cases ω-oxidation is of importance. In this type of degradation the end methyl group is hydroxylated by a mixed function oxygenase (cf. C.2.6.4) to form an ω-hydroxy fatty acid. Further oxidation forms a dicarboxylic acid.

Long-chain alkanes are also attacked by the enzymes of ω-oxidation. The fatty acids formed as a result can react further in primary metabolism. Microorganisms which oxidize alkanes are therefore of great economic importance in the production of organic products (e.g. fodder) from petroleum, natural gas and petroleum residues. Recently an infection by petroleum bacteria has been reported to be the cause of a mysterious destruction of roads in some districts of Western Australia. The road surface became brittle and broken within a short time due to degradation of the petroleum portion of the asphalt by the bacteria.

References for further reading

Stoffel, W. Über Biosynthese und biologischen Abbau hochungesättigter Fettsäuren, *Naturwiss.* **53** (1965), pp. 621–30.

Mudd, J. B. Fat Metabolism in Plants, *Ann. Rev. Plant Physiol.* **18** (1967), pp. 229–52.

Green, D. E. and Allmann, D. W. 'Fatty Acid Oxidation', *Metabolic Pathways*, Vol. II, ed. D. M. Greenberg (Academic Press, New York, 1968), pp. 1–36.

Stumpf, P. K. Metabolism of Fatty Acids, *Ann. Rev. Biochem.* **38** (1969), pp. 691–712.

2.2.7 Formation of fatty acid esters

Triglycerides, designated as fats or fatty oils according to their consistency, are the most important forms of storage of fatty acids in living organisms.

They are synthesized by acylation of one mole of glycerol-1-phosphate by two moles of a CoA ester of the fatty acid. L-α-Phosphatidic acid is first formed, which is then dephosphorylated to a D-α,β-diglyceride and phosphate. The diglyceride thus formed then reacts with a further molecule of fatty acid CoA to yield the triglyceride (fig. 51).

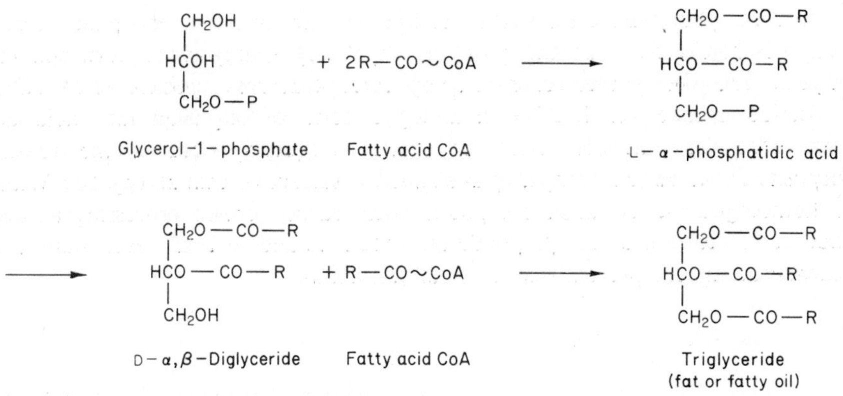

$$
\begin{array}{c}
CH_2OH \\
| \\
HCOH \\
| \\
CH_2O-P
\end{array}
\quad + \; 2R-CO\sim CoA \quad \longrightarrow \quad
\begin{array}{c}
CH_2O-CO-R \\
| \\
HCO-CO-R \\
| \\
CH_2O-P
\end{array}
$$

Glycerol–1–phosphate   Fatty acid CoA   L–α–phosphatidic acid

$$
\longrightarrow \quad
\begin{array}{c}
CH_2O-CO-R \\
| \\
HCO-CO-R \\
| \\
CH_2OH
\end{array}
\quad + \; R-CO\sim CoA \quad \longrightarrow \quad
\begin{array}{c}
CH_2O-CO-R \\
| \\
HCO-CO-R \\
| \\
CH_2O-CO-R
\end{array}
$$

D–α,β–Diglyceride   Fatty acid CoA   Triglyceride
(fat or fatty oil)

Fig. 51. Formation of fats and fatty oils.

The naturally occurring triglycerides are mixtures. They occur as a solid if there is a low proportion of unsaturated fatty acids (e.g. oleic acid, linoleic acid and linolenic acid), and in liquid form if there is a higher proportion. They can be degraded to glycerol and fatty acids by means of lipases which are secreted by the pancreas of mammals and human beings, but also occur in all other organisms.

D-α,β-Diglycerides may be substituted by other compounds besides fatty-acid CoA esters. Thus condensation with cytidine diphosphate choline leads to lecithin (fig. 52), while condensation with cytidine diphosphate ethanolamine leads to phosphatidyl ethanolamine (cephalin). In this compound the choline group of lecithin is replaced by an ethanolamine grouping.

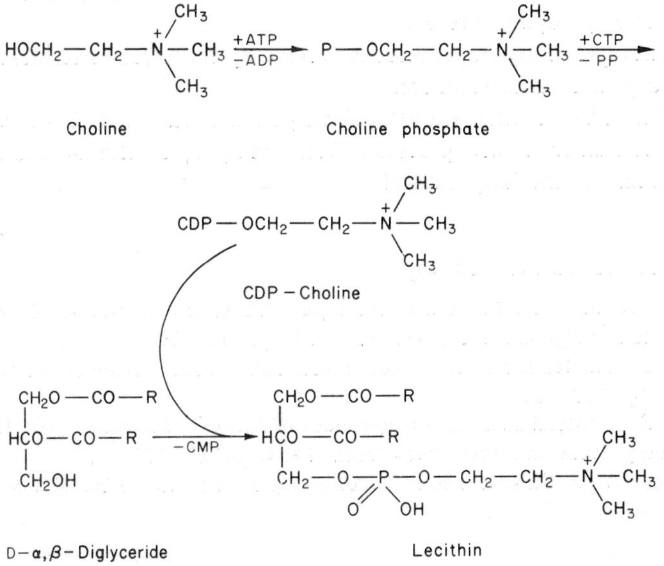

$$
HOCH_2-CH_2-\overset{+}{N}\!\!\begin{array}{c}CH_3\\ \\ CH_3\end{array}\!\!-CH_3 \xrightarrow[-ADP]{+ATP} P-OCH_2-CH_2-\overset{+}{N}\!\!\begin{array}{c}CH_3\\ \\ CH_3\end{array}\!\!-CH_3 \xrightarrow[-PP]{+CTP}
$$

Choline   Choline phosphate

$$
CDP-OCH_2-CH_2-\overset{+}{N}\!\!\begin{array}{c}CH_3\\ \\ CH_3\end{array}\!\!-CH_3
$$

CDP – Choline

$$
\begin{array}{c}
CH_2O-CO-R \\
| \\
HCO-CO-R \\
| \\
CH_2OH
\end{array}
\xrightarrow{-CMP}
\begin{array}{c}
CH_2O-CO-R \\
| \\
HCO-CO-R \\
| \\
CH_2-O-P-O-CH_2-CH_2-\overset{+}{N}\!\!\begin{array}{c}CH_3\\ \\ CH_3\end{array}\!\!-CH_3
\end{array}
$$

D–α,β– Diglyceride   Lecithin

Fig. 52. Biosynthesis of lecithin.

In these phosphate diesters the free hydroxyl group of the phosphate residue may be substituted by another group, e.g. by N-acyl sphingosine to form sphingo-myelin. Thus the synthesis of more complicated structures is possible (cf. D.2.2.8).

Besides alkanes (cf. D.2.2.9), homologous series of long-chain fatty acids and their aldehydes, alcohols, esters and acetals frequently occur in the wax-like deposits of the cuticle of higher plants, and the surface of animals (fig. 53). Within a homologous series, those compounds with carbon chains possessing an even number of carbon atoms preponderate (while alkanes with an even number of carbon atoms are present only in small amounts).

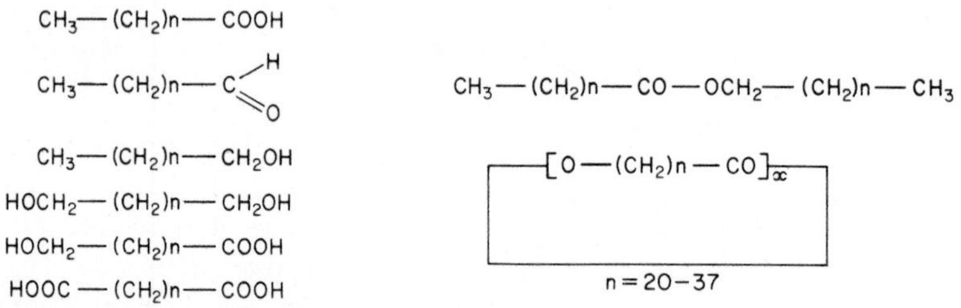

$$CH_3\!-\!(CH_2)n\!-\!COOH$$

$$CH_3\!-\!(CH_2)n\!-\!C\!\!\stackrel{H}{\underset{O}{\diagdown}}$$

$$CH_3\!-\!(CH_2)n\!-\!CH_2OH$$

$$HOCH_2\!-\!(CH_2)n\!-\!CH_2OH$$

$$HOCH_2\!-\!(CH_2)n\!-\!COOH$$

$$HOOC\!-\!(CH_2)n\!-\!COOH$$

$$CH_3\!-\!(CH_2)n\!-\!CO\!-\!OCH_2\!-\!(CH_2)n\!-\!CH_3$$

$$[O\!-\!(CH_2)n\!-\!CO]_{\overline{x}}$$

$$n = 20-37$$

Fig. 53. Fatty acids and derivatives of fatty acids from epicuticular wax.

Compounds with very long carbon chains are synthesized by the fatty-acid-elongation complex described in D.2.2.2. The long-chain fatty acids thus formed are probably reduced to the corresponding aldehydes and alcohols after con-version to the CoA-ester.

Three reaction mechanisms have been shown to be involved in the formation of these ester groupings present.

(a) The reversal of saponification involving the direct esterification of free fatty acids and free alcohols.

(b) The same reaction with the participation of fatty-acid CoA-esters.

(c) The transfer of acid groupings from sphingolipids, lecithin and compounds similar to the fatty alcohols.

References for further reading

Strickland, K. P. 'The Biogenesis of the Lipids', *Biogenesis of Natural Compounds*, ed. P. Bernfeld (Pergamon Press, Oxford, 1967), pp. 103–205.

Eglinton, E. and Hamilton, R. J. Leaf Epicuticular Waxes, *Science (Washington)* **156** (1967), pp. 1322–35.

Rossiter, R. J. 'Metabolism of Phosphatides', *Metabolic Pathways*, Vol. II, ed. D. M. Greenberg (Academic Press, New York, 1968), pp. 69–115.

Kolattukudy, P. E. Biosynthesis of Cuticular Lipids, *Ann. Rev. Plant Physiol.* **21** (1970), pp. 163–92.

2.2.8 Biosynthesis of sphingolipids and plasmalogens

Sphingolipids are those compounds which possess sphingosine, *trans*-D-*erythro*-1,3-dihydroxy-2-amino-octadec-4-ene, or a similar substance as the common structural element. They are further differentiated into sphingomyelins, cerebrosides and gangliosides according to the substituents present on the basic skeleton. Representatives of these classes of substances occur in all organisms. They are constituents of the lipid portion of the cell membrane and the membrane of cell organelles, and as such are of very great importance in the metabolic processes taking place in the cell. The gangliosides participate in the important functions of the central nervous system of human beings and animals.

Sphingosine originates through a reaction involving palmityl CoA and serine, in which the carboxyl group of serine is eliminated. In animal tissues the reduc-

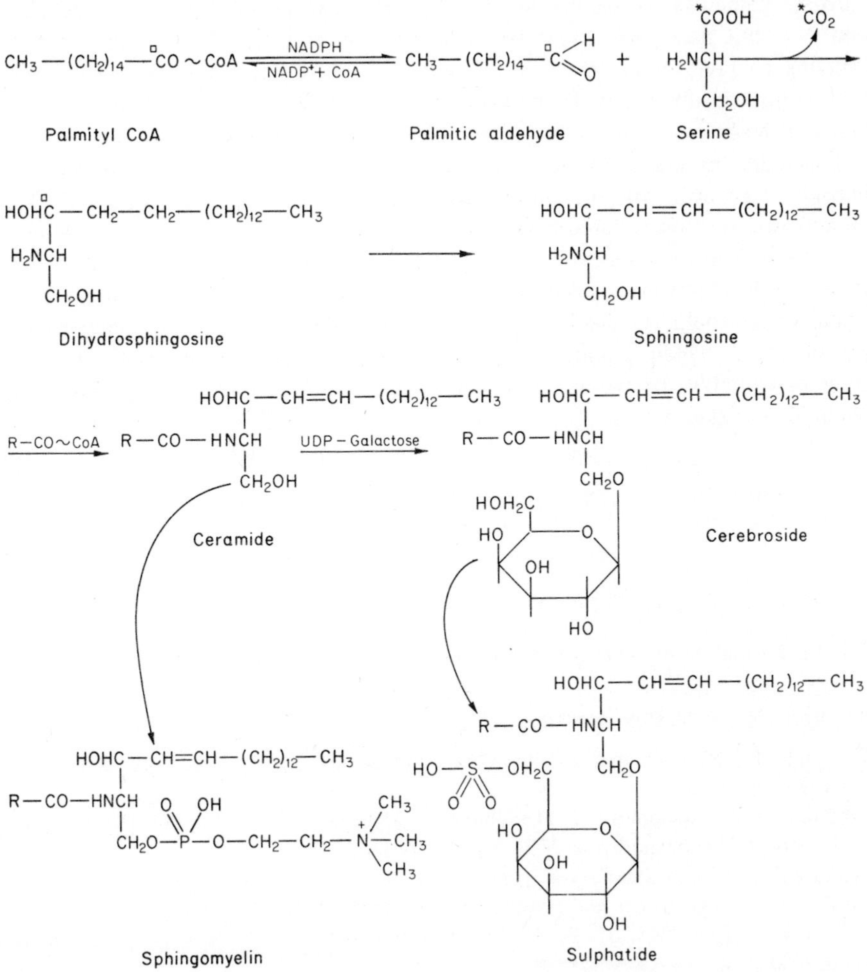

$$CH_3-(CH_2)_{14}-\overset{o}{C}O \sim CoA \xrightleftharpoons[\text{NADP}^+ + \text{CoA}]{\text{NADPH}} CH_3-(CH_2)_{14}-\overset{o}{C}\overset{H}{\underset{O}{\diagdown}} + \quad H_2N\overset{*COOH}{\underset{CH_2OH}{\overset{|}{C}H}} \xrightarrow{\quad *CO_2 \quad}$$

Palmityl CoA Palmitic aldehyde Serine

Fig. 54. Formation of sphingomyelins, cerebrosides and sulphatides.

tion of palmityl CoA to palmitic aldehyde by means of NADPH as well as the condensation of this compound with serine to form dihydrosphingosine have been demonstrated. The latter compound is then converted to sphingosine in a dehydrogenation step catalysed by a flavoenzyme.

In the formation of sphingolipids the sphingosine is probably acylated at the β-position by a fatty-acid CoA-ester, forming an N-acyl sphingosine (ceramide). This may further be substituted at the α-position by CDP-choline to form sphingomyelin (fig. 54).

In the formation of cerebrosides the ceramide is glycosidically linked with a sugar (galactose or glucose) which in the case of sulphatides is further esterified with a sulphate group at position 6. In the case of mucolipids (e.g. the gangliosides) the molecule of ceramide is linked at the α-position with several sugars (e.g. glucose, galactose, N-acetyl glucosamine, N-acetyl galactosamine and fucose) and N-acetyl neuraminic acid (sialic acid). The mucolipids can combine with proteins to give lipoproteins, proteolipids and phosphatidolipids and can form very complex substances. If the synthesis of gangliosides is disturbed in human beings it leads to the so-called sphingolipidoses, e.g. amaurotic idiocy.

Plasmalogens are substances in which glycerol is linked with a substituted phosphate group in the α-position, with a fatty acid in the β-position and with an aliphatic α,β-unsaturated alkane through an ether grouping in the α'-position.

Little is known about the biosynthesis of the compounds belonging to this group. It has, however, been shown in the case of choline plasmalogen that hexadecanol (palmityl alcohol) serves as the precursor for the α,β-unsaturated carbon chain, which is linked by an ether bridge at the α'-position. A saturated ether is probably formed as an intermediate which is dehydrogenated in the last reaction step (fig. 55).

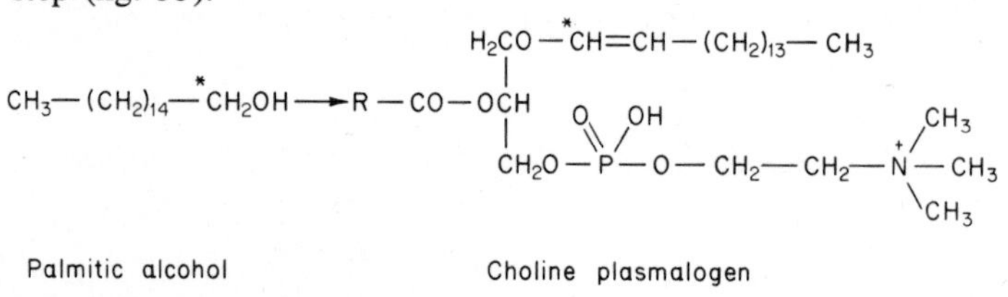

Palmitic alcohol   Choline plasmalogen

Fig. 55. Formation of choline plasmalogen.

References for further reading

Wiegandt, H. Struktur und Funktion der Ganglioside, *Angew Chem.* **80** (1968), pp. 89–98.

Rossiter, R. J. 'Metabolism of Phosphatides', *Metabolic Pathways*, Vol. II, ed. D. M. Greenberg (Academic Press, New York, 1968), pp. 69–115.

Lennarz, W. J. Lipid Metabolism, *Ann. Rev. Biochem.* **39** (1970), pp. 359–88.

Kanfer, J. N. 'Biosynthesis and Hydrolysis of Sphingolipids', *Metabolic Conjugation and Metabolic Hydrolysis*, ed. W. H. Fishman (Academic Press, New York, 1970), pp. 603–28.

Stoffel, W. Sphingolipids, *Ann. Rev. Biochem.* **40** (1971), pp. 57–82.

2.2.9 Formation of n-alkanes from fatty acids

The exterior of higher plants is covered by a cuticle on which a waxy layer of variable thickness is usually deposited. These deposits which have quite a complicated composition (cf. D.2.2.7 also) play an important role in wetting, and in the regulation of the water economy of plants. Some of these waxes, e.g. the carnauba wax from *Copernicia cerifera*, are of commercial interest.

In a number of higher plants the major portion of these deposits consist of alkanes with about thirty carbon atoms. In *Brassica oleracea*, for example, nonacosane, 15-nonacosanol and 15-nonacosanone occur in large quantities. *Iso*alkanes and *anteiso*alkanes preponderate in *Nicotiana tabacum*. The alkanes usually contain an odd number of carbon atoms. They are accompanied by fatty acids and fatty-acid derivatives of approximately the same chain length and in most cases with an even number of carbon atoms. The alkanes originate by decarboxylation from these very long-chain fatty acids (cf. D.2.2.2). Thus a straight-chain fatty acid with thirty carbon atoms is the precursor for the unbranched alkanes nonacosane, nonacosanol and nonacosanone. The formation of *iso*alkanes and *anteiso*alkanes starts from fatty acids in whose biosynthesis isobutyryl CoA, isovaleryl CoA and α-methylbutyryl CoA serve as the starter molecules (cf. D.2.2.1) (fig. 56).

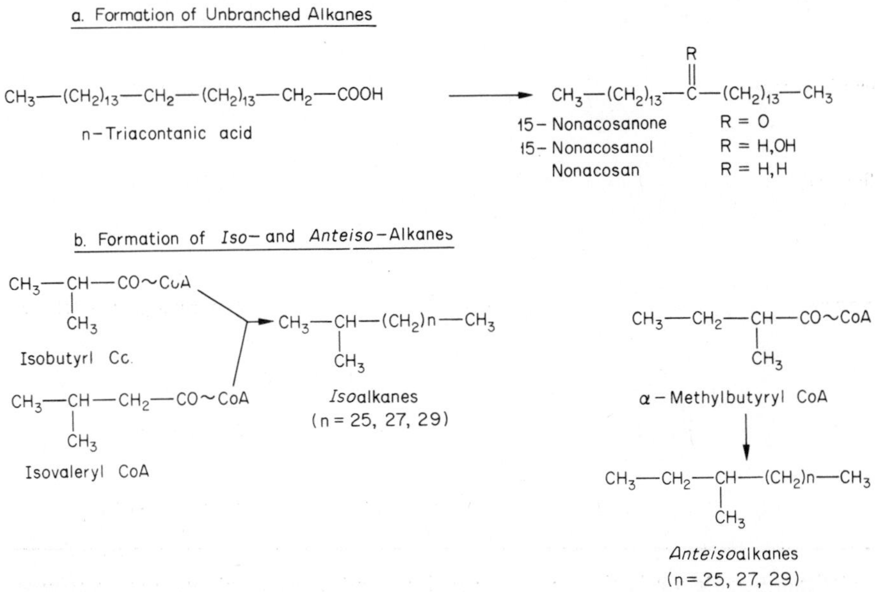

a. Formation of Unbranched Alkanes

CH_3—$(CH_2)_{13}$—CH_2—$(CH_2)_{13}$—CH_2—$COOH$ $\longrightarrow$ CH_3—$(CH_2)_{13}$—$\overset{\overset{R}{\|}}{C}$—$(CH_2)_{13}$—$CH_3$

n–Triacontanic acid

15– Nonacosanone	R = O
15– Nonacosanol	R = H,OH
Nonacosan	R = H,H

b. Formation of *Iso–* and *Anteiso–*Alkanes

CH_3—CH—$CO{\sim}CoA$
      |
     CH_3
Isobutyrl Cc.

CH_3—CH—CH_2—$CO{\sim}CoA$
      |
     CH_3
Isovaleryl CoA

$\longrightarrow$ CH_3—CH—$(CH_2)n$—CH_3
                       |
                      CH_3
*Iso*alkanes
(n = 25, 27, 29)

CH_3—CH_2—CH—$CO{\sim}CoA$
             |
            CH_3
α – Methylbutyryl CoA

$\downarrow$

CH_3—CH_2—CH—$(CH_2)n$—CH_3
             |
            CH_3
*Anteiso*alkanes
(n = 25, 27, 29)

Fig. 56. Formation of alkanes.

References for further reading

Douglas, A. G. and Eglinton, G. 'The Distribution of Alkanes', *Comparative Phytochemistry*, ed. T. Swain (Academic Press, London, 1966), pp. 57–77.

Eglinton, G. and Hamilton, R. J. Leaf Epicuticular Waxes, *Science (Washington)* **156** (1967), pp. 1322–35.

Kolattukudy, P. E. Biosynthesis of Surface Lipids, *Science (Washington)* **159** (1968), pp. 498–505.

2.2.10 Biosynthesis of acetylenic derivatives

Until now acetylenic derivatives have only been found in fungi (e.g. Basidiomycetes) and a number of higher plants. They either are unbranched aliphatic compounds or compounds with alicyclic and/or heterocyclic rings in the molecule. A great variety of functional groups may be present. A few representatives of this class of compound are given in table 8.

Table 8. A few naturally occurring acetylene derivatives

Name	Structure	Occurrence
Diatrene	$HOOC-CH{=}CH-(C{\equiv}C)_3-CH_2OH$	Fungal cultures
Artemesia ketone	$CH_3-(C{\equiv}C)_3-CH{\overset{t}{=}}CH-CH_2-CH_2-CO-CH_2-CH_3$	Anthemideae
Falcarinone	$CH_3-(CH_2)_6-CH{\overset{c}{=}}CH-CH_2-(C{\equiv}C)_2-CO-CH_2-CH_3$	Araliaceae, Ammiaceae
Matricaria lactone	$CH_3-CH{\overset{c}{=}}CH-C{\equiv}C-CH{\overset{t}{=}}$	Asteraceae
Artemisia lactone	$CH_3-C{\equiv}C-C{\overset{t}{=}}CH-CH{\overset{t}{=}}$	Asteraceae
Carlina oxide		*Carlina acaulis*
Laurencine		*Laurentia glandulifera* (alga
		Echinops species
		Echinops species

It has been shown both in micro-organisms and in higher plants that acetylenic compounds are formed from one molecule of acetyl CoA, which serves as the starter molecule, and several malonyl CoA units. Biogenesis proceeeds in most cases via oleic acid, and polyunsaturated fatty acids (cf. D.2.2.5) whose double bonds are converted by dehydrogenation to triple bonds (such dehydrogenation reactions possibly proceed at a later stage of the reaction sequence also). The chains are shortened by conversion to vinyl compounds as well as by α-oxidation and β-oxidation (cf. D.2.2.6).

The metabolic pathway outlined in fig. 57 appears to be involved in the formation of dehydromatricaria acid and other polyenes in the case of fungi. Oleic acid is first converted via linoleic acid to crepis acid and dehydrocrepis acid. This compound is then converted to dehydromatricaria acid by means of further dehydrogenation and oxidation steps. Acetylenes with eleven to fourteen carbon atoms may originate from the intermediates of this reaction. Those with odd numbers of carbon atoms are probably formed from the corresponding acids by decarboxylation.

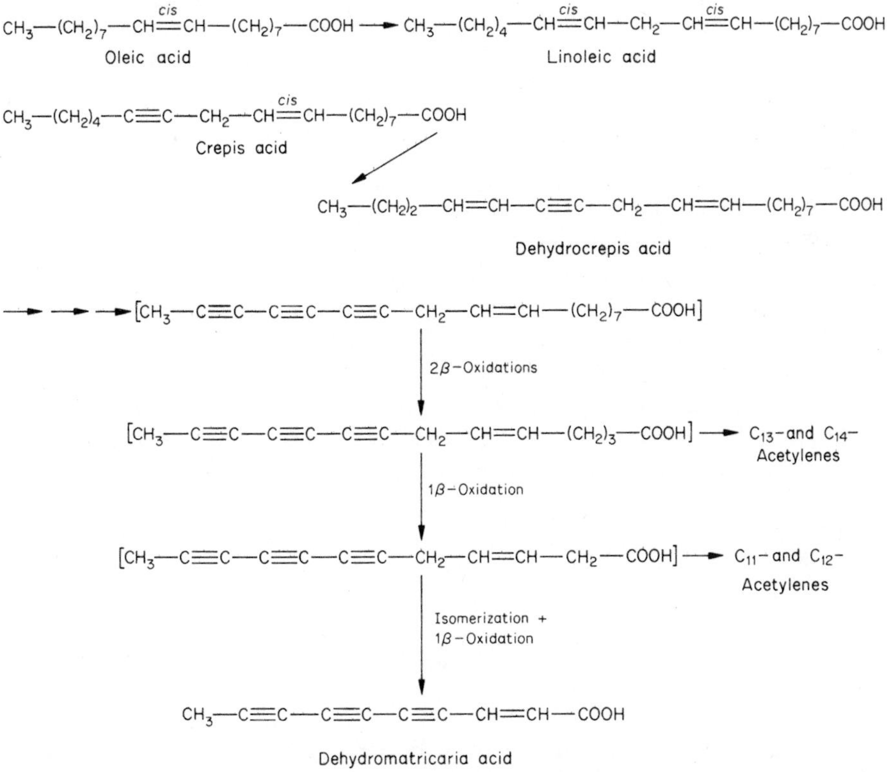

$$CH_3\text{—}(CH_2)_7\text{—}\overset{cis}{CH\text{=}CH}\text{—}(CH_2)_7\text{—}COOH \longrightarrow CH_3\text{—}(CH_2)_4\text{—}\overset{cis}{CH\text{=}CH}\text{—}CH_2\text{—}\overset{cis}{CH\text{=}CH}\text{—}(CH_2)_7\text{—}COOH$$

Oleic acid   Linoleic acid

$$CH_3\text{—}(CH_2)_4\text{—}C\text{≡}C\text{—}CH_2\text{—}\overset{cis}{CH\text{=}CH}\text{—}(CH_2)_7\text{—}COOH$$

Crepis acid

$$CH_3\text{—}(CH_2)_2\text{—}CH\text{=}CH\text{—}C\text{≡}C\text{—}CH_2\text{—}CH\text{=}CH\text{—}(CH_2)_7\text{—}COOH$$

Dehydrocrepis acid

$$\longrightarrow \longrightarrow \longrightarrow [CH_3\text{—}C\text{≡}C\text{—}C\text{≡}C\text{—}C\text{≡}C\text{—}CH_2\text{—}CH\text{=}CH\text{—}(CH_2)_7\text{—}COOH]$$

$\downarrow$ 2β-Oxidations

$$[CH_3\text{—}C\text{≡}C\text{—}C\text{≡}C\text{—}C\text{≡}C\text{—}CH_2\text{—}CH\text{=}CH\text{—}(CH_2)_3\text{—}COOH] \longrightarrow C_{13}\text{- and } C_{14}\text{- Acetylenes}$$

$\downarrow$ 1β-Oxidation

$$[CH_3\text{—}C\text{≡}C\text{—}C\text{≡}C\text{—}C\text{≡}C\text{—}CH_2\text{—}CH\text{=}CH\text{—}CH_2\text{—}COOH] \longrightarrow C_{11}\text{- and } C_{12}\text{- Acetylenes}$$

$\downarrow$ Isomerization + 1β-Oxidation

$$CH_3\text{—}C\text{≡}C\text{—}C\text{≡}C\text{—}C\text{≡}C\text{—}CH\text{=}CH\text{—}COOH$$

Dehydromatricaria acid

Fig. 57. Probable pathway for the formation of dehydromatricaria acid and other polyacetylenes from oleic acid.

The majority of naturally occurring acetylenic compounds are biosynthesized from the compounds mentioned above. Some of the reaction mechanisms involved which are clearly understood are shown below:

Isomerization reactions

(a) The formation of allene groupings

The —C≡C—CH$_2$—group can be converted to a —CH=C=CH—

grouping by means of an isomerase. Depending on the particular enzyme both possible absolute configurations may be formed.

(b) Cis-trans isomerism

The double bonds present in most acetylenic compounds may be isomerized by means of special enzymes. *Cis-trans* isomers almost always occur together in nature.

Addition reactions

Additions to the triple bond of acetylene derivatives is easy to observe *in vitro*. *In vivo*, this reaction is probably responsible for the formation of several different compounds.

(a) Formation of phenyl rings

In a number of naturally occurring acetylene derivatives the molecule contains a phenyl ring which does not carry hydroxyl groups and does not arise from a polyketide (cf. D.2.3). It is thought that these rings are formed by addition reactions to the triple bonds.

Though experimental proof of this supposition is awaited, the experimental conversion of matricaria ester into the aromatic compound I indicates the possibility of such a mechanism. A scheme, outlined in fig. 58, is postulated for the rearrangement, which assumes the initial formation of a thioether followed by the formation of an aromatic ring with the shift of a methyl group.

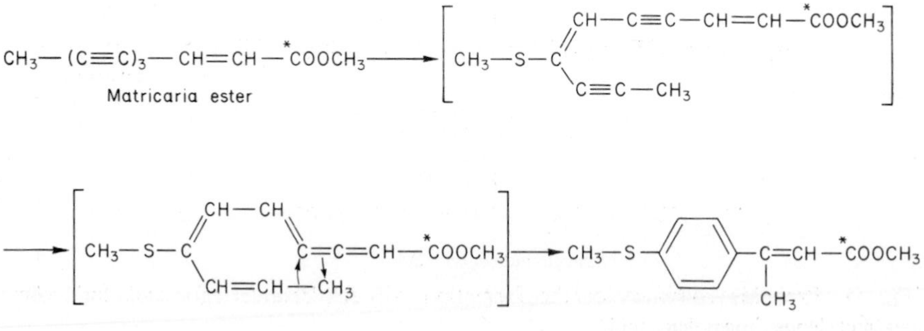

Fig. 58. Synthesis of aromatic compounds from aliphatic acetylene derivatives.

(b) Formation of thioethers

The synthesis of thioethers, takes place formally by the addition of H_2S to a triple bond with or without subsequent alkylation. Cyclic thio derivatives are possibly formed from dithio compounds by the elimination of sulphur

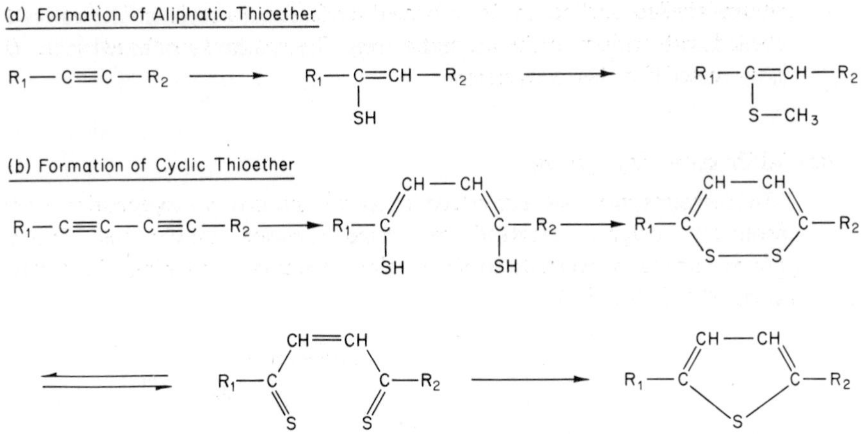

(a) Formation of Aliphatic Thioether

(b) Formation of Cyclic Thioether

Fig. 59. Formation of thio derivatives from acetylenic compounds.

(fig. 59). The actual sulphur containing precursor is unknown, but feeding experiments have shown that sulphur from sulphate and cysteine may be incorporated.

(c) Formation of furan derivatives

These are formed by addition of oxygen across the triple bonds. Furan derivatives are formed from inenoles via dihydrofuran derivatives. Thus

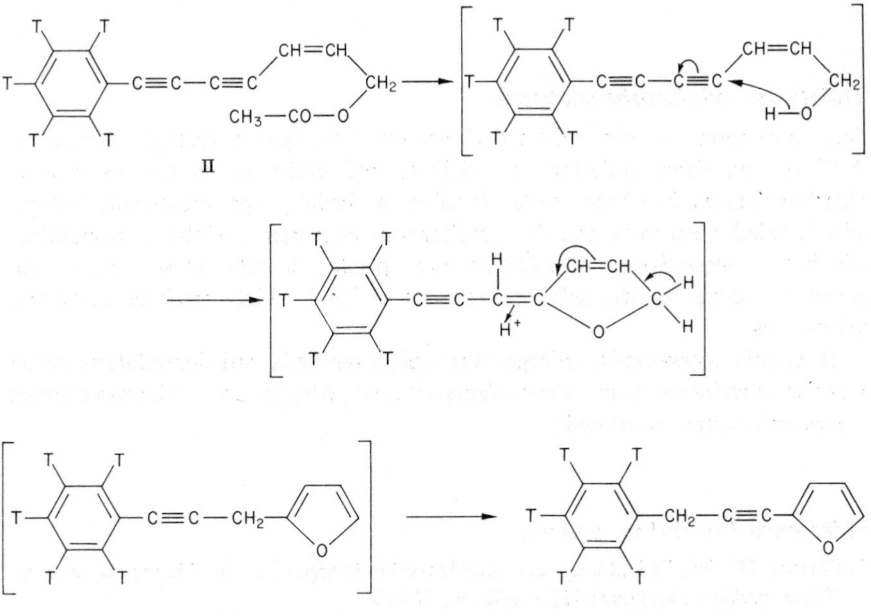

Carlina oxide

Fig. 60. Formation of furan derivatives from acetylenic compounds.

tritium-labelled carlina oxide is formed by feeding acetylene derivative II labelled with tritium in the aromatic ring. The pathway outlined in fig. 60 is suggested for this conversion.

(d) Addition to vinyl groups

Vinyl groups may be converted to epoxy groups by oxygenation with molecular oxygen catalysed by mixed function oxygenases. Epoxy groups can be converted to the terminal functional groupings illustrated in fig. 61 (cf. C.2.6.5).

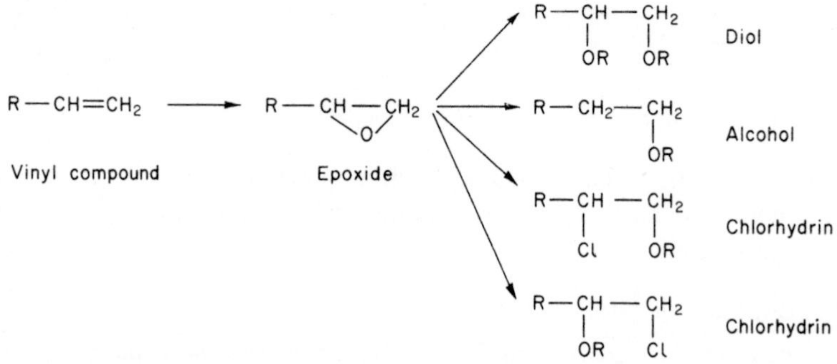

Fig. 61. Compounds deriving from substances with terminal vinyl groups.

Oxidations and decarboxylations

The conversion of terminal methyl groups into hydroxymethyl groups (cf. D.2.2.6), and their oxidation to aldehyde and carboxyl groups, as well as decarboxylations, has been made plausible by feeding experiments with isotopically labelled precursors and the simultaneous occurrence of the corresponding substituted derivatives. In addition the specific decarboxylases have been shown to occur in the cell-free extracts of fungi which produce acetylenic compounds.

Acetylenic compounds undergo very quick synthesis and degradation in the organisms producing them. During degradation, hydrogenation of the triple bonds occurs and acetate is formed.

References for further reading

Bohlmann, F. Die Bedeutung der Acetylenverbindungen für die Pflanzensystematik, *Planta medica (Stuttgart)* **12** (1964), pp. 384–9.

Jantsen, E. F. 'Ethylene and Polyacetylenes', *Plant Biochemistry*, ed. J. Bonner and J. E. Varner (Academic Press, New York 1965), pp. 641–64.

Bohlmann, F. Natürlich vorkommende Acetylen-Verbindungen, *Fortschr. chem. Forsch.* **6** (1966), pp. 65–100.

Bu'Lock, J. D. 'The Biogenesis of Natural Acetylenes', *Comparative Phytochemistry*, ed. T. Swain (Academic Press, London, 1966), pp. 79–95.

Bu'Lock, J. D. 'Biosynthesis of Polyacetylenes in Fungi', *Biosynthesis of Antibiotics*, ed. J. F. Snell (Academic Press, New York, 1966), pp. 141–57.

Bohlmann, F. Biogenetische Beziehungen der natürlichen Acetylenverbindungen. *Fortschr. Chem. org. Naturstoffe* **25** (1967), pp. 1–62.

Anchel, M. 'Biogenesis and Biological Activity of Polyacetylenes', *Antiobotics*, Vol. II, ed. D. Gottlieb and P. D. Shaw (Springer Verlag, Berlin, 1967), pp. 189–215, 441–5.

2.3 Formation of polyketides

A great number of natural products arise from polyketo acids in which methylene and carbonyl groups alternate with each other in regular sequence. These compounds have been termed polyketides, since they were considered by earlier workers to be polymers of ketene. Polyketides are produced especially by lower fungi, living free or in symbiosis with algae. Compounds of this type, however, are also found in higher plants, e.g. special anthraquinone derivatives (cf. D.2.3.1). The flavonoids, occurring in great variety (cf. D.20.5.2), may also be included with polyketides, though other ('non-ketide') groupings are present in the molecule.

Polyketides originate by means of multi-enzyme complexes from one molecule of acetyl CoA serving as the starter (cinnamoyl CoA in case of the flavonoid compounds) and several molecules of malonyl CoA, with the loss of the latter's free carboxyl group. In contrast to fatty-acid biosynthesis (cf. D.2.2.1), reduction steps do not take place with the result that the above-mentioned polyketo acid is formed. These polyketo acids are unstable and have not yet been detected in the free form. The polyketo acid is probably attached to a subunit of the multi-enzyme complex, and is only released if it is stabilized by one or more ring closures (by elimination of water from a carbonyl and methylene group). In the cyclic compounds thus formed, the carbonyl groups occur mostly in the enol form which is stablized by resonance.

The regular substitution pattern of hydroxyl groups which might be expected

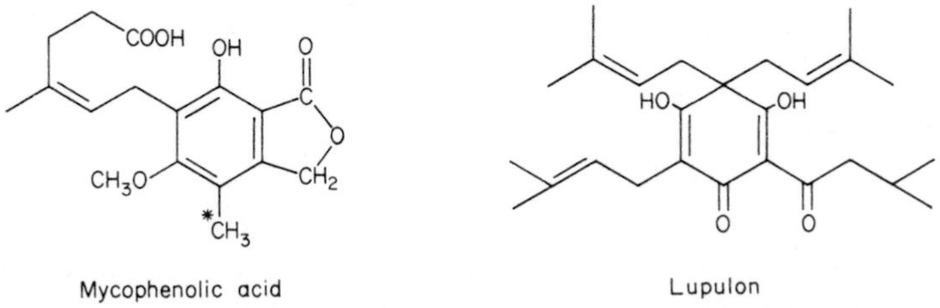

Mycophenolic acid   Lupulon

Fig. 62. The structure of the polyketides, mycophenolic acid and lupulon.

in polyketides due to enolization may be considerably changed by further reactions. In these reactions oxidation and reduction steps as well as methylation of hydroxyl groups are of special importance. In addition the enolates often undergo electrophilic substitutions by methyl groups (cf. C.3.3) and isopentenyl groups (cf. D.5.1) (fig. 62). In this instance the methyl group which originates from methionine is marked by '*'.

References for further reading

Birch, A. J. Biosynthesis of Polyketides and Related Compounds, *Science (Washington)* **156** (1967), pp. 202–6.

Dimroth, P., Walter, H. and Lynen, F. Biosynthese von 6-Methylsalicylsäure, *Europ. J. Biochem.* **13** (1970), pp. 98–110.

2.3.1 Formation of anthracene derivatives and seco-anthraquinones

Anthracene compounds usually occur in nature at the oxidation level of anthraquinone or of anthrone. In most of these compounds the anthracene nucleus is substituted by a number of hydroxyl groups and by a one-carbon side chain. In micro-organisms the fungal genera *Penicillium* and *Aspergillus* especially are characterized by the widespread occurrence of the compounds of this class. Similar substances also occur in Polygonaceae and Rhamnaceae.

Feeding experiments with ^{14}C and ^{18}O labelled acetate or malonate have shown that all carbon atoms and the oxygen atoms marked in fig. 63 originate from both these precursors (cf. in contrast the formation of anthraquinones not substituted in ring A in higher plants, cf. D.6.2.1). An anthrone substituted in position 2 by a carboxyl group and in position 3 by a methyl group is first formed by cyclization. In the formation of most anthracene derivatives the carboxyl group is then eliminated, and is present only in the case of endocrocin, an anthraquinone from the lichen *Cetralia endocrocea*, found in Japan. In the case of many compounds the methyl group is oxidized to an alcohol or carboxyl group.

Linkage of several anthraquinone units is possible by means of phenol oxidase reactions (cf. C.2.3). Many dimeric anthraquinones are found, especially in *Penicillium islandicum*.

Besides anthraquinone derivatives, the so-called ergochromes occur as pigments in *Claviceps purpurea*. These compounds consist of two xanthone derivatives linked at positions 2,2' or 4,4'. The experiments carried out on the formation of these compounds (especially the distribution of radioactivity in the molecule after feeding of emodines) suggest that they be considered as secoanthraquinones and that they originate from anthraquinone derivatives as shown in fig. 64. It appears that the ring system of anthraquinone is first cleaved to form benzophenone derivatives which are then converted in several steps to xanthone derivatives, e.g. xanthone derivative A, the basic units of the ergochromes. The dimerization may follow after oxidation catalysed by phenol oxidases (cf. C.2.3).

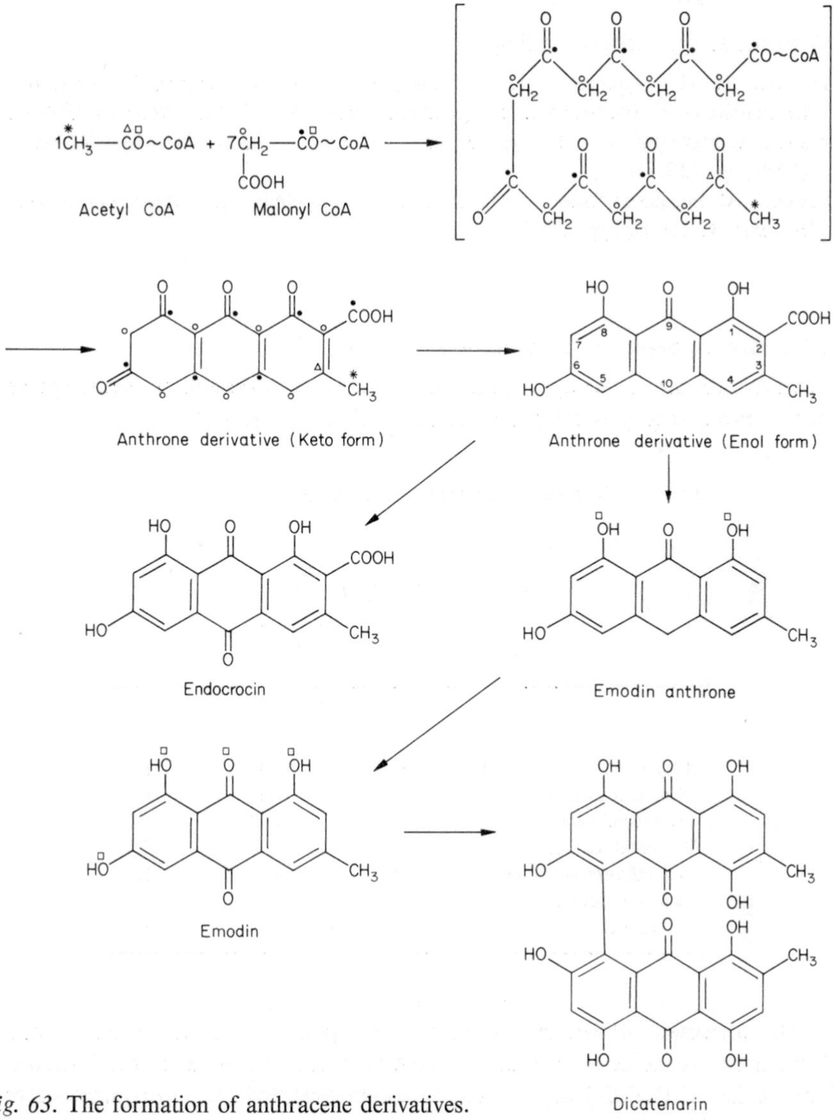

$$1\overset{*}{C}H_3\!-\!\overset{\triangle\,\square}{C}O\!\sim\!CoA \;+\; 7\overset{\bullet}{C}H_2\!-\!\overset{\bullet\,\square}{C}O\!\sim\!CoA \longrightarrow$$

Acetyl CoA   Malonyl CoA

Anthrone derivative (Keto form)

Anthrone derivative (Enol form)

Endocrocin

Emodin anthrone

Emodin

Dicatenarin

Fig. 63. The formation of anthracene derivatives.

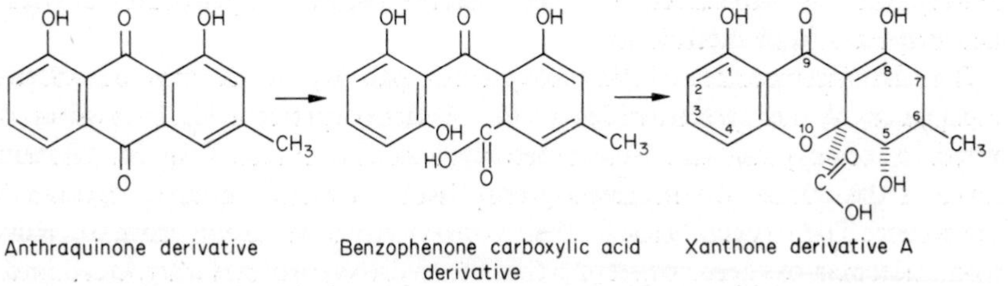

Anthraquinone derivative

Benzophenone carboxylic acid derivative

Xanthone derivative A

Fig. 64. Possible pathway of formation of secoanthraquinones.

References for further reading

Thomson, R. H. 'Quinones: Nature, Distribution and Biosynthesis', *Chemistry and Biochemistry of Plant Pigments* (Academic Press, New York, 1965), pp. 309–32.

Franck, B. Struktur und Biosynthese der Mutterkorn-Farbstoffe, *Angew. Chem.* **81** (1969), pp. 269-78.

Luckner, M. Die Biosynthese von Hydrochinon- und *p*-Chinonderivaten, *Fortschritte der Botanik* **31** (1969), pp. 110–22.

2.3.2 Biosynthesis of tetracyclines

Tetracyclines are a group of antibiotics which are produced by some species of *Streptomyces* and possess the structure outlined in table 9.

Table 9. Naturally occurring tetracyclines

	R_1	R_2	R_3
Tetracycline	H	CH_3	H
7-Chlorotetracycline	Cl	CH_3	H
5-Oxytetracycline	H	CH_3	OH
7-Bromotetracycline	Br	CH_3	H
6-Demethyltetracycline	H	H	H
7-Chloro-6-demethyl- tetracycline	Cl	H	H

The formation of the ring skeleton takes place by condensation of malonyl CoA units. A molecule of malonyl CoA (not acetyl CoA as in the formation of fatty acids, cf. D.2.2.1) or a molecule of malonamoyl CoA probably serves as the starter. The methyl group at carbon atom 6, which is found in some of the tetracyclines, is introduced into the molecule before ring closure (cf. D.2.3.4) and originates from methionine.

The first intermediates of this biosynthetic pathway which may be isolated are compounds of the pretetramide group. 6-Methyl-pretetramide is possibly converted to tetracycline and 7-chlorotetracycline as outlined in fig. 65. Glutamine serves as the donor of the amino group. Both *N*-methyl groups originate from methionine. The oxygen atom of the hydroxyl group at carbon atom 6 originates from molecular oxygen, suggesting that the hydroxyl group is introduced into the molecule through a mixed function oxygenase reaction.

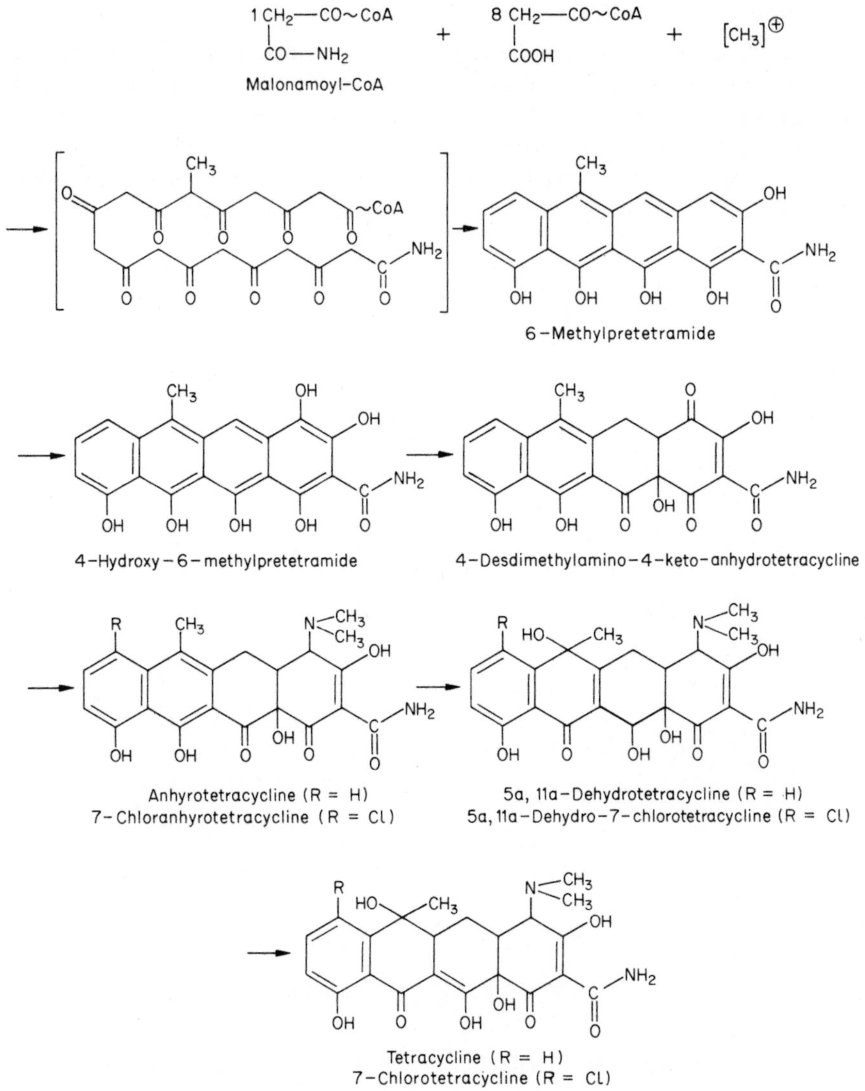

Fig. 65. Biosynthesis of tetracycline and 7-chlorotetracycline.

References for further reading

Turley, R. H. and Snell, J. F. 'Biosynthesis of Tetracycline Antibiotics', *Biosynthesis of Antibiotics*, ed. J. F. Snell (Academic Press, New York, 1966), pp. 95–120.

McCormick, J. R. D. 'Tetracyclines', *Antibiotics*, Vol. II, ed. D. Gottlieb and P. D. Shaw (Springer Verlag, Berlin, 1967), pp. 113–22.

2.3.3 Formation of griseofulvin

Griseofulvin, an antibiotic active against fungi, and obtained from *Penicillium griseofulvum*, is probably formed as outlined in fig. 66. This scheme was worked out from feeding experiments. From one molecule of acetyl CoA and six molecules of malonyl CoA, the polyketo compound I is first formed. This then cyclizes to the benzophenone derivative II. Through methylation by means of *S*-adenosylmethionine (cf. C.3.3) griseophenone C is formed, and is then converted to

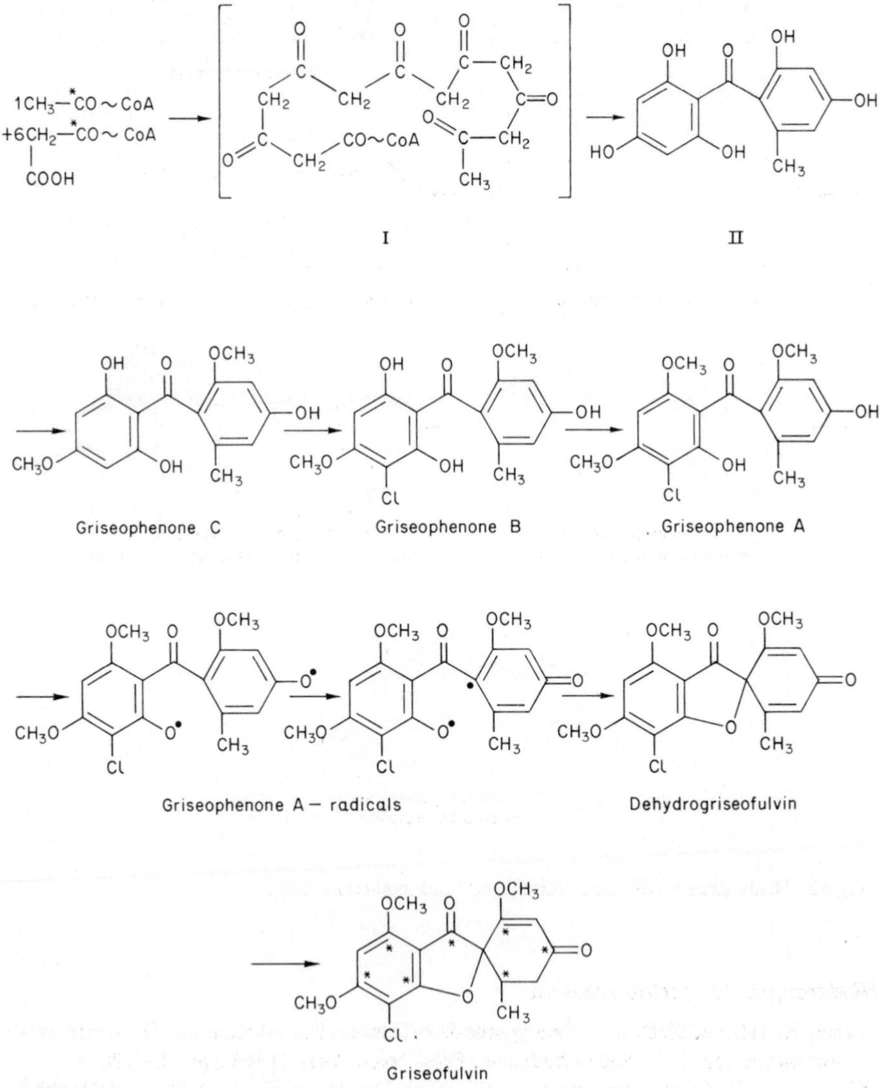

Griseophenone C

Griseophenone B

Griseophenone A

Griseophenone A – radicals

Dehydrogriseofulvin

Griseofulvin

Fig. 66. Biosynthesis of griseofulvin.

griseophenone B by chlorination (cf. C.2.4.2), and to griseophenone A by renewed methylation. Attack by a phenol oxidase (cf. C.2.3) on the latter compound forms a double radical, which changes spontaneously to dehydrogriseofulvin. In the last step griseofulvin is formed from this compound by hydrogenation.

References for further reading

Grove, J. F. Griseofulvin and Some Analogues, *Fortschritte Chem. org. Naturstoffe* 22 (1964), pp. 203–64.

Grove, J. F. 'Griseofulvin', *Antibiotics*, Vol. II, ed. D. Gottlieb and P. D. Shaw (Springer Verlag, Berlin, 1967), pp. 123–33 and 440.

2.3.4 The biosynthesis of polyketide phenol carboxylic acids and phenol carboxylic acid derivatives in fungi and lichens

Monocyclic polyketides with a carboxyl group are produced in large number and many varieties by fungi, living either in symbiosis with algae, as lichens, or as the independently occurring lower fungi. Orsellinic acid, occurring alone or as a constituent of the depside, lecanoric acid, is one of the simplest compounds of this group. The thallus of the lichen *Parmelia tinctorum* contains, besides lecanoric acid, the depside, atranorin, which differs from lecanoric acid in having two additional carbon atoms.

Feeding of the corresponding radioactively labelled precursors has established that orsellinic acid and formyl orsellinic acid originate from one molecule of acetate and three molecules of malonate. One oxygen atom of the carboxyl group is derived from water during the hydrolytic elimination of the terminal CoA-residue. All the other oxygen atoms and the carbon atoms are derived from these precursor acids. Since orsellinic acid cannot be converted to formyl orsellinic acid, the incorporation of the one-carbon fragment must take place before aromatization.

Lecanoric acid and atranorin are derived from two molecules of orsellinic acid and formyl orsellinic acid respectively by the action of an esterase (fig. 67). Decarboxylation of orsellinic acid to 1-methyl-3,5-dihydroxybenzene is possible through the action of a specific decarboxylase, detectable in cell-free extracts of the lichen *Umbillicaria pustulata*.

The cyclization of the polyketide (as yet not detected in the free state) formed from one molecule of acetate and three molecules of malonate may also take place in the manner shown in fig. 68, besides that outlined in fig. 67, with the formation of a substituted phloroacetophenone derivative. By the incorporation of a one-carbon unit before aromatization, 2,4,6-trihydroxy-5-methylphloroacetophenone is formed, which may be converted to usnic acid (fig. 68) by oxidative coupling (cf. D.2.3) via several intermediate stages. 2,4,6-Trihydroxyphloroacetophenone itself is not a precursor of usnic acid.

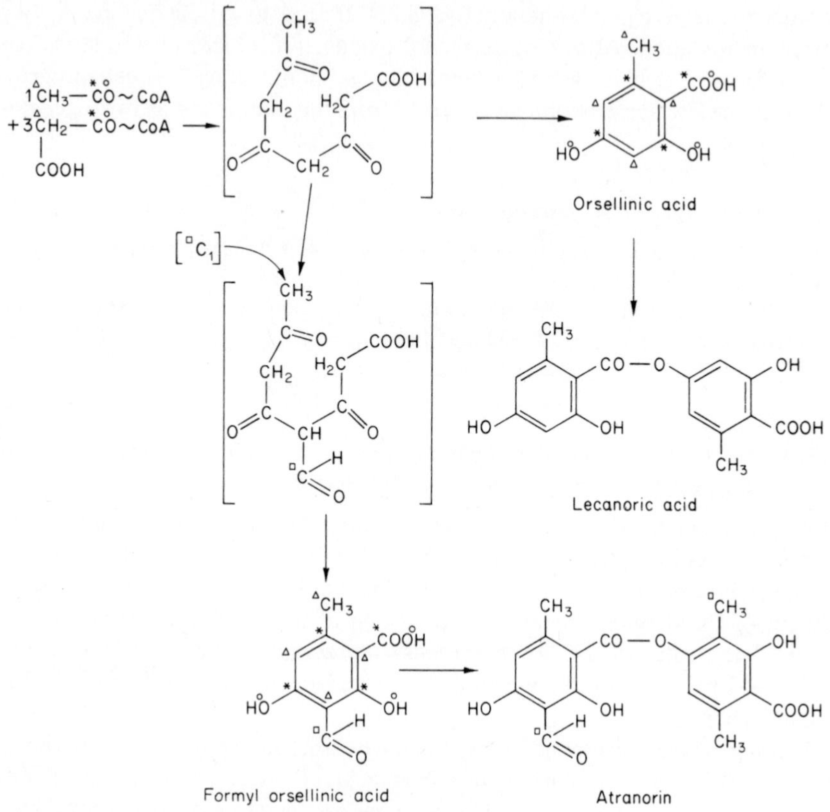

Fig. 67. Formation of lecanoric acid and atranorin.

A number of micro-organisms, especially soil fungi, convert the phenol carboxylic acids synthesized by them in the above way to a number of further partly autoxidizable phenols by means of decarboxylation, oxidation and hydroxylation reactions. The products originating from orsellinic acid in *Epicoccum nigrum* are illustrated in fig. 69. Gallic acid is among them. In other organisms this compound arises either from shikimic acid (cf. D.6.2.2) or from trihydroxy cinnamic acid (cf. D.20.6). There are, therefore, three biosynthetic ways for the formation of this compound. Orcinol is a precursor of fumigatin, a benzoquinone derivative found in *Aspergillus fumigatus*.

The trihydroxy benzenes of the pyrogallol type are particularly autoxidizable at suitable pH values and may be converted to the corresponding quinones. These can undergo nucleophilic substitutions and add other phenols, amines or amino acids. The resulting product may undergo further oxidations or substitutions. The high molecular weight, extremely complex, brown compounds formed in this way are termed humic acids. They represent the major proportion of the organic substance of most soils.

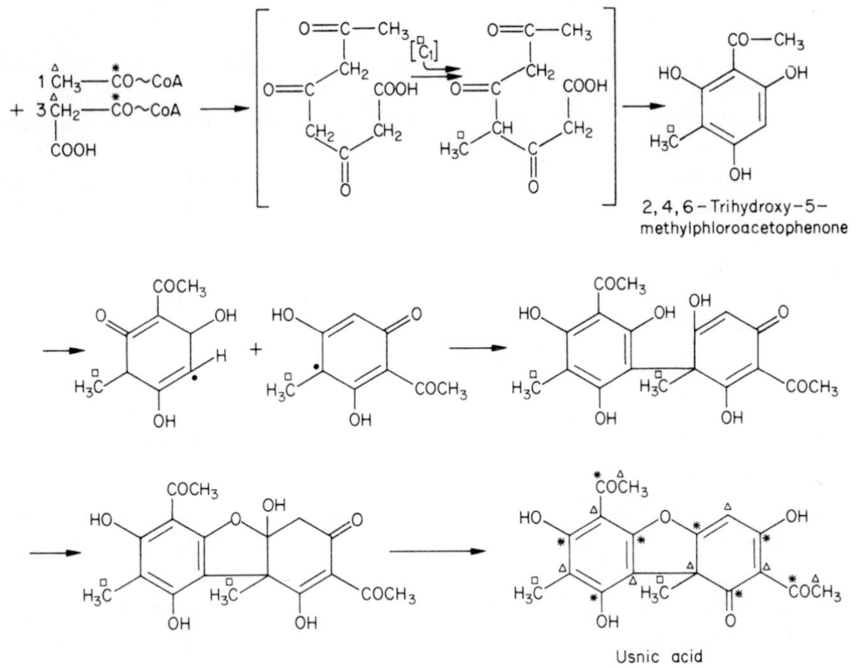

Fig. 68. Formation of usnic acid.

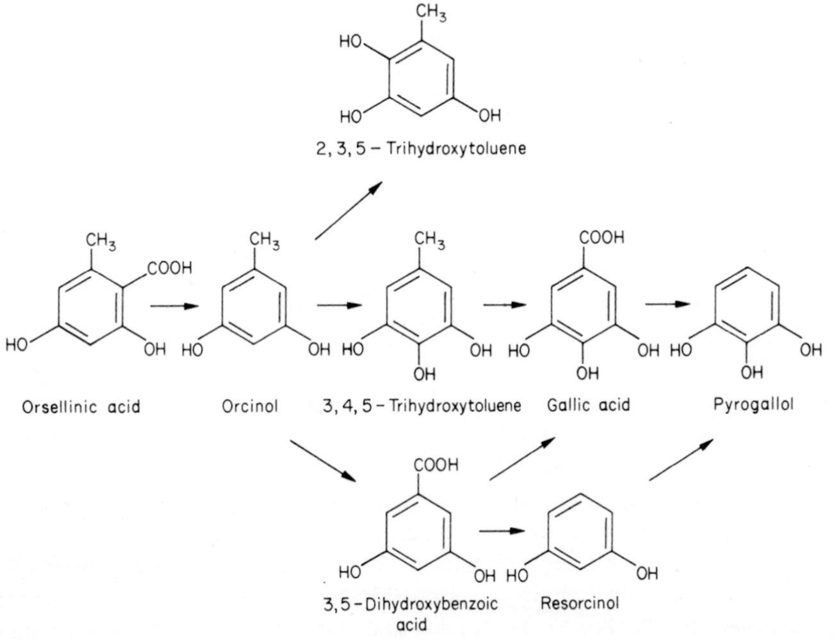

Fig. 69. Products of conversion of orsellinic acid in *Epicoccum nigrum*.

References for further reading

Shibata, S. Chemistry and Biosynthesis of Some Fungal Metabolites, *Chem. in Britain* **3** (1967), pp. 110–21.

Birch, A. J. Biosynthesis of Polyketides and Related Compounds, *Science (Washington)* **156** (1967), pp. 202–6.

Martin, J. P., Richards, S. J. and Haider, K. Properties and Decomposition and Binding Action in Soil of 'Humic Acid' Synthesized by *Epicoccum nigrum*, *Soil Sci. Soc. of America, Proc.* **31** (1967), pp. 657–62.

Tanenbaum, S. W. 'Some Acetate Derived Antibiotics', *Antibiotics*, Vol. II, ed. D. Gottlieb and P. D. Shaw (Springer Verlag, Berlin, 1967), pp. 82–112.

Mosbach, K. Zur Biosynthese von Flechtenstoffen, Produkten einer symbiotischen Lebensgemeinschaft, *Angew. Chem.* **81** (1969), pp. 233–44.

2.3.5 Formation of hemlock alkaloids, cycloheximide and other piperidine derivatives

Alkaloids of the coniine type have been found up until now in a few higher plants, besides *Conium maculatum*, the hemlock. In contrast to the Solanaceae, the aerial parts of *Conium* are the main sites of alkaloid synthesis. The *Conium* alkaloids probably originate from four molecules of acetate via an intermediate polyketide. After feeding acetate-1-^{14}C, carbon atoms 2,4,6 and 2' of γ-coniceine

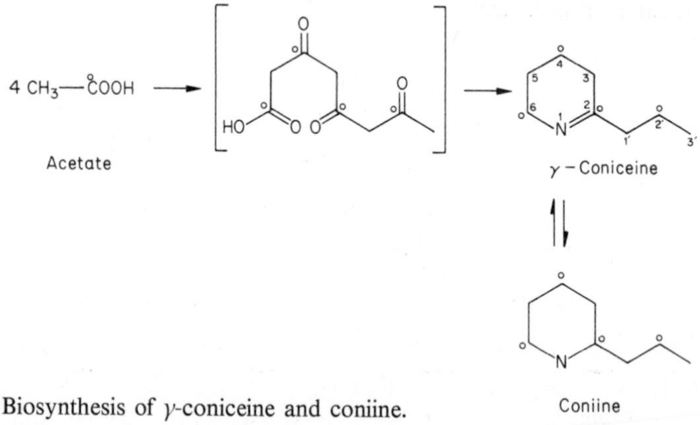

Fig. 70. Biosynthesis of γ-coniceine and coniine.

and coniine were radioactive (fig. 70). 5-Oxooctanoic acid and 5-oxo-octanol are the immediate precursors of γ-coniceine. γ-Coniceine is easily converted to D-coniine and vice versa (fig. 70).

In the case of the piperidine alkaloids, carpaine, cassiine and prosopine (fig. 71), which have been isolated from species of *Carica, Azima, Cassia* and *Prosopis*, synthesis via the polyketides is postulated. No proof has, as yet, been put forward.

In the case of the antibiotic cycloheximide, on the other hand, formation from acetate has been confirmed. By feeding malonic acid-1,3-^{14}C the distribution of radioactivity shown in fig. 72 was obtained. This is consistent with the pattern

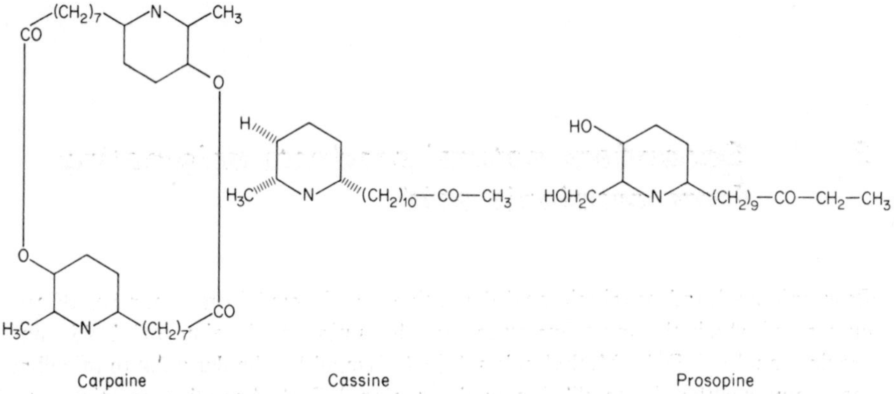

Carpaine   Cassine   Prosopine

Fig. 71. The structural formulae of some piperidine alkaloids.

expected after incorporation of acetate-1-^{14}C with the exception of the portion of the molecule boxed off which must have originated from a molecule of malonate, including both carboxyl groups. This is further supported by the fact that $^{14}CO_2$ is incorporated to a greater extent in the carbon atoms of the carbonyl groups of the cycloheximide ring (malonyl CoA is synthesized from acetyl CoA and CO_2 by means of acetyl CoA carboxylase, cf. D.2.1). The origin of the nitrogen is unknown. Both the *C*-methyl groups originate from methionine.

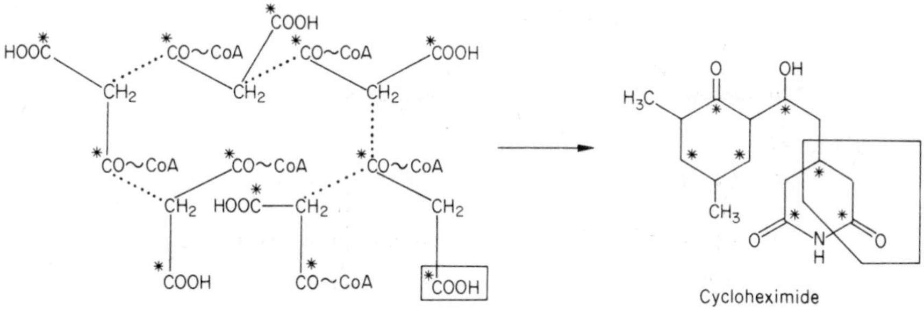

Cycloheximide

Fig. 72. Formation of cycloheximide.

References for further reading

Vaněk, Z., Cudlin, J. and Vondráček, M. 'Cycloheximide and Other Glutarimide Antibiotics', *Antibiotics*, Vol. II, ed. D. Gottlieb and P. D. Shaw (Springer Verlag, Berlin, 1967), pp. 222–7.

Liebisch, H. W. 'Piperidinalkaloide', *Biosynthese der Alkaloide*, ed. K. Mothes and H. R. Schütte (VEB Deutscher Verlag der Wissenschaften, Berlin, 1969), pp. 275–311.

Leete, E. Biosynthesis of the Hemlock and Related Piperidine Alkaloids, *Accounts chem. Res.* **4** (1971), pp. 100–7.

3. Secondary natural products originating from propionic acid

Propionic acid and propionic acid derivatives are formed from a number of precursors of which the most important are the amino acid valine (cf. D.11), and succinic acid (cf. D.4). Methylmalonyl CoA formed by the degradation of valine may remain either in equilibrium with propionyl CoA through a reaction catalysed by a biotin enzyme (cf. C.3.1) on the one hand or may be converted to succinyl CoA. Since both reactions are reversible, succinyl CoA can be converted to propionyl CoA via methylmalonyl CoA (fig. 73). Propionyl CoA may also be formed by activation of propionic acid (cf. C.1.2) and originates during β-oxidation of fatty acids possessing an odd number of carbon atoms (cf. D.2.2.6). In *Escherichia coli* propionic acid is degraded to pyruvic acid via acrylic acid and lactic acid (fig. 73).

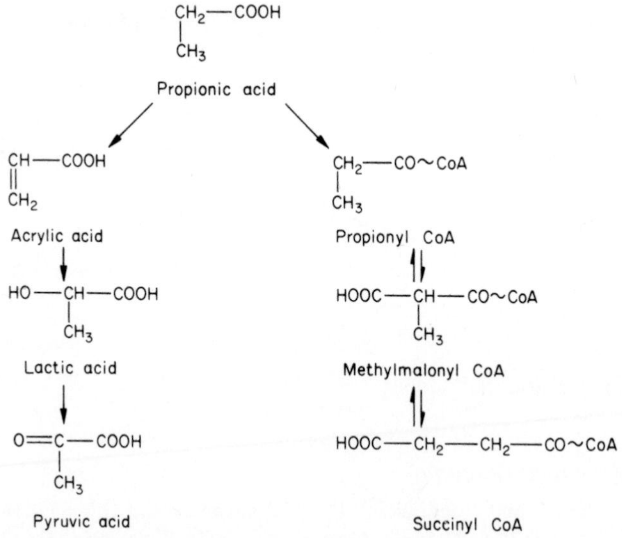

$$
\begin{array}{c}
CH_2\text{---}COOH \\
| \\
CH_3
\end{array}
$$

Propionic acid

$$
\begin{array}{c}
CH\text{---}COOH \\
\| \\
CH_2
\end{array}
$$
Acrylic acid

$$
HO\text{---}CH\text{---}COOH \\
| \\
CH_3
$$
Lactic acid

$$
O\text{=}C\text{---}COOH \\
| \\
CH_3
$$
Pyruvic acid

$$
\begin{array}{c}
CH_2\text{---}CO\sim CoA \\
| \\
CH_3
\end{array}
$$
Propionyl CoA

$$
HOOC\text{---}CH\text{---}CO\sim CoA \\
| \\
CH_3
$$
Methylmalonyl CoA

$$
HOOC\text{---}CH_2\text{---}CH_2\text{---}CO\sim CoA
$$
Succinyl CoA

Fig. 73. Formation of propionyl CoA and methylmalonyl CoA.

References for further reading

Meister, A. *Biochemistry of the Amino Acids*, Vol. II. (Academic Press, New York, 1965).

3.1 Biosynthesis of methyl fatty acids and macrolide antibiotics

Propionyl CoA behaves like acetyl CoA in many reactions. From the electronic structure given in C.1.2, it is obvious that the α-carbon atom can be easily substituted by electrophilic groups (with CO_2, methylmalonyl CoA is formed, cf. fig. 73) or the carbon atom of the carbonyl group can react with nucleophilic groupings, e.g. with the α-carbon atom of methylmalonyl CoA and malonyl CoA. The carbon chains of various secondary natural products may therefore contain acetate as well as propionate units.

Table 10. Fatty acids which are fully or partly synthesized from propionate units

Name/Formula	Occurrence	Precursors
$-$Methyl valeric acid $H_2-CH_2-CH-COOH$ CH_3	*Ascaris lumbricoides*	2 molecules of propionate
$_{32}-$Mycocerosic acid $H_3-(CH_2)_{19}-CH-CH_2-CH-CH_2-CH-CH_2-CH-COOH$ CH_3 CH_3 CH_3 CH_3	Mycobacteria	10 molecules of acetate and 4 molecules of propionate
$_{27}-$Phthienoic acid $H_3-(CH_2)_{17}-CH-CH_2-CH-CH=C-COOH$ CH_3 CH_3 CH_3	Mycobacteria	9 molecules of acetate and 3 molecules of propionate
$,4,6,8-$Tetramethyldecanoic acid $H_3-CH_2-CH-CH_2-CH-CH_2-CH-CH_2-CH-COOH$ CH_3 CH_3 CH_3 CH_3	*Anser anser domesticus*	1 molecule of acetate and 4 molecules of propionate
$,4,6,8-$Tetramethylundecanoic acid $H_2-CH_2-CH-CH_2-CH-CH_2-CH-CH_2-CH-COOH$ H_3 CH_3 CH_3 CH_3 CH_3	*Anser anser domesticus*	5 molecules of propionate

A number of fatty acids, synthesized in this way, are found in animals and micro-organisms. A few characteristic representatives of this group of compounds are given in table 10. In some cases the exact mode of formation is uncertain. It is, however, thought to be similar to the one described for the unbranched fatty acids in D.2.2.1.

In the case of a number of macrolide antibiotics, carbon chains occur with every alternate carbon atom more or less regularly substituted by a methyl group. By the feeding of isotopically labelled precursors it has been shown that propionic acid participates in the synthesis of these compounds. The erythromycins A, B and C occurring in *Streptomyces erythreus* are formed in this way. Erythronolide B is first formed and is subsequently converted to erythromycin by glycosylation (fig. 74). By analogy to fatty-acid biosynthesis (cf. D.2.2.1) it is supposed that one molecule of propionyl CoA and six molecules of methylmalonyl CoA are involved in the synthesis of erythronolide B and all the reaction steps proceed on one multi-enzyme complex.

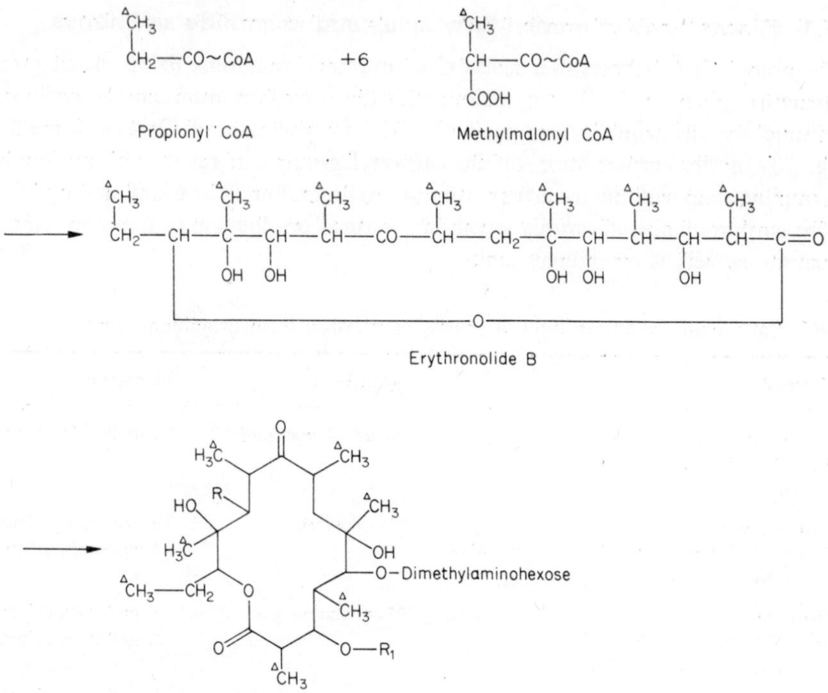

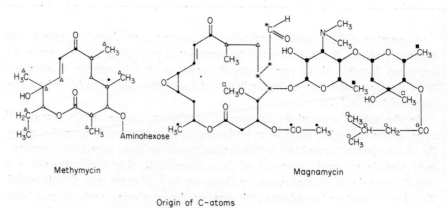

Propionyl CoA +6 Methylmalonyl CoA

Erythronolide B

Erythromycin A R = OH, R₁ = L-Cladinose
Erythromycin B R = H, R₁ = L-Cladinose
Erythromycin C R = OH, R₁ = L-Mycarose

Fig. 74. The biosynthesis of erythromycins.

Methymycin

Magnamycin

Origin of C-atoms

• Acetate, △ Propionate, ∗ Succinate,
■ Glucose, □ Methionine, ○ Valine

Fig. 75. Biosynthesis of methymycin and magnamycin.

Methymycin, synthesized by *Streptomyces venezuela*, is formed from one molecule of acetate and five molecules of propionate. Acetate, propionate, succinate, glucose, L-valine and methyl groups from methionine are involved in the formation of magnamycin, found in *Streptomyces halstedii* (fig. 75).

References for further reading

Corcoran, J. W. and Chick, M. 'Biochemistry of the Macrolide Antibiotics', *Biosynthesis of Antibiotics*, ed. J. F. Snell (Academic Press, New York, 1966), pp. 159–201.

Grisebach, H. *Biosynthetic Pathways in Microorganisms and Higher Plants* (John Wiley, New York, 1967).

Vaněk, Z. and Majer, J. 'Macrolide Antibiotics', *Antibiotics*, Vol. II, ed. D. Gottlieb and P. D. Shaw (Springer Verlag, Berlin, 1967), pp. 154–88.

Birch, A. J. 'Nystatin', *Antibiotics*, Vol. II, ed. D. Gottlieb and P. D. Shaw (Springer Verlag, Berlin, 1967), pp. 228–30.

4. Biosynthesis of secondary products from the acids of the tricarboxylic acid and glyoxylic acid cycles

The tricarboxylic acid cycle (fig. 76) is not only of significance in the degradation of acetyl CoA with the resulting formation of CO_2 and the reduced coenzymes used substantially in cell respiration. It also supplies intermediates for the synthesis of a number of amino acids, e.g. glutamic acid and aspartic acid, as well as other natural products. Replacement of the compounds taken out of the cycle is provided for by the degradation of isocitric acid to glyoxylic acid and succinic acid (cf. D.4.2). Malic acid is formed from glyoxylic acid and acetyl CoA. The tricarboxylic acid cycle 'short-circuited' by these reactions is known as the glyoxylic acid cycle.

4.1 Formation of substituted citric acid derivatives from oxaloacetic acid

A number of higher and lower fungi synthesize acids of the agaricic acid type which probably originate from oxaloacetic acid and a fatty-acid CoA-ester (e.g. stearyl CoA) in a reaction corresponding to the formation of citric acid in the tricarboxylic acid cycle (cf. figs. 76 and 77).

Derivatives of such alkyl citric acids are found in the nonadrides, a group of fungal metabolites, consisting of three representatives (glaucanic acid, glauconic acid and byssochlamic acid).

The biosynthesis of these products, which are characterized by a nine-membered alicyclic ring and two five-membered anhydride rings, has been investigated with radioactively labelled compounds. The results obtained have led to the conclusion that oxalacetic acid and caproyl CoA condense to form n-butylcitric acid. The nonadrides are formed from two molecules of this compound (fig. 78).

References for further reading

Miller, M. W. *The Pfizer Handbook of Microbial Metabolites* (McGraw-Hill Book Company, New York, 1961).

Sutherland, J. K. The Nonadrines, *Fortschr. Chem. org. Naturstoffe* **25** (1967), pp. 131–49.

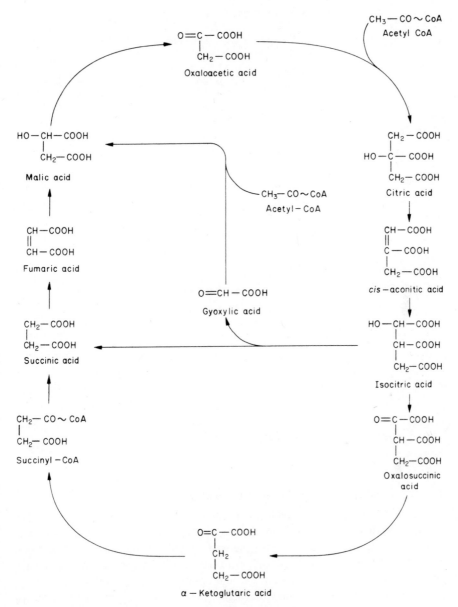

Fig. 76. Tricarboxylic acid cycle and glyoxylic acid cycle.

CH₃—(CH₂)₁₅—CH₂—CO ∼ CoA + O=C—COOH ⟶ CH₃—(CH₂)₁₅—CH—COOH

$$CH_3-(CH_2)_{15}-CH_2-CO \sim CoA \ + \ \overset{\displaystyle O=C-COOH}{\underset{\displaystyle CH_2-COOH}{|}} \longrightarrow \overset{\displaystyle CH_3-(CH_2)_{15}-CH-COOH}{\underset{\displaystyle CH_2-COOH}{\underset{\displaystyle HO-C-COOH}{|}}}$$

Stearyl — CoA   Oxaloacetic acid   Agaricic acid

Fig. 77. Possible way of formation of agaricic acid.

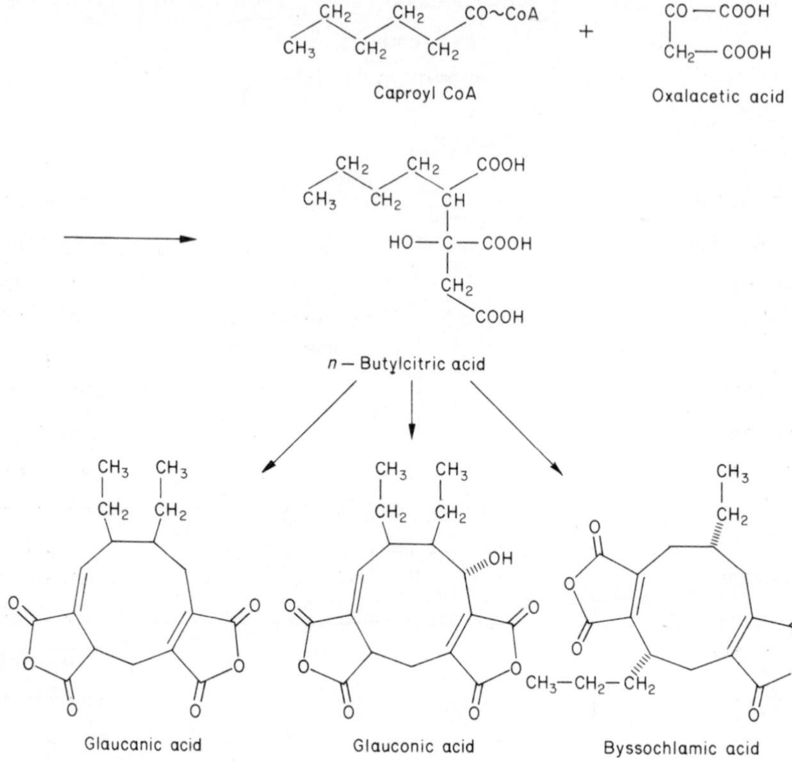

Fig. 78. Possible way of biosynthesis of nonadrides.

4.2 Conversion of glyoxylic acid to oxalic acid

Oxalic acid and calcium oxalate are synthesized by higher plants in particular. Calcium oxalate is formed from the excess calcium present in the plant. The amount of calcium oxalate stored is therefore usually proportional to the calcium content of the nutrient media. Oxalic acid has long been considered a secondary product which is of no significance to the organism. It is, however, now known that in certain micro-organisms and higher plants *O*-oxalylhomoserine is an important intermediate in the formation of methionine. Enzymes of spinach catalyse the condensation of this compound with cysteine to give cystathionine (fig. 79).

The oxalic acid produced in this reaction, after conversion to the CoA-ester, can either be transferred to homoserine or converted by decarboxylation to formic acid which enters one-carbon metabolism (cf. C.3.2). Cystathionine may be converted to methionine as described in D.10.

Oxalic acid is formed from glyoxylic acid by the action of the enzyme glycolate oxidase in the presence of oxygen, and catalase which destroys the hydrogen

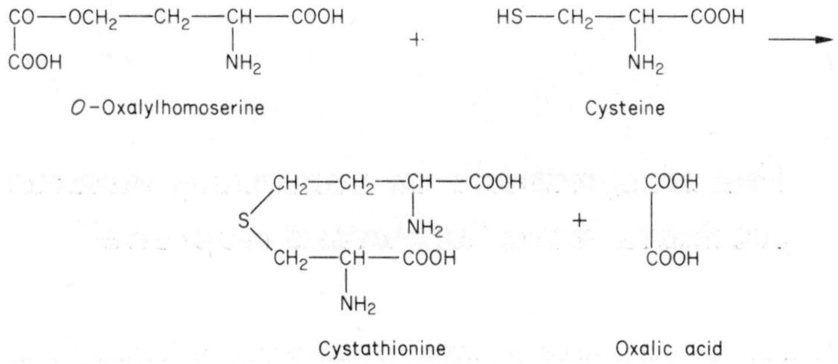

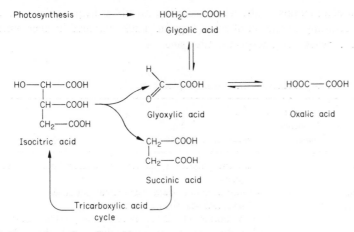

Fig. 79. Conversion of *O*-oxalylhomoserine to cystathionine.

peroxide formed as a side product. Glyoxylic acid is synthesized from isocitric acid by the action of the enzyme isocitrate lyase (cf. fig. 76), but may also be produced from glycolic acid by the action of glycolate oxidase, and is thus derived from the metabolic products of photosynthesis (fig. 80). Glycolate oxidase, catalase and a NAD$^+$-dependent glyoxylic acid-reductase are localized in special particles in the cells, the glyoxysomes.

Fig. 80. Formation of oxalic acid.

References for further reading

Ranson, S. L. 'Plant Acids', *Biosynthetic Pathways in Higher Plants*, ed. J. B. Pridham and T. Swain (Academic Press, London, 1965), pp. 179–98.

Zenk, M. H. Biochemie und Physiologie sekundärer Pflanzenstoffe, *Ber. dtsch. Bot. Ges.* **80** (1967), pp. 573–91.

Bornkamm, R. Typen des Oxalatstoffwechsels grüner Blätter bei einigen Familien höherer Pflanzen, *Flora Abt. A*, **160** (1969), pp. 317–36.

Tolbert, N. E. Microbodies – Peroxysomes and Glyoxysomes, *Ann. Rev. Plant Physiol.* **22** (1971), pp. 45–74.

5. The biosynthesis of secondary natural products from 'activated isoprene'

A great number of compounds occurring in nature may be formally considered to have originated from fragments whose carbon skeletons correspond to isoprene (methylbutadiene, fig. 81) (the so-called isoprene rule). Table 11 illustrates the

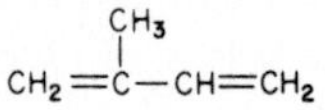

$$CH_2 = \overset{\overset{\displaystyle CH_3}{|}}{C} - CH = CH_2$$

Fig. 81. Isoprene.

variety of compounds synthesized according to this principle. They are differentiated according to the number of isoprene units contained in their molecules into monoterpenes, sesquiterpenes, diterpenes etc.

Table 11. Natural products formed from isoprene units.

Class	Number of isoprene units	Occurrence
Hemiterpenes	1	Isoprene, emitted by the leaves of different species of higher plants
Monoterpenes	2	Constituents of volatile oils, iridoid substances
Sesquiterpenes	3	Constituents of volatile oils and resins
Diterpenes	4	Constituents of volatile oils and resins, phytol, vitamin A
Triterpenes	6	Squalene, steroids, pentacyclic triterpenes
Tetraterpenes	8	Carotenes, xanthophylls
Polyterpenes	n	Rubber, gutta-percha, balata

Isoprene units are, in addition, constituents of complex molecules. The ubiquinones (cf. D.6.3.2), vitamins of the K-group (cf. D.6.2.1) and plastoquinones (cf. D.20.2) have side chains which consist of up to ten isoprene units, and the ring system of coumarins (cf. D.20.4.2) as well as the quinoline ring (cf. D.6.4.2) may be substituted by isopentenyl groups. The terpenes pristane and phytane which have been detected even in pre-cambrian fossils are probably derived from the diterpene alcohol phytol, a constituent of the chlorophyll molecule (cf. D.8.1).

References for further reading

Goodwin, T. W. 'The Biological Significance of Terpenes in Plants', *Terpenoids in Plants*, ed. J. B. Pridham (Academic Press, London, 1967), pp. 1–23.

Rasmussen, R. A. Isoprene, that Missing Link in Terpene Biosynthesis, Identified as a Foliage Volatile, *Plant Physiol.* **44** (1969), p. 40.

5.1 Formation of isopentenyl pyrophosphate ('activated isoprene') from acetyl CoA

All isoprenoid compounds originate from isopentenyl pyrophosphate, which is also known as 'activated isoprene'. Isopentenyl pyrophosphate is synthesized from acetyl CoA in the same manner by both plants and animals

Acetoacetyl CoA is first formed from two molecules of acetyl CoA by head-to-tail condensation. This reaction is catalysed by the enzyme thiolase (cf. D.2.2.6).

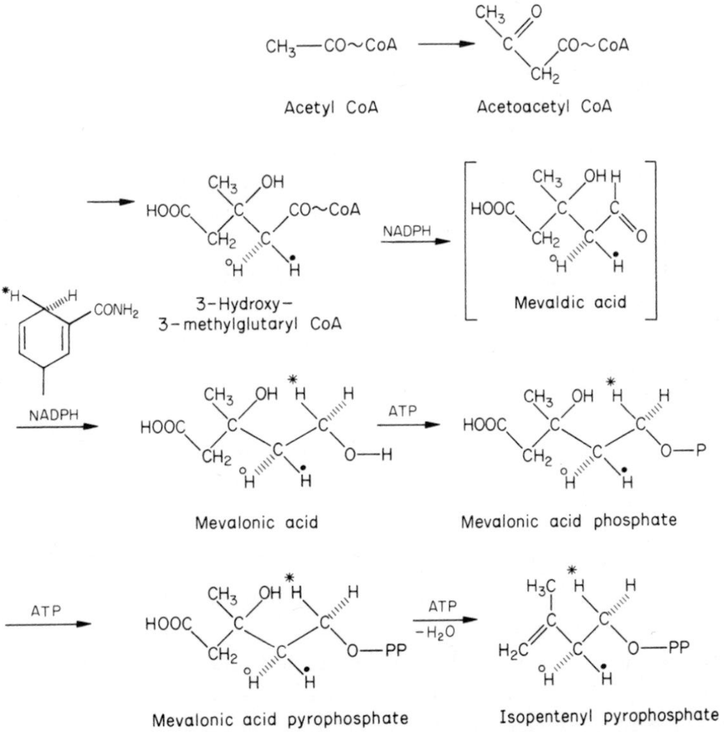

Fig. 82. Biosynthesis of isopentenyl pyrophosphate from acetyl CoA.

A third molecule of acetyl CoA adds to the carbonyl group at position three of acetoacetyl CoA to form 3-hydroxy-3-methylglutaryl CoA (fig. 82) (for the biosynthesis of this compound from leucine, cf. D.12.1).

3-Hydroxy-3-methylglutaryl CoA is then reduced to mevalonic acid. This reduction probably proceeds via mevaldic acid as an intermediate, although this compound does not occur in the free form. The reaction is practically irreversible and mevalonic acid is the compound which may be regarded as the direct precursor for isoprenoid compounds.

The enzyme mevaldate reductase transfers the hydrogen stereospecifically from the 'A' side of NADH or NADPH to the substrate (cf. C.2.2.1), $5R$-[^{3}H]-Mevalonic acid is thus formed from H_A-[^{3}H]-NADH or -NADPH (the tritium atom is marked '*' in fig. 82). Mevalonic acid thus formed is then phosphorylated at the primary alcoholic group to form mevalonic acid monophosphate and then in a second reaction step mevalonic acid pyrophosphate is formed. Isopentenyl pyrophosphate is obtained from the latter compound by decarboxylation and elimination of a molecule of water. The reaction requires the presence of ATP and results in the production of ADP and inorganic phosphate. The exact mechanism of this reaction is unknown.

References for further reading

Bonner, J. 'The Isoprenoids', *Plant Biochemistry*, ed. J. Bonner and J. E. Varner (Academic Press, New York, 1965), pp. 665–92.

Goodwin, T. W. 'The Biosynthesis of Carotenoids', *Chemistry and Biochemistry of Plant Pigments*, ed. T. W. Goodwin (Academic Press, London, 1965), pp. 143–73.

Goodwin, T. W. 'Regulation of Terpenoid Synthesis in Higher Plants', *Biosynthetic Pathways in Higher Plants*, ed. J. B. Pridham and T. Swain (Academic Press, London, 1965), pp. 57–71.

Donnings, C. and Popják, G. Studies on the Biosynthesis of Cholesterol, XVIII: The Stereospecificity of Mevaldate Reductase and the Biosynthesis of Asymmetrically Labelled Farnesyl Pyrophosphate, *Proc. Royal. Soc.* B **163** (1966), pp. 465–91.

Porter, J. W. and Anderson, D. G. Biosynthesis of Carotenes, *Ann. Rev. Plant Physiol.* **18** (1967), pp. 197–228.

Ryback, G. 'Specifically Labelled Substrates of Terpenoid Biosynthesis', *Terpenoids in Plants*, ed. J. B. Pridham (Academic Press, London, 1967), pp. 47–58.

5.2 Polymerization of isopentenyl pyrophosphate

The synthesis of terpenes takes place by the condensation of several molecules of isopentenyl pyrophosphate. Dimethylallyl pyrophosphate, which is formed from the former compound by the shift of the double bond catalysed by isopentenyl pyrophosphate isomerase, serves as a starter molecule for this polymerization both in animals and plants. The elimination of a hydrogen atom in this reaction at carbon atom 2 is strictly stereospecific. The hydrogen atom marked with 'o' is always eliminated, except in the case of rubber biosynthesis discussed below, while that marked '•' is retained (cf. fig. 83). The radioactivity in the terpenes is incorporated only from $4R$-[^{3}H]-mevalonic acid but not from $4S$-[^{3}H]-mevalonic acid.

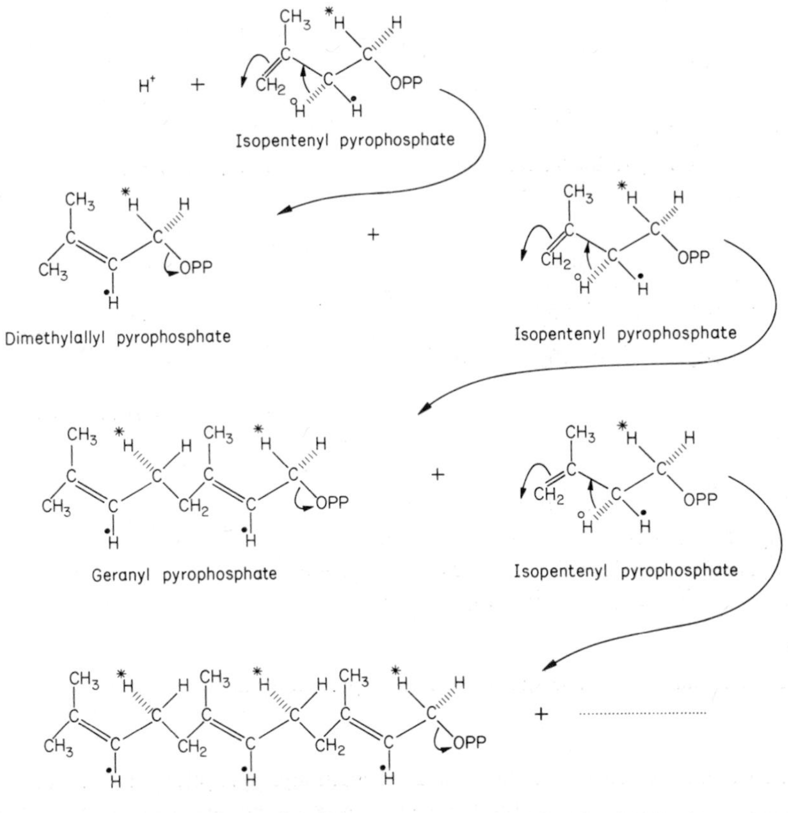

Fig. 83. Formation of compounds with all-*trans* configuration from isopentenyl pyrophosphate.

One molecule of dimethylallyl pyrophosphate then serves as an acceptor for one molecule of isopentenyl pyrophosphate. The pyrophosphate group is then lost from the starter molecule or from the head of the chain that is being elongated. The condensation may be considered as a nucleophilic substitution by the CH_2 group of isopentenyl pyrophosphate. The substitution causes an inversion of configuration at carbon atom 1 of the starter molecule (or of the elongating chain when further isopentyl pyrophosphate molecules are added) since the CH_2 group of isopentenyl pyrophosphate opposite the pyrophosphate group enters the molecule from the side in a concerted reaction. The inversion of configuration has been shown by replacement of a hydrogen atom at carbon atom 1 by deuterium. This reaction resembles the well-known S_N2 reaction of organic chemistry. During the resulting shift of the double bond, occurring simultaneously with the new C–C bonding, the hydrogen atom marked 'o' at carbon atom 2 is lost. The resulting monoterpene, geranyl pyrophosphate, that is first formed in the above

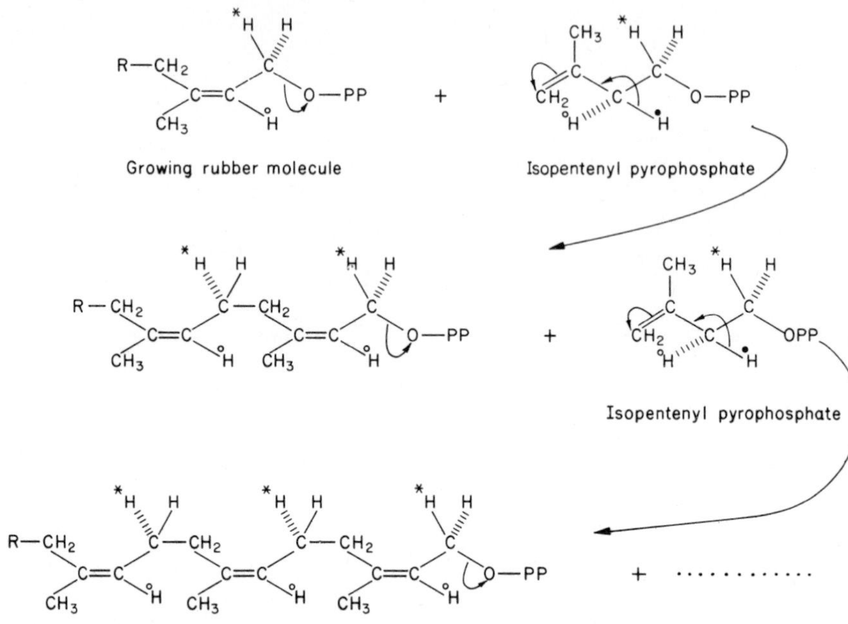

Fig. 84. Formation of rubber from isopentenyl pyrophosphate.

reaction is converted by addition of an isopentenyl pyrophosphate molecule by a similar mechanism, to farnesyl pyrophosphate, and this is then converted to geranyl-geranyl pyrophosphate. The long-chain terpene derivatives (with the exception of rubber) are most probably synthesized in a similar manner. Configurations around all the double bonds in these compounds are *trans*.

Rubber differs from other polyterpenes (e.g. gutta-percha and balata) in the fact that the double bonds are *cis*. Rubber is obtained from the latex of *Hevea brasiliensis* where it occurs in certain connected cell systems, the latex vessels. Latex contains particles, the majority of which are made up of rubber and have a diameter of up to 3 μm. These are suspended in an aqueous solution which is a mixture of cytoplasm of latex cells which contains soluble enzymes, ribosomes, mitochondria etc. (cf. B.3). Latex is in certain cases capable of synthesizing rubber even when removed from the latex vessels.

While formation of isopentenyl pyrophosphate proceeds in the latex fluid, polymerization to rubber takes place mostly on the surface of pre-existing rubber particles. It is catalysed by an enzyme which is present on the boundary between the particles and the aqueous phase and, unlike the synthesis of all other terpenes, the hydrogen atom marked '•' is eliminated from carbon atom 2 of isopentenyl pyrophosphate. This confers *cis* configuration on the carbon chain (fig. 84). The rubber molecules formed are of very different sizes and are made up of about

500 to more than 5000 isoprene units. The inversion of configuration at carbon atom 1 of isopentenyl pyrophosphate (or the growing chain during polymerization) during rubber biosynthesis has not yet been investigated, but it appears to take place in the same manner as in the formation of *trans* terpenes.

References for further reading

Porter, J. W. and Anderson, D. G. Biosynthesis of Carotenes, *Ann. Rev. Plant Physiol.* **18** (1967), pp. 197–228.

Hemming, F. W. 'Polyisoprenoid Alcohols (Prenols)', *Terpenoids in Plants*, ed. J. B. Pridham (Academic Press, London, 1967), pp. 223–39.

Bonner, J. 'Rubber Biogenesis', *Biogenesis of Natural Compounds*, ed. P. Bernfeld (Pergamon Press, Oxford, 1967), pp. 941–52.

Popják, G. 'Conversion of Mevalonic Acid into Prenyl Hydrocarbons as Exemplified by the Synthesis of Squalene', *Natural Substances Formed Biologically from Mevalonic Acid*, ed. T. W. Goodwin (Academic Press, New York, 1970), pp. 17–33.

Hemming, F. W. 'Polyprenols', *Natural Substances Formed Biologically from Mevalonic Acid*, ed. T. W. Goodwin (Academic Press, New York, 1970), pp. 105–17.

5.3 The head-to-head condensation of farnesyl pyrophosphate and geranylgeranyl pyrophosphate with the formation of phytoene and squalene

Although in the formation of most terpenoid compounds the isoprene groups are linked by means of the head-to-tail condensation discussed in an earlier section, head-to-head condensation also occurs in certain cases (e.g., in the synthesis of carotenoid compounds and of squalene). This has been most extensively investigated in the synthesis of squalene.

It is interesting to note that the reaction proceeds stereospecifically, since a hydrogen atom at carbon atom 1 from one of the two farnesyl groups is replaced by a hydrogen atom originating from NADPH. The condensation may proceed according to the mechanism outlined in fig. 85.

By substitution of one of the two farnesyl groups by a negatively charged group (X in fig. 85) and elimination of the pyrophosphate group, a derivative of nerolidol may be formed first (a pyrophosphate group, for example may act as a substituent in place of X). This derivative may undergo nucleophilic substitution by the second molecule of farnesyl pyrophosphate with the elimination of the diphosphate group and inversion of configuration at carbon atom 1 in a similar way to that discussed under D.5.2. Since the double bond in the nerolidyl derivative is lost, a positive charge is formed at carbon atom 2 of this molecule which is at first covered by the substituent X and is then removed from the molecule by the elimination of a proton from carbon atom 2 of the nerolidyl portion of the molecule. Of the two hydrogen atoms at carbon atom 2 the one above the plane

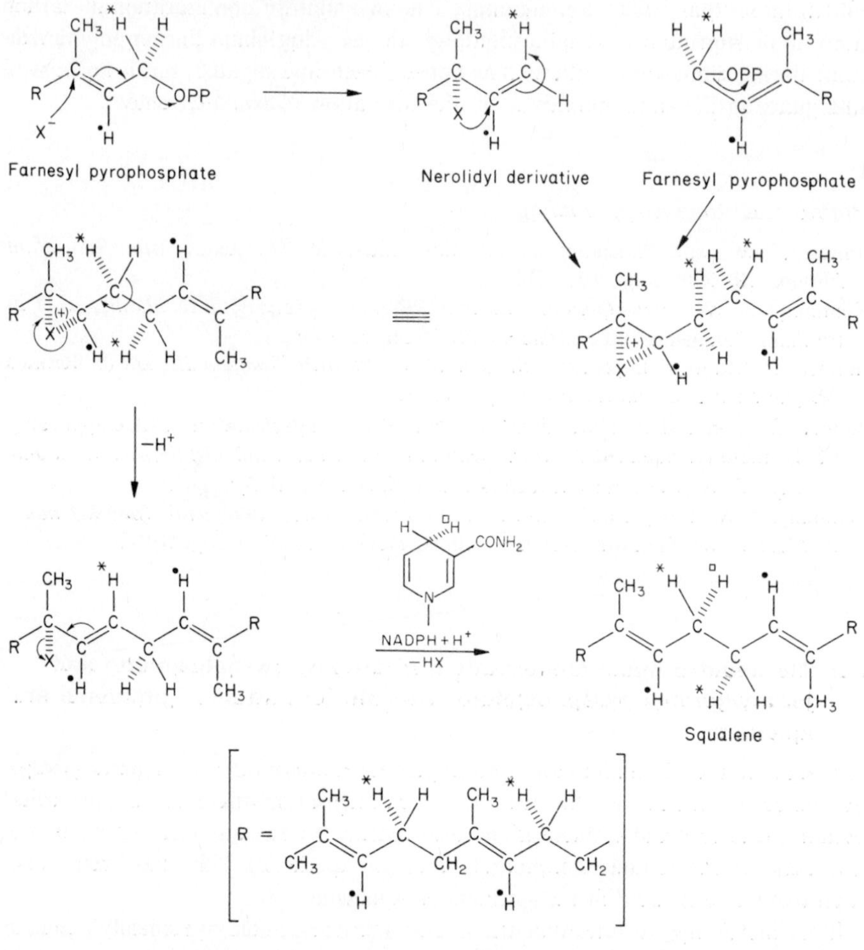

Fig. 85. Formation of squalene from farnesyl pyrophosphate.

of the paper in the formula in fig. 85 is lost while the one marked '*' is retained in squalene. During the subsequent shift of the double bond from carbon atoms 2 and 3 to carbon atoms 3 and 4 of the nerolidyl portion associated with the elimination of group X, a hydrogen atom from NADPH is introduced into the molecule and squalene is formed. By using $4S$-[^{3}H]-NADPH it has been shown that this hydrogen atom originates from the 'B' side of the reduced nicotinamide (cf. C.2.1.1).

Carotenoids are probably formed in a similar manner from geranylgeranyl pyrophosphate. However, the last step of the reaction in this case is not a reduction but the formation of a *cis* double bond between the two carbon atoms 1 and 1'. Phytoene is formed with the retention of both the hydrogen atoms marked

'*'. The additional double bond prevents the folding of the molecule which is prerequisite for the cyclization process as described for squalene under D.5.7.1.

References for further reading

Porter, J. W. and Anderson, D. G. Biosynthesis of Carotenes, *Ann. Rev. Plant Physiol.* **18** (1967), pp. 197–228.

Frantz, I. D. and Schoepfer, G. J. Sterol Biosynthesis, *Ann. Rev. Biochem.* **36** (1967), pp. 691–726.

Goodwin, T. W. 'Recent Investigations on the Biosynthesis of Carotenoids and Triter-penes', *Perspectives in Phytochemistry*, ed. J. B. Harborne and T. Swain (Academic Press, London, 1969), pp. 75–90.

Popják, G. 'Conversion of Mevalonic Acid into Prenyl Hydrocarbons as Exemplified by the Synthesis of Squalene', *Natural Substances Formed Biologically from Mevalonic Acid*, ed. T. W. Goodwin (Academic Press, New York, 1970), pp. 17–33.

5.4 Formation of monoterpenes, sesquiterpenes and diterpenes from geranyl-, farnesyl- and geranylgeranyl pyrophosphates

(a) Monoterpenes

The aliphatic monoterpene hydrocarbons of volatile oils (e.g. myrcene and oci-mene) originate from geranyl pyrophosphate probably by way of neryl pyrophos-phate (*cis* geranyl pyrophosphate) with the elimination of a phosphate group

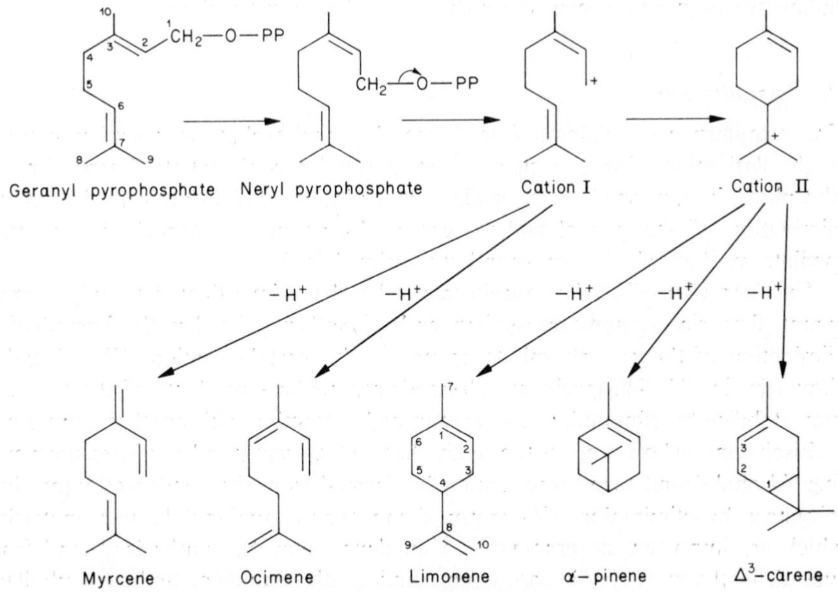

Fig. 86. Possible ways for the formation of terpene hydrocarbons from geranyl pyrophosphate.

(cation I) and a proton (shifts of double bonds are not excluded) (fig. 86). By the addition of the positively charged carbon atom 1 to the double bond between carbon atoms 6 and 7, a six-membered ring (cation II) is probably formed first before the molecule stabilizes by the elimination of a proton or by reaction with another compound. According to the position at which elimination of the proton takes place, the compound will remain monocyclic (e.g. limonene) or may form a second ring (e.g. α-pinene and Δ^3-carene). It has been found in the case of *Pinus pinea* that carbon atom 10 of limonene was free from radioactivity after feeding geraniol-4,8-C^{14}. The formation of the double bond lying outside the ring must therefore be strictly stereospecific.

Aliphatic compounds with an oxygen atom at carbon atom 1 are formed by hydrolytic elimination of the pyrophosphate group. Geraniol is formed from geranyl pyrophosphate by this reaction. Nothing is yet known, however, about the exact pathway of formation of oxygen-containing cyclic compounds (e.g. menthol and carvone).

Radioactivity could be detected in the constituents of the volatile oil of peppermint plants as soon as three minutes after exposure to $^{14}CO_2$. Monoterpene hydrocarbons especially were strongly labelled. Since radioactivity was incorporated to the same extent in all the compounds of this group it was assumed that the terpene hydrocarbons are formed in parallel reactions from only one or a few precursors. The oxygen-containing monoterpenes were labelled more slowly. These and other findings indicate that the formation of oxygenated compounds is a more complicated process. After the conclusion of exposure to $^{14}CO_2$ the radioactive terpenes formed are partly degraded in most plants.

(b) Sesquiterpenes

The sesquiterpenes originate from farnesyl pyrophosphate in a similar manner to the derivation of monoterpenes from geranyl pyrophosphate. Farnesyl pyrophosphate is converted to the widely occurring alcohol farnesol by hydrolytic elimination of the pyrophosphate group. Farnesene is formed via an intermediate product which corresponds to cation I in fig. 86.

The formation of cyclic sesquiterpenes has been investigated to only a small extent. The ring systems most frequently found may be directly formed after elimination of the pyrophosphate group via the unstable cations III, IV and V shown in fig. 87. The positively charged intermediate products of the type a–f may stabilize by elimination of a proton or by reaction with another compound.

Bisabolene, whose ring system is the basis of a great number of sesquiterpenes (e.g. bisabolol and turmerones), may be formed from the condensation product of type a by elimination of a proton. From type c, guajanolides (e.g. matricin), which are important as precursors of azulenes, may be synthesized, and from type d, in the most simple case, compounds of the humulene and caryophyllene types may originate.

From the cation IV, longifolene as well as α- and β-himalachene may be

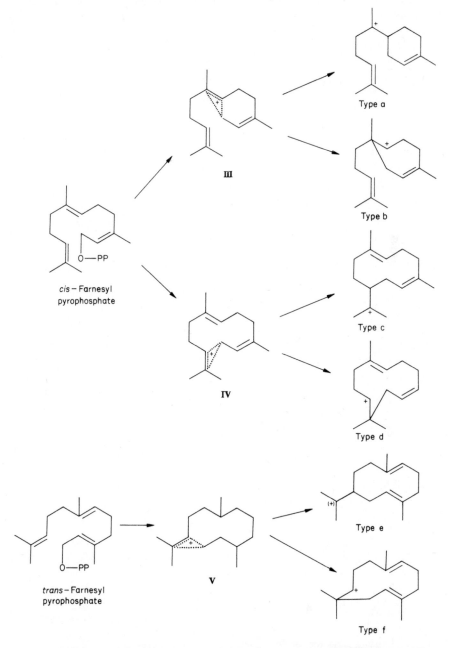

Fig. 87. Possible ways for the formation of different types of cyclic sequiterpenes from farnesyl pyrophosphate.

formed after several rearrangements. The formation of longifolene in this manner is in agreement with the radioactivity distribution obtained after feeding acetate to *Pinus longifolia*.

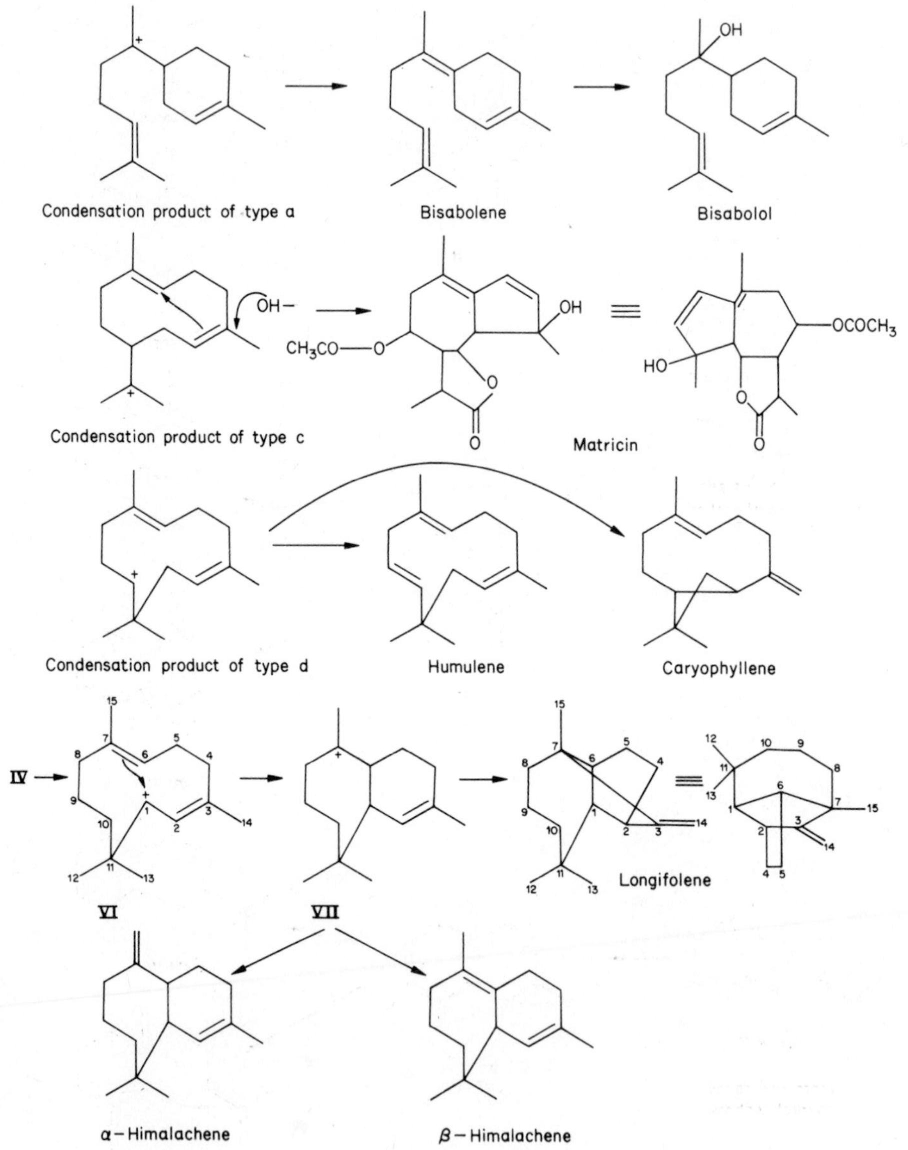

Fig. 88. Possible ways for the formation of some sesquiterpenes.

(c) Diterpenes

The starting material for diterpene synthesis is geranylgeranyl pyrophosphate. This substance may be converted to geranylgeraniol as well as to the aliphatic C_{20}-hydrocarbon by the elimination of pyrophosphate. Phytol is one of the highly hydrogenated alcohols derived from geranylgeraniol.

Geranylgeranyl pyrophosphate is also easily converted to bicyclic and tricyclic derivatives. Cyclization is catalysed by enzymes and is probably initiated by the action of a proton. In the case of tetracyclic and pentacyclic triterpenes, however, cyclization proceeds in a different way than in the case of squalene (cf. D.5.7.1).

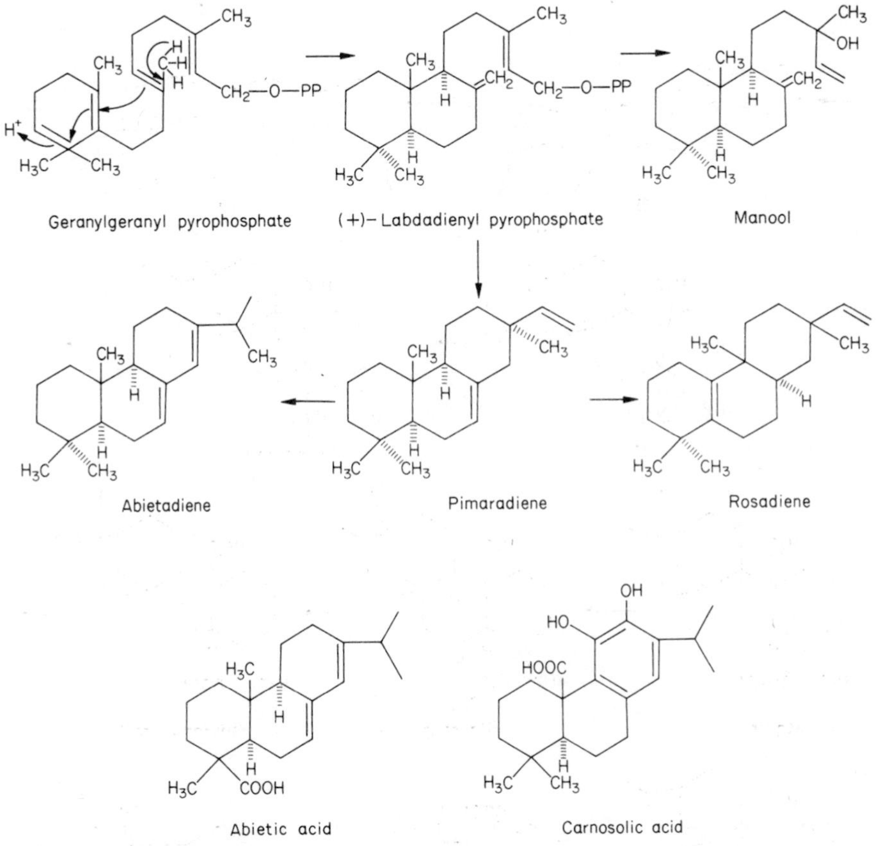

Geranylgeranyl pyrophosphate   (+)-Labdadienyl pyrophosphate   Manool

Abietadiene   Pimaradiene   Rosadiene

Abietic acid   Carnosolic acid

Fig. 89. Formation of cyclic diterpenes of the (+)-labdadienol series.

A bicyclic compound, corresponding to labdadienol, which, depending on its configuration, belongs to the (+)- or (−)-series, is first formed during cyclization. Compounds of the type outlined in fig. 89 may be formed from (+)-labdadienyl pyrophosphate by the elimination of the pyrophosphate group. The carbon

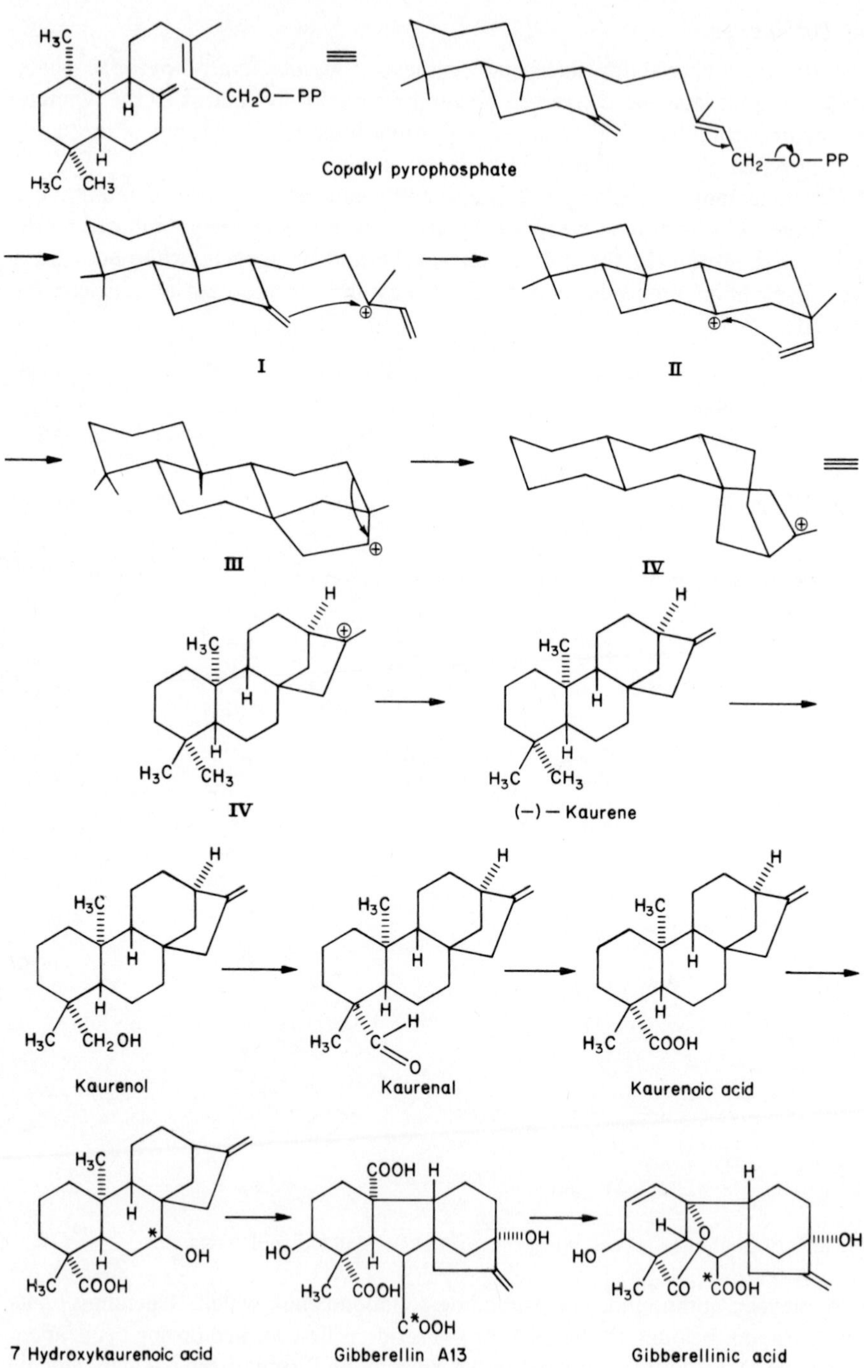

Copalyl pyrophosphate

I

II

III

IV

IV

(−) − Kaurene

Kaurenol

Kaurenal

Kaurenoic acid

7 Hydroxykaurenoic acid

Gibberellin A13

Gibberellinic acid

Fig. 90. Biosynthesis of gibberellins.

skeleton of abietadiene, which differs from that of other compounds, is formed by the 1,2-shift of a methyl group to the vinyl side chain. (Abietadiene is the precursor of abietic acid, which occurs in large amounts in the resins of conifers, and of carnosolic acid, the characteristic bitter principle of the Labiatae.

(−)-Labdadienyl pyrophosphate (copalyl pyrophosphate) is the precursor of gibberellins which are active as hormones in higher plants (fig. 90). More than twenty representatives of this group are known today, of which gibberellic acid (GA_3) is the most important.

In gibberellin formation, copalyl pyrophosphate is first converted to cation I with the elimination of a pyrophosphate group and a shift of a double bond. This cyclizes further to cations II and III and by the shift of a bond changes to cation IV which stabilizes to yield (−)-kaurene with the elimination of a proton. The latter compound is the first stable intermediate of the synthetic chain. From it, through mixed function oxidations (cf. C.2.6), first kaurenol, and then kaurenal, kaurenoic acid and 7-hydroxy kaurenoic acid are formed. The conversion of 7-hydroxy kaurenoic acid to gibberellins is accompanied by a contraction of one ring, from which the carbon atom marked '*' is eliminated. Thus C_{20}-gibberellins are first formed, e.g. gibberellin A_{13}, which by further reactions may be changed to C_{19}-gibberellins, for example.

(d) Terpene alkaloids

Nitrogen-containing sesquiterpene and diterpene alkaloids are found in a number of higher plants (cf. fig. 91). Feeding experiments with *Dendrobium* sp. and with

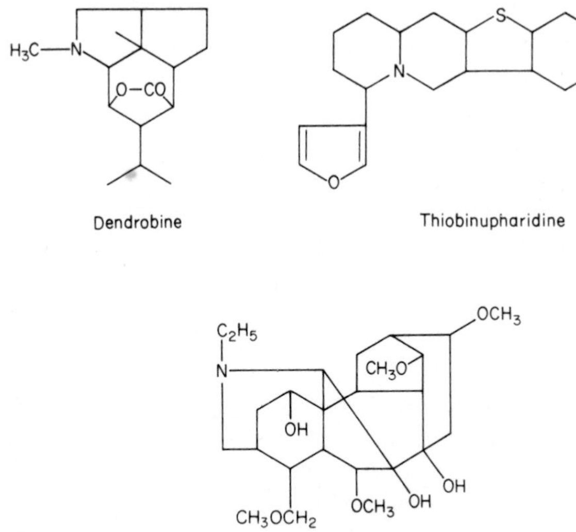

Dendrobine

Thiobinupharidine

Lycoctonine

Fig. 91. The stuctural formulae of some terpene alkaloids.

Nuphar luteum have demonstrated the incorporation of mevalonic acid into dendrobine and thiobinupharidine. Mevalonic acid is incorporated into lycoctonine and related alkaloids by species of *Delphinium*. Details of the formation of these compounds are not known.

References for further reading

Bates, R. B. 'Terpenoid Antibiotics', *Antibiotics*, Vol. II, ed. D. Gottlieb and P. D. Shaw (Springer Verlag, Berlin, 1967), pp. 134–51.

Nicholas, H. J. 'The Biogenesis of Terpenes in Plants', *Biogenesis of Natural Compounds*, ed. P. Bernfeld (Pergamon Press, Oxford, 1967), pp. 829–901.

Schütte, H. R. 'Isoprenoide Alkaloide', *Biosynthese der Alkaloide*, ed. K. Mothes and H. R. Schütte (VEB Deutscher Verlag der Wissenschaften, Berlin, 1969), pp. 601–15.

Lang, A. Gibberellins: Structure and Metabolism, *Ann. Rev. Plant Physiol.* **21** (1970), pp. 537–70.

Francis, M. J. O. 'Monoterpene Biosynthesis', *Aspects of Terpenoid Chemistry and Biochemistry*, ed. T. W. Goodwin (Academic Press, London, 1971), pp. 29–51.

Herout, V. 'Biochemistry of Sesquiterpenoids', *Aspects of Terpenoid Chemistry and Biochemistry*, ed. T. W. Goodwin (Academic Press, London, 1971), pp. 53–94.

Macmillan, J. 'Diterpenes – The Gibberellins', *Aspects of Terpenoid Chemistry and Biochemistry*, ed. T. W. Goodwin (Academic Press, London, 1971), pp. 153–80.

5.5 Biosynthesis of iridoid compounds

Iridoid compounds are cyclic monoterpenes structurally derived from iridodial (fig. 92). Iridodial, as well as substances of similar structure, occurs in the defensive secretions of the ants of the *Iridomyrmex* sp. and is named after these organisms. The iridoid compounds were formerly known as pseudo indicans since a few of them polymerize in the presence of acids and oxygen to blue coloured substances and thus behave like indican, which under similar conditions produces the blue dye indigo (cf. D.19.6).

Iridoid substances are especially widespread in higher plants. They may be divided into compounds of the loganin group which possess ten carbon atoms, those of the aucubin group with nine carbon atoms, and those of the unedoside group with eight carbon atoms. In the case of aucubin and related compounds, carbon atom 10, and in the case of unedoside, the carbon atoms in positions 10 and 11 have been lost (fig. 92).

The iridoid compounds are derived from geranyl pyrophosphate. The pathway outlined in fig. 93 has been made probable by feeding experiments. Geranyl pyrophosphate appears at first to be hydroxylated in position 8 and then isomerized to a nerol derivative. 8-Hydroxynerol is then formed after hydrolytic elimination of the pyrophosphate group. 8-Hydroxygeraniol and 8-hydroxynerol as well as the pyrophosphate of these compounds are incorporated into loganin with good incorporation rates.

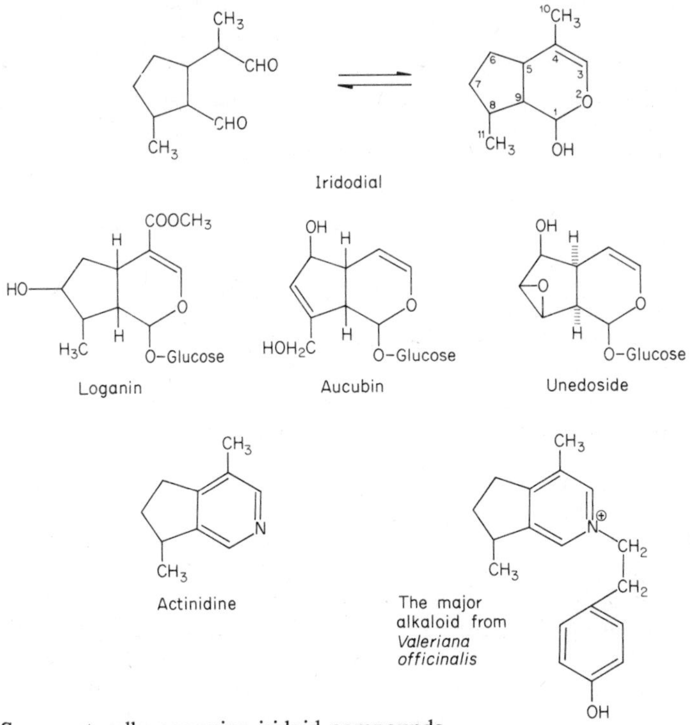

Fig. 92. Some naturally occurring iridoid compounds.

A further hydroxylation of 8-hydroxynerol and a subsequent oxidation may lead to the, as yet unidentified, trialdehyde I which then is transformed to the enol compound II. The occurrence of both these compounds might explain why, in contrast to the formation of all other terpenes, an isomerization takes place during the biosynthesis of the iridoids, in which carbon atoms 8 and 10 of geranyl pyrophosphate lose their identity. As established in many species geranyl pyrophosphate labelled radioactively in position 8 as well as in position 10 leads to the formation of iridoid compounds uniformly labelled in positions 3 and 10 (fig. 93).

A reductive cyclization should lead in one step from compound I to the iridoid III which may be converted to loganin via 7-deoxyloganic acid.

The present state of our knowledge indicates that all iridoid compounds are derived from 7-deoxyloganic acid and loganin. The molecule may be changed further in many ways, e.g. be oxidized or reduced, hydroxylated and decarboxylated.

An important pathway leads to the so-called secologanins. The cyclopentane ring is first broken between positions 7 and 8 (fig. 94). The 11-hydroxyloganin derivative is probably the actual starting point for the cleavage (X could be, for example, a phosphate group). Secologanin is formed first and is then converted to gentiopicroside probably via IV and sweroside and swertiamarin.

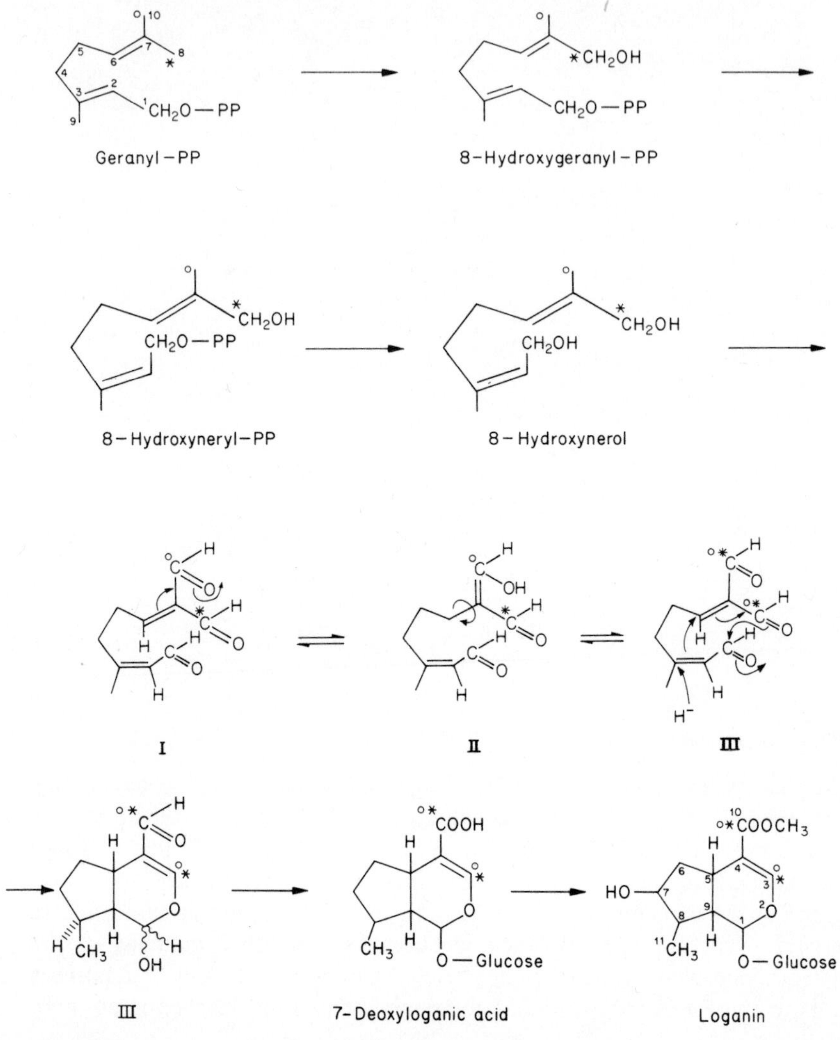

Fig. 93. Formation of loganin from geranyl pyrophosphate.

A number of other alkaloids originate from iridoids and secoiridoids. Only compounds of simple structure of the actinidine and loganin type are discussed at this juncture. The iridoid indole and isoquinoline alkaloids are dealt with in sections D.19.3 and D.20.1.2.

Since the iridoid glycosides are precursors for the iridoid indole and isoquinoline alkaloids and gentiopicroside is converted *in vitro* in the presence of ammonia to the iridoid alkaloid gentianin (fig. 95), it is probable that the iridoid glycosides are also important precursors for the biosynthesis of the simple iridoid alkaloids *in vivo*. Actinidine and the major alkaloid of *Valeriana officinalis* can be syn-

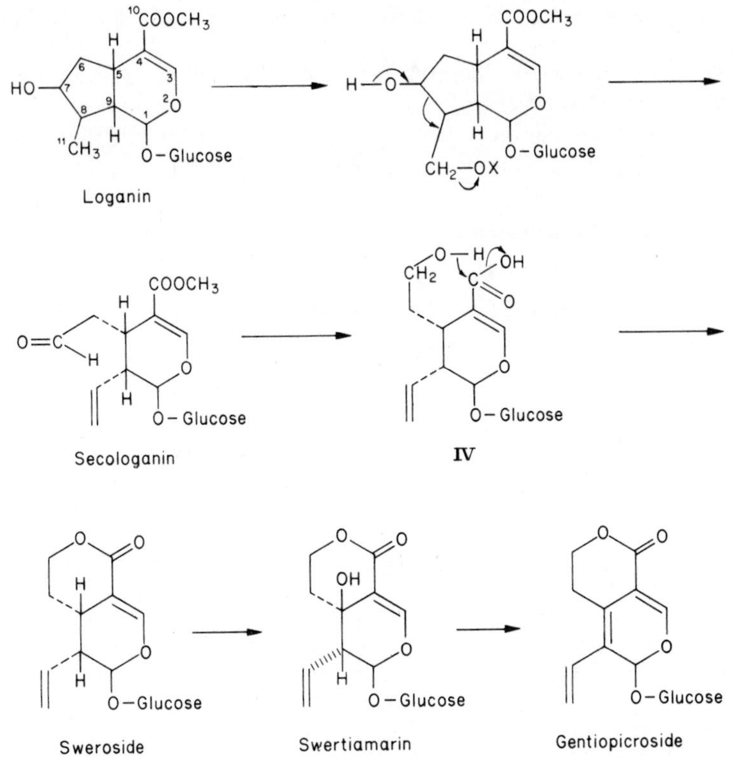

Fig. 94. Formation of secologanin and glycosides of the secologanin type from loganin.

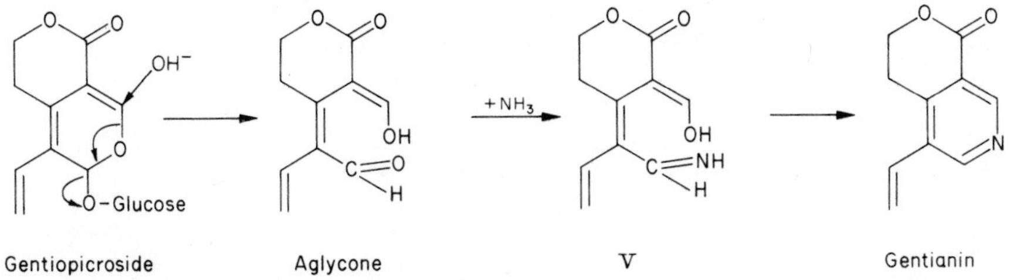

Fig. 95. Conversion of gentiopicroside to gentianin

thesized from a derivative of loganin with the participation of ammonia or tyramine respectively. More detailed experiments are, however, awaited.

References for further reading

Schütte, H. R. 'Isoprenoide Alkaloide', *Biosynthese der Alkaloide*, ed. K. Mothes and H. R. Schütte (VEB Deutscher Verlag der Wissenschaften, Berlin, 1969), pp. 601–15.
Leete, E. Alkaloid Biosynthesis, *Adv. in Enzymology* **32** (1969), pp. 373–422.

Gross, D. Die Biosynthese iridoider Naturstoffe, *Fortschritte der Botanik* **32** (1970), pp. 93–108.

Gross, D. Naturstoffe mit Pyridinstruktur und ihre Biosynthese, *Fortschritte Chemie org. Naturstoffe* **28** (1970), pp. 109–61.

Plouvier, V. and Favre-Bonvin, J. Les iridoïdes et seco-iridoïdes: répartition, structure, propriétés, biosynthèse, *Phytochemistry* **10** (1971), pp. 1697–1722.

5.6 Formation of carotenoid compounds and vitamin A

The yellow to red colouring matter of the carotenoid type occurs in lower plants as well as in the tissues of higher plants and is found especially in the chloroplasts

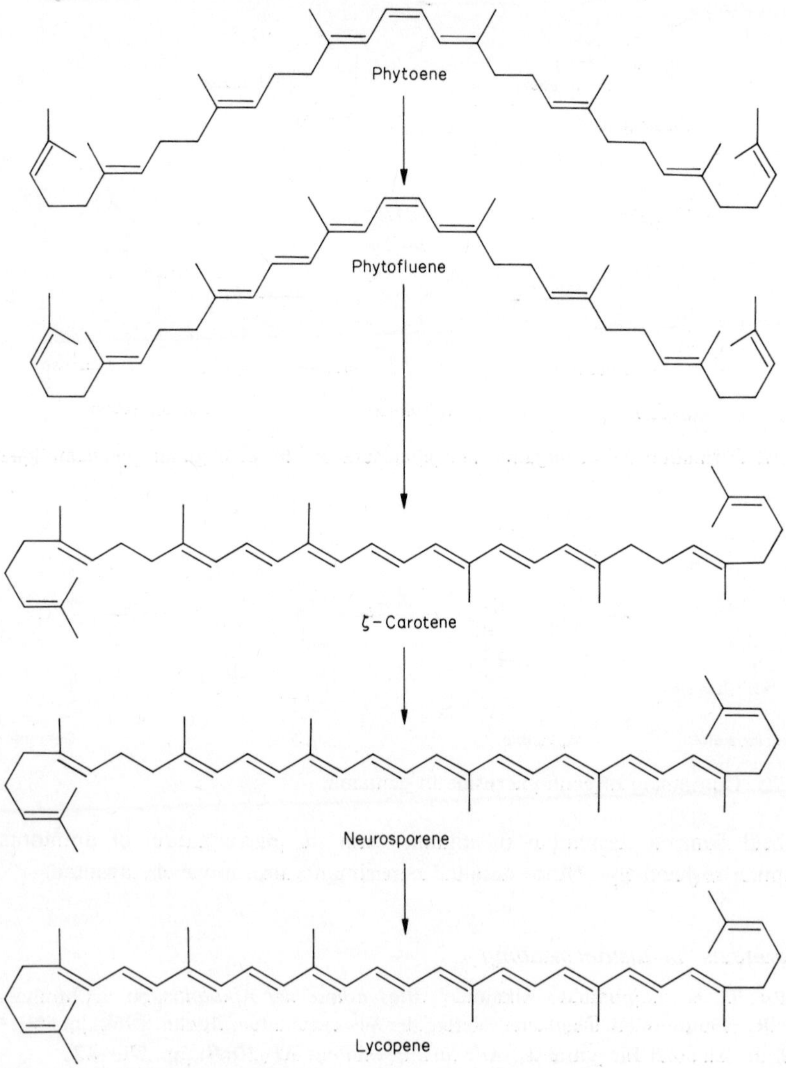

Fig. 96. Conversion of phytoene to lycopene.

and chromoplasts. Carotenoid compounds cannot be synthesized by animals. The colour of the carotenoids is due to the great number of *trans*-double bonds found in these compounds. They may be either free of oxygen (carotenes in the strict sense) or may contain hydroxyl-, methoxyl-, epoxide- or keto-groups and are then known as xanthophylls. Carotenoid compounds may either be aliphatic or may have one six-membered carbocyclic ring at one or both ends of the molecule.

The carotenoid compounds originate from a colourless hydrocarbon, phytoene (cf. D.5.3). Phytoene is reduced stepwisely in stereospecific reactions via phytofluene, ζ-carotene and neurosporene to lycopene (fig. 96). During the conversion

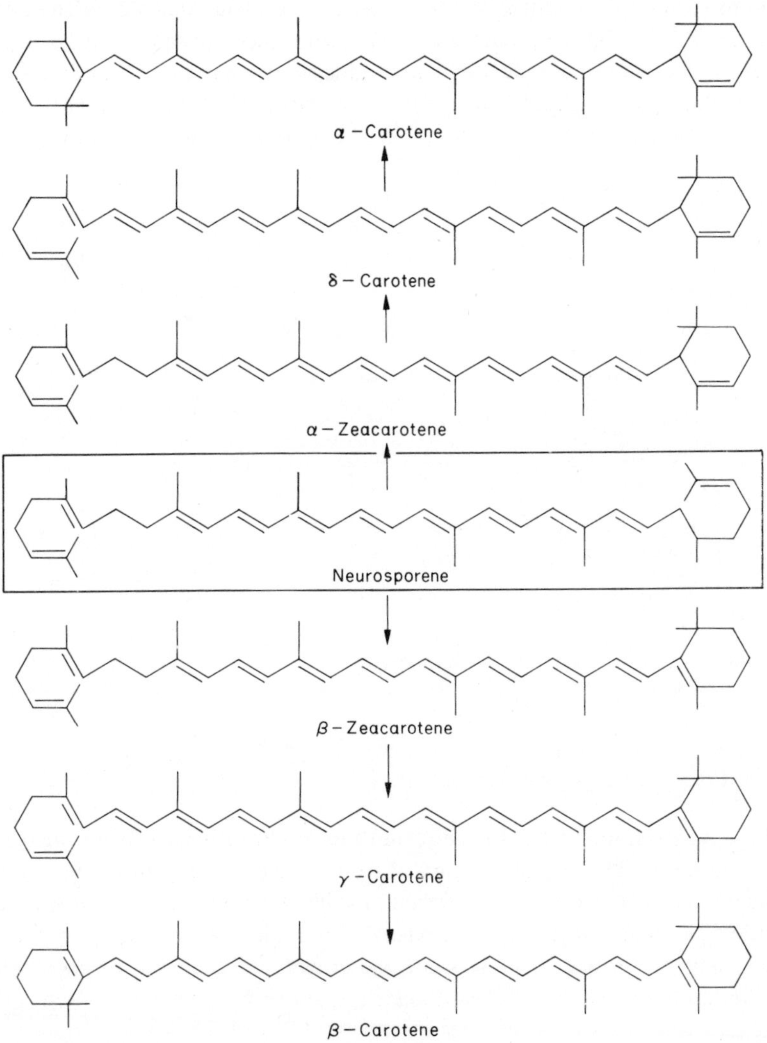

Fig. 97. Formation of cyclic carotenes from neurosporene.

of phytofluene to ξ-carotene the central *cis*-double bond is converted to a *trans*-double bond as a result of which the all-*trans* configuration of the carotenes is formed. The formation of the carotenoids probably takes place mostly in the chloroplasts.

Several pathways appear to exist for the biosynthesis of cyclic carotenes. They all probably start from aliphatic carotenoids. Thus from neurosporene, on the one hand, β-zeacarotene, γ-carotene and β-carotene may be formed and on the other, α-zeacarotene, δ-carotene and α-carotene (fig. 97). In a few organisms, however, β-carotene and γ-carotene appear to be formed from lycopene.

The cyclization occurring during the formation of the ionone rings, of which the β-ionone ring is present in β-zeacarotene, γ-carotene and β-carotene and the α-ionone ring in α-zeacarotene and δ-carotene, most probably takes place as outlined in fig. 98. Here a proton is first added to carbon atom 3 and a bond between carbon atoms 2 and 7 is formed. Then either a hydrogen atom from position 7 or 5 is eliminated, resulting in the β-ionone ring and α-ionone ring, respectively.

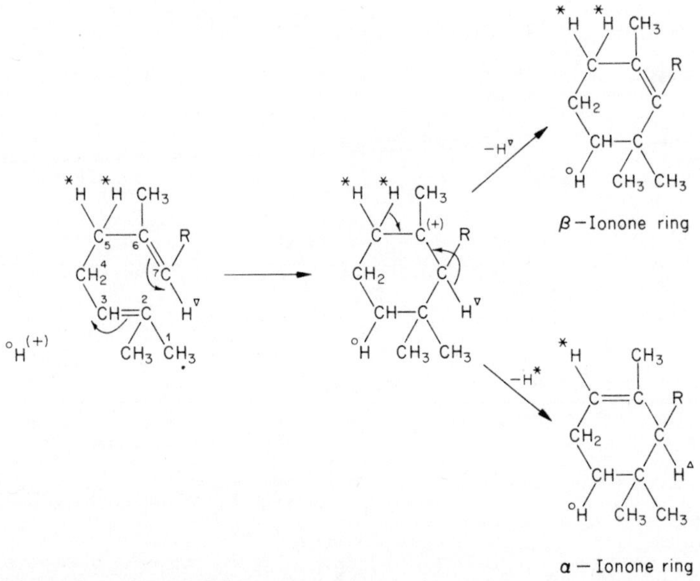

Fig. 98. Formation of α- and β-ionone rings.

The oxygen-containing carotenes (xanthophylls) are formed from the oxygen-free compounds. The oxygen originates from molecular oxygen so that it may be assumed that an epoxide ring is formed in the first step from a double bond by means of a mixed function oxygenase. From this, various oxygen-containing groups may be formed in subsequent reactions (cf. C.2.6.5). It has been shown that the xanthophylls can also be reduced again to oxygen-free carotenes in the plants.

The hydroxylated and methoxylated carotenes in *Rhodospirillum rubrum* are probably synthesized from lycopene as outlined in fig. 99. Synthesis can also proceed from neurosporene.

Similar pathways of formation are assumed for the hydroxylated and methoxylated cyclic carotenes.

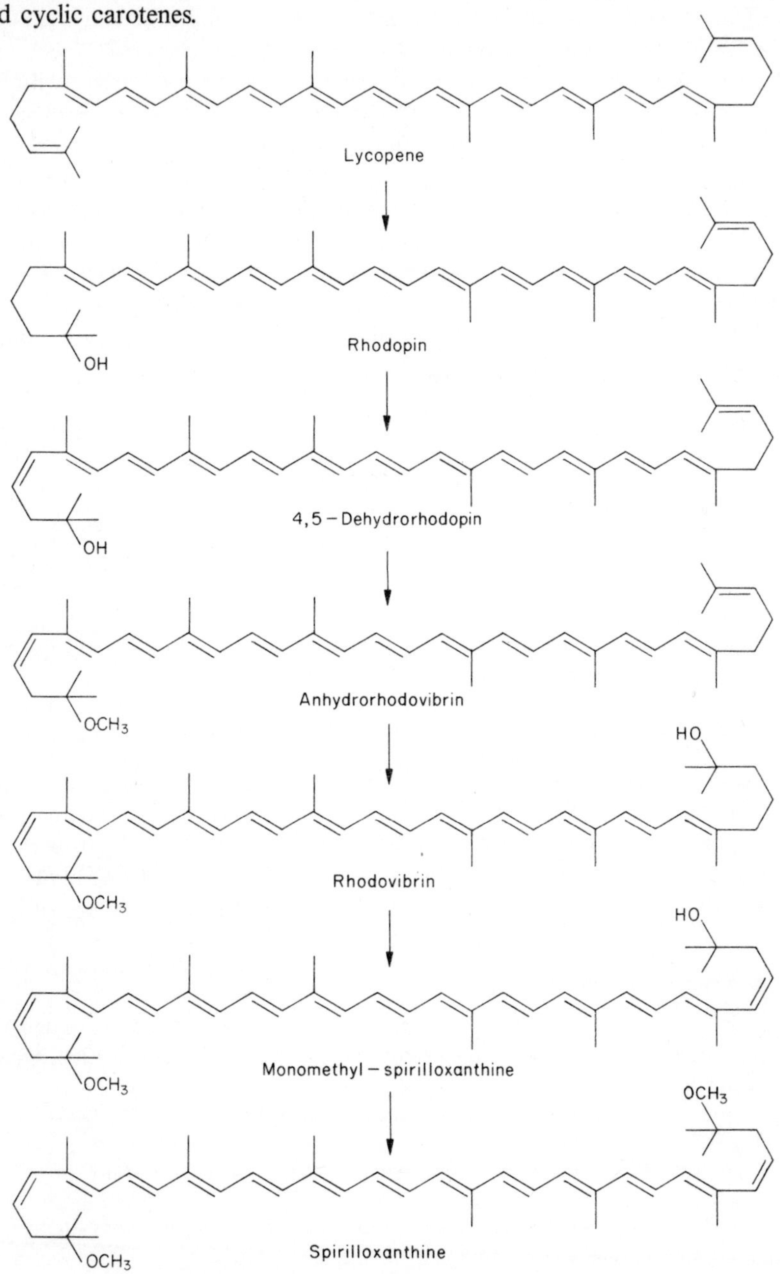

Fig. 99. Formation of spirilloxanthine from lycopene.

Carotenes which possess a β-ionone ring may be converted to vitamin A in animals, The molecule is degraded by β-oxidation from one end to the central double bond. In this manner retinal is first formed via retinolic acid, and is subsequently reduced to vitamin A_1 (retinol). The metabolite actually active in metabolism is still unknown. Possibly it is related to retinolic acid.

The activity of vitamin A is very diverse, for example it interferes with hormonal regulation in an unknown manner, influencing the biosynthesis of steroid hormones in the adrenal cortex. It is in addition essential for the perception of the optical stimulation. The visual pigment rhodopsin in the retina of the human eye is an enzymatically active iron-containing chromoprotein which consists, partly, of 11-*cis*-retinal. Chromoprotein is bleached by the action of light. The first step consists of a photoisomerization with the formation of all-*trans*-retinal followed by changes in the secondary structure of the protein. The individual stages of bleaching and regeneration of the visual purple are extremely complicated and cannot be discussed here in detail.

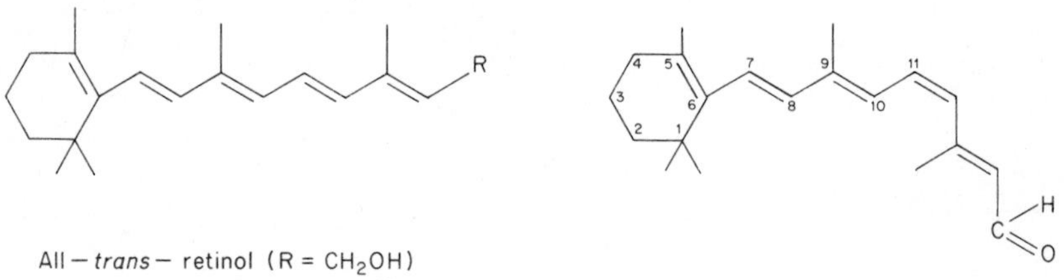

All − *trans* − retinol (R = CH₂OH)

All − *trans* − retinal (R = C<H / =O)

All − *trans* − retinolic acid (R = COOH)

11 − *cis* − retinol

Fig. 100. The structural formulae of vitamin A, and its derivatives.

Perception of smell, according to recent investigations, is based on complex formation between the gas molecules causing the smell and carotenoid compounds, including vitamin A. The latter compounds have some properties of semiconductors, which are changed by absorption of gas molecules. A model built on the basis of these findings could distinguish between various types of smell.

References for further reading

Chichester, C. O. and Nakayama, T. O. M. 'The Biosynthesis of Carotenoids and Vitamin A', *Biogenesis of Natural Compounds*, ed. P. Bernfeld (Pergamon Press, Oxford, 1967), pp. 641–78.

Porter, J. W. and Anderson, D. G. Biosynthesis of Carotenes, *Ann. Rev. Plant Physiol.* **18** (1967), pp. 197–228.

Sebrell, W. H. and Harris, R. S. *The Vitamins*, Vol. 1 (Academic Press, New York, 1967).

Czygen, F. C. Biosynthese der Carotinoide, *Ber. dtsch. Bot. Ges.* **80** (1967), pp. 627–44.

Mackinney, G. 'Carotenoids and Vitamin A', *Metabolic Pathways*, Vol. II, ed. D. M. Greenberg (Academic Press, New York, 1968), pp. 221–73.

Williams, J. O. and Rosenberg, B. Halbleiter 'riechen', *Umschau in Wissenschaft und Technik* **69** (1969), p. 348.

Goodwin, T. W. 'Recent Investigations on the Biosynthesis of Carotenoids and Triterpenes', *Perspectives in Phytochemistry*, ed. J. B. Harborne and T. Swain (Academic Press, London, 1969), pp. 75–90.

Britton, G. 'General Aspects of Carotenoid Biosynthesis', *Aspects of Terpenoid Chemistry and Biochemistry*, ed. T. W. Goodwin (Academic Press, London, 1971), pp. 225–89.

5.7 Biosynthesis of tetracyclic and pentacyclic triterpenes

Besides aliphatic compounds of the squalene type a large number of cyclic triterpenes with four or five rings occur in nature.

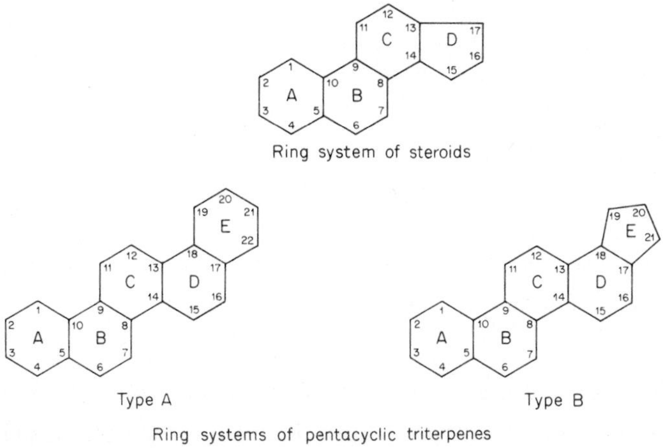

Ring system of steroids

Type A   Type B

Ring systems of pentacyclic triterpenes

Fig. 101. Ring systems of steroids and pentacyclic triterpenes.

Steroids are the most important tetracyclic triterpenes. They originate from cyclopentanoperhydrophenanthrene (sterane), and in most cases carry a hydroxyl group at position 3 and are often substituted by methyl groups at positions 10 and 13 and by a side chain at position 17. In addition, further methyl groups, hydroxyl groups, double bonds etc. may be present.

Steroids are probably synthesized by all living organisms (a probable exception being the bacteria). Certain representatives often occur in plants as well as animals. The plant steroids, however, are mostly glycosides while those in animals occur almost exclusively in the free form.

In the case of pentacyclic triterpenes, ring D of the steroid system is enlarged to a six-membered ring by an additional carbon atom. An additional six-membered ring (type A) or a five-membered ring (type B) is joined to this. Pentacyclic triperpenes of type A are widespread, especially in dicotyledons. Position 3 of the ring system of the pentacyclic triterpenes is always substituted by a hydroxyl group. Further, a number of methyl groups, and in the case of type B, an isopropyl side chain at carbon atom 19, are present (fig. 101).

Since the molecules of the cyclic triterpenes consist either completely or largely of cyclohexane or cyclopentane rings, the structural formulae illustrated above do not satifactorily represent their spatial configuration. The cyclohexane rings can exist fundamentally in two forms, of which the chair conformation is the more favourable with regard to energy than the boat conformation (fig. 102).

Different conformations are also possible for the cyclopentane ring. This and the various modes of coupling give rise to a large number of conformations (cf. the formulae of cholestane and coprostane in fig. 102).

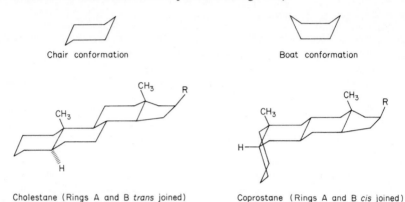

Chair conformation   Boat conformation

Cholestane (Rings A and B *trans* joined)   Coprostane (Rings A and B *cis* joined)

Fig. 102. The conformation of cholestane and coprostane.

5.7.1 Formation of cyclic triterpene ring systems from squalene.

The formation of cyclic triterpenes takes place both in animals as well as plants from the aliphatic compound squalene. This isoprenoid compound is first oxidized to squalene-2, 3-epoxide (fig. 103). The double bond between carbon atoms 2 and 3 is thus lost. In certain cases the formation of a second epoxide group at the other end of the molecule is also possible (see page 153).

The epoxide ring is broken by the action of protons forming a hydroxyl group at carbon atom 3 and a positive charge remains at carbon atom 2. The 3-hydroxysqualene cation thus formed cyclizes spontaneously. The number and conformation of the rings formed are probably dependent on the folding of the squalene chain. This is possibly determined by the attachment to the surface of the enzyme catalysing this conversion. The positive charge that remains in the molecule after cyclization is lost by the elimination of a proton. The positive charge may, however, wander in the molecule, causing the substituents on the carbon skeleton (H-atoms and methyl groups) to be so shifted that they retain their spatial orientation above and below the plane of the paper (*cis*-anionotrophy of the hydride and methide groups). The ring system itself on the contrary must change its conformation during the shifts, or else *cis*-anionotropy would be impossible. The cyclization as well as the anionotropy is a single concerted mechanism. The intermediates formulated above do not occur in the free form.

In the formation of tetracyclic triterpenes with the cholestane conformation (fig. 102), squalene must be so folded that it possesses a chair-boat-chair-boat conformation. The prosterol cation IV which carries a positive charge at carbon atom 20 is probably the first compound with a steroid skeleton to be formed by cyclization of the corresponding 3-hydroxysqualene cation via compounds I, II and III. By the shift of two hydrogen atoms and two methyl groups, steroid VIII is probably formed via compounds V, VI and VII. It can then be converted to lanosterol by the elimination of a proton from carbon atom 9 and the formation of a double bond between carbon atoms 8 and 9. Lanosterol is particularly widespread in animals and is an important intermediate in the formation of other naturally occurring steroids. The positive charge reaches carbon atom 9 in a further shift forming compound IX, from which cycloartanol

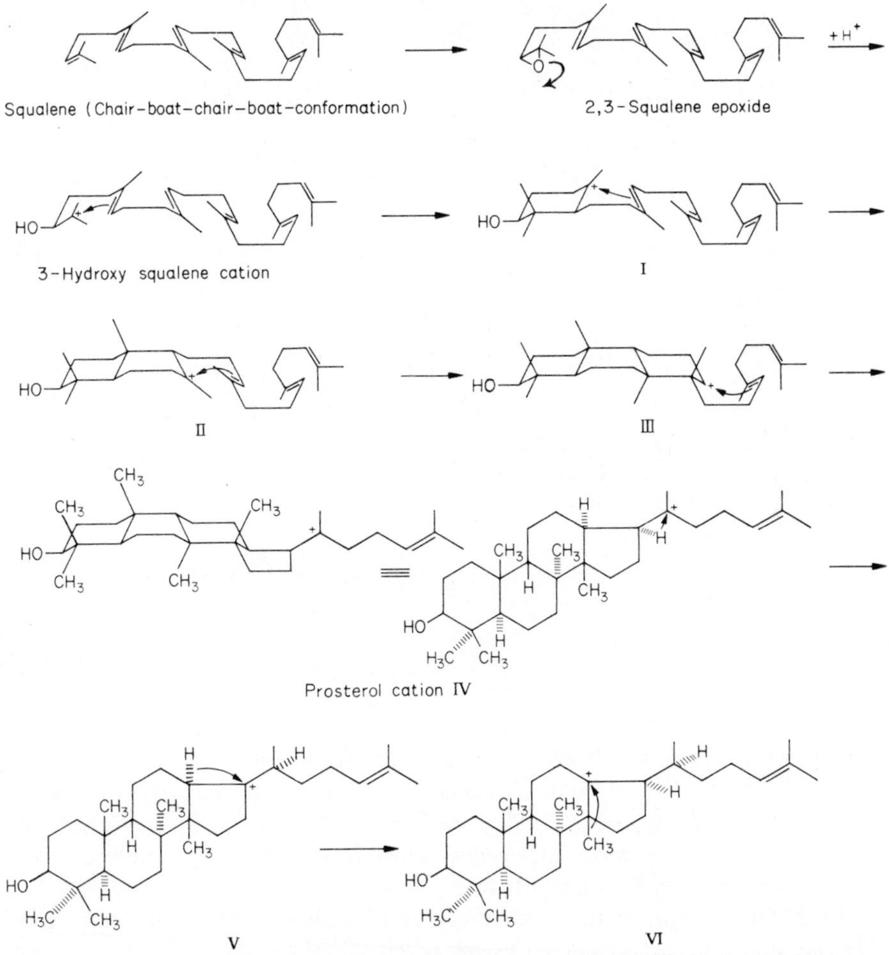

Squalene (Chair-boat-chair-boat-conformation)   2,3-Squalene epoxide

3-Hydroxy squalene cation   I

II   III

Prosterol cation IV

V   VI

Fig. 103. Possible biosynthetic pathway of lanosterol, cycloartanol and cucurbitacin E.

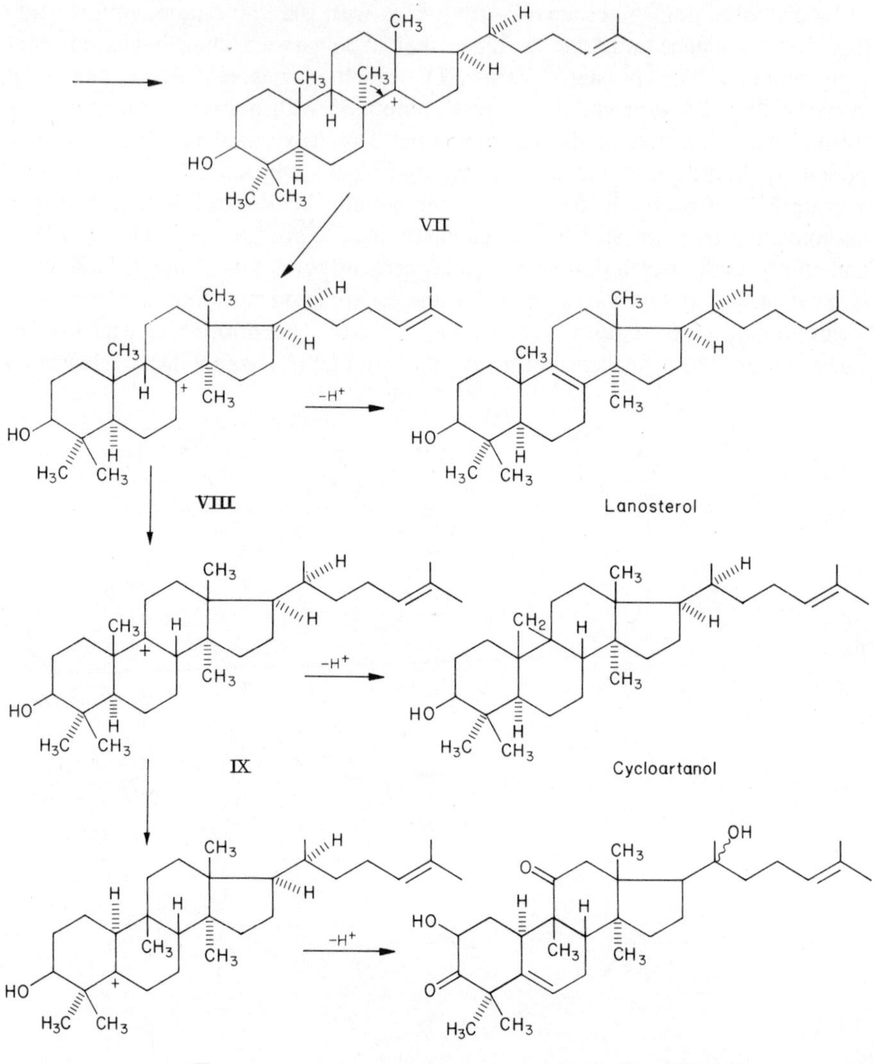

Fig. 103 (contd.).

may be obtained by the elimination of a proton from carbon atom 19 (fig. 103). Cycloartanol occurs frequently in plants and is the key product from which all other sterane derivatives are synthesized. Thus in the case of *Euphorbia* species, the lanosterol present in the latex is not formed by the pathway mentioned above but by isomerization from cycloartanol.

By a further shift of the positive charge to carbon atom 5 compound X is formed, and may be considered as the precursor of the cucurbitacins. In these very bitter compounds, which have been found in Cucurbitaceae and in the genus

Iberis of Crucifereae, the steroid ring system has the same configuration as in lanosterol and cycloartanol.

Compounds such as dammaradienol and euphol in which rings C and D of the steroid nucleus are linked differently may originate by a similar mechanism, but are derived from squalene which is in the chair-chair-chair-boat conformation (fig. 104).

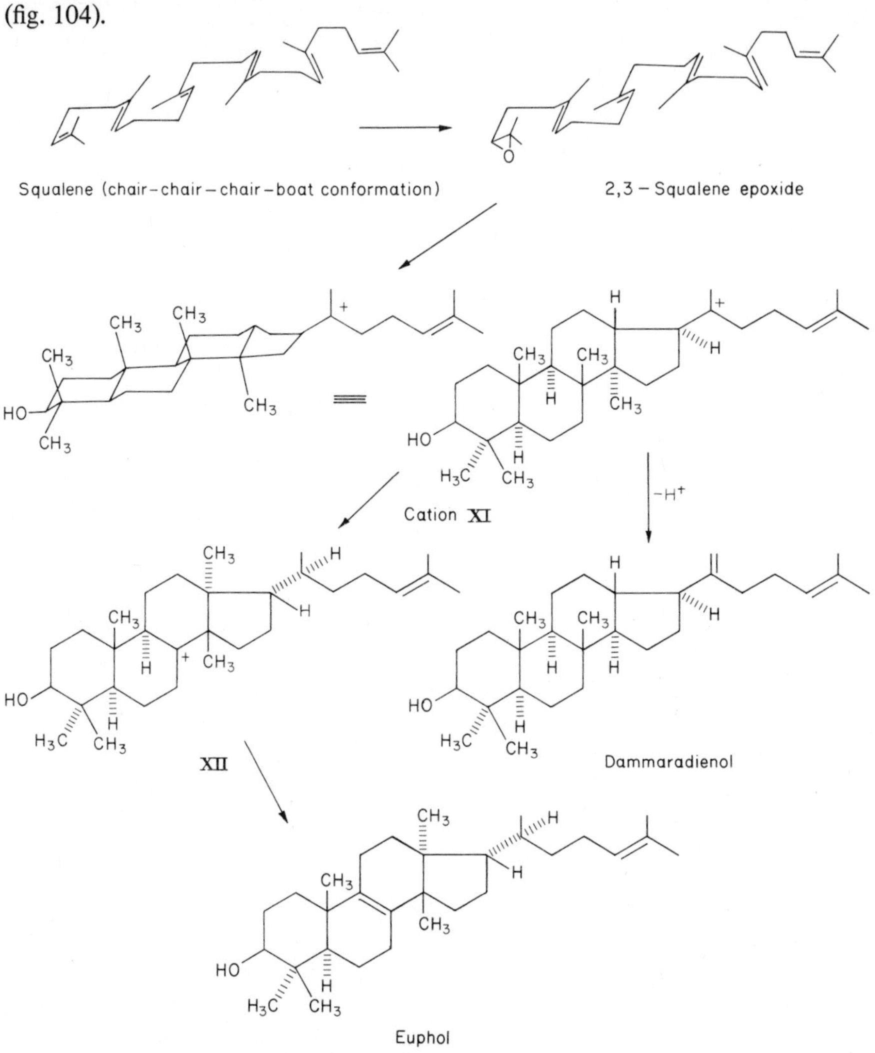

Fig. 104. Formation of dammaradienol and euphol.

The steroid cation XI thus formed may either directly eliminate a proton with the formation of a double bond between carbon atoms 20 and 21 forming dammaradienol, or the shift of the above-mentioned substituents may result in cation XII, which may change to euphol.

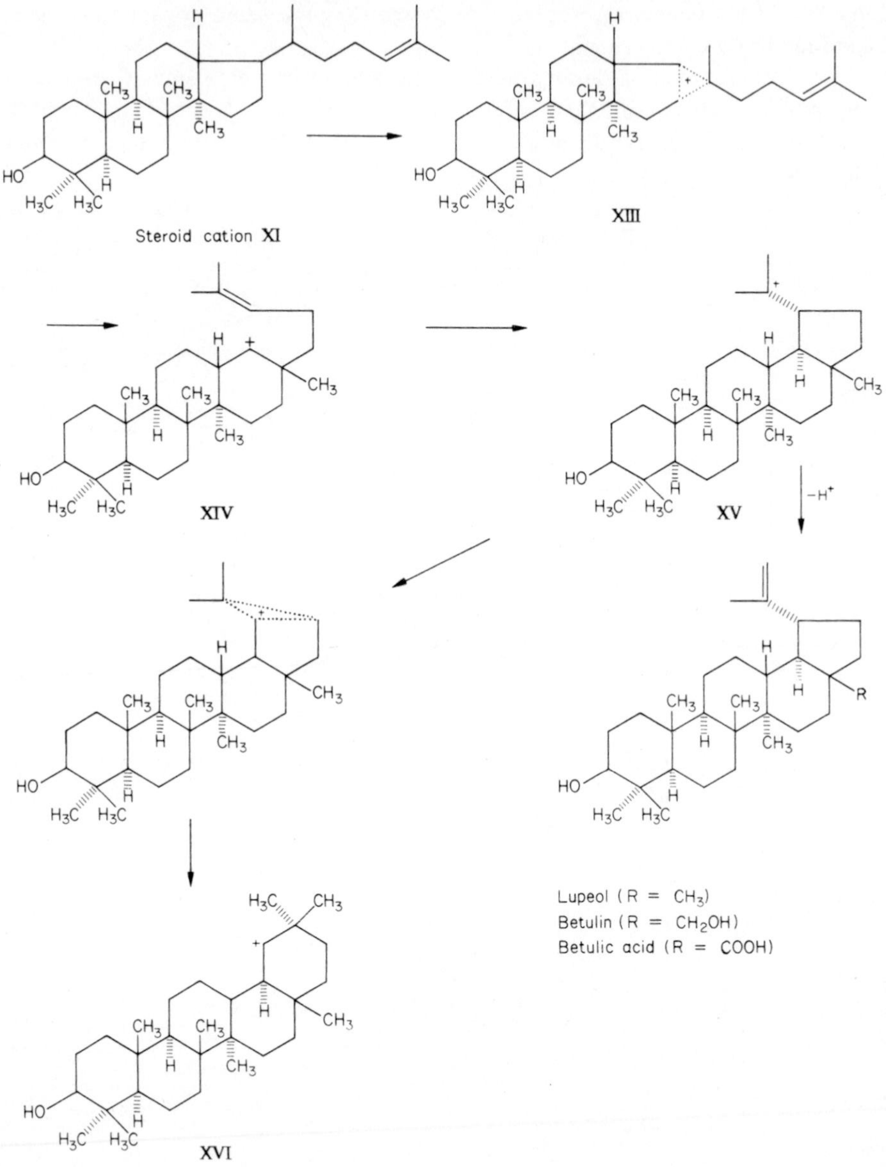

Fig. 105. Formation of pentacyclic triterpenes, part I.

The steroid cation **XI** shown in fig. 104 may also be that intermediate from which the pentacyclic triterpenes originate (fig. 105 and 106). By means of a Wagner-Meerwein rearrangement an enlargement of ring D (compound **XIV**) may first take place via compound **XIII**. A further ring closure to form com-

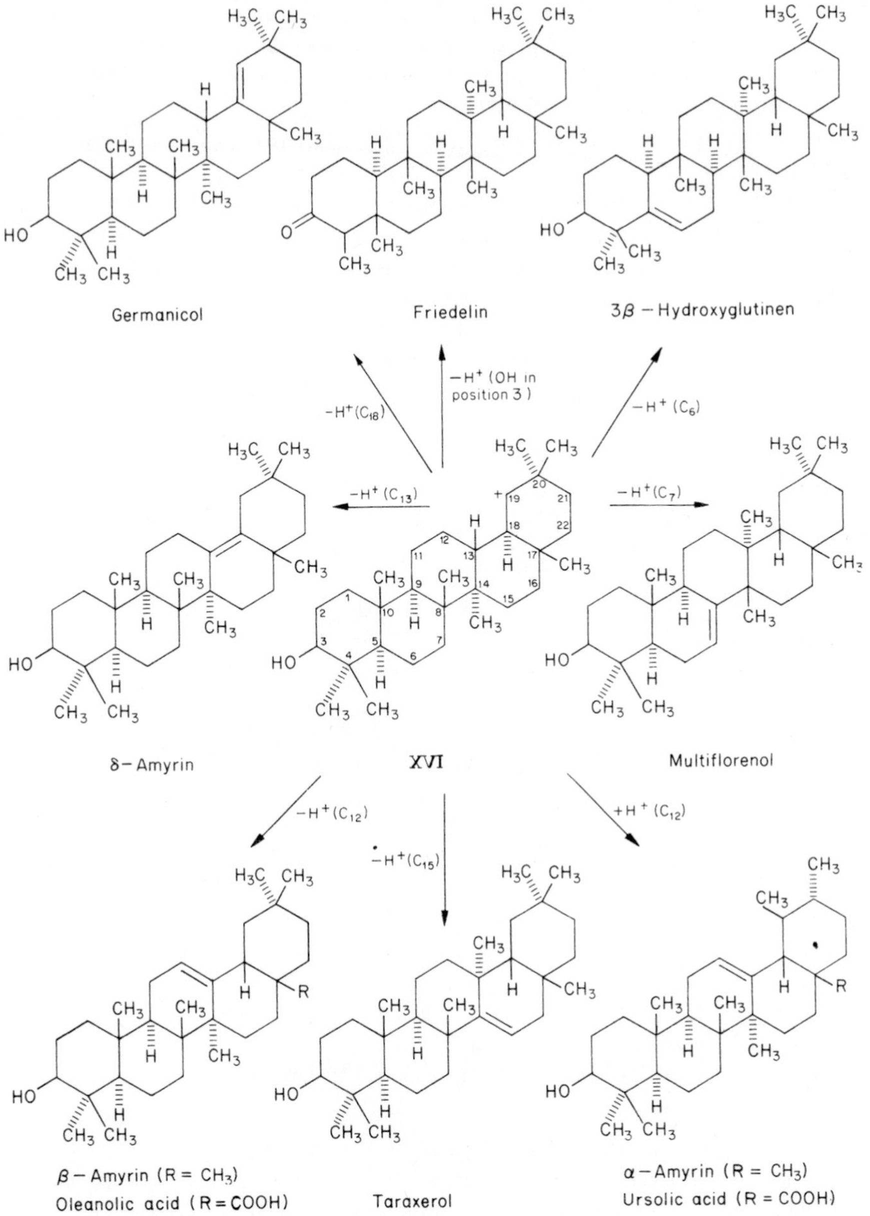

Fig. 106. Formation of pentacyclic triterpenes, part II.

pound XV may then follow. If a proton is eliminated from carbon atom 30 of the latter, compounds of the lupeol type may originate, a representative of which is the triterpene, betulin, which occurs in large quantities in birch bark. A further

Wagner-Meerwein rearrangement leads to compound XVI, through an enlargement of the newly built ring E. This may be regarded as the starting material for the synthesis of a large number of pentacyclic triterpenes (fig. 106). Individual compounds may arise from compound XVI by *cis*-anionotropy of substituents and shift of the positive charge, as well as by the elimination of a proton, as described above.

Some pentacyclic triterpenes are termed acidic sapogenins since they have properties similar to the steroid sapogenins (cf. D.5.7.2). They are the active principles of the sapogenin drugs used in therapeutics. The aescigenin glycosides occur in the seeds of *Aesculus hippocastanum*, quillaic acid in the bark of *Quillaia saponaria* and the glyzyrrhetic acid glycosides in the underground parts of *Glyzyrrhiza glabra* (cf. fig. 107). The compounds illustrated in fig. 106 often occur in latex and resins.

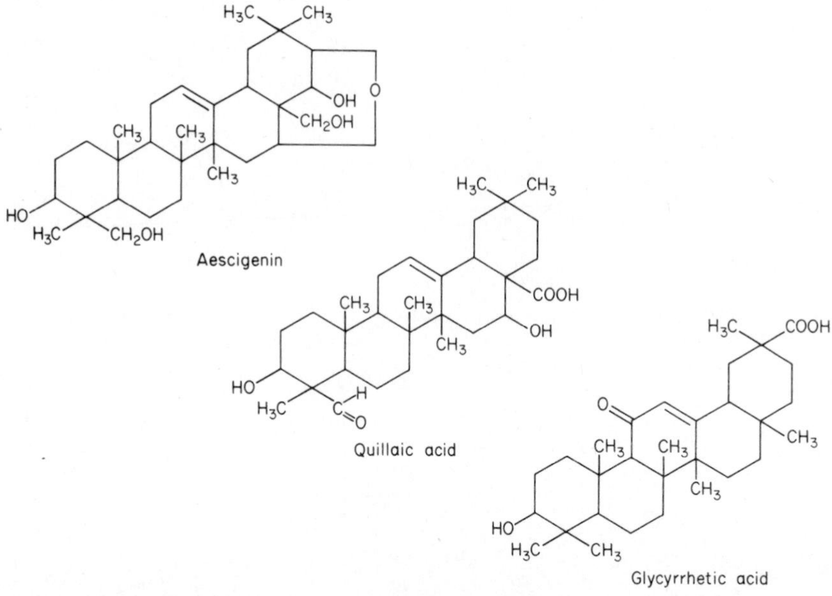

Fig. 107. Formulae of aglycones of some acidic sapogenins.

If oxygenation of squalene takes place on both the terminal double bonds of the molecule (rather than on one) then 2,3;27,28-squalenediepoxide is formed. This compound can yield α-onocerin (onocol) by cyclization and elimination of two protons and has been found in the roots of *Ononis spinosa* (fig. 108).

References for further reading

See D.5.7.2.

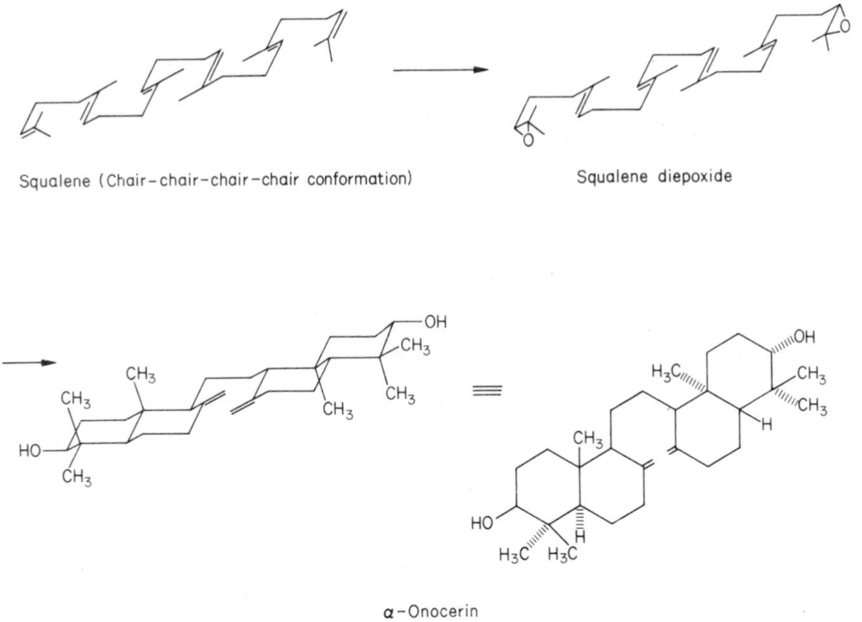

Squalene (Chair–chair–chair–chair conformation)   Squalene diepoxide

α-Onocerin

Fig. 108. Formation of α-onocerin.

5.7.2 Biosynthesis of various groups of steroids

(a) Sterols

Steroids which are characterized by a long isoprenoid side chain at carbon atom 17 are called sterols. These compounds which usually consist of twenty-seven to twenty-nine carbon atoms are ubiquitous both in plants and animals. In contrast to lanosterol and cycloartanol, the most important naturally occurring sterols possess methyl groups only in positions 10 and 13.

In the case of animals, lanosterol is probably the starting material from which the other sterols are synthesized. It is first converted to cholesterol via zymosterol. It is probable that the methyl group at carbon atom 14 is first removed by oxidation, followed by both the methyl groups at carbon atom 4. In plants, on the contrary, sterols are found which, for example, like cycloleucanol and macdougallin, have a methyl group at carbon atom 14, and in the case of the former compound, also at carbon atom 4. This may be regarded as a variation of demethylation in which the methyl groups at carbon atom 4 are first removed. It is assumed that before elimination the methyl groups are hydroxylated by a mixed function oxygenase (cf. C.2.6.4) and then oxidized to carboxyl groups via the aldehyde. In the isolated liver, for example, one molecule of carbon dioxide appears per eliminated methyl group in the formation of cholesterol (fig. 109).

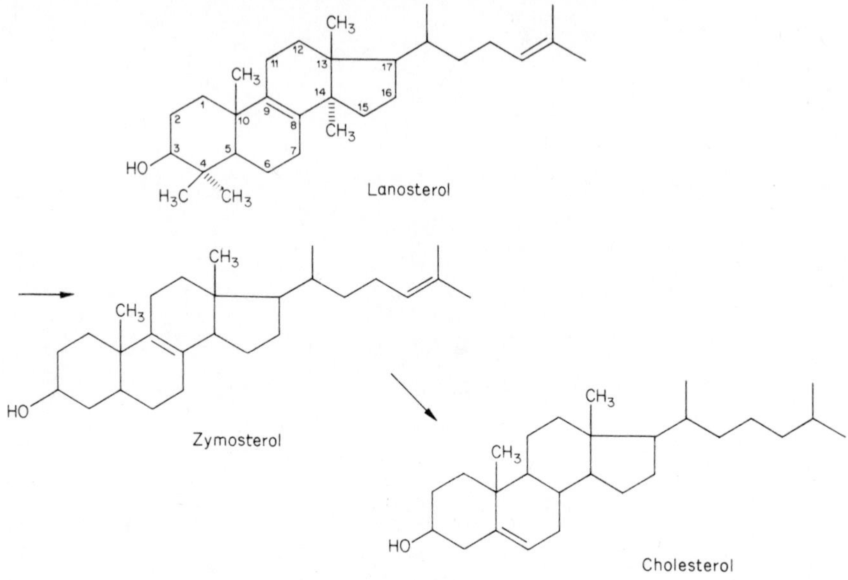

Fig. 109. Conversion of lanosterol to cholesterol.

Cholesterol is an important intermediate in the synthesis of other steroids, and is synthesized by plants as well as animals.

A number of sterols have been found which, in comparison to lanosterol, have one or two additional carbon atoms in the side chain (cf. the formulae of ergosterol, eburicoic acid, stigmasterol and β-sitosterol in fig. 110).

It has been shown that these carbon atoms originate from the methyl group of methionine. In the case of derivatives such as stigmasterol and β-sitosterol, both the carbon atoms of the ethyl group substituted at position 24 come from this source. The C-methylations probably take place at the lanosterol level, since with eburicoic acid a compound is known which has an additional methyl group, but is structurally little different from lanosterol. One hydrogen atom of the methyl group of methionine is lost during the methylation. The mechanism therefore resembles that of the synthesis of tuberculostearic acid from oleic acid (cf. D.2.2.3).

(b) Bile acids

The bile acids, in contrast to most other steroids, have the configuration of coprostane (cf. fig. 102). The side chain at carbon atom 17 carries a terminal carboxyl group and has five carbon atoms in the compounds isolated from higher animals and human beings (cf. the formula of cholic acid in fig. 111). These compounds are synthesized by the liver and excreted into the bile. They help in the emulsification of fats during digestion.

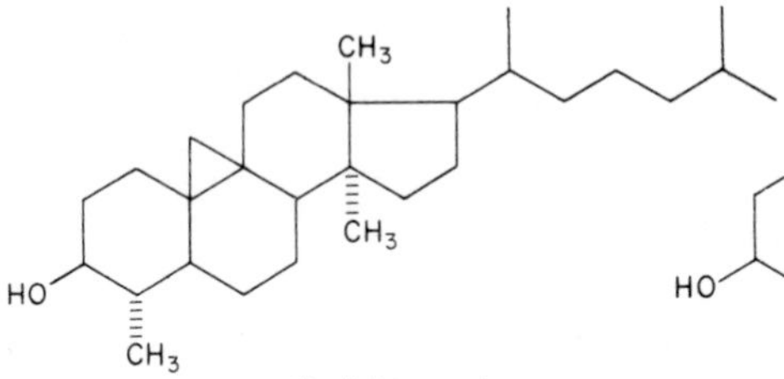

Cycloleucanol

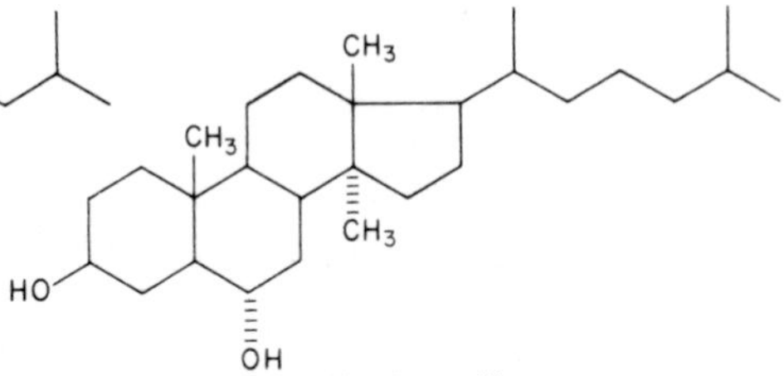

Macdougallin

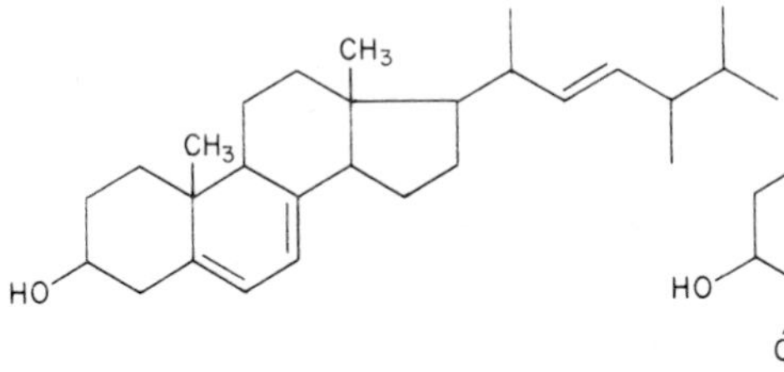

Ergosterol

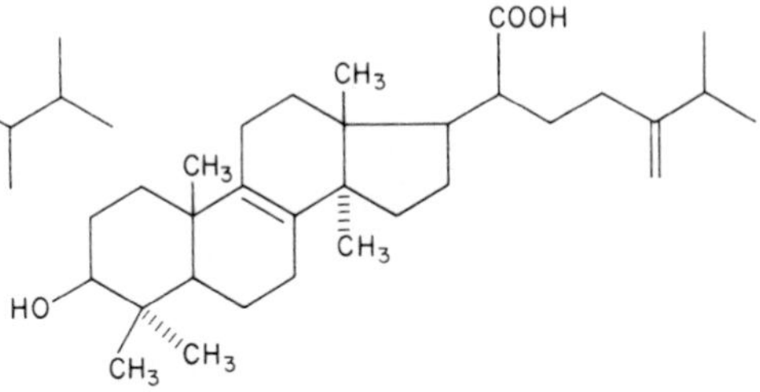

Eburicoic acid

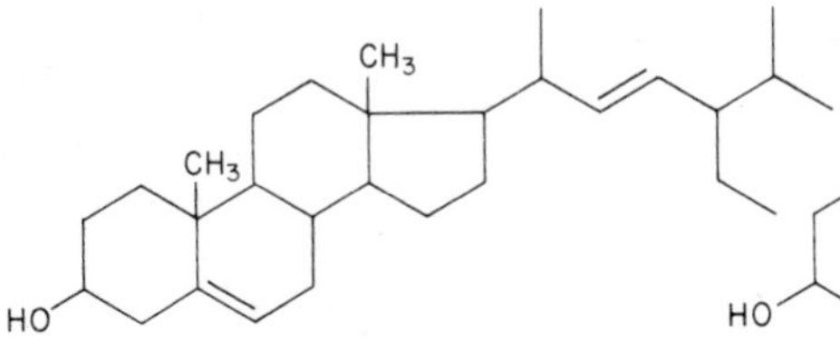

Stigmasterol

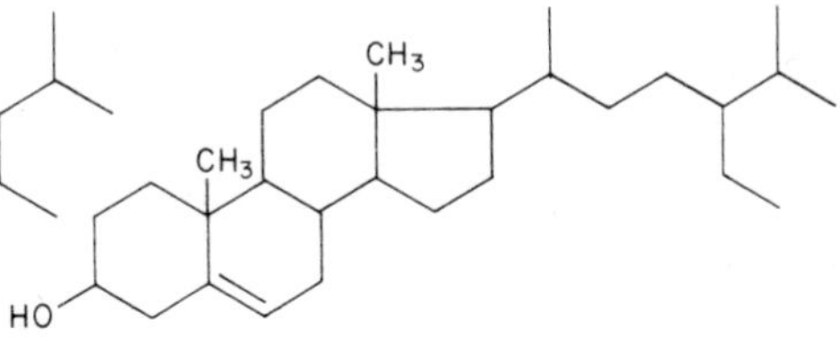

β-Sitosterol

Fig. 110. Naturally occurring sterols.

It has been shown without exception that the bile acids originate from choles-terol. The hydroxylation pattern is first laid down in the steroid ring system, and afterwards the side chain is cleaved off. One of the terminal methyl groups is oxidized to a carboxyl group before the shortening begins. 3,7,12-Trihydroxy coprostane and the corresponding trihydroxycoprostanic acid are easily converted to cholic acid by rats. The shortening of the side chain is observed only in higher animals. Bile acids of the coprostanic acid type are found in crocodiles, turtles and primitive fishes.

(c) Steroid hormones

The carbon skeleton characteristic of the steroid hormones is formed either by a further shortening to two carbon atoms, or by the complete removal, of the side

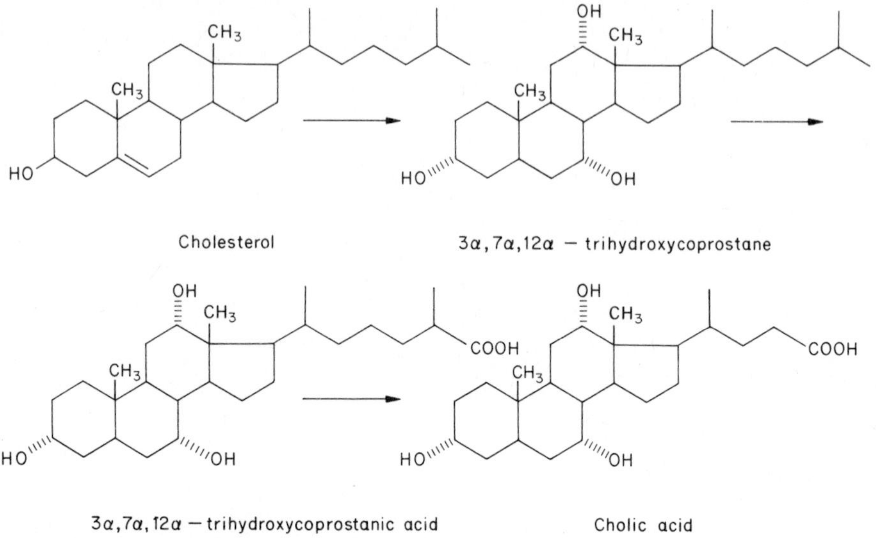

Cholesterol

$3\alpha,7\alpha,12\alpha$ — trihydroxycoprostane

$3\alpha,7\alpha,12\alpha$ — trihydroxycoprostanic acid

Cholic acid

Fig. 111. Biosynthesis of cholic acid.

chain. Thus pregnenolone and progesterone which occur in plants and animals, and the hormones cortisone and cortisol synthesized by the adrenal cortex of higher animals, carry a side chain of two carbon atoms which is replaced either by a hydroxyl or a carbonyl group in the hormones, testosterone and oestrone respectively, which are synthesized by the gonads. Androstane and oestrane derivatives have been recently found in plants also. Testosterone is the active principle of the defensive secretion of beetles of the genus *Ilybius*.

The synthesis of these compounds has been investigated both in plants as well as in higher animals, and it has been shown that they originate from cholesterol (cf. fig. 112). Enzyme preparations have been isolated from adrenal cortex, testicles and ovaries which are capable of converting cholesterol to pregnenolone and isocaproic acid. 20,22-Dihydroxycholesterol is formed first in this conversion. Pregnenolone is converted to progesterone by oxidation and isomerization.

Progesterone is the parent substance for the formation of both the adrenal cortex and sex hormones (fig. 112). The synthesis of aldosterone which is characterized by a formyl grouping at carbon 18 proceeds via 18-hydroxy corticosterone which is probably formed from progesterone by mixed function oxidation (cf. C.2.6.4). The side chain is eliminated as acetate in the biosynthesis of sex hormones. The reaction is initiated by hydroxylation at position 17 (cf. C.2.6.4).

The hydroxylation of the methyl group at position 19 (cf. C.2.6.4) and the subsequent oxidation of the hydroxymethyl group to an aldehyde group are prerequisites both for the aromatization of ring A and loss of the carbon atom at position 19, during the conversion of testosterone to oestrone. The latter is

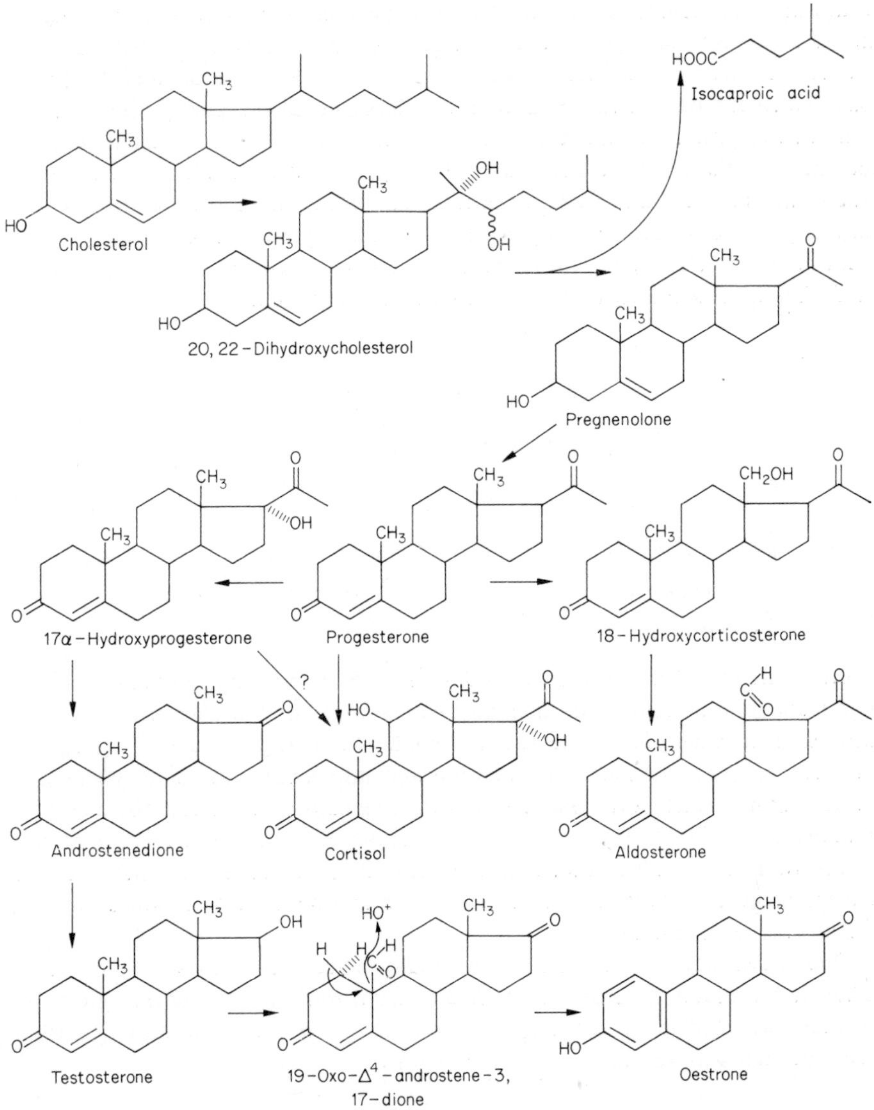

Fig. 112. Biosynthesis of steroid hormones.

eliminated as formic acid, the 1-β-H-atom also being lost. The reaction is probably initiated by attack by an OH^+ ion (cf. fig. 112) or a hydroxide radical (cf. C.2.4).

(d) Cardiac glycosides

Cardiac glycosides in which the steroid ring system is substituted at position 17 either by an unsaturated five-membered lactone ring (cardenolide type) or an unsaturated six-membered lactone ring (bufadienolide type) (cf. fig. 113) occur almost exclusively in higher plants. Cardenolides have until now been found in fourteen families (e.g. Liliaceae, Ranunculaceae, Asclepiadaceae and Apocynaceae) which are not related to one another. Bufadienolides have only been isolated from the two genera *Scilla* and *Bowiea* of Liliaceae, *Helleborus* of Ranunculaceae and the skin glands of various toads (*Bufo* sp.). A large number of rare sugars which have not yet been found in other places participate in the synthesis of the cardiac glycosides occurring in plants, as well as of the pregnane derivatives also found there (cf. D.1). The steroid molecules are linked via an ester grouping with suberylarginine in the case of the bufadienolide compounds of toads.

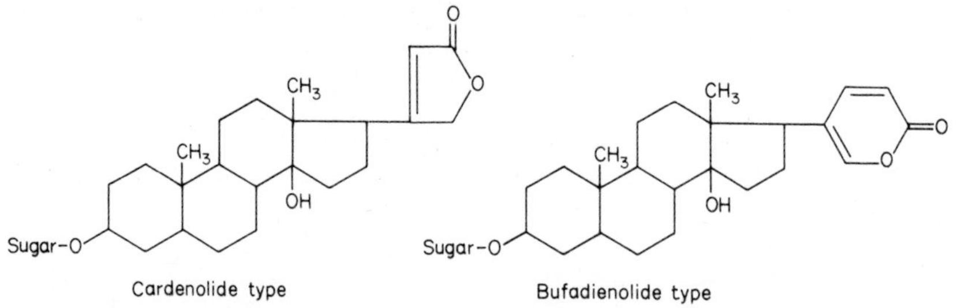

Cardenolide type   Bufadienolide type

Fig. 113. Cardiac glycosides.

Biosynthesis of the cardiac glycosides starts from pregnane derivatives. These compounds have been found in many plants containing cardiac glycosides but also occur in other plants which do not contain cardiac glycosides. Progesterone has been shown to be specifically incorporated into cardenolides. The two missing carbon atoms which form the lactone ring originate from acetate (or acetyl or malonyl CoA) (fig. 114). It is not clear at what stage of the synthesis the hydroxyl group with the β-configuration at carbon atom 14 is introduced into the molecule. This is associated with reversal of the configuration at this carbon atom and fusion of rings C and D.

Direct conversion of cholesterol to bufadienolides has been shown in toads. Like the formation of helleborin in *Helleborus atrorubens*, biosynthesis probably proceeds via a pregnane derivative which in this case reacts with a three-carbon acid and not with acetate.

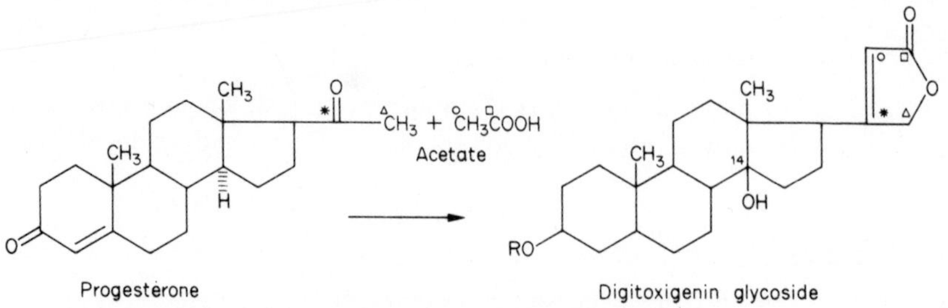

Progesterone   Digitoxigenin glycoside

Fig. 114. Formation of cardinolides from pregnane derivatives.

(e) Steroid sapogenins

Steroid sapogenins have till now been found only in higher plants. They are particularly abundant in Liliaceae, Solanaceae and Scrophulariaceae. These compounds are characterized by oxygen-containing rings, which are linked to the steroid ring system at carbon atom 17 and carbon atom 16. The fusion of rings A and B of the steroid ring system is either *cis* (e.g. in smilagenin) or *trans* (e.g. in gitogenin) (cf. fig. 102). The compounds lower the surface tension of aqueous solutions and permit the formation of stable foams. The term sapogenin is derived from this property.

The conversion of cholesterol to diosgenin has been shown by feeding experiments. Compounds such as kryptogenin and dihydrokryptogenin are assumed to be the intermediates from which the oxygen-containing rings of the sapogenins, linked by means of a spiroketal grouping, may be easily formed (fig. 115).

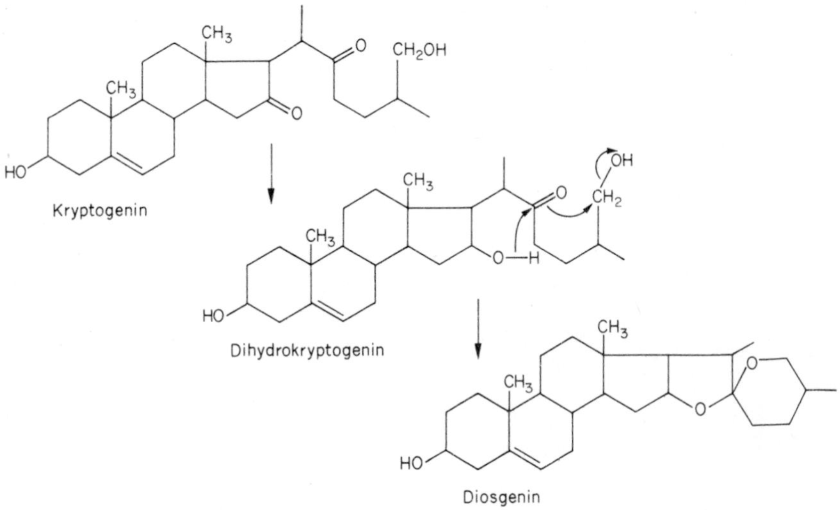

Fig. 115. Formation of steroid sapogenins.

(f) Steroid alkaloids

Steroid alkaloids directly derived from cholesterol may be differentiated from those which are derivatives of pregnane and possess a two-carbon side chain at carbon atom 17. Compounds of the former type have been found only in a few higher plants. They are always accompanied by steroid sapogenins, but are less widespread than these. Alkaloids of the tomatidine and veratramine type occur in the genera *Solanum*, *Veratrum* and *Holarrhena*, and compounds with the structure of solanidine only within the genus *Solanum*. In almost all cases alkaloids are glycosidically bound to sugars at position 3. It is thought that alkaloids of the tomatidine and solanidine type are formed via intermediates which are similar to tetrahydrokryptogenin (fig. 116). After replacement of the keto group at position 22 by an imino group (compound I), the cation II may first be formed by reaction of this grouping with the primary alcoholic group. It later stabilizes either with the formation of compounds of the tomatidine group, or is converted to compounds of the solanidine type via intermediate III.

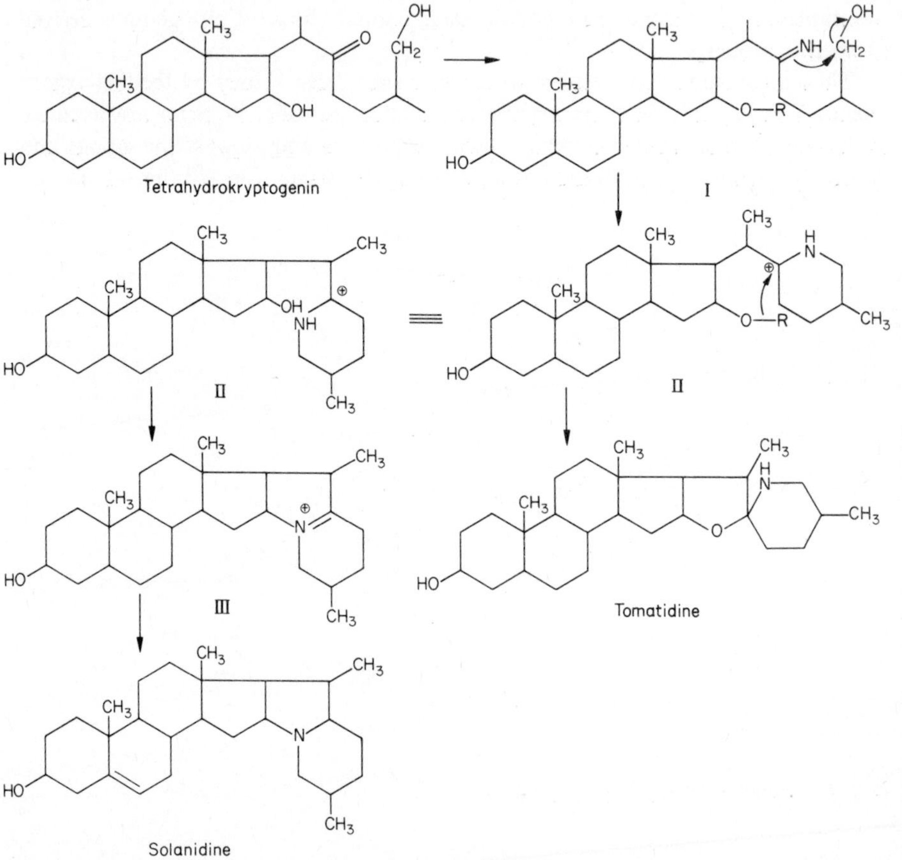

Fig. 116. Possible ways of biosynthesis of the alkaloids of the tomatidine and solanidine types.

Those steroid alkaloids in which the side chain at carbon atom 17 consists of only two carbon atoms may be formed from progesterone (progesterone has been detected in those plants which possess the corresponding alkaloids). Thus, in the case of alkaloids of the funtumafrine type, the oxo group at carbon atom 21, and in the case of funtumine, the carbonyl group at carbon atom 3 must be replaced by an amino group (fig. 117).

It has recently been discovered that the alkaloids holaphylline and holaphylla-mine, found in *Holarrhena floribunda*, originate from pregnenolone (see page 156). Holaphyllamine is probably a precursor of holaphylline (fig. 117).

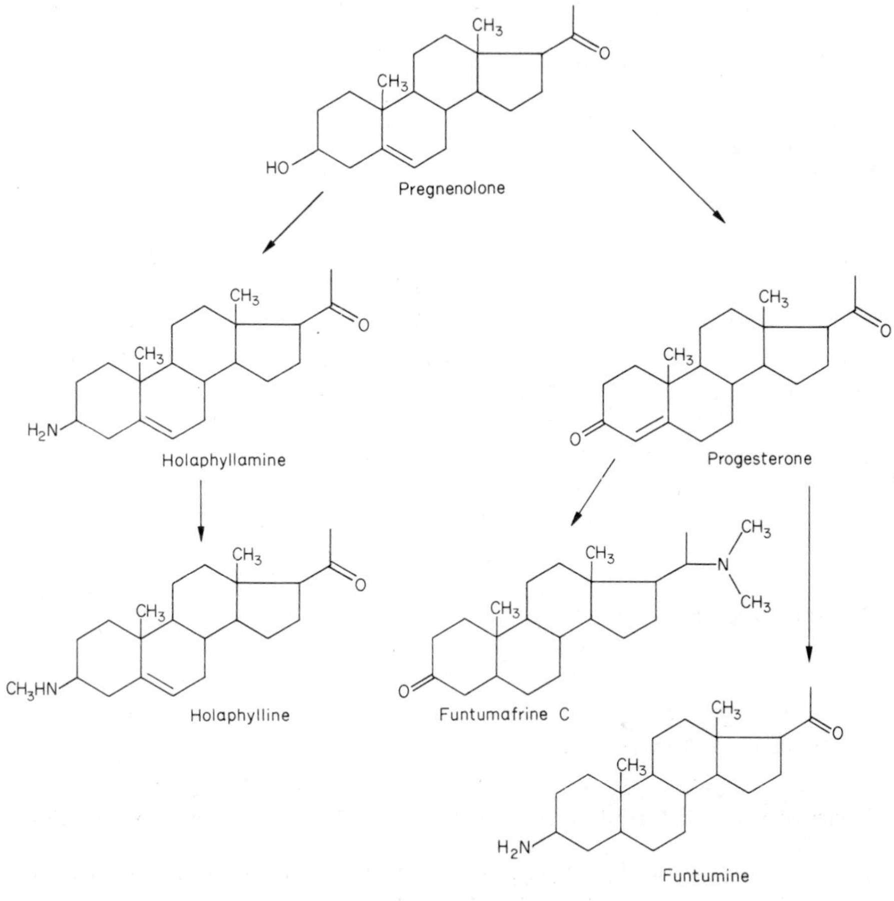

Fig. 117. Formation of steroid alkaloids from progesterone and pregnenolone.

The salamander alkaloids such as samandarine (fig. 118), which have a slightly altered steroid ring system, also originate from cholesterol. The nitrogen atom in ring A of these compounds is derived from glutamine.

Experiments on the biosynthesis of the *C-Nor-D*-homosteroid alkaloids of the veratramine and the veracevine types (cf. fig. 118) have not yet been carried out. It may, however, be concluded from their chemical structure that these com-pounds are derived from steroid precursors.

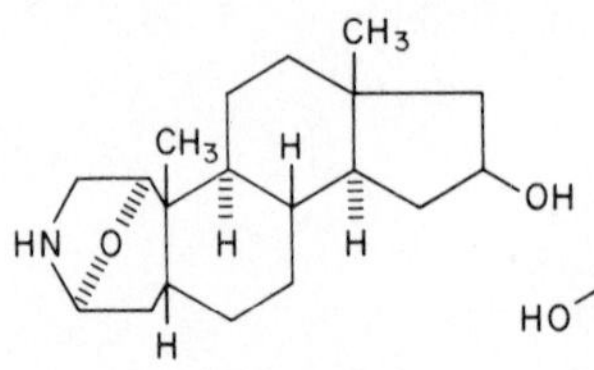

Samandarine

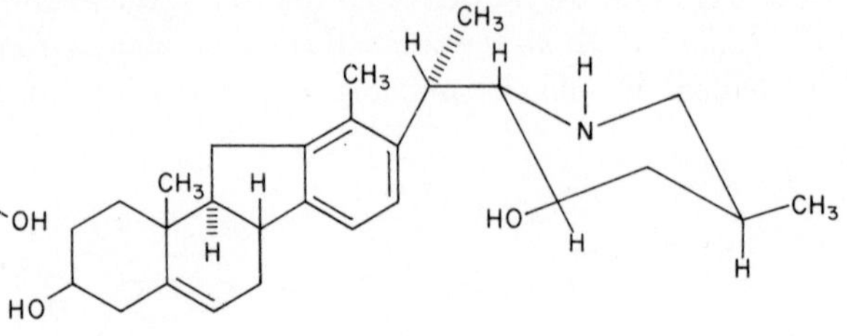

Veratramine

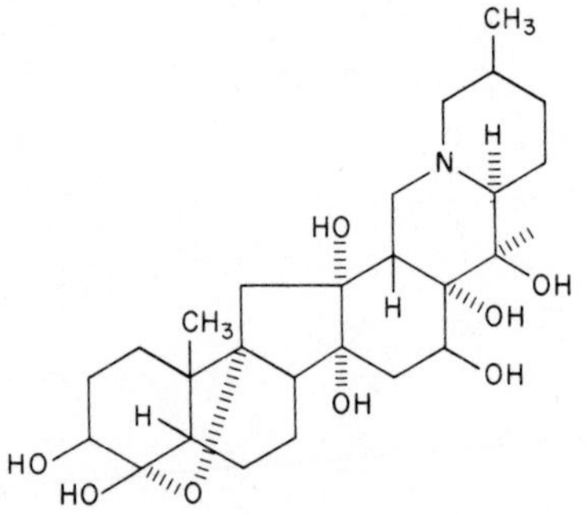

Veracevine

Fig. 118. Representatives of various types of steroid alkaloids.

References for further reading

Fieser, L. F. and Fieser, M. *Steroide.* (Verlag Chemie, Weinheim, 1961).

Baumgarten, G. *Die herzwirksamen Glykoside* (VEB Georg Thieme Verlag, Leipzig, 1963).

Boiteau, P., Pasich, B. and Ratsimamanga, A. R. *Le Triterpénoides en physiologie végétale et animale* (Gauthier Villars, Paris, 1964).

Habermehl, G. Chemie und Toxikologie der Salamanderalkaloide, *Naturwiss.* **53** (1966), pp. 123–28.

Staple, E. 'The Biosynthesis of Steroids', *Biogenesis of Natural Compounds*, ed P. Bernfeld (Pergamon Press, Oxford, 1967), pp. 207–45.

Reichstein, T. Cardenolid- und Pregnanglykoside, *Naturwiss.* **54** (1967), pp. 53–67.

Holker, J. S. E. Conformational Analysis of Terpenes and Steroids, *Terpenoids in Plants,* ed. J. B. Pridham (Academic Press, London, 1967), pp. 25–45.

Goad, L. J. 'Aspects of Phytosterol Biosynthesis', *Terpenoids in Plants*, ed. J. B. Pridham (Academic Press, London, 1967), pp. 159–90.

Frantz, I. D. and Schroepfer, G. J. Sterol Biosynthesis, *Ann. Rev. Biochem.* **36** (1967), pp. 691–726.

Leete, E. 'Alkaloid Biogenesis', *Biogenesis of Natural Compounds*, ed. P. Bernfeld (Pergamon Press, Oxford, 1967), pp. 953–1023.

Morand, P. and Lyall, J. The Steroidal Estrogens, *Chem. Reviews* **68** (1968), pp. 85–124.

Schütte, H. R. 'Steroidalkaloide', *Biosynthese der Alkaloide*, ed. K. Mothes and H. R. Schütte (VEB Deutscher Verlag der Wissenschaften, Berlin, 1969), pp. 616–44.

Danielsson, H. and Tschen, T. T. 'Steroid Metabolism', *Metabolic Pathways*, Vol. II, ed. D. M. Greenberg (Academic Press, New York, 1968), pp. 117–68.

Samuels, L. T. and Eik-Nes, K. B. 'Metabolism of Steroid-Hormones', *Metabolic Pathways*, Vol. II, ed. D. M. Greenberg (Academic Press, New York, 1968), pp. 169–220.

Heftmann, E. *Steroid Biochemistry* (Academic Press, New York, 1970).

Yamamoto, S. and Bloch, K. 'Enzymatic Studies on the Oxidative Cyclization of Squalene', *Natural Substances Formed Biologically from Mevalonic Acid*, ed. T. W. Goodwin (Academic Press, London, 1970), pp. 35–43.

Wiss, O. and Gloor, U. 'Sterol Biosynthesis', *Natural Substances Formed Biologically from Mevalonic Acid*, ed. T. W. Goodwin (Academic Press, London, 1970), pp. 45–77.

Tschesche, R. Zur Biosynthese der Cardenolid- und Bufadienolidglykoside, *Planta Medica (Stuttgart)*. Supplement 4/1971, pp. 34–39.

5.8 Formation of vitamin D_2 and D_3

The vitamins of the D-group are of special importance in the transport of calcium ions in the animal organism. Their absence results in considerable damage due to calcium deficiency and is characterized by rachitis. Too high a dosage of the vitamin is, however, also harmful.

The D-vitamins originate by opening of ring B of 7-dehydrocholesterol which is formed from cholesterol in the animal body, and by the same process from the plant sterol, ergosterol. The conversion of both these compounds takes place non-enzymatically by the action of UV-light (260–85 nm). Provitamin D_3 or D_2 formed via the unstable intermediate I, is converted to vitamins D_3 or D_2 by isomerization. A number of other processes take place after ultra-violet exposure which are of little importance in the synthesis of D-vitamins, and are therefore not discussed here.

Exposure of the surface of the body to the sun's rays produces a sufficient amount of vitamin D_3 in animals and human beings through this series of reactions. The rate of formation is considerably dependent on the extent of hairiness and the colour, and hence the 'transparency', of the skin. The dark skin of human races living in the proximity of the equator thus affords a protection against increased formation of the vitamin, while the white or light-skinned races living nearer the polar regions better utilize the smaller amount of UV-radiation falling there.

Infants usually only expose the face to UV-radiation, but although the surface area is so small it is sufficient, since the rate of synthesis of the necessary vitamin D is extraordinarily rapid. This is possible because the skin is still soft and has

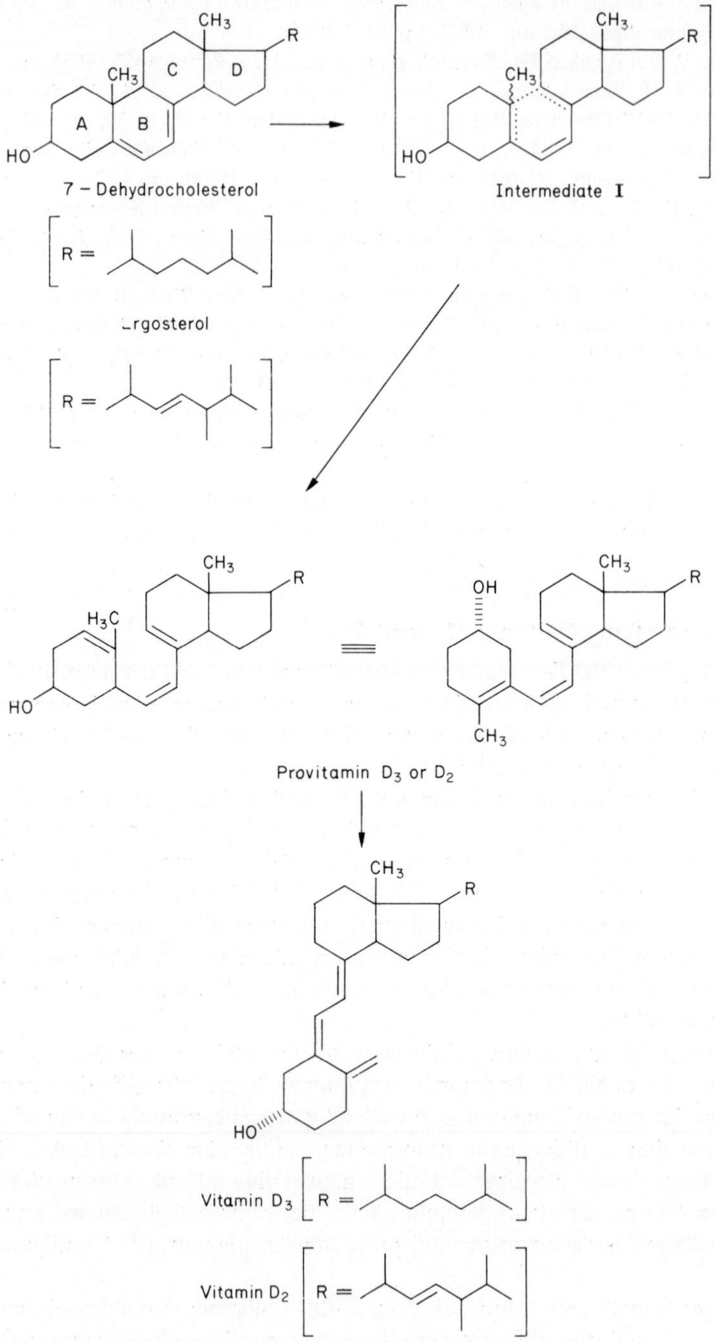

7 – Dehydrocholesterol

$$R = \text{/\\/\\/}$$

rgosterol

$$R = \text{/\\=/\\/}$$

Intermediate I

Provitamin D_3 or D_2

Vitamin D_3 $\left[R = \text{/\\/\\/} \right]$

Vitamin D_2 $\left[R = \text{/\\=/\\/} \right]$

Fig. 119. Formation of D-vitamins from sterols.

little pigment (the redness of the cheeks is due to the fact that there is no pigment to mask the blood vessels).

Vitamin D_3, also known as cholecalciferol, is present in large amounts in fish liver oils and can be taken into the human body in this form. Commercially synthesized vitamin D_2 (calciferol) is somewhat less active in human beings than vitamin D_3 but is used to a great extent in therapy.

References for further reading

Fieser, L. F. and Fieser, M. *Steroide* (Verlag Chemie, Weinheim, 1961).

Bersin, T. *Biochemie der Vitamine* (Akademische Verlagsgesellschaft, Frankfurt, 1966).

Farnsworth Loomis, W. Skin Pigment Regulation of Vitamin D Biosynthesis in Man, *Science (Washington)* **157** (1967), pp. 501–6.

Sanders, G. M., Pot, J. and Havinga, E. Some Recent Results in the Chemistry and Stereochemistry of Vitamin D and its Isomers. *Fortschr. Chem. org. Naturstaffe* **27** (1969), pp. 131–57.

Heftmann, E. *Steroid Biochemistry* (Academic Press, New York, 1970).

6. Secondary products originating from dehydroquinic acid, shikimic acid, chorismic acid and anthranilic acid

Phosphoenolpyruvic acid and erythrose-4-phosphate (fig. 120) serve as precursors for the biogenetically related carbocyclic compounds, quinic acid, shikimic acid, chorismic acid and anthranilic acid. Addition of the carbonyl group of the sugar to the double bond of the enol pyruvate leads to the formation of 3-deoxy-D-*arabino*-heptulosonic acid-7-phosphate via intermediate I. The phosphate group of the phosphoenolpyruvic acid is lost in the process.

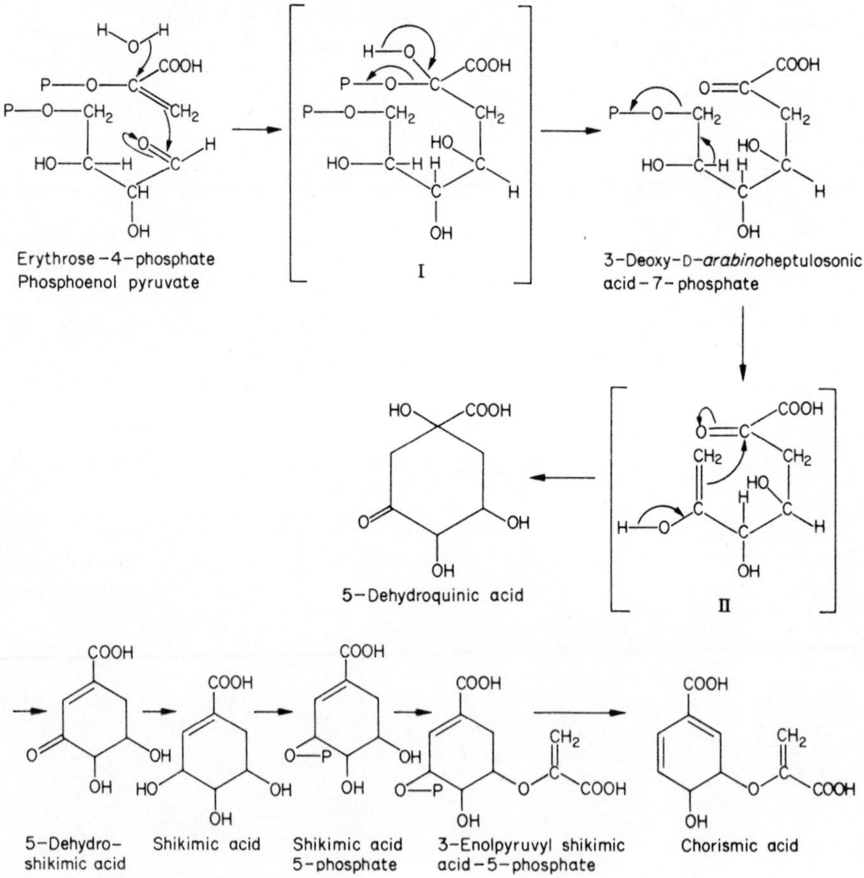

Erythrose-4-phosphate
Phosphoenol pyruvate

I

3-Deoxy-D-*arabino*heptulosonic acid-7-phosphate

5-Dehydroquinic acid

II

5-Dehydroshikimic acid   Shikimic acid   Shikimic acid 5-phosphate   3-Enolpyruvyl shikimic acid-5-phosphate   Chorismic acid

Fig. 120. Formation of chorismic acid.

3-Deoxy-D-*arabino*-heptulosonic acid-7-phosphate cyclizes to 5-dehydro-quinic acid by loss of the phosphate group. 5-Dehydro-shikimic acid is formed from this compound by loss of water, and is then reduced to shikimic acid.

Shikimic acid occurs in high concentrations in the young branches of conifers and can easily be isolated in large quantities in the spring from the apex of the shoots of *Ginkgo biloba*.

After phosphorylation at position 5, shikimic acid reacts with phosphoenol pyruvate with the formation of 3-enolpyruvylshikimic acid-5-phosphate. This compound yields chorismic acid by elimination of the phosphate group and a proton. Chorismic acid is the starting point for the synthesis of anthranilic acid, *p*-aminobenzoic acid and *p*-hydroxybenzoic acid (cf. D.6.3) as well as for the amino acids, phenylalanine and tyrosine (cf. D.20).

References for further reading

Meister, A. *Biochemistry of the Amino Acids*, Vol. II. (Academic Press, New York, 1965).

6.1 The formation of quinic acid and chlorogenic acid

Quinic acid is widespread in plants and is formed by reduction of 5-dehydro-quinic acid. The reaction is catalysed by quinic acid dehydrogenase and is reversible. Quinic acid thus, for example, is converted to shikimic acid.

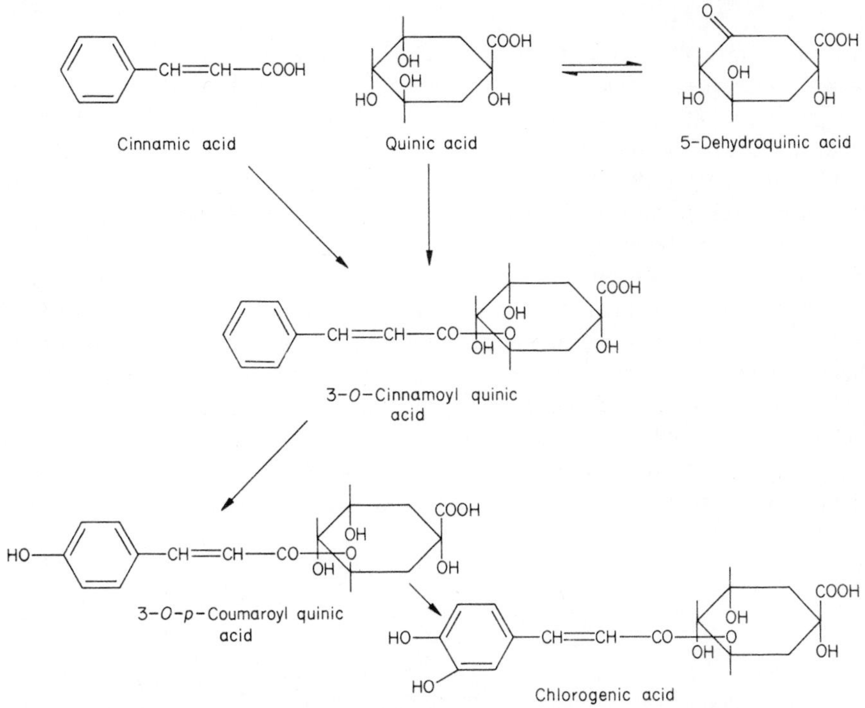

Fig. 121. Formation of quinic acid and chlorogenic acid.

Quinic acid can also react with cinnamic acid with the formation of 3-*O*-cinnamoyl quinic acid. Chlorogenic acid is formed from this compound by hydroxylation, via 3-*O*-*p*-coumaroyl quinic acid (fig. 121). Chlorogenic acid occurs in many higher plants.

References for further reading

Towers, G. H. N. 'Metabolism of Phenolics in Higher Plants and Microorganisms', *Biochemistry of Phenolic Compounds*, ed. J. B. Harborne (Academic Press, London, 1964), pp. 249–94.

Neish, A. C. 'Major Pathways of Biosynthesis of Phenols', *Biochemistry of Phenolic Compounds*, ed. J. B. Harborne (Academic Press, London, 1964), pp. 295–359.

6.2 Shikimic acid as precursor of secondary natural products

6.2.1 The biosynthesis of naphthoquinone and anthraquinone derivatives

The most important naturally occurring naphthoquinones belong to the group of K-vitamins. Vitamin K_2 (fig. 122), and vitamin K_1 (phylloquinone) which is derived from it by partial hydrogenation of the side chain, are especially widespread. The speed of biosynthesis of some blood-coagulating factors, e.g. of the Stuart factor X, depends on the quantity of K-vitamins contained in the liver mitochondria. Coagulation time is lengthened by an insufficiency of vitamin K.

Naphthoquinone derivatives which lack a side chain or possess only a short one are also found in higher plants and micro-organisms. Representatives of this group are 2-hydroxy-naphthoquinone (lawsone) synthesized by *Impatiens balsamina* and 5-hydroxy-naphthoquinone (juglone) occurring in *Juglans regia*.

While naphthoquinones of the javanicin type found in fungi are polyketides and are synthesized from acetate and malonate, the naphthoquinone nucleus in bacteria and higher plants originates in almost all cases from shikimic acid and a three-carbon unit, through a series of poorly understood reactions (an exception, however, is chimaphilin, see below). The three-carbon unit adds to carbon atom 6 of shikimic acid, with the formation of naphthohydroquinone and naphthoquinone via α-naphthol. Both these compounds are easily interconvertible and are

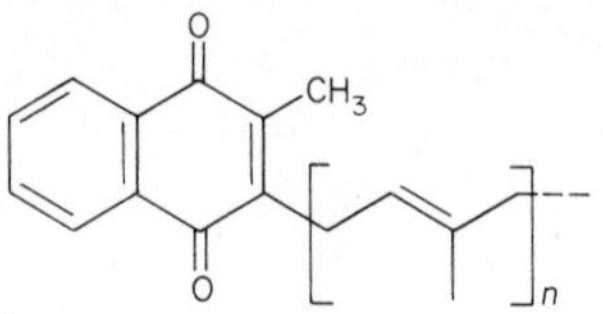

Fig. 122. Vitamin K_2 ($n = 6$–9)

the parent compounds in the biosynthesis of other naphthoquinone derivatives. For example, in the formation of lawsone and juglone they are hydroxylated, and in the biosynthesis of K-vitamins, first methylated and then probably substituted at the oxidation level of naphthohydroquinone by a prenyl pyrophosphate with the elimination of pyrophosphate as described in D.6.3.2 and D.20.2.

Anthraquinone derivatives are formed in a similar manner in the Rubiaceae. Shikimic acid is incorporated into ring A, and 1,4-naphthoquinone into rings A and B of alizarin, purpurin-3-carboxylic acid and purpurin, which occur in *Rubia tinctorum*, the madder. The rest of the molecule originates from mevalonic acid. Prenylnaphthoquinones of the desoxylapachol type appear to be intermediates (fig. 123).

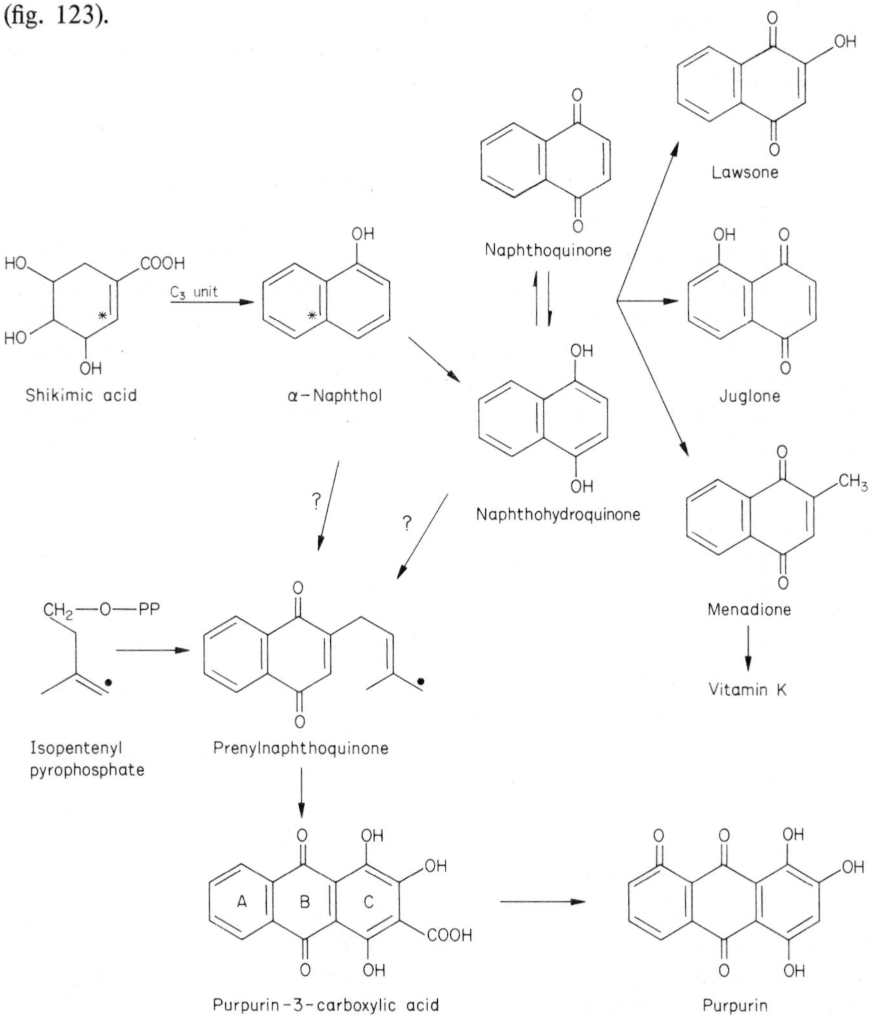

Fig. 123. The formation of naphthoquinone and anthraquinone derivatives from shikimic acid.

The formation of the anthraquinone ring system in *Rubia tinctorum* thus takes place in a manner different to that of anthraquinone derivatives substituted in ring A, which occur in micro-organisms and other higher plants, e.g. Rhamnaceae and Polygonaceae (cf. page 104).

References for further reading

Zenk, M. H. and Leistner, E. Biosynthesis of Quinones, *Lloydia* **31** (1968), pp. 275–92.
Luckner, M. Die Biosynthese von Hydrochinon- and *p*-Chinonderivaten, *Fortschritte der Botanik* **31** (1969), pp. 110–22.
Threlfall, D. R. and Whistance, G. R. 'Biosynthesis of Isoprenoid Quinones and Chromanols', *Aspects of Terpenoid Chemistry and Biochemistry*, ed. T. W. Goodwin (Academic Press, London, 1971), pp. 357–404.

6.2.2 The conversion of shikimic acid to protocatechuic and gallic acids

As well as the pathway described in D.20.6, in which gallic acid is formed from 3,4,5-trihydroxycinnamic acid, and the pathway from orsellinic acid discussed in D.2.3.4, direct conversion of shikimic acid to gallic acid has been detected in a few higher plants (e.g. *Rhus typhina*, *Acer saccharinum*, *Camellia sinensis* and *Vaccinium vitis-idaea*) as well as in micro-organisms (*Phycomyces blakesleeanus*). In *Phycomyces*, protocatechuic acid also is formed in this way. Like 5-dehydroshikimic acid this compound is probably an intermediate in the formation of gallic acid.

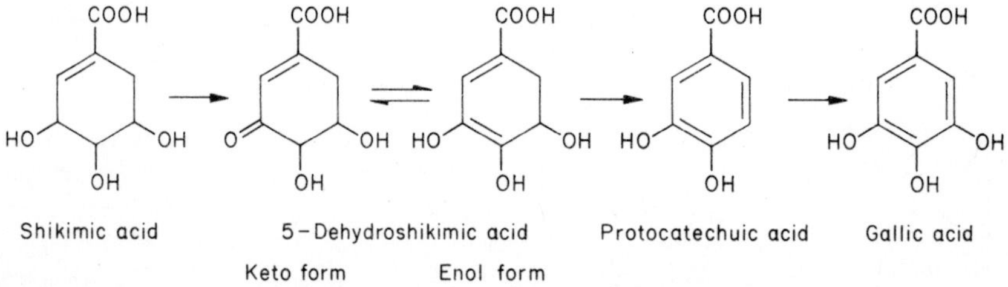

Fig. 124. Possible pathway for the formation of protocatechuic acid and gallic acid from shikimic acid.

References for further reading

Humphries, S. G. 'The Biosynthesis of Tannins', *Biogenesis of Natural Compounds*, ed. P. Bernfeld (Pergamon Press, Oxford, 1967), pp. 801–27.
Swain, T. 'The Tannins', *Plant Biochemistry*, ed. J. Bonner and J. E. Varner (Academic Press, New York, 1965), pp. 552–80.
Cornthwaite, D. and Haslam, E. Gallotannins. Part IX. The Biosynthesis of Gallic Acid in *Rhus typhina*, *J. Chem. Soc. (London)* (1965), pp. 3008–11.

6.3 Secondary natural products formed from chorismic acid

6.3.1 The conversion of chorismic acid to anthranilic acid, *p*-aminobenzoic acid, salicylic acid and *p*-hydroxybenzoic acid

Chorismic acid is not only an intermediate in the synthesis of prephenic acid and thus of phenylpropane units (cf. D.20), but may also be converted to *o*- and *p*-amino- or hydroxybenzoic acids after elimination of the pyruvate group (for the formation of anthranilic acid from trypthophan and *o*- and *p*-hydroxybenzoic acids from hydroxycinnamic acids cf. D.19.5.1 and D.20.6).

In the synthesis of aminobenzoic acids from shikimic acid, catalysed by the enzymes anthranilate synthetase and *p*-aminobenzoate synthetase, the amino group originates from the amide group of glutamine. Intermediates of the reaction have not been detected and probably do not occur in the free form. However, the pathway outlined in fig. 125 is considered likely.

The first step is probably the conversion of chorismic acid to the amide I, the amino group of which then is linked to the carbocyclic ring (compounds II and

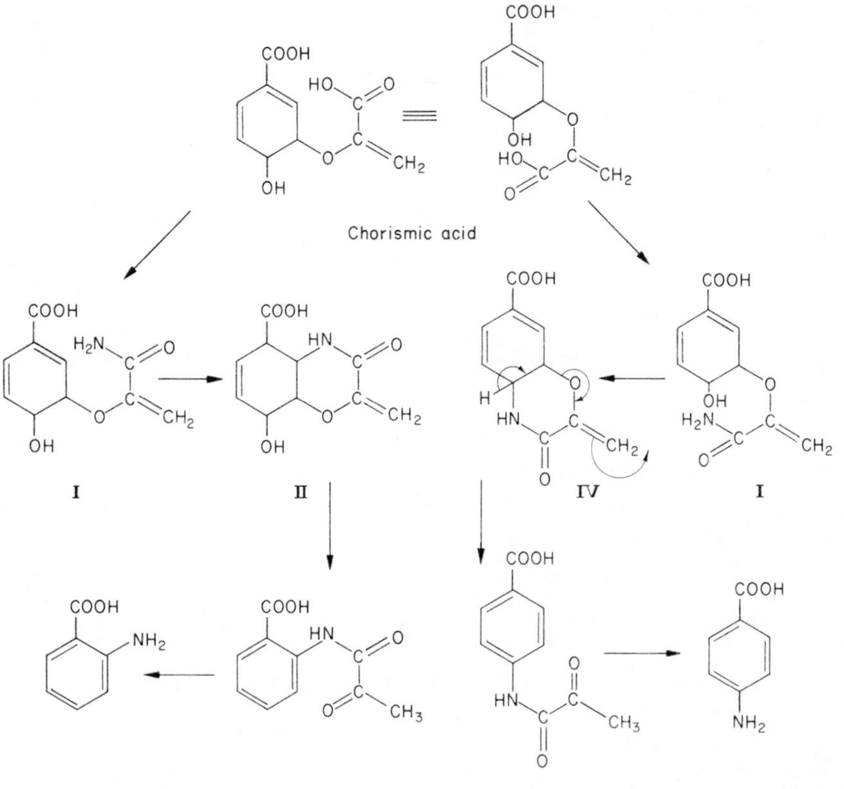

Anthranilic acid III V *p*-Aminobenzoic acid

Fig. 125. Possible mechanism of formation of anthranilic acid and *p*-aminobenzoic acid from chorismic acid.

IV). Opening of the heterocyclic ring forms substances III and V. They may be considered to be amides of pyruvic acid and may be decomposed hydrolytically to pyruvic acid and anthranilic acid or *p*-aminobenzoic acid.

p-Hydroxybenzoic acid originates directly from chorismic acid by elimination of pyruvic acid. Little is known about the formation of salicylic acid from chorismic acid, which occurs in tubercle bacilli under conditions of iron deficiency. However the biosynthesis may proceed via the cyclic product VI and salicylic acid-pyruvic acid ester (fig. 126).

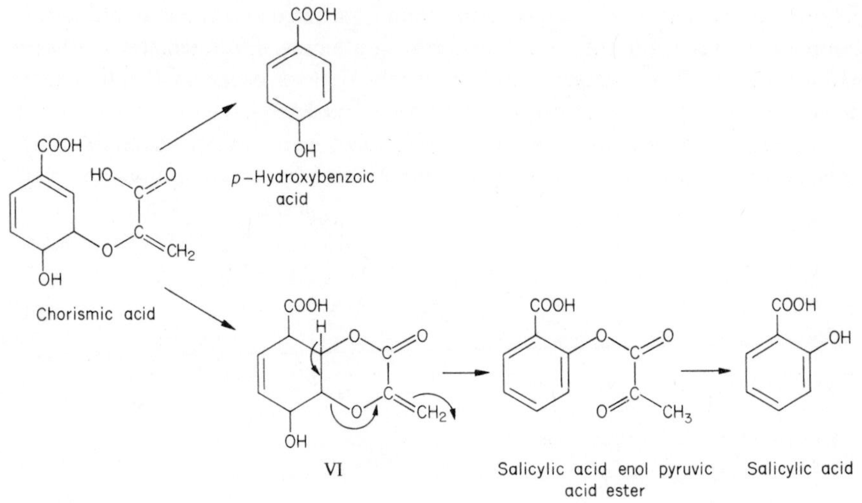

Fig. 126. Formation of *p*-hydroxybenzoic acid and salicylic acid from chorismic acid.

References for further reading

Ratledge, C. 'Biosynthesis of Salicylic Acid by *Mycobacterium smegmatis*', *Biosynthesis of Aromatic Compounds*, ed. G. Billek (Pergamon Press, Oxford, 1966), pp. 61–6.

Metzner, C. 'Evolution des Recherches en Phytochimie de 1962 à 1964', *Actualités de Phytochimie Fondamentale*, ed. C. Metzner (Masson et Cie, Paris, 1966), pp. 1–34.

Gröger, D. Anthranilic Acid as Precursor of Alkaloids, *Lloydia* **32** (1969), pp. 221–46.

6.3.2 Formation of ubiquinones from *p*-hydroxybenzoic acid

The ubiquinones (coenzymes Q) belong to the group of *iso*prenoid *p*-benzoquinones (cf. also D.20.2). The side chains of individual compounds consist of

5 to 10 isoprene residues (or twenty-five to fifty carbon atoms). Thus the molecule of ubiquinone 50 (coenzyme Q_{10}) has a chain of fifty carbon atoms or ten isoprene units.

The ubiquinones participate in cell respiration and are probably attached with its hydrophobic side chains to the lipid component of the membranes in such a way that the quinoid ring can undergo oxidoreductions. Most of them are found in the mitochondria. p-Hydroxybenzoic acid and a prenyl pyrophosphate corresponding to the length of the side chain serve as precursors in all organisms investigated so far. In bacteria, p-hydroxybenzoic acid is formed from chorismic acid (cf. D.6.3.1) and in higher plants from p-coumaric acid (cf. D.20.6).

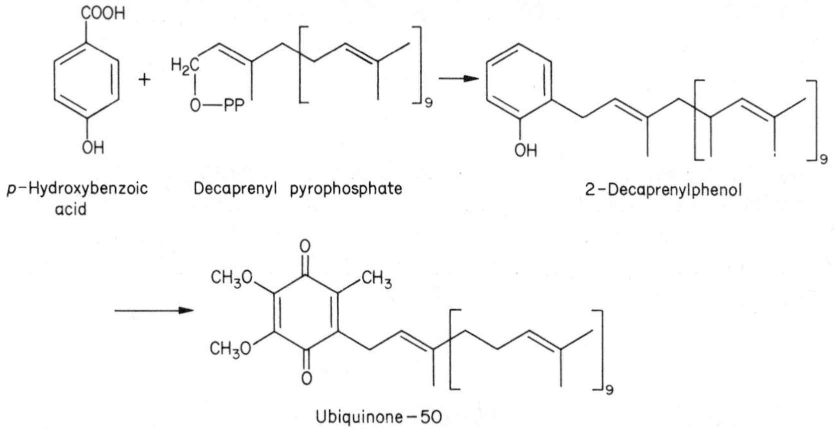

| p-Hydroxybenzoic acid | Decaprenyl pyrophosphate | | 2-Decaprenylphenol |

Ubiquinone-50

Fig. 127. The formation of ubiquinones.

The carboxyl group of p-hydroxybenzoic acid is lost after coupling the acid with prenyl pyrophosphate and 2-decaprenylphenol is observed as an intermediate during the formation of ubiquinone-50. No proof is as yet available for an oxidative decarboxylation as has been detected in the formation of hydroquinone (cf. D.20.6).

The prenylhydroquinone that is formed is hydroxylated, oxidized, methylated etc. by further reaction steps. the O-methyl and C-methyl groups originate from methionine (fig. 127) (cf. in contrast D.20.2). All three hydrogen atoms of the transferred methyl groups are retained in the product (cf. C.3.3).

References for further reading

Thomson, R. H. 'Quinones: Nature, Distribution and Biosynthesis', *Chemistry and Biochemistry of Plant Pigments*, ed. T. W. Goodwin (Academic Press, London, 1965), pp. 309–32.

Zenk, M. H. and Leistner, E. Biosynthesis of Quinones, *Lloydia* **31** (1968), pp. 275–92.

Rudney, H. 'The Biosynthesis of Terpenoid Quinones', *Natural Substances Formed Biologically from Mevalonic Acid*, ed. T. W. Goodwin (Academic Press, London, 1970), pp. 89–103.

Threlfall, D. R. and Whistance, G. R. 'Biosynthesis of Isoprenoid Quinones and Chromanols', *Aspects of Terpenoid Chemistry and Biochemistry*, ed. T. W. Goodwin (Academic Press, London, 1971), pp. 357–404.

6.4 Formation of secondary natural products from anthranilic acid

Anthranilic acid may be formed from chorismic acid by the pathway described in D.6.3.1 as well as from the amino acid tryptophan as described in D.19.5 and D.19.5.1. Since anthranilic acid is also a precursor of tryptophan, the enzymes for the following cycle are present in most living organisms (especially in microorganisms and animals):

$$\text{Anthranilic acid} \longrightarrow \text{Indole}$$
$$\uparrow \qquad\qquad \downarrow$$
$$\text{Tryptophan}$$

It is not clear whether this cycle proceeds to any considerable extent *in vivo*.

Anthranilic acid behaves frequently as an α-amino acid in the synthesis of secondary natural products. Thus, real alkaloids or, after condensation with other amino acids, cyclic peptides which are benzodiazepine rather than piperazine derivatives (cf. D.21.1) are formed from it.

6.4.1 Formation of 3-hydroxyanthranilic acid and its derivatives

3-Hydroxyanthranilic acid is synthesized by most organisms from 3-hydroxykynurenine by the action of the enzyme kynureninase (cf. D.19.5.1). Direct hydroxylation of anthranilic acid has also been shown in a few cases.

Anthranilic acid is the precursor of the alkaloid, damascenine, which occurs in *Nigella damascena*, and which is localized in certain epidermal cells of the seeds. The methyl groups of the alkaloid originate from methionine (fig. 128) (cf. C.3.3).

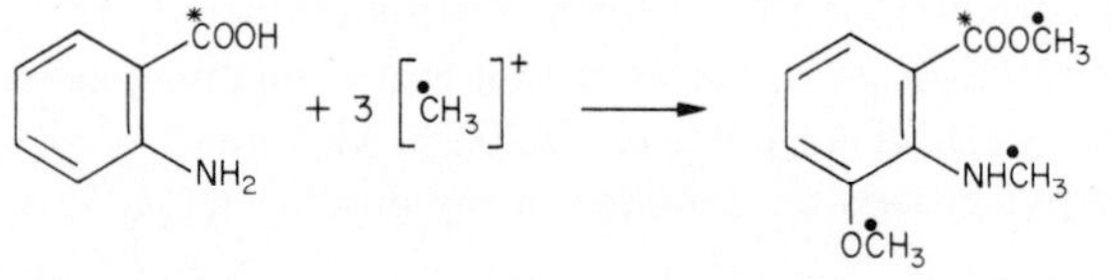

Anthranilic acid   Damascenine

Fig. 128. Formation of damascenine from anthranilic acid.

Recently an enzyme system has been found in the chloroplasts of *Tecoma stans*, which oxidizes anthranilic acid to catechol via 3-hydroxyanthranilic acid and *o*-aminophenol. Since the enzyme catalysing the hydroxylation requires molecular oxygen, tetrahydrofolic acid and NADPH, it is a mixed function oxygenase (cf. C.2.6.1).

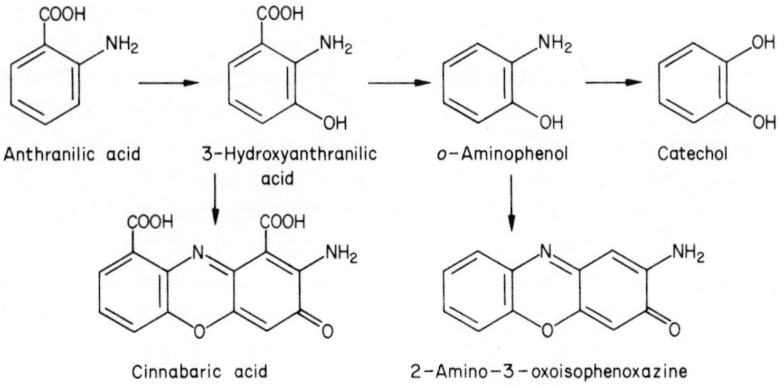

Anthranilic acid   3-Hydroxyanthranilic   o–Aminophenol   Catechol
                           acid

Cinnabaric acid   2–Amino–3–oxoisophenoxazine

Fig. 129. Secondary conversions of anthranilic acid in *Tecoma stans*.

3-Hydroxyanthranilic acid and *o*-aminophenol may be oxidatively dimerized to phenoxazines in *Tecoma stans*. Cinnabaric acid is formed from the former, and 2-amino-3-oxoisophenoxazine from the latter (fig. 129) (cf. also D.19.5.2). Three atoms of oxygen are required in the condensation of two molecules of 3-hydroxyanthranilic acid to one molecule of cinnabaric acid.

The actinomycins, a group of antibiotics consisting of a phenoxazine chromophore, actinocin, and a peptide side chain (e.g. actinomycin D, cf. fig. 130), are synthesized by *Streptomyces antibioticus* and a few other micro-organisms.

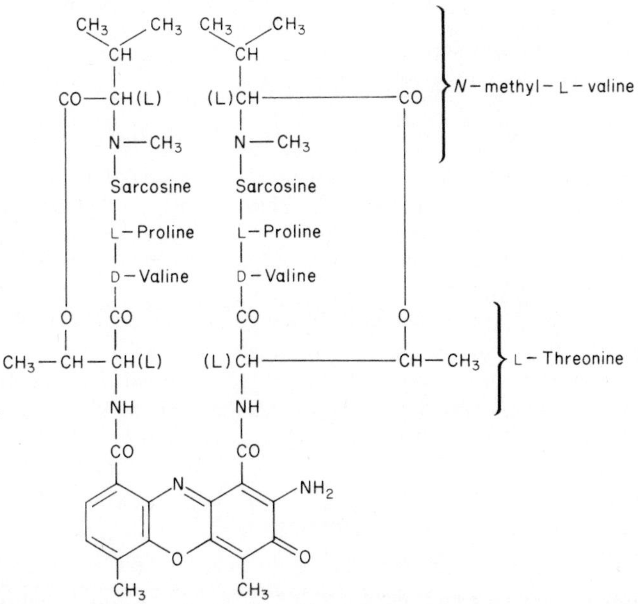

Fig. 130. Actinomycin D.

Actinocin is formed from tryptophan, 3-Hydroxyanthranilic acid is a direct precursor. This compound is first methylated at position 4 and is then converted to the peptide lactone corresponding to the particular actinomycin. The methyl group originates from methionine (cf. C.3.3). The pathway of synthesis of the peptide portion is unknown.

Two molecules of 3-hydroxy-4-methylanthranilic acid peptide lactone condense oxidatively to yield phenoxazinone (fig. 131). The enzyme, phenoxazinone synthetase, which catalyses this reaction has been isolated from *Streptomyces antibioticus*. In the synthesis of one molecule of actinomycin three atoms of oxygen are consumed and six hydrogen atoms are ultimately eliminated from the substrate. The reaction probably proceeds via quinonimines. Dimerization of these compounds is a spontaneous reaction (cf. D.7.7).

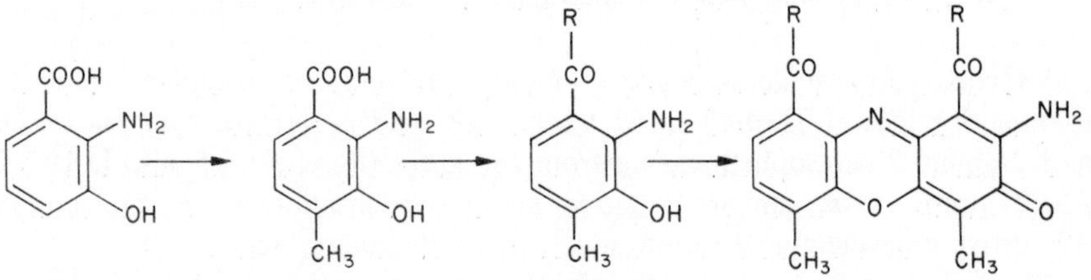

| 3 – Hydroxyanthranilic acid | 3 – Hydroxy – 4 – methylanthranilic acid | 3 – Hydroxy – 4 – methylanthranilic acid pentapeptide lactone | Actinomycin |

Fig. 131. Formation of actinomycins.

References for further reading

Rao, P. V. S. and Vaidyanathan, C. S. Enzymic Conversion of 3-Hydroxyanthranilic Acid to Cinnabaric Acid by the Leaves of *Tecoma stans*, *Arch. Biochem. Biophysics* **115** (1966), pp. 27–34.

Mothes, K. Die Anthranilsäure als Vorstufe von Alkaloiden, *Deutsche Apotheker-Ztg.* **106** (1966), pp. 1409–15.

Katz, E. 'Actinomycin', *Antibiotics*, Vol. II, ed. D. Gottlieb and P. D. Shaw (Springer Verlag, Berlin, 1967), pp. 276–341.

Schütte, H. R. 'Verschiedenes', *Biosynthese der Alkaloide*, ed. K. Mothes and H. R. Schütte (VEB Deutscher Verlag der Wissenschaften, Berlin, 1969), pp. 645–77.

Gröger, D. Anthranilic Acid as Precursor of Alkaloids, *Lloydia* **32** (1969), pp. 221–46.

6.4.2 Biosynthesis of quinoline alkaloids

Some quinoline alkaloids originate from anthranilic acid (cf. in contrast D.19.4.1, D.19.4.2 and D.19.5.1). Examples from higher plants include echinorine, occurring in the genus *Echinops* of the family Compositae and a large number of quinoline, furoquinoline and benzoquinoline alkaloids (acridones) synthesized by the Rutaceae.

In the biosynthesis of the alkaloids shown in fig. 132 *o*-aminobenzoylacetyl CoA is probably first formed by the condensation of the CoA-ester of anthranilic acid and acetyl CoA. This compound may either be reduced to *o*-aminobenzoyl acetaldehyde and converted to echinorine via 4-hydroxy-quinoline or decomposed by hydrolysis to *o*-aminobenzoylacetic acid and coenzyme A. *o*-Aminobenzoylacetic acid exists in a pH dependent equilibrium with 2,4-dihydroxy-quinoline.

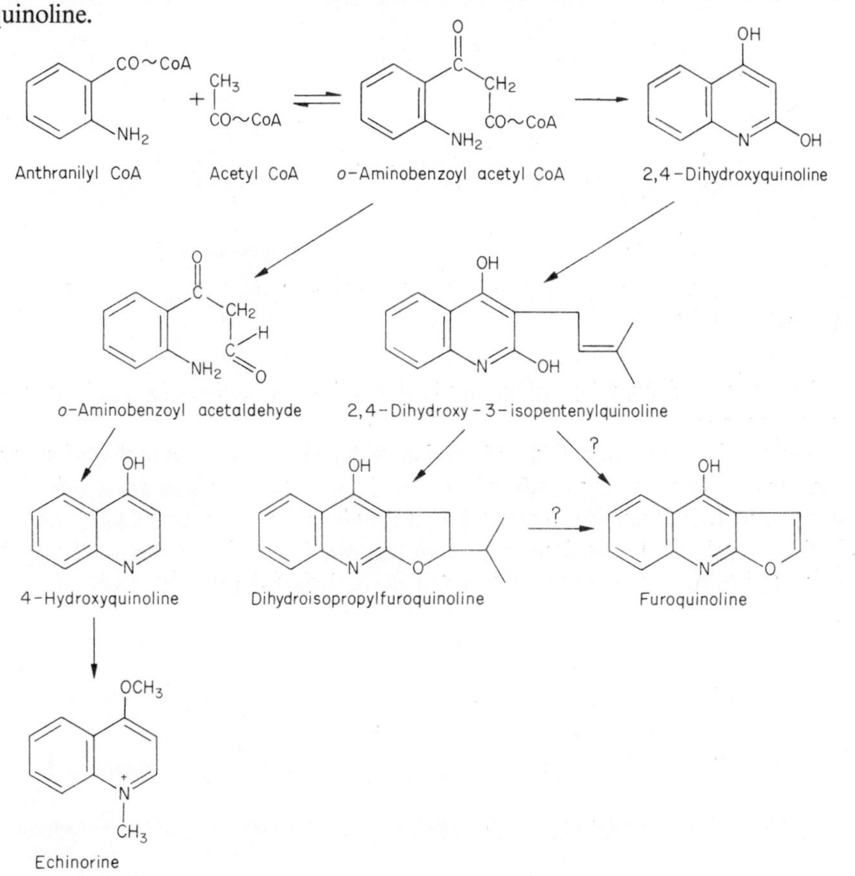

Fig. 132. Possible pathway of formation of quinoline and furoquinoline alkaloids from anthranilyl CoA, acetyl CoA and isopentenyl pyrophosphate.

This quinoline derivative is substituted by isopentenyl pyrophosphate in a later biosynthetic step. 2,4-Dihydroxy-3-isopentenyl quinoline so formed is a precursor for dihydroisopropylfuroquinoline as well as for the furoquinoline alkaloids (cf. the formation of furocoumarins from dihydroisopropylfuro-coumarins, D.20.4.2). It is not clear whether the isopropyl group lost during formation of the furoquinoline alkaloids is eliminated before or after ring closure.

The acridone alkaloids may be formed from anthranilyl CoA and three molecules of malonyl CoA. The polyketo compound that is formed first (cf. D.2.3) stabilizes to form hydroxylated acridones by ring closure and enolization (fig. 133).

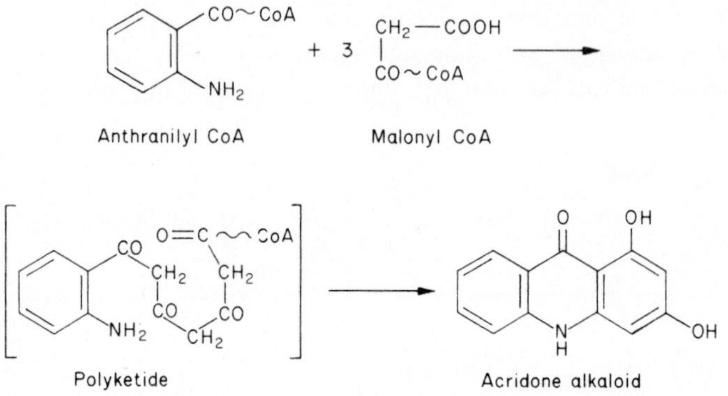

Anthranilyl CoA   Malonyl CoA

Polyketide   Acridone alkaloid

Fig. 133. Possible pathway of formation of acridone alkaloids from anthranilyl CoA and malonyl CoA.

Two types of quinoline compounds derived from anthranilic acid occur in micro-organisms:

Various 2-*n*-alkyl-quinolones-(4) and their *N*-oxides are found in *Pseudomonas aeruginosa*. These compounds which are known as pseudans are synthesized from activated anthranilic acid and a molecule of an activated β-keto acid. A 2-alkyl-3-carboxy-4-quinolone is probably formed first. This compound may be converted to the corresponding pseudan by decarboxylation (fig. 134).

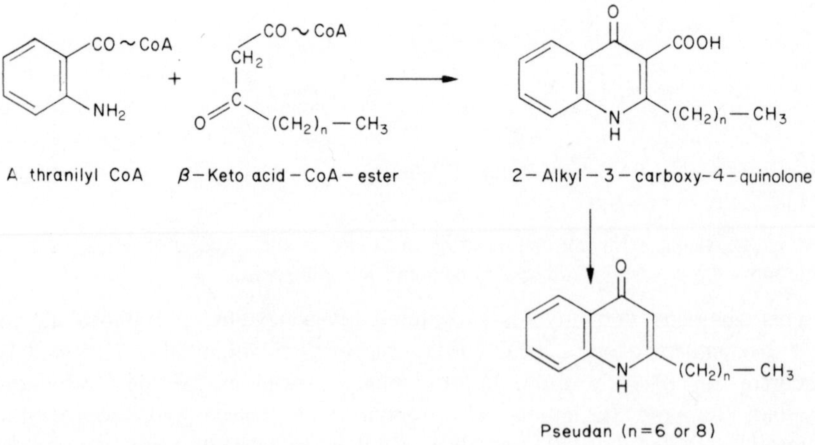

A thranilyl CoA β-Keto acid – CoA – ester   2 – Alkyl – 3 – carboxy-4 – quinolone

Pseudan (n = 6 or 8)

Fig. 134. Formation of pseudanes from activated anthranilic acid and β-keto acid.

The 2-phenyl-4-quinolone derivatives occurring in the Rutaceae seem to be formed from anthranilyl CoA and a benzoylacetyl CoA ester (cf. D.20.6) in a similar way.

The quinoline derivatives viridicatine and viridicatol synthesized by moulds of the genus *Penicillium* originate from anthranilic acid and phenylalanine (fig. 135). The benzodiazepine derivatives, cyclopeptine, dehydrocyclopeptine, cyclopenine and cyclopenol, are formed first.

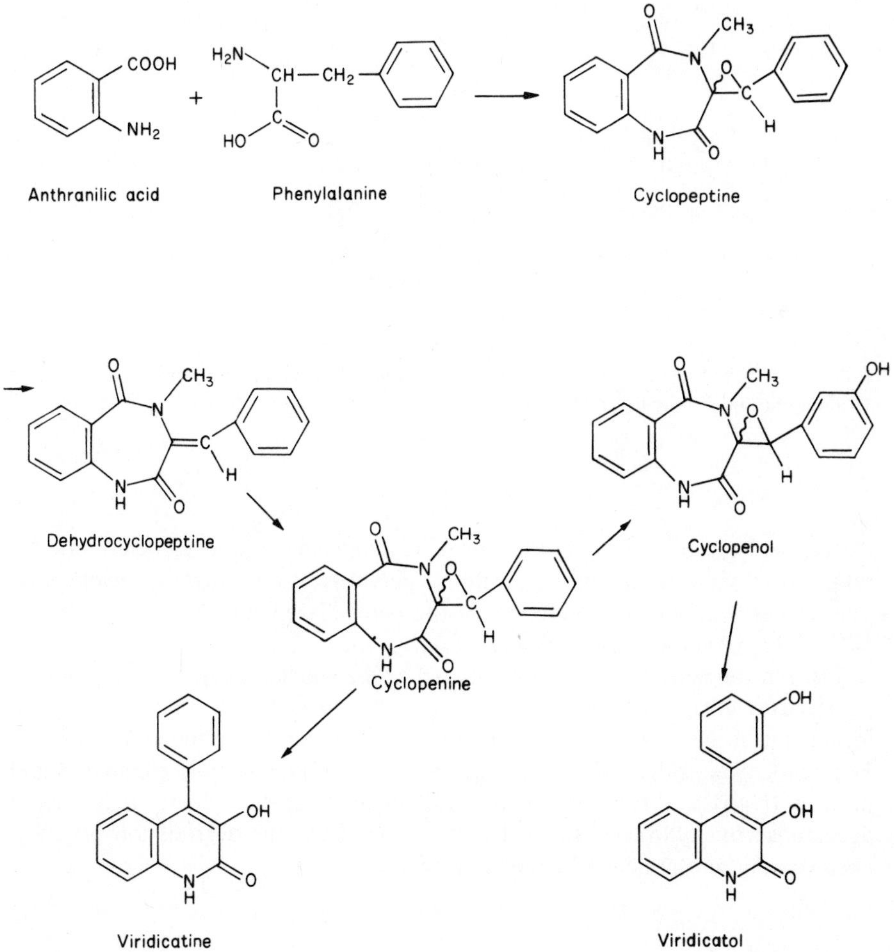

Anthranilic acid   Phenylalanine   Cyclopeptine

Dehydrocyclopeptine   Cyclopenine   Cyclopenol

Viridicatine   Viridicatol

Fig. 135. Formation of viridicatine and viridicatol.

These substances may be regarded as cyclic peptide derivatives. The phenyl ring of cyclopenol and viridicatol carries a hydroxyl group in the *m*-position. Natural products with this type of hydroxylation pattern are quite rare. The hydroxyl group is probably introduced into the molecule by a mixed function

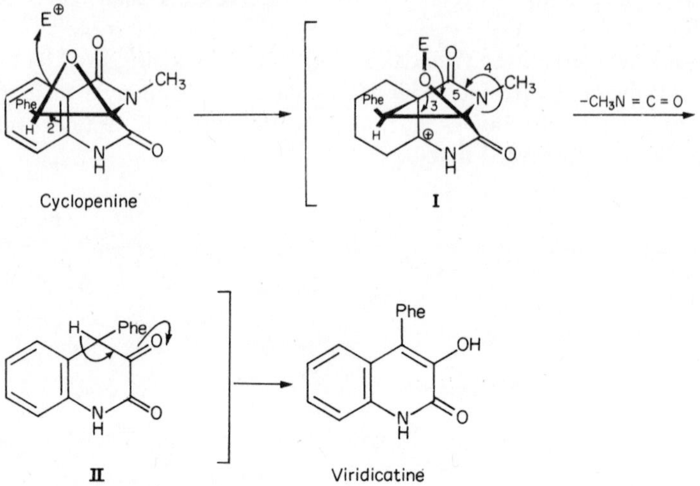

Cyclopenine

I

$-CH_3N = C = O$

II

Viridicatine

Fig. 136. Possible pathway for the conversion of cyclopenine and cyclopenol to viridicatine and viridicatol respectively.

oxygenase (cyclopenine-*m*-hydroxylase). Cyclopenine and cyclopenol are re-arranged to viridicatine and viridicatol respectively by the enzyme cyclopenase. The carbonyl group at position 5 of the benzodiazepine ring system and the $\geq N-CH_3$ group are lost in this reaction.

The rearrangement is probably initiated at the epoxide oxygen by addition of an electron-attracting grouping. In a concerted mechanism (the intermediates formulated in fig. 136 are not really found) the tricyclic compound I is formed. During decomposition of this compound methyl isocyanate is eliminated and product II arises, which enolizes to viridicatine. Methylisocyanate may further decompose by hydrolysis to methylamine and CO_2, the degradation products observed during the action of the enzyme cyclopenase.

References for further reading

Leete, E. 'Alkaloid Biogenesis', *Biogenesis of Natural Compounds*, ed. P. Bernfeld Pergamon Press, Oxford, 1967, pp. 953–1023.

Gröger, D. 'Acridinalkaloide', *Biosynthese der Alkaloide*, ed. K. Mothes and H. R. Schütte (VEB Deutscher Verlag der Wissenschaften, Berlin, 1969), pp. 562–7.

Luckner, M 'Chinoline', *Biosynthese der Alkaloide*, ed. K. Mothes and H. R. Schütte (VEB Deutscher Verlag der Wissenschaften, Berlin, 1969), pp. 510–50.
Gröger, D. Anthranilic Acid as Precursor of Alkaloids, *Lloydia* **32** (1969), pp. 221–46.

6.4.3 Formation of quinazoline alkaloids

The quinazoline alkaloids are a small group of secondary natural products. Representatives are found in micro-organisms, e.g. in *Pseudomonas aeruginosa* (cf. D.19.5.4), as well as in higher plants, e.g. Acanthaceae, Palmaceae, Papilionaceae, Rutaceae, Saxifragaceae and Zygophyllaceae, and in animals e.g. in the defence secretions of the millipede, *Glomeris marginata* (cf. A.1).

In higher plants and animals, the aromatic ring, the carbon atom joined to it and probably one of the two nitrogen atoms originate from anthranilic acid. A one-carbon or a two-carbon unit and frequently a third molecule also take part in the formation of quinazoline alkaloids. There is as yet no information as to the nature of the one-carbon or two-carbon unit.

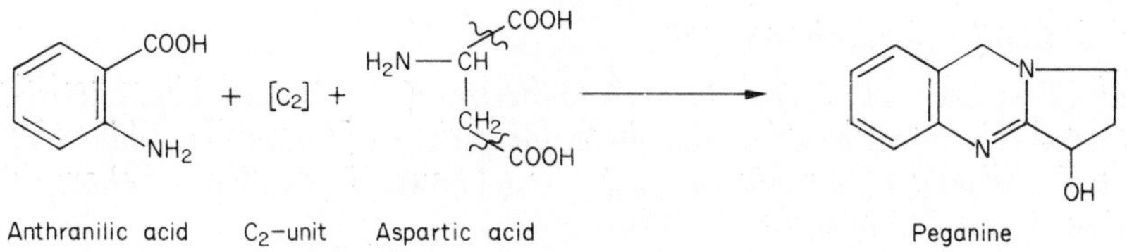

Anthranilic acid C_2–unit   Aspartic acid   Peganine

Fig. 137. Biosynthesis of peganine.

In the case of peganine, which occurs in *Adhadota vasica*, the third precursor is aspartic acid. Both the carboxyl groups of the amino acid are lost during incorporation (fig. 137). Deoxypeganine is not formed during the synthesis.

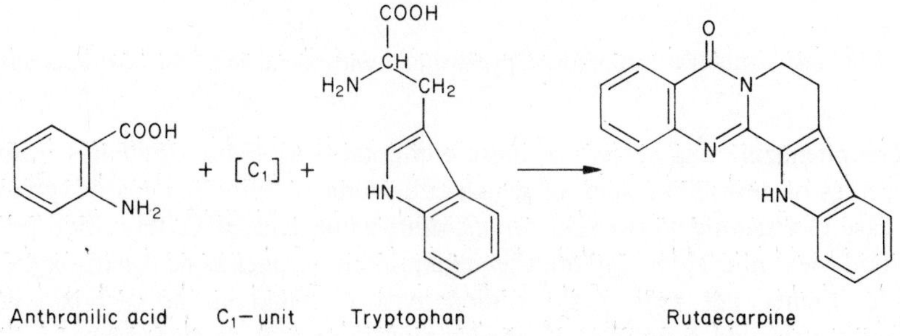

Anthranilic acid C_1–unit   Tryptophan   Rutaecarpine

Fig. 138. Formation of rutaecarpine.

During the biosynthesis of rutaecarpine in *Evodia rutaecarpa* tryptophan is incorporated into the β-carboline portion (fig. 138). Tryptamine, however, is probably the actual precursor (cf. D.19.3).

References for further reading

Mothes, K. Die Anthranilsäure als Vorstufe von Alkaloiden, *Deutsche Apotheker-Zeitung* **106** (1966), pp. 1409–45.

Schildknecht, H. and Wenneis, W. F. Über Arthropoden-Abwehrstoffe XXV; Anthranilsäure als Precursor der Arthropoden-Alkaloide Glomerin und Homoglomerin, *Tetrahedron. Letters* (1967), pp. 1815–18.

Gröger, D. 'Chinazolin-Alkaloide', *Biosynthese der Alkaloide*, ed. K. Mothes and H. R. Schütte (VEB Deutscher Verlag der Wissenschaften, Berlin, 1969), pp. 551–61.

Gröger, D. Anthranilic Acid as Precursor of Alkaloids, *Lloydia* **32** (1969), pp. 221–46.

6.4.4 Synthesis of phenazines

Various species of *Pseudomonas* synthesize a number of characteristically coloured substances based on the phenazine skeleton. Pyocyanine, a blue pigment from *Pseudomonas aeruginosa*, is the best known. Derivatives of phenazine-1-carboxylic acid also occur (fig. 139).

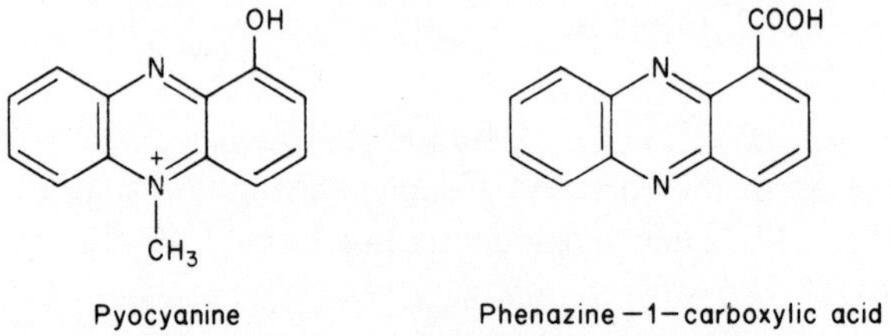

Pyocyanine   Phenazine —1— carboxylic acid

Fig. 139. The structural formulae of pyocyanine and phenazine-1-carboxylic acid.

The phenazine ring system of these compounds probably originates from two molecules of anthranilic acid, of which either one or both are decarboxylated. In the case of *Pseudomonas chlororaphis*, anthranilic acid-[14]COOH is incorporated in good yield into chlororaphin, the semiquinone of phenazine-1-carboxylic acid amide. Eighty per cent of the radioactivity is found in the carboxyl group. Shikimic acid and a number of smaller molecules such as glycerine and alanine are incorporated in higher amounts too. It is supposed that these substances are incorporated via anthranilic acid, but definitive proof is not yet available.

References for further reading

MacDonald, J. C. 'Pyocyanine', *Antibiotics*, Vol. II, ed. D. Gottlieb and P. D. Shaw (Springer Verlag, Berlin, 1967), pp. 52–65.

Schütte, H. R. 'Verschiedenes', *Biosynthese der Alkaloide*, ed. K. Mothes and H. R. Schütte (VEB Deutscher Verlag der Wissenschaften, Berlin, 1969), pp. 645–77.

7. General pathways for the formation of secondary natural products from L-amino acids

Amino acids are compounds which have at least one carboxyl group and one amino group in the molecule. In most of the naturally occurring amino acids the amino group is situated in the α-position relative to the carboxyl group (α-amino acids, general formula R—CHNH$_2$—COOH). If R does not represent H, then compounds of this structural type have a centre of asymmetry at the α-carbon atom. Amino acids with the L-configuration are especially widespread.

Twenty-three amino acids occur as constituents of proteins in all living organisms. They may thus be considered compounds of primary metabolism. There are, however, in addition, more than 200 non-protein amino acids known which are only synthesized by a few organisms and the majority of these may be regarded as secondary natural products. The naturally occurring D-amino acids are also included here.

A large number of secondary natural products are derived from the protein amino acids. One may differentiate between those groups of substances, the individual representatives of which are similar to each other and are synthesized from a large number of primary amino acids by similar reactions (D.7.1–D.7.7) and those substances which originate by special metabolic pathways which are characteristic for one or a few of the protein amino acids (D.8–D.20).

7.1 Formation of D-amino acids

D-Amino acids have been found in recent years in many organisms in relatively large quantities. They constitute a major portion of the free amino acids of the body fluid of some insects during certain developmental stages and are also found in micro-organisms as well as in human urine (cf. D.21.3, D.21.4 and D.21.5). D-Alanine and D-serine, as well as D-cysteine, D-glutamic acid and D-ornithine are especially widespread. D-Cysteine is a constituent of luciferin (cf. D.9.3).

D-Amino acids usually originate from L-amino acids by the action of amino acid racemases. Some of these enzymes contain pyridoxal phosphate as the coenzyme (cf. C.4). It is supposed that the Schiff base (fig. 18), formed by the reaction of this enzyme with the amino acid, is changed by the elimination of a proton into an imino compound I. The Schiff base may be formed again by a reverse reaction involving the addition of a proton, when, due to the specificity of the enzyme, the L- as well as the D-form of the amino acids is formed. Other racemases are flavin enzymes (cf. C.2.1.2). Racemization appears to proceed via the imino acids which are formed by reversible dehydrogenation.

Transaminases with high optical specificity participate in the synthesis of several D-amino acids (cf. C.4). Thus an enzyme from *Bacillus subtilis* catalyzes the following reaction:

D-alanine + α-ketoglutaric acid $\rightleftharpoons$ pyruvic acid + D-glutamic acid

L-Amino acids do not react with these transaminases.

The conversion of the L-form to the D-form is also possible in the case of linked amino acid groups (cf. formation of penicillin N and cephalosporin C, D.21.3, as well as the biosynthesis of peptide antibiotics, D.21.5). The D-alanine unit of octopine is formed by the stereospecific hydrogenation of a Schiff base composed of arginine and pyruvic acid (cf. D.17.1). The degradation of D-amino acids is catalysed by specific oxidases. Ammonia and an α-keto acid are formed in this reaction (cf. C.2.1.2).

References for further reading

Meister, A. *Biochemistry of the Amino Acids*, Vol. I (Academic Press, New York, 1965).
Corrigan, J. J. D-Amino Acids in Animals, *Science (Washington)* **164** (1969), pp. 142–9.

7.2 Biosynthesis of acylated amino acids

The acylated amino acids, in all cases investigated so far, originate from an unactivated amino acid and the CoA-ester of the acylating acid (cf. C.1.2) (fig. 5). In the case of formylation, formic acid is transferred to the acceptor from the tetrahydrofolic acid derivative (cf. C.3.2).

The following acids are of special significance as constituents of acylated amino acids:

Formic acid (N-formylmethionine is the starter molecule in the synthesis of polypeptide chains in bacteria).

Acetic acid (N-acetylphenylalanine and other acylated amino acids are equally effective as starter molecules in certain protein synthetic systems. They play a role in the metabolism of D-amino acids, especially in fungi, and also in the synthesis of cysteine and methionine (cf. D.9 and D.10). δ-N-Acetylornithine constitutes 10% of the dry weight of the roots of *Corydalis ochotensis*).

Malonic acid (N-malonyl derivatives of D-amino acids are synthesized by a number of higher plants).

Oxalic acid and succinic acid (the corresponding homoserine derivatives participate in the biosynthesis of methionine, cf. D.10).

References for further reading

Thompson, J. F., Morris, C, and Smith, I. K. New Naturally Occurring Amino Acids, *Ann. Rev. Biochem.* **38** (1969), pp. 137–58.

7.3 Synthesis of amines

Amines are synthesized by animals as well as plants and micro-organisms. Low molecular weight volatile compounds of this class of substances are present to a great extent in the aroma of many flowers, for example of the genus *Crataegus*, and in the smell of the foetid morel (*Phallus impudicus*) and other fungi.

Amines originate in nature in several independent ways:

(a) The most important is the enzymic decarboxylation of amino acids.

Amino acid decarboxylases (cf. C.4) occur in micro-organisms as well as in plants and animals. A number of important amines which are formed by decarboxylation are given in table 12.

(b) The second reaction is catalysed by aldehyde transaminases (cf. C.4) which transfer ammonia between amino acids and aldehydes (fig. 140).

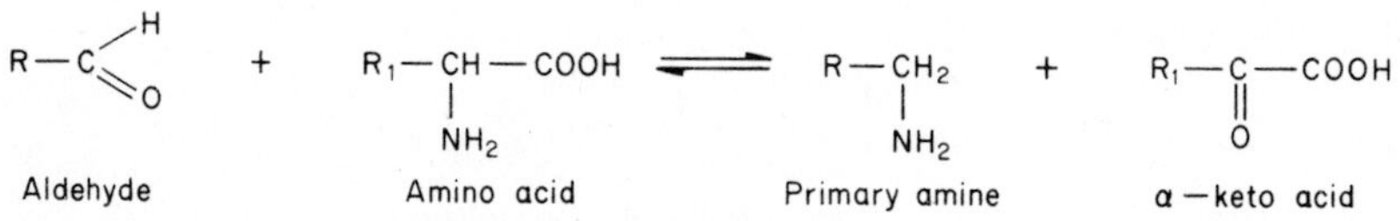

Fig. 140. Formation of amines by transamination.

Table 12. Some amines formed by decarboxylation of amino acids

Formula	Amino acid $\longrightarrow$ Amine		Occurrence of corresponding amino acid decarboxylases
	R = COOH	R = H	
CH_2-CH-R $\mid$ $\mid$ OH NH_2	Serine	Ethanolamine (cholamine)	Animals, higher plants
CH_3-CH_2-CH-R $\mid$ NH_2	α-Aminobutyric acid	n-Propylamine	Micro-organisms
$\begin{matrix} H_3C \\ H_3C \end{matrix}\!\!>\!CH-CH-R$ NH_2	Valine	i-Butylamine	Micro-organisms, higher plants
$\begin{matrix} H_3C \\ H_3C \end{matrix}\!\!>\!CH-CH_2-CH-R$ NH_2	Leucine	i-Amylamine	Micro-organisms, higher plants
$CH_3-(CH_2)_4-CH-R$ $\mid$ NH_2	α-Aminoheptanoic acid	n-Hexylamine	Micro-organisms
CH_2-CH_2-CH-R $\mid$ $\mid$ $S-CH_3$ NH_2	Methionine	3-Methylmer-captopropyl-amine	Higher plants
$HOOC-CH_2-CH-R$ $\mid$ NH_2	Aspartic acid	β-Alanine	Micro-organisms, higher plants, animals

Formula	Amino acid ⟶ Amine		Occurrence of corresponding amino acid decarboxylases
	R = COOH	R = H	
$HOOC-CH_2-CH_2-CH-R$ $\quad\quad\quad\quad\quad\quad NH_2$	Glutamic acid	γ-Aminobutyric acid	Micro-organisms, animals, higher plants
$CH_2-(CH_2)_3-CH-R$ $NH_2 \quad\quad\quad\quad NH_2$	Lysine	Cadaverine	Micro-organisms, higher plants
$CH_2-(CH_2)_2-CH-R$ $NH_2 \quad\quad\quad\quad NH_2$	Ornithine	Putrescine	Micro-organisms, higher plants
$CH_2-(CH_2)_2-CH-R$ $NH \quad\quad\quad\quad NH_2$ $\quad C=NH$ H_2N	Arginine	Agmatine	Micro-organisms, higher plants
⬡$-CH_2-CH-R$ $\quad\quad\quad\quad NH_2$	Phenylalanine	Phenylethyl-amine	Micro-organisms, higher plants
$HO-$⬡$-CH_2-CH-R$ $\quad\quad\quad\quad\quad\quad NH_2$	Tyrosine	Tyramine	Micro-organisms, animals, higher plants
⬡$-CH_2-CH-R$ $\quad\quad\quad\quad NH_2$	Tryptophan	Tryptamine	Micro-organisms, animals, higher plants
$HO-$⬡$-CH_2-CH-R$ $\quad\quad\quad\quad\quad NH_2$	5-Hydroxy-tryptophan	Serotonin	Micro-organisms, animals, higher plants
⬡$-CH_2-CH-R$ $\quad\quad\quad NH_2$	Histidine	Histamine	Micro-organisms, animals

The amines illustrated in table 13, as well as others, are synthesized in this manner by micro-organisms and higher plants.

(c) Methylamine is formed from N-methyl amino acids by the action of amino acid oxidases (cf. C.2.1.2). Thus a glycine-oxidase has been found in animal tissues which converts glycine to glyoxylic acid and ammonia, and sarcosine to glyoxylic acid and methylamine. In higher plants N-methylamino ethanol (cf. D.7.4) is converted to glycolic aldehyde and methylamine in a similar reaction (fig. 141). For the degradation of adrenaline to 3-methoxy-4-hydroxymandelic acid aldehyde and methyl-amine cf. D.20.6.

Table 13. Some primary amines originating from aldehydes by transamination

Aldehyde	Amine	Aldehyde	Amine
Formaldehyde	Methylamine	*i*-Butyraldehyde	*i*-Butylamine
Acetaldehyde	Ethylamine	*i*-Valeraldehyde	*i*-Amylamine
Propionaldehyde	*n*-Propylamine	Hexanal	Hexylamine

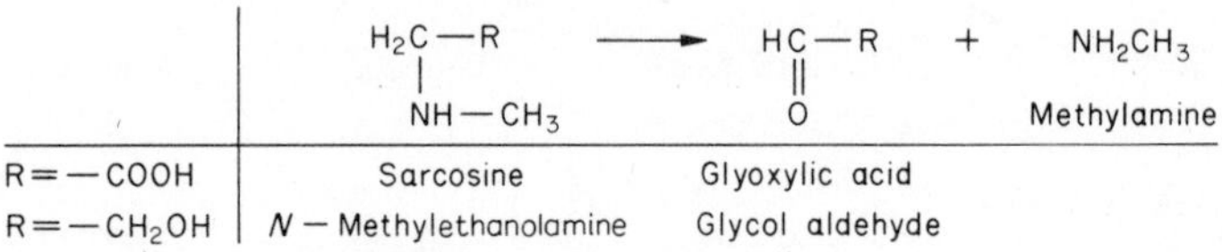

R = — COOH	Sarcosine	Glyoxylic acid	
R = — CH₂OH	*N* — Methylethanolamine	Glycol aldehyde	

Fig. 141. Formation of methylamine.

(*d*) Trimethylamine is formed during the degradation of ergothioneine to thiolurocanic acid (cf. D.18) or during the bacterial degradation of choline. In the case of the latter reaction which is catalysed by a cobalamin-containing enzyme, acetaldehyde is formed as well (fig. 142).

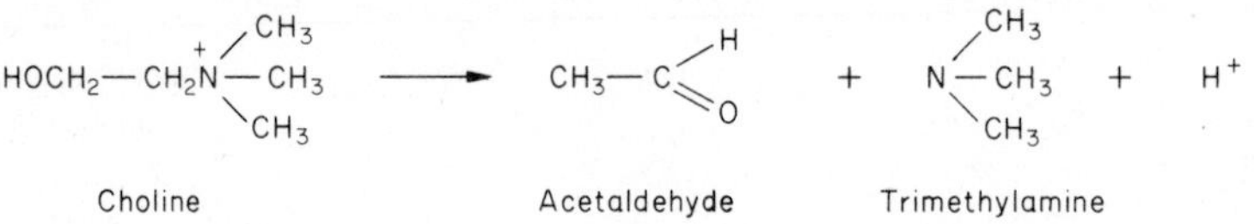

Choline   Acetaldehyde   Trimethylamine

Fig. 142. Degradation of choline to trimethylamine and acetaldehyde.

References for further reading

Meister, A. *Biochemistry of the Amino Acids*, Vol. I (Academic Press, New York, 1965).

Steiner, M., Hartmann, T., Dönges, D. and Bast, E. Biosynthese flüchtiger Amine durch einer Amine-Transaminase in Blütenpflanzen, *Naturwiss.* **54** (1967), pp. 370–1.

Schütte, H. R. 'Einfache Amine', *Biosynthese der Alkaloide*, ed. K. Mothes and H. R. Schütte (VEB Deutscher Verlag der Wissenschaften, Berlin, 1969), pp. 168–82.

Smith, T. A. The Occurrence, Metabolism and Functions of Amines in Plants, *Biol. Rev.* **46** (1971), pp. 201–41.

7.4 Formation of methylated amino acids and betaines

Methylated amino acids are widespread in the animal and plant kingdoms as well as in micro-organisms. If the nitrogen atom is made quarternary by methylation then they are called betaines. Some of the more important methylated amino acids and betaines as well as their mode of formation are outlined in fig. 143. All the methyl groups contained in these compounds originate from methionine (cf. C.3.3).

Two pathways exist for the synthesis of betaine (glycine betaine). As well as the stepwise methylation of glycine (fig. 143), biosynthesis from serine is of

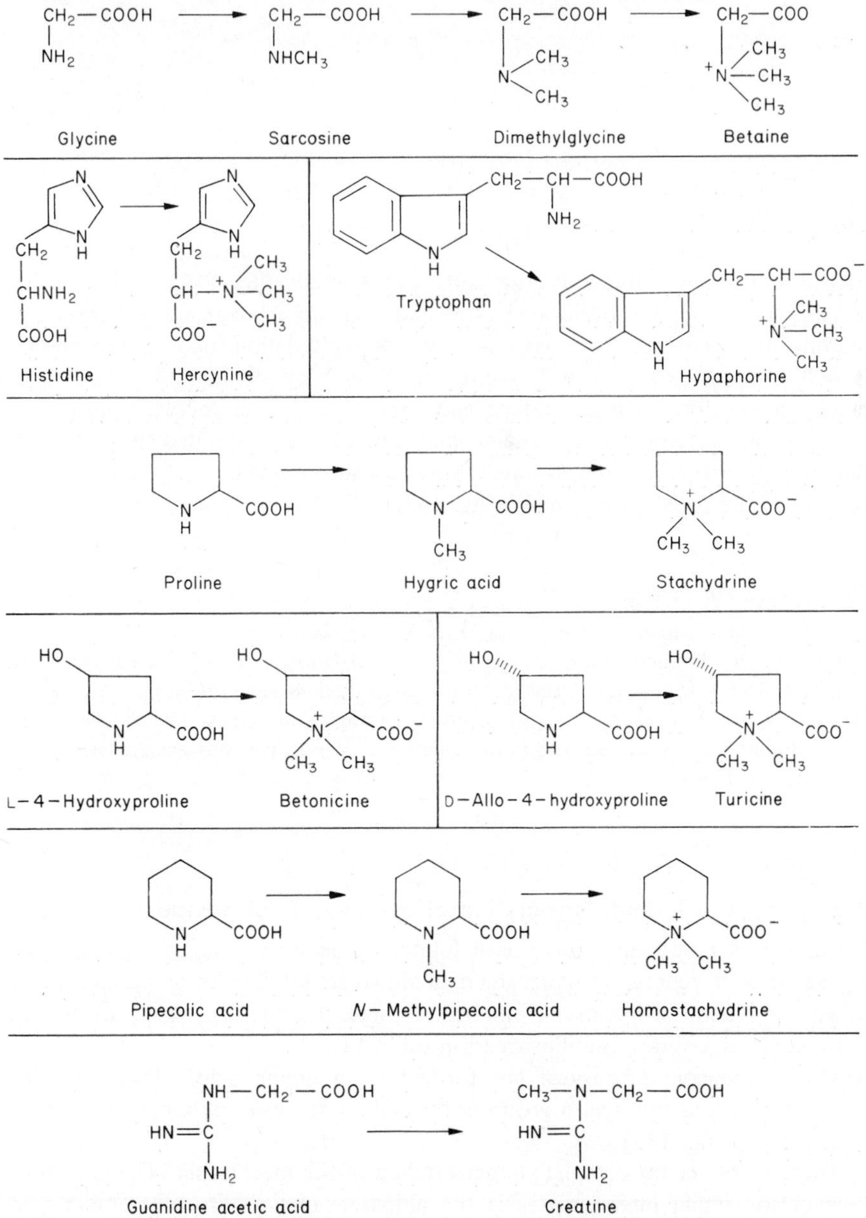

Glycine → Sarcosine → Dimethylglycine → Betaine

Histidine → Hercynine

Tryptophan → Hypaphorine

Proline → Hygric acid → Stachydrine

L-4-Hydroxyproline → Betonicine

D-Allo-4-hydroxyproline → Turicine

Pipecolic acid → N-Methylpipecolic acid → Homostachydrine

Guanidine acetic acid → Creatine

Fig. 143. Formation of some methylated amino acids and betaines.

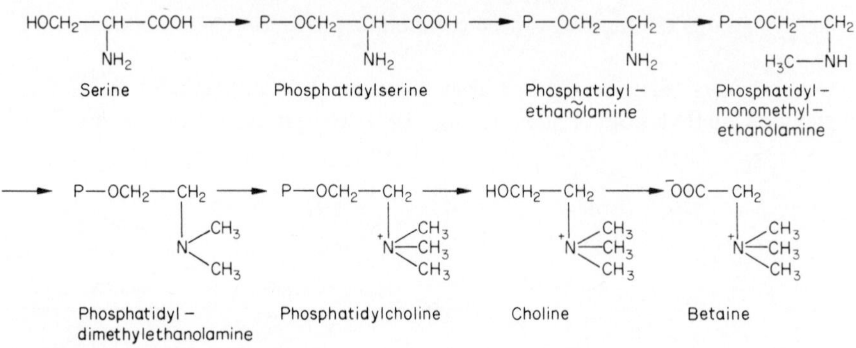

HOCH₂—CH—COOH → P—OCH₂—CH—COOH → P—OCH₂—CH₂ → P—OCH₂—CH₂

Serine   Phosphatidylserine   Phosphatidyl–ethanolamine   Phosphatidyl–monomethyl–ethanolamine

→ P—OCH₂—CH₂ → P—OCH₂—CH₂ → HOCH₂—CH₂ → ⁻OOC—CH₂

Phosphatidyl–dimethylethanolamine   Phosphatidylcholine   Choline   Betaine

Fig. 144. Formation of betaine.

special importance in micro-organisms and animals (fig. 144). Serine is first converted to the phosphoric acid ester and then decarboxylated to form phosphatidylethanolamine. By means of stepwise methylation, phosphatidylcholine is first formed, and is then converted to choline by elimination of phosphate (acetylcholine formed from choline and acetyl CoA is an important hormone). Choline can be oxidized to betaine and can also be degraded to glycine via dimethylglycine and sarcosine by means of transmethylases (cf. C.3.3) by the reverse of the pathway outlined in fig. 143.

References for further reading

Meister, A. *Biochemistry of the Amino Acids*, Vol. I (Academic Press, New York, 1965).

Schütte, H. R. 'Einfache Amine', *Biosynthese der Alkaloide*, ed. K. Mothes and H. R. Schütte (VEB Deutscher Verlag der Wissenschaften, Berlin, 1969), pp. 168–82.

Schütte, H. R. 'Methylierung und Transmethylierung', *Biosynthese der Alkaloide*, ed. K. Mothes and H. R. Schütte (VEB Deutscher Verlag der Wissenschaften, Berlin, 1969), pp. 123–67.

7.5 Biosynthesis and degradation of cyanogenic glycosides

A number of compounds have been found, particularly in higher plants, which are capable of generating hydrogen cyanide when acted upon by certain glycosidases or chemical agents. A number of these compounds, which are termed cyanogenic glycosides, are illustrated in table 14.

The cyanogenic glycosides are formed from amino acids. Their carboxyl group is lost and the cyano group is formed from the α-carbon atom and the amino group (fig. 145).

There is as yet no complete understanding of the mechanism of the reaction. One of the initial intermediates is the aldoxime of the particular amino acid. This appears to be first dehydrated to the nitrile, hydroxylated to the cyano-

Table 14. Some frequently occurring cyanogenic glycosides

Name	R_1	R_2	Sugar
Linamarin	—CH$_3$	—CH$_3$	Glucose
Lotaustralin	—CH$_2$—CH$_3$	—CH$_3$	Glucose
Prunasin	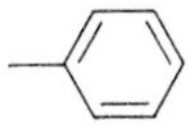 ⬡	—H	Glucose
Amygdalin	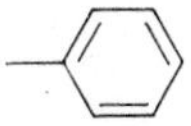 ⬡	—H	Gentiobiose
Dhurrin	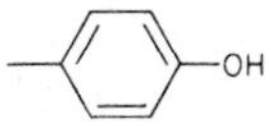 ⬡—OH	—H	Glucose

hydrin and then glycosylated (fig. 146). *N*-Hydroxyamino acids (cf. D.21.2) and the α-keto acid oximes originating from them by dehydrogenation are possibly intermediates in the synthesis of the aldoximes.

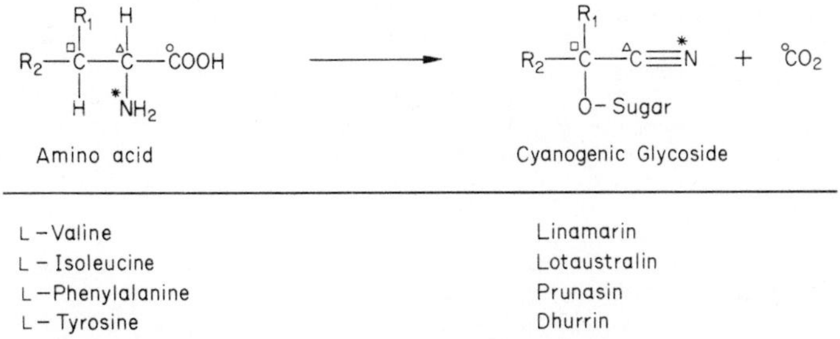

$$R_2-\overset{\overset{R_1}{|}}{\underset{\underset{H}{|}}{C}}-\overset{\overset{H}{|}}{\underset{\underset{NH_2}{|}}{C}}-\overset{O}{\overset{||}{C}}OOH \longrightarrow R_2-\overset{\overset{R_1}{|}}{\underset{\underset{O-\,Sugar}{|}}{C}}-C≡N \;+\; CO_2$$

Amino acid	Cyanogenic Glycoside
L –Valine	Linamarin
L – Isoleucine	Lotaustralin
L –Phenylalanine	Prunasin
L – Tyrosine	Dhurrin

Fig. 145. Amino acids as precursors of cyanogenic glycosides.

The cyanogenic glycosides formed by the above pathway are not necessarily the end products of metabolism and may be degraded by the plants synthesizing them. Such degradation takes place very quickly on destruction of the tissue structure. For instance, free hydrocyanic acid is quickly detectable in large quantities when seeds of *Linum usitatissimum* which contain the cyanogenic glycoside linamarin are ground in the presence of water. By feeding isotopically labelled compounds it was found that continuous synthesis and degradation of the cyanogenic glycosides take place in intact organisms, and that the balance of these reactions determines the concentration of the compounds in the tissue.

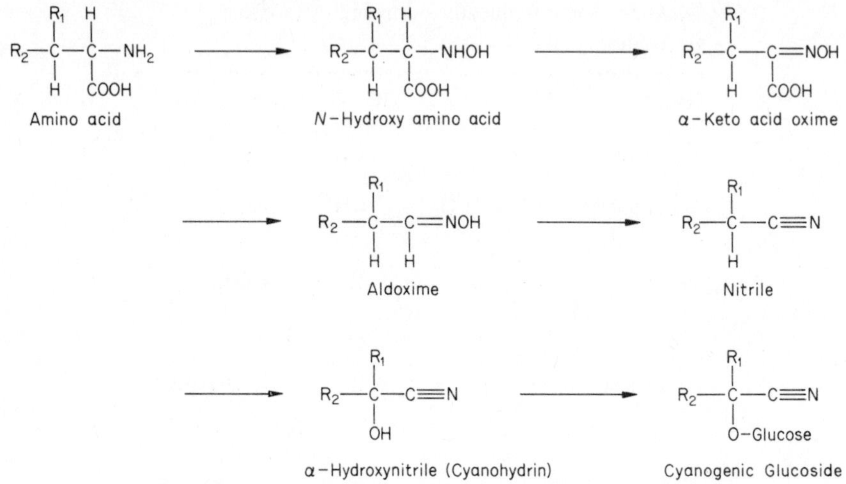

Fig. 146. Probable pathway for the formation of cyanogenic glycosides from amino acids.

Two groups of enzymes take part in degradation of cyanogenic glycosides. The degradation of the sugar moeity takes place by means of β-glycosidases. The cyanohydrins formed by this reaction are degraded in a second reaction to the aldehyde or the ketone and hydrocyanic acid by means of oxynitrilases (fig. 147).

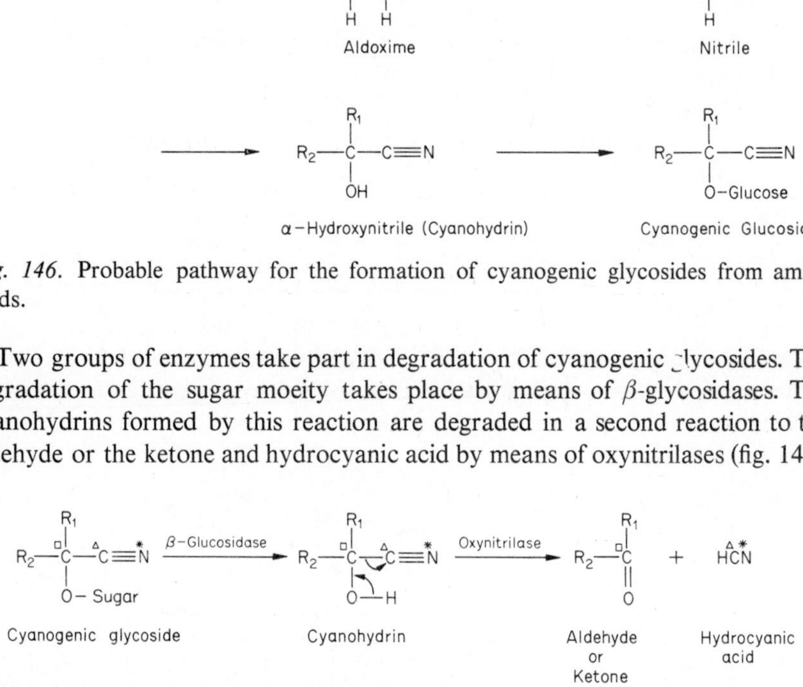

Fig. 147. The degradation of cyanogenic glycosides.

The best known of the β-glycosidases is emulsin, which occurs in seeds of *Prunus amygdalus* and whose natural substrate is amygdalin. The substrate specificity of emulsin is, however, poor, since besides amygdalin a large number of other β-glycosides are also degraded.

While hydrocyanic acid is very toxic to most organisms due to reaction with enzyme systems containing heavy metals such as copper and iron, plants, especially those containing cyanogenic glycosides, not only tolerate hydrocyanic acid, but even utilize it in metabolism. In the case of higher plants, hydrocyanic acid may react with serine or cysteine to form β-cyanoalanine with the elimination of either water or H_2S. The β-cyanoalanine thus formed may be converted to asparagine by reaction with water (fig. 148). In this way the carbon atom as well as the nitrogen atom of hydrocyanic acid are converted to a compound of

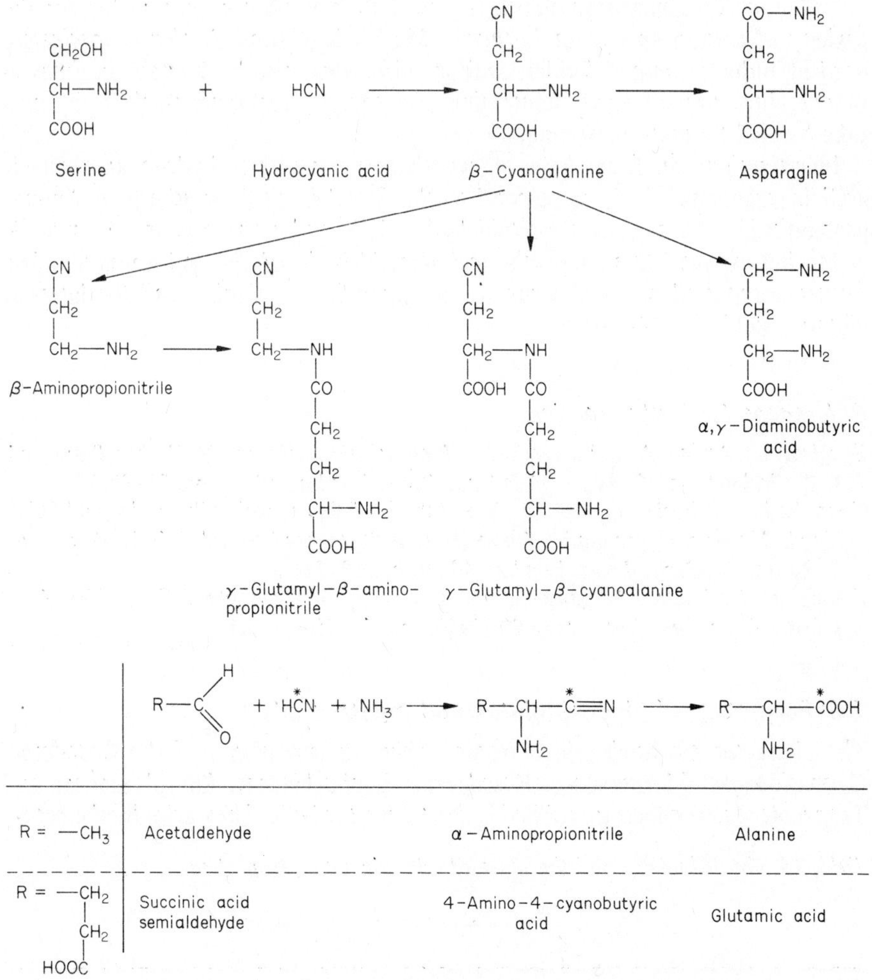

Fig. 148. The assimilation of hydrocyanic acid by higher plants and fungi.

primary metabolism and may be further utilized by the organism. Since the conversion of β-cyanoalanine to asparagine takes place very quickly, β-cyanoalanine is detectable only in minute quantities in most plants assimilating hydrocyanic acid.

In *Vicia* species where β-cyanoalanine may be accumulated, this compound is a precursor of other secondary plant products (fig. 148). On the one hand, decarboxylation with the formation of β-aminopropionitrile occurs, and on the other, linkage of glutamic acid to the amino group of β-cyanoalanine or of β-aminopropionitrile is possible. β-Cyanoalanine is also probably a precursor of the α,γ-diaminobutyric acid found in *Vicia* seedlings; this has not yet been proved experimentally, however.

γ-Glutamyl-β-aminopropionitrile is the lathyrism factor responsible for the toxicity of certain species of *Lathyrus*. The seeds of some *Lathyrus* species are used in India as animal fodder, and are also consumed by humans in times of famine. Ingestion of seeds containing γ-glutamyl-β-aminopropionitrile by mistake has led to acute poisoning.

Investigations on fungi have shown that in some cases hydrocyanic acid is used in reactions which correspond to the Strecker nitrile synthesis. α-Aminopropionitrile is formed from acetaldehyde, HCN and ammonia, and 4-amino-4-cyano butyric acid from succinic acid semialdehyde, HCN and ammonia. The former compound may subsequently be saponified to alanine and the latter to glutamic acid (fig. 148).

References for further reading

Fowden, L. 'Amino Acid Biosynthesis', *Biosynthetic Pathways in Higher Plants,* ed. J. B. Pridham and T. Swain (Academic Press, London, 1965), pp. 73–99.

Conn, E. E. and Butler, G. W. 'The Biosynthesis of Cyanogenic Glucosides and Other Simple Nitrogen Compounds', *Perspectives in Phytochemistry*, ed. J. B. Harborne and T. Swain (Academic Press, London, 1969), pp. 47–74 .

Eyjólfsson, R. Recent Advances in the Chemistry of Cyanogenic Glycosides, *Fortschritte Chem. org. Naturstoffe* **28** (1970), pp. 74–108.

7.6 Formation of glucosinolates and mustard oils

The characteristic smell which occurs when various plants of the Cruciferae, Capparidaceae, Moringaceae, Resedaceae, Euphorbiaceae, Phytolaccaceae and Tropaeolaceae families are rubbed is due to mustard oils. They arise by the action

Table 15. Glucosinolates and the mustard oils originating from them,

$$R-C\begin{array}{l}{}^{S-\text{Glucose}}\\{}_{N-SO_4^-}\end{array} \longrightarrow R-N{=}C{=}S \; + \; \text{Glucose} + SO_4^{--}$$

Glucosinolate	Mustard oil	R
Sinigrin	Allyl mustard oil	$CH_2{=}CH{-}CH_2{-}$
Gluconapin	3-Butenyl mustard oil	$CH_2{=}CH{-}CH_2{-}CH_2{-}$
Glucobrassicanapin	4-Pentenyl mustard oil	$CH_2{=}CH{-}CH_2{-}CH_2{-}CH_2{-}$
Glucoibervirin	ω-Methylthiopropyl mustard oil	$CH_3{-}S{-}CH_2{-}CH_2{-}CH_2{-}$
Sinalbin	*p*-Hydroxybenzyl mustard oil	$HO{-}\langle\text{benzene}\rangle{-}CH_2{-}$
Glucotropaeolin	Benzyl mustard oil	$\langle\text{benzene}\rangle{-}CH_2{-}$
Gluconasturtiin	Phenylethyl mustard oil	$\langle\text{benzene}\rangle{-}CH_2{-}CH_2{-}$

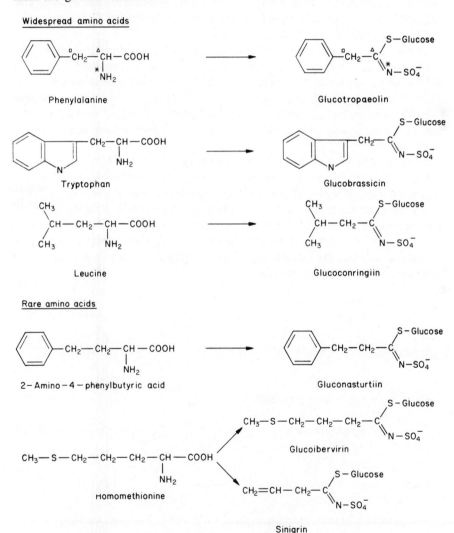

5,5-Dimethyl-2-oxazolidinethione

5-Phenyl-2-oxazolidinethione

Fig. 149. Oxazolidinethiones originating from mustard oils.

of enzymes on the sulphur-containing glucosides, the glucosinolates (mustard oil glucosides). Some of these compounds and the mustard oils originating from them are given in table 15.

Widespread amino acids

Phenylalanine

Glucotropaeolin

Tryptophan

Glucobrassicin

Leucine

Glucoconringiin

Rare amino acids

2-Amino-4-phenylbutyric acid

Gluconasturtiin

Glucoibervirin

Homomethionine

Sinigrin

Fig. 150. Formation of glucosinolates from amino acids.

The mustard oil originating from those glucosinolates in which group R is substituted by a hydroxyl group in the β-position cyclizes spontaneously to yield oxazolidinethiones. Thus 5,5-dimethyl-2-oxazolidinethione is formed from gluco-conringiin (R = CH$_3$—COH(CH$_3$)—CH$_2$—) and 5-phenyl-2-oxazolidinethione from glucobarbarin (cf. fig. 151).

The glucosinolates are synthesized from amino acids. The carboxyl group of the amino acid is lost, while the amino group is retained in the glucoside. Ald-oximes, corresponding to the particular amino acid, and which possibly originate via N-hydroxy amino acid and the α-keto acid oxime (cf. D.7.5), are inter-mediates in the biosynthetic pathway. Other intermediates are not known. Thus glucotropaeolin is formed from phenylalanine, glucobrassicin from tryptophan and glucoconringiin from leucine (fig. 150).

In the case of certain glucosinolates, derivation from the protein amino acids is not possible. Thus, in those plants which contain such compounds, the secon-dary amino acids which have been detected may be considered to be precursors. Thus 2-amino-4-phenylbutyric acid, the precursor of gluconasturtiin, has been found in *Nasturtium officinale*. 2-Amino-4-phenylbutyric acid probably origin-ates in a similar way to that described under D.12 for the formation of leucine from valine.

In the case of sinigrin, it has been shown that the amino acid homomethionine is the precursor. This amino acid differs from methionine in that it possesses an additional CH$_2$-group, and probably originates in the way described in D.12. The stage at which sulphur and the methyl group are eliminated during the forma-tion of sinigrin is still unclear. It has been found, however, that conversion does not proceed via allylglycine and 2-amino-5-hydroxyvaleric acid. The CH$_3$S-grouping of homomethionine is retained in glucoibervirin, which occurs together with sinigrin in *Iberis sempervirens*.

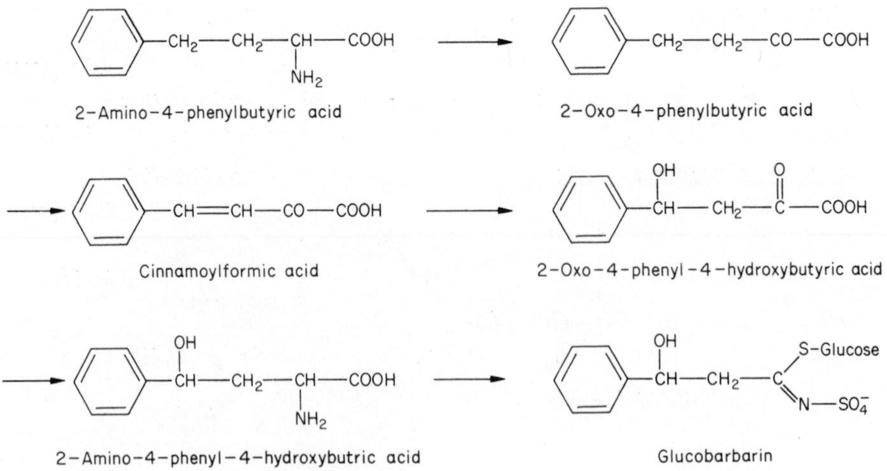

2-Amino-4-phenylbutyric acid   2-Oxo-4-phenylbutyric acid

Cinnamoylformic acid   2-Oxo-4-phenyl-4-hydroxybutyric acid

2-Amino-4-phenyl-4-hydroxybutric acid   Glucobarbarin

Fig. 151. Biosynthesis of glucobarbarin.

2-Amino-4-phenyl-4-hydroxybutyric acid, which is probably a precursor for glucobarbarin, may be formed from 2-amino-4-phenylbutyric acid as shown in fig. 151.

The enzyme myrosinase, which is active in the cleavage of glucosinolates, is localized in special cells. Destruction of tissue structure liberates the enzyme and the glucosides are cleaved. An isomerase is also present in certain plants, which isomerizes the mustard oils (isothiocyanates) to rhodanids (thiocyanates) (fig. 152).

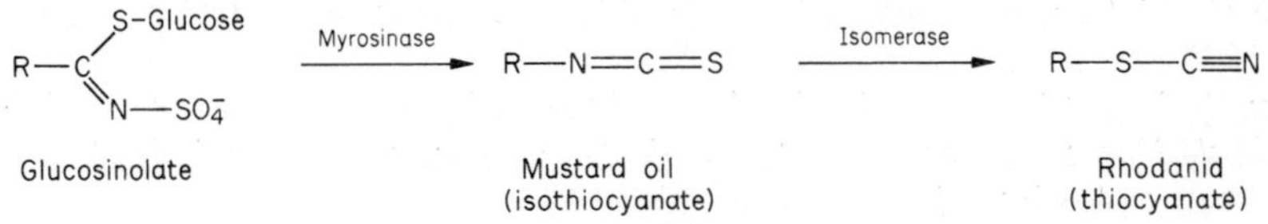

Fig. 152. Formation of mustard oils and thiocyanates from glucosinolates.

Glucobrassicin, which occurs in some species of *Brassica*, may be enzymatically (probably by myrosinase) converted to indolylacetonitrile at low pH. The formation of hydrocyanic acid in the Cruciferae probably occurs through decomposition of nitriles formed from glucosinolates. In neutral media glucobrassicin is decomposed to glucose, sulphate, thiocyanic acid and hydroxymethylindole.

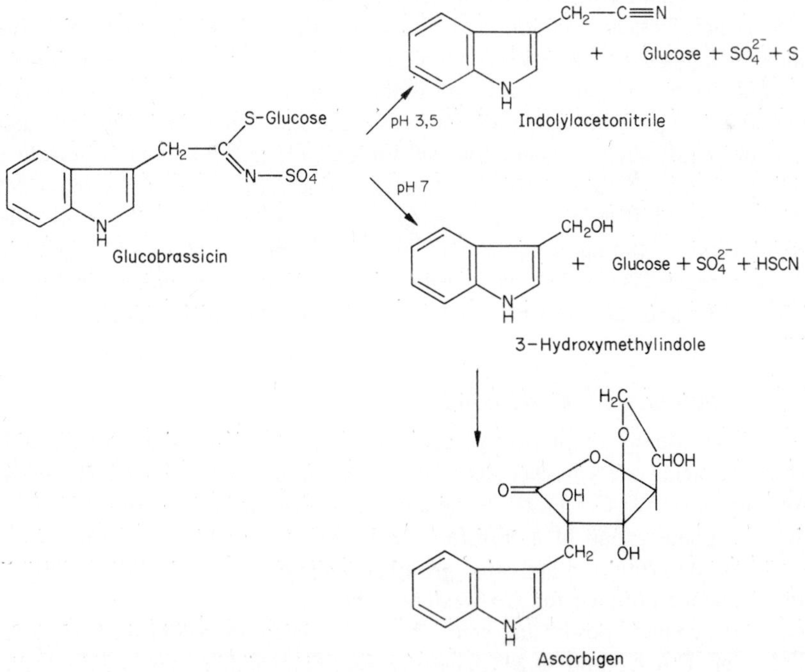

Fig. 153. Formation of indolylacetonitrile and ascorbigen from glucobrassicin.

The latter compound reacts with ascorbic acid to form ascorbigen, a substance which was isolated from cabbages many years ago.

References for further reading

Kjaer, A. Naturally Derived *iso*-Thiocyanates (Mustard Oils) and Their Parent Glucosides, *Fortschr. Chem. org. Naturstoffe* **18** (1960), pp. 122–76.

Hegnauer, R. *Chemotaxonomie der Pflanzen*, Vol. III (Birkhäuser-Verlag, Basel, 1964), pp. 587–92.

Underhill, E. W. and Wetter, L. R. 'Biosynthesis of Mustard Oil Glucosides', *Biosynthesis of Aromatic Compounds*, ed. G. Billek (Pergamon Press, Oxford, 1966), pp. 129–37.

Conn, E. E. and Butler, G. W. 'The Biosynthesis of Cyanogenic Glucosides and Other Simple Nitrogen Compounds', *Perspectives in Phytochemistry*, ed. J. B. Harborne and T. Swain (Academic Press, London, 1969).

7.7. General reactions in the formation of alkaloids

The term 'alkaloid' was coined in 1819 by C. F. W. Meissner, a pharmacist from Halle. Meissner suggested that the known alkaline plant products should be classified not as alkalis, but as alkaloids, since they differ from alkalis in many respects. Like most groups of natural products classified according to external properties (e.g. tannins, sapogenins, bitter substances, antibiotics) the alkaloids are chemically and biochemically heterogeneous.

The *N*-heterocyclic compounds which originate from amino acids are usually termed 'real alkaloids' nowadays, while substances which originate from amino acids but possess the nitrogen in an aliphatic linkage are called 'protoalkaloids' (cf. D.20.1.1). The pseudoalkaloids formed by the addition of ammonia to nitrogen-free compounds such as isoprenoid substances (cf. D.5.5 and D.5.7.2) are different from both the above groups. The biosynthesis of real alkaloids requires the formation of an *N*-heterocyclic ring system, including the aliphatically bound nitrogen atom of the amino acid. This involves formation of a C–N bond. Intramolecular reactions as well as intermolecular reactions in which several compounds participate are possible. The following are the most important reaction types:

(a) Formation of acid amide bonds

Acid amides usually originate by reaction of an activated acid with an amine. Carboxylic acids are involved as their CoA-esters (cf. C.1.2), amino acids as pyrophosphates (cf. C.1.1), or in the case of protein synthesis, bound to the terminal phosphate group of a transfer ribonucleic acid. During the formation of the acid amide group the activated group is eliminated. The reaction corresponds to the sequence outlined for CoA-esters in fig. 5.

Natural products possessing acid amide groupings are found more frequently in fungi (cf. D.6.4.2 and D.21). The grouping may undergo secondary changes such as oxidation and reduction (cf. D.21.1 and D.21.2).

It has been shown in certain cases that the acid amide linkage present in the alkaloid molecule does not originate by reaction of an acid with an amine, but by later oxidation (e.g. in the conversion of sparteine to lupanine, cf. D.16.3).

(b) Formation of azomethines (Schiff bases)

Azomethines are frequently formed spontaneously (cf. C.) from compounds with amino- and carbonyl groups.

$$R_1-C{\overset{R_2}{\underset{O}{}}} \quad + \quad H_2N-R_3 \quad \longrightarrow \quad R_2-\overset{\overset{R_2}{|}}{C}=N-R_3 \quad + \quad H_2O$$

Carbonyl   Primary   Azomethine
compound   amine

If secondary amines take part in this reaction, compounds with a quarternary nitrogen atom are formed. These, due to the strong polarization of the molecule, are particularly reactive and easily undergo further reactions.

$$R_1-C{\overset{R_2}{\underset{O}{}}} \quad + \quad HN{\overset{R_3}{\underset{R_4}{}}} \quad \longrightarrow \quad R_1-\overset{\overset{R_2}{|}}{C}=\overset{\oplus}{N}{\overset{R_3}{\underset{R_4}{}}} \quad + \quad OH^-$$

Carbonyl   Secondary   Quarternary
compound   amine   azomethine

The amines which take part in the formation of Schiff bases are formed by decarboxylation of amino acids (cf. D.7.3). They are frequently methylated (cf. D.7.4).

The carbonyl compounds are formed in many cases from amines by transamination (cf. C.4) and oxidative deamination (cf. C.2.1.2). In some reactions, however, keto acid derivatives originating from acetate metabolism (cf. D.6.4.2) also take part.

(c) Mannich condensation

The reaction of a carbonyl compound, usually an aldehyde, with an amine and a compound with an acidic hydrogen is known in the narrower sense as a Mannich reaction. The β-substituted ethyl amines such as phenylethylamine and its derivatives (cf. D.20.1.2) as well as tryptamine and its derivatives (cf. D.19.3) are amines which possess an acidic hydrogen atom.

In this reaction the carbon atom of the carbonyl group is present together with a nitrogen atom and a carbon atom with a negative fractional charge, and it reacts first with the more negative grouping. Only when it is the amine does a Mannich condensation occur (fig. 154).

The Mannich condensation in the above-mentioned form has often been observed *in vitro*, but also occurs under cellular conditions (i.e. at room temperature, physiological pH etc.). *In vivo*, an acyl CoA-ester probably acts as the carbonyl compound, and an acid amide is formed as an intermediate instead of the *N-*

hydroxymethyl derivative. Thus in the biosynthesis of harman, acetyl CoA and N-acetyl tryptamine (cf. D.19.3) have been shown to be the actual precursors.

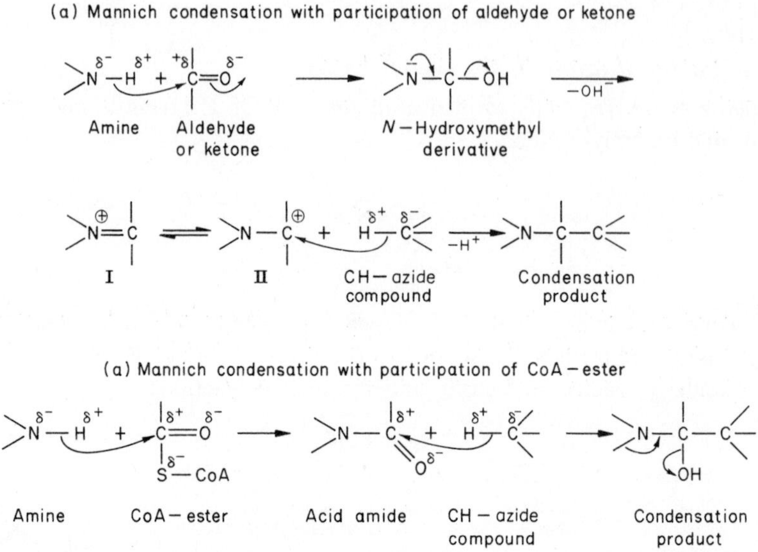

(a) Mannich condensation with participation of aldehyde or ketone

(a) Mannich condensation with participation of CoA − ester

Fig. 154. Mannich condensation.

(d) Addition of amines to quinoid substances

As well as the three reaction types described above, the addition of amines to o- or p-quinoid systems is of importance in a few cases (cf. the formation of melanins, D.20.1.6, the ommochromes, D.19.5.2, the phenoxazines, D.6.4.1, and the humic acids, D.2.4.4). p-Quinoid and o-quinoid intermediates are formed from o-quinones and p-quinones respectively, and subsequently rearranged to form aromatic compounds (fig. 155).

(e) Formation of N-glycosides

The formation of N-glycosidic bonds (cf. D.1.1) plays a role in the biosynthesis of the indole ring system (cf. D.19) and the pteridine ring system (cf. D.8.2.3). This type of linkage is, however, less important than the others mentioned above.

References for further reading

Bessler, O. Über Wilhelm Meissner, den Schöpfer des Begriffs 'Alkaloid'. Abh. dtsch. Akad. Wiss., Berlin, Kl. Chem., Geol., Biol. (1963), no. 4, pp. 13–15.
Leete, E. Alkaloid Biosynthesis, Ann. Rev. plant. Physiol. 18 (1967), pp. 179–96.

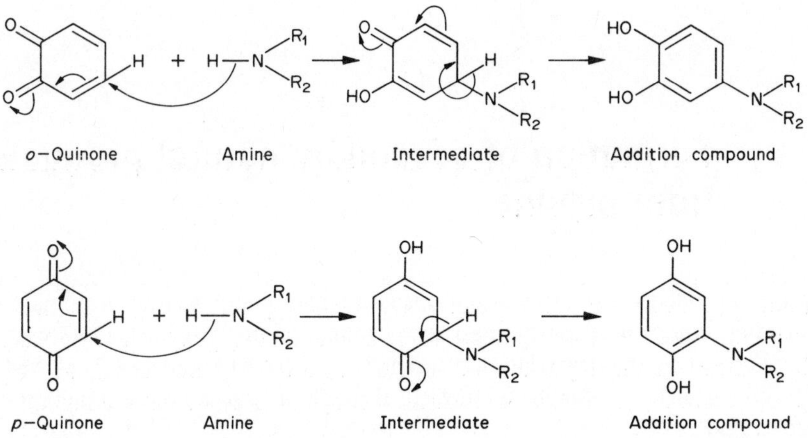

Fig. 155. Addition of amines to quinones.

Mothes, K. 'Biologie der Alkaloide', *Biosynthese der Alkaloide*, ed. K. Mothes and H. R. Schütte (VEB Deutscher Verlag der Wissenschaften, Berlin, 1969), pp. 1–20.

Liebisch, H. W. 'Cyclisierungsmechanismen bei der Alkaloidbiosynthese', *Biosynthese der Alkaloide*, ed. K. Mothes and H. R. Schütte (VEB Deutscher Verlag der Wissenschaften, Berlin, 1969), pp. 101–22.

Liebisch, H. W. Cyclisierungsmechanismen bei der Alkaloidbiosynthese, *Fortschritte chem. Forsch.* **9** (1968), pp. 534–604.

8. Formation of secondary natural products from glycine

Glycine is produced by micro-organisms and higher plants as well as animals. Of special importance in its biosynthesis is the amination of glyoxylic acid. Glyoxylic acid is formed by the degradation of isocitric acid (cf. D.4 and D.4.2) as well as from glycolaldehyde which is formed through a transketolase reaction via glycolic acid.

Glycine is also formed in smaller quantities by a number of other reactions, e.g. the decomposition of serine (cf. C.3.2) and threonine, by demethylation of sarcosine (cf. D.7.4), as well as during the degradation of purines via imidazole derivatives (cf. D.8.7).

The amino acid in its turn is a precursor for a few groups of important natural products.

References for further reading

Meister, A. *Biochemistry of the Amino Acids*, Vol. II (Academic Press, New York, 1965).

Reinbothe, H. 'Metabolismus der in die Alkaloidbiogenese eintretenden Aminosäuren', *Biosynthese der Alkaloide*, ed. K. Mothes and H. R. Schütte (VEB Deutscher Verlag der Wissenschaften, Berlin, 1969), pp. 40–100.

8.1 Formation of porphyrins, bile pigments and cobalamin derivatives

(a) Formation of compounds with the porphyrin ring system

Iron and magnesium chelates of porphyrins are widespread in nature. Examples of iron-containing porphyrin derivatives are haemoglobin, which mediates oxygen transport in the blood, and the prosthetic groups of cytochromes, peroxidases and catalases. The porphyrin ring system of the chlorophylls synthesized by green plants contains magnesium.

Porphobilinogen is the basic skeleton of all naturally occurring porphyrin derivatives. This compound is excreted in large quantities in the urine of patients with acute porphyria.

Porphobilinogen originates from succinyl CoA and glycine. These compounds condense with the elimination of coenzyme A to form α-amino-β-keto adipic acid. This compound then undergoes spontaneous decarboxylation to form δ-aminolaevulinic acid. By elimination of the elements of water a molecule of porphobilinogen is formed from two molecules of this acid (fig. 156). The enzymes catalysing these reactions have been isolated from erythrocytes and liver cells.

Porphobilinogen polymerizes with the elimination of ammonia to form the symmetrical tetrapyrrole, uroporphyrinogen I. This reaction is catalysed by the enzyme uroporphyrinogen-I-synthetase. An enzyme preparation catalysing this reaction has been isolated from spinach leaves.

Uroporphyrinogen-I-synthetase and uroporphyrinogen-III-cosynthetase act together in the formation of uroporphyrinogen III. This compound differs from uroporphyrinogen I in that the acetyl and propionyl groups in ring D are reversed. The cosynthetase probably reacts with an intermediate of uroporphyrinogen-I-synthetase. No intermediates have as yet been found, however.

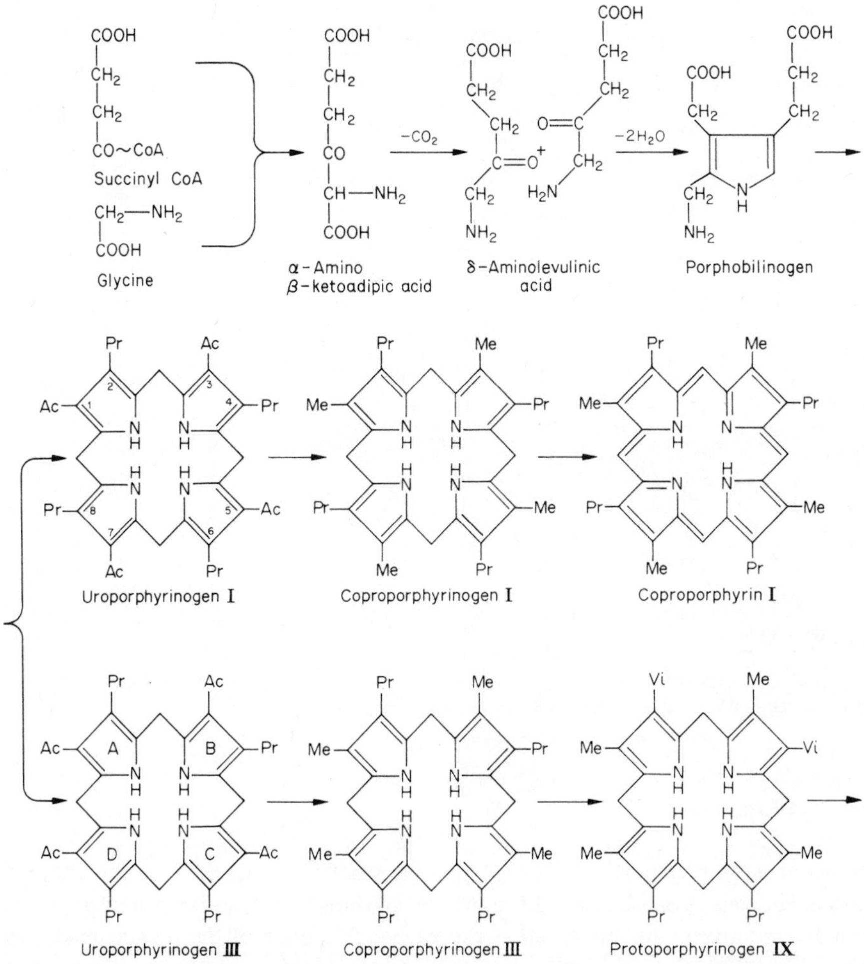

Fig. 156. Formation of porphobilinogen, haem and chlorophyll a.

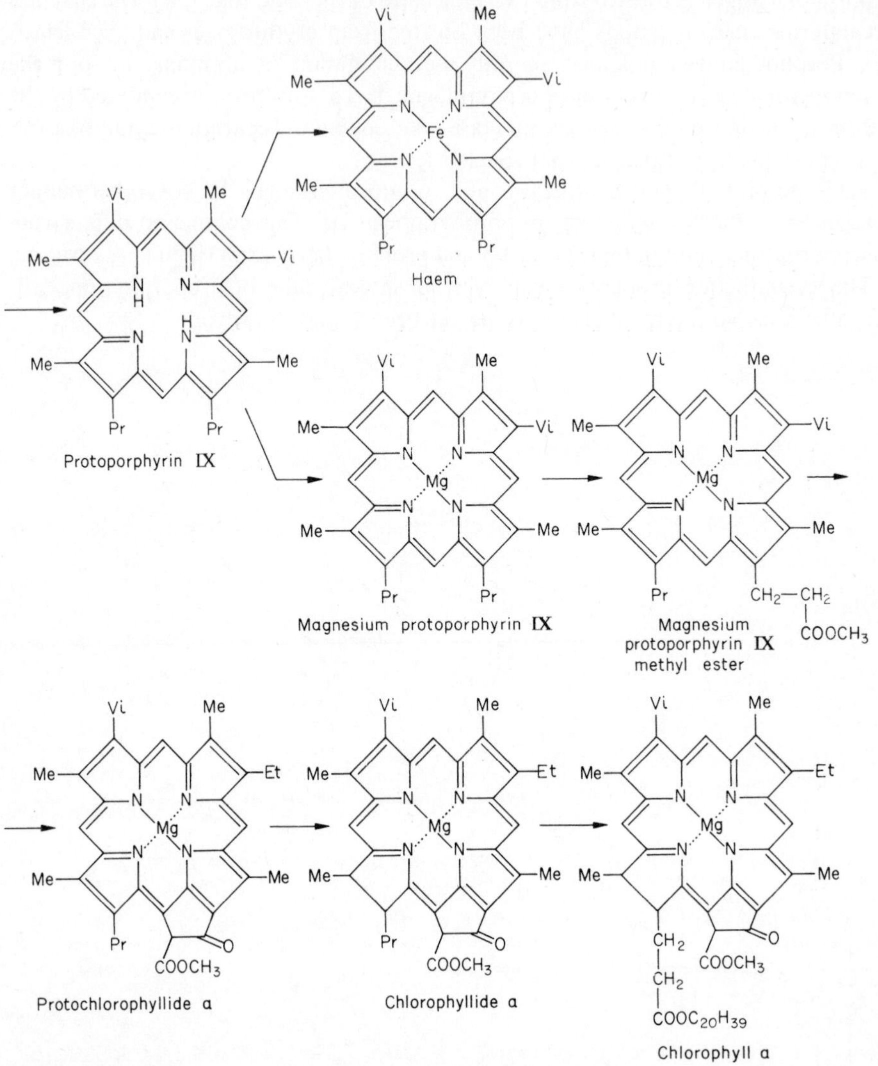

Protoporphyrin IX

Haem

Magnesium protoporphyrin IX

Magnesium protoporphyrin IX methyl ester

Protochlorophyllide a

Chlorophyllide a

Chlorophyll a

Ac = —CH₂—COOH   Me = —CH₃   Et = —CH₂—CH₃
Pr = —CH₂—CH₂—COOH   Vi = —CH=CH₂

Fig. 156 (contd.).

Uroporphyrinogen III and uroporphyrinogen I are converted by the enzyme porphyrinogen decarboxylase to coproporphyrinogen III and coproporphyrinogen I respectively. In this reaction the carboxyl groups of the four acetate side chains in positions 1, 3, 5 and 8 or 7 are removed leaving methyl groups in each of these positions. Early termination of this reaction has enabled intermediates retaining three, two or one acetate groups to be found. In the case of iron deficiency, coproporphyrinogen III is accumulated by certain micro-organisms.

Protoporphyrinogen IX is formed from coproporphyrinogen III by the action of the enzyme coproporphyrinogen-III-oxidase. The propionic acid side chains are decarboxylated at positions 2 and 4 and converted to vinyl groups by dehydrogenation. A further oxidation takes place to yield protoporphyrin IX. The latter reaction corresponds to the oxidation of coproporphyrinogen I to coproporphyrin I (fig. 156). This and similar coproporphyrins are constituents of human faeces. The quantity excreted may be very large in certain diseases.

Protoporphyrin IX is the starting material for synthesis of iron-containing porphyrins of the haem type as well as those of the chlorophyll type which contain magnesium. The introduction of iron is mediated by the so-called 'protoporphyrin-iron-chelating-enzyme' (PICE) which has been isolated from rat liver mitochondria (cf. D.21.2 for the origin of the iron). The enzyme which introduces magnesium into protoporphyrin IX to form magnesiumprotoporphyrin IX has not been isolated.

The monomethyl ester of magnesiumprotoporphyrin IX is formed first in the synthesis of protochlorophyllide a. Then a number of reactions take place whose sequence is not clear. The esterified propionic acid side chain is oxidized at the β-position and reacts with the neighbouring carbon atom, linking rings C and D, to form a cyclopentenone ring, while the vinyl group in position 4 is reduced to an ethyl group.

In the synthesis of chlorophyll a, protochlorophyllide a is first reduced to chlorophyllide a and then esterified with phytol (fig. 156). The reaction proceeds on a specific protein complex, the so-called holochrome.

Besides haem and chlorophyll a, a number of structurally related compounds are known, e.g. chlorophylls b, c, d and e as well as the bacterial chlorophylls, haematoporphyrin IX and haem a, and the cytochromes. The pathways of synthesis leading to these compounds are similar to the one described above and will therefore not be discussed individually.

(b) Cleavage of porphyrin ring system with the formation of bile pigments

The bile pigments arise by degradation of haemoglobin and other protoporphyrin-IX-derivatives.

Haemoglobin is first oxidatively converted to the green pigment, choleglobin (verdoglobin). In this reaction the porphyrin ring system is cleaved between rings A and B. The blue-green biliverdin is formed from this by the loss of iron and the protein component, globin. Biliverdin is easily reduced to the orange-red pigment, bilirubin. Albumin-bound bilirubin occurs in small quantities in the blood stream. This bilirubin-protein complex is decomposed in the liver, and bilirubin itself, or glycosidically linked with glucuronic acid (cf. D.1.1.2), is excreted into the bile juice. Bilirubin and bilirubin glucuronide are the most important bile pigments,

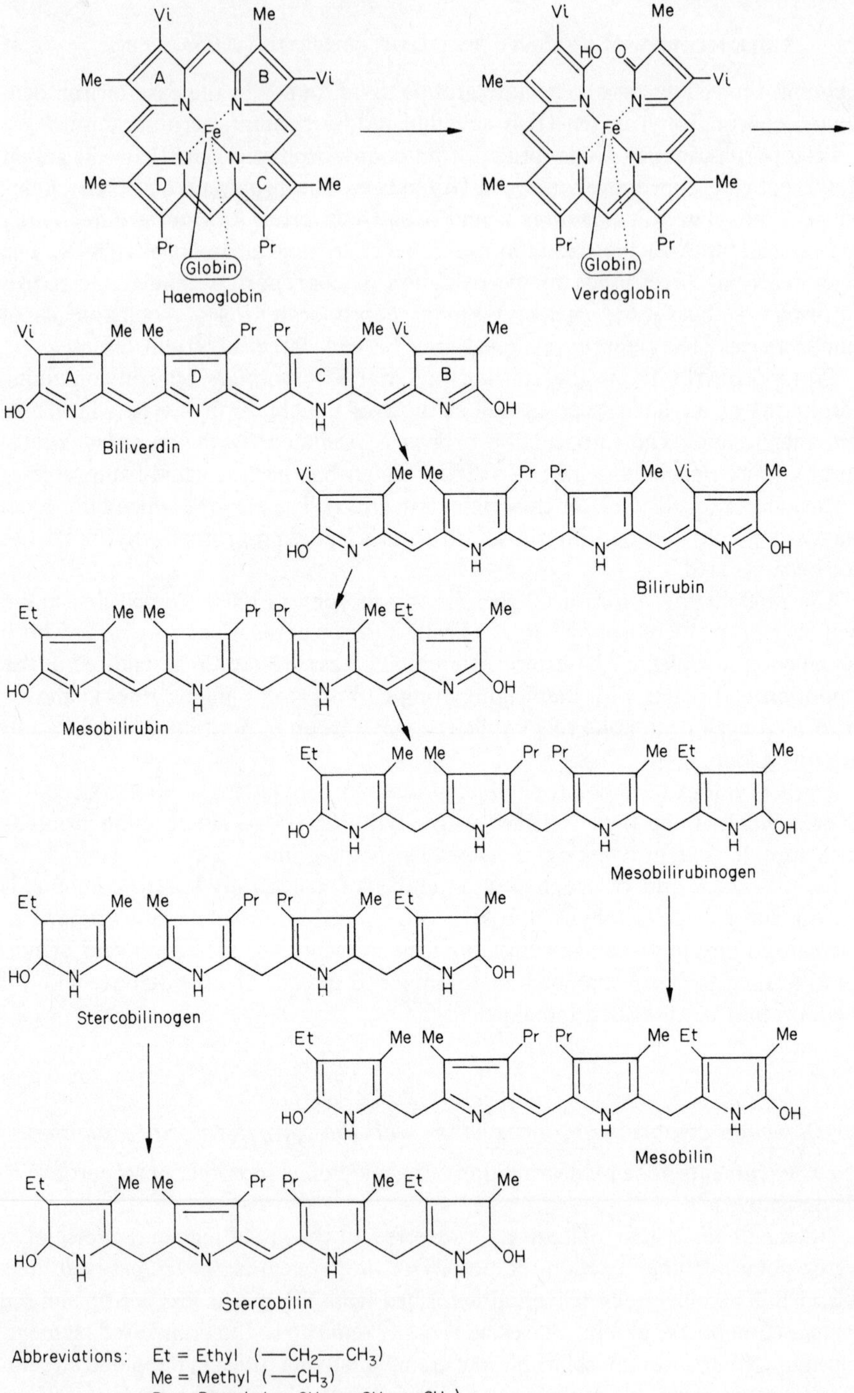

Vi — Vinyl

Haemoglobin

Verdoglobin

Biliverdin

Bilirubin

Mesobilirubin

Mesobilirubinogen

Stercobilinogen

Mesobilin

Stercobilin

Abbreviations: Et = Ethyl (—CH₂—CH₃)
                 Me = Methyl (—CH₃)
                 Pr = Propyl (—CH₂—CH₂—CH₃)
                 Vi = Vinyl (—CH══CH₂)

Fig. 157. Formation and metabolism of bile pigments.

and impart a yellow colour to the faeces if chyme passes into the intestine and is excreted.

Usually, however, the glucuronide compound is degraded in the intestine, and bilirubin is further changed both by body enzymes and by the action of the intestinal bacteria. Mesobilirubin is formed by reduction of the vinyl groups. Further reduction yields mesobilirubinogen and stercobilinogen. The dark brown pigments of the faeces, mesobilin and stercobilin, originate from the two latter compounds by spontaneous, non-enzymic, oxidation (fig. 157). The degradation of bile pigments to coloured compounds containing only two pyrrol nuclei has also been detected.

Some plants, especially the algae and cryptomonads as well as some animals, e.g. certain butterflies, contain bile pigment-protein complexes, the phycobili-proteins. Pigments of the phycoerythrin and phycocyanin type belong to this group and participate in light absorption during photosynthesis in Protophyceae and Cyanophyceae.

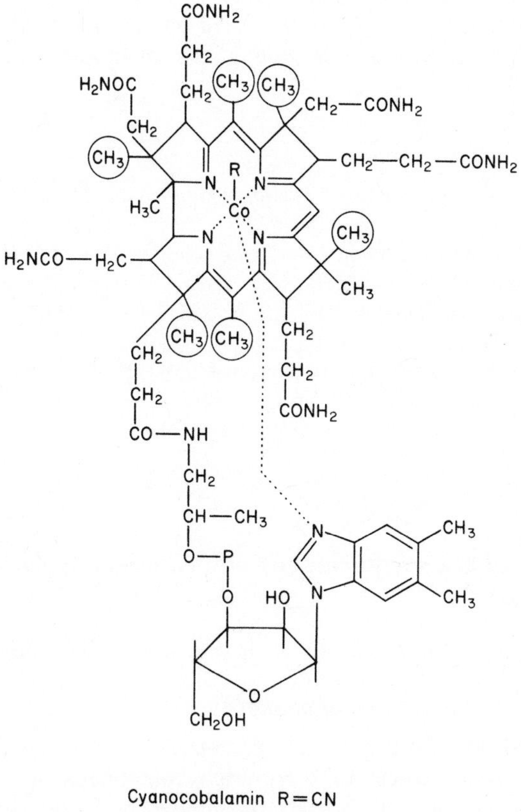

Cyanocobalamin $R = CN$
Methylcobalamin $R = CH_3$

Fig. 158. Structural formulae of cyanocobalamin (vitamin B_{12}) and methylcobalamin.

Nothing certain is known about the formation of phycobiliproteins. It has not yet been clarified as to whether the protein is already linked to the porphyrin which serves as the precursor, or whether a free bile pigment is formed first which reacts at a later stage with the protein component.

(c) Biosynthesis of cobalamins

Corrin, the porphyrin-like ring system of cobalamin derivatives, closely resembles uroporphyrin III in the arrangement and structure of the side chains. In contrast to the porphyrins, however, the pyrrol rings are extensively reduced and two of them are directly linked to one another (fig. 158).

The pathway of formation of corrin has been investigated in the case of cyanocobalamin, vitamin B_{12}. It has been shown that δ-aminolaevulinic acid is a precursor. The methyl groups within the circles in fig. 158 originate from methionine (cf. C.3.3). The exact biosynthetic pathway is still unknown. Human beings and mammals are not capable of synthesizing corrin derivatives.

Cobalamin derivatives are coenzymes of the enzymes of 'one-carbon' metabolism. Methylcobalamin (fig. 158) is formed as an intermediate (cf. C.3.2.) in the action of N^5-methyltetrahydrofolic acid-homocysteine-methyl transferase on N^5-methyl-THF.

References for further reading

Goodwin, T. W. (Ed.). *Porphyrins and Related Compounds* (Academic Press, London, 1968).

Schersten, T. 'Bile Acid Conjugation', *Metabolic Conjugation and Metabolic Hydrolysis*, ed. W. H. Fishman (Academic Press, New York, 1970), pp. 75–121).

Rüdiger, W. Neues aus Chemie und Biochemie der Gallenfarbstoffe, *Angew. Chem.* **82** (1970), pp. 527–34.

Friedmann, H. C. and Cagen, L. M. Microbial Biosynthesis of B_{12}-like Compounds, *Ann. Rev. Microbiology* **24** (1970), pp. 159–208.

Shlyk, A. A. Biosynthesis of Chlorophyll b, *Ann. Rev. Plant Physiol.* **22** (1971), pp. 169–84.

8.2 Formation of purine derivatives and compounds derived from purines

8.2.1 Biosynthesis of compounds with the purine ring system

(a) Formation of inosine monophosphate

Purine glycosides are constituents of the nucleic acids of all living organisms. Their phosphate esters, so-called 'energy-rich' compounds, are of particular importance in metabolism (cf. C.1.1). A number of methylated purines (e.g. caffeine, theobromine, theophylline and herbipoline) occurring in higher plants and sponges are secondary products and may be regarded as alkaloids. Methylated purines have also been found in the nucleic acids of all living organisms investigated till now. Purines substituted by isopentenyl groups in place of methyl groups, such as zeatin isolated from *Zea mays*, are active in plants as growth hormones, and are termed kinins.

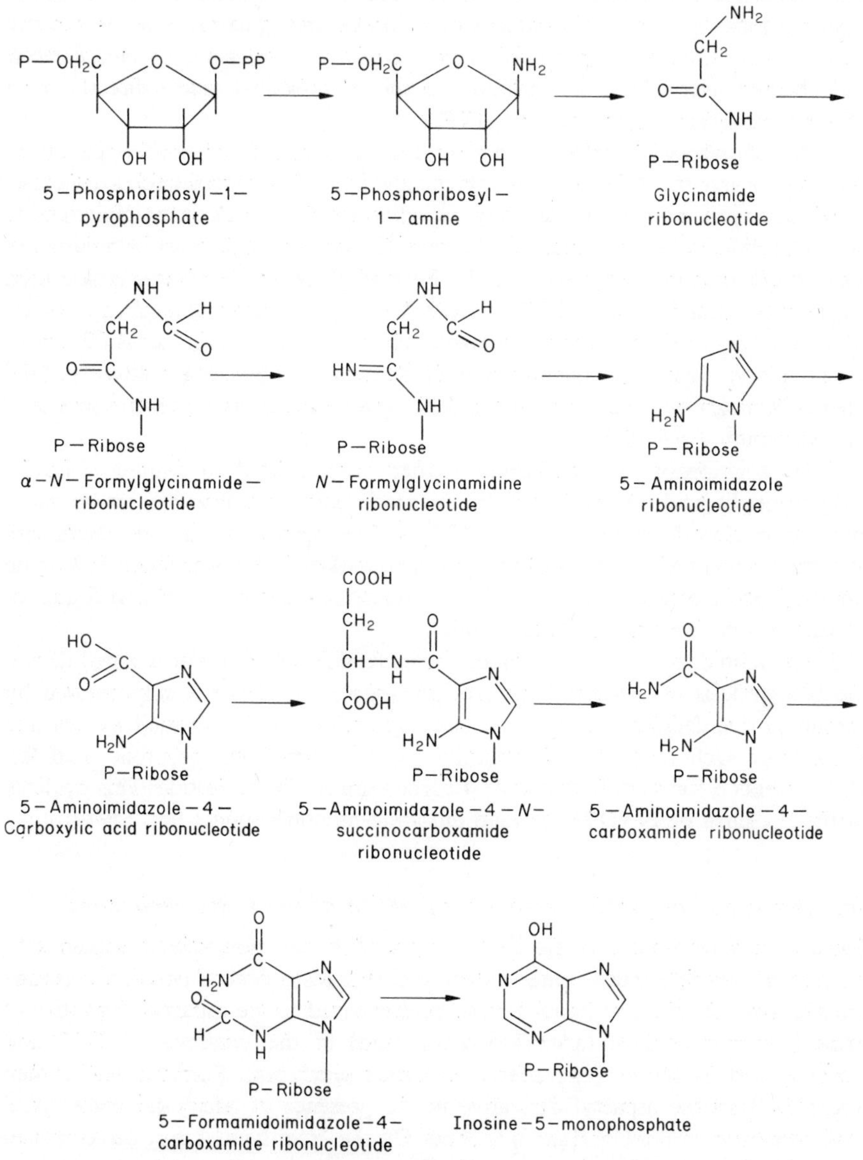

Fig. 159. Biosynthesis of inosine monophosphate.

The synthesis of the purine ring system probably proceeds in nearly the same manner in all organisms. The sequence of reactions (fig. 159) explained below relates to micro-organisms and animals, but recent evidence suggests that it may also hold for higher plants.

The biosynthesis begins with the reaction of 5-phosphoribosyl-1-pyrophosphate with glutamine in the case of bacteria and mammals, or asparagine in the case of higher plants, to form 5-phosphoribosyl-1-amine and glutamic acid or aspartic acid. Certain enzyme preparations from *Aerobacter aerogenes*, *Escherichia coli* and chicken liver also can synthesize 5-phosphoribosyl-1-amine directly from ribose-5-phosphate, ammonia and ATP.

5-Phosphoribosyl-1-amine is converted to glycinamide ribonucleotide in the presence of glycine, ATP and the enzyme glycinamide ribonucleotide synthetase.

Glycinamide ribonucleotide may be converted to α-N-formylglycinamide ribonucleotide in the presence of glycinamide ribonucleotide transformylase and anhydrocitrovorum factor (cf. C.3.2), Formyl glycinamide ribonucleotide then reacts with glutamine and ATP to yield N-formyl glycinamidine ribonucleotide which cyclizes to form 5-aminoimidazole ribonucleotide in a further ATP dependent reaction. This imidazole compound is accumulated by certain mutants of the fungus *Neurospora crassa*, and in the dephosphorylated form by certain strains of the bacterium *Escherichia coli*.

5-Aminoimidazole ribonucleotide is then carboxylated to 5-amino-4-imidazolecarboxylic acid ribonucleotide by the biotin-containing enzyme, aminoimidazole ribonucleotide carboxylase (cf. C.3.1). This compound further reacts with aspartic acid and ATP yielding 5-aminoimidazole-4-N-succinocarboxamide ribonucleotide, and the latter is then converted to fumaric acid and 5-aminoimidazole-4-carboxamide ribonucleotide.

5-Aminoimidazole-4-carboxamide ribonucleotide which is also formed during the biosynthesis of imidazole glycerol phosphate (cf. D.18) is accumulated by certain purine-deficient mutants in large quantities, and is formed by cell-free extracts of such mutants. Formylation by N^{10}-formyltetrahydrofolic acid (cf. C.3.2) yields 5-formamidoimidazole-4-carboxamide ribonucleotide. This cyclizes in the presence of NADPH to yield inosine-5-monophosphate (fig. 159).

(b) Conversion of inosine monophosphate to other purine derivatives

Inosine monophosphate is the first compound in this biosynthetic sequence to contain a complete purine ring system and it holds a central position in purine metabolism. On the one hand it may be converted to the aspartyl derivative of inosine monophosphate (adenylosuccinic acid) in the presence of GTP and aspartic acid by the enzyme adenyl succinate synthetase. Fumaric acid is then liberated from the aspartyl derivative in the presence of adenylsuccinate lyase, and adenosine monophosphate is formed. On the other hand it may be converted to xanthosine monophosphate in an NAD^+ dependent reaction. Adenosine monophosphate may be converted to inosine monophosphate via adenosine triphos-

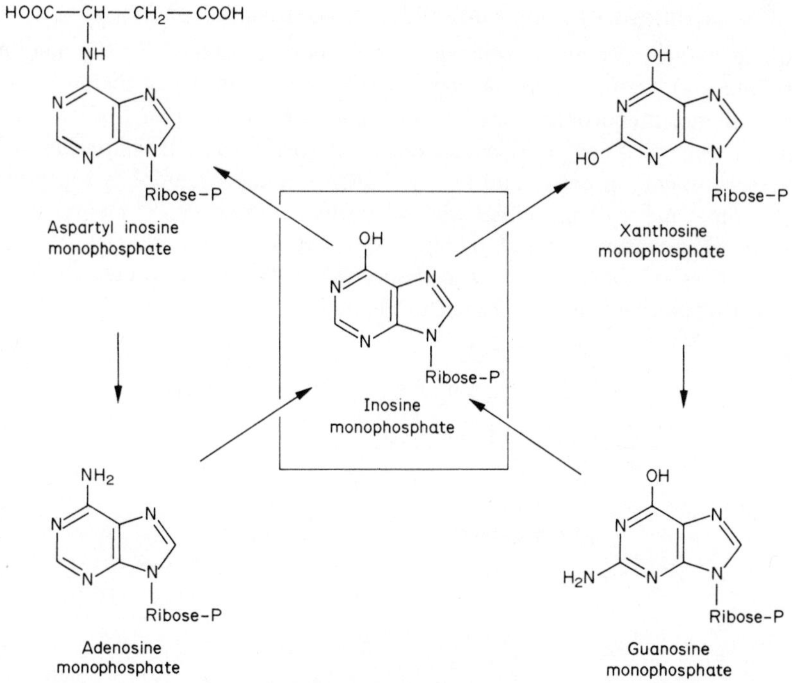

Fig. 160. Formation of adenosine-, guanosine- and xanthosine monophosphate.

phate and 1-(5′-phosphoribosyl)-ATP by reactions similar to those involved in the formation of the imidazole ring of histidine (cf. D.18). Direct conversion of adenosine monophosphate to inosine monophosphate is also possible by the enzyme adenosine aminohydrolase (= adenosine deaminase). Guanosine monophosphate is formed from xanthosine monophosphate in the presence of ATP and ammonia. In the presence of NADPH, guanosine monophosphate may be deaminated to inosine monophosphate (fig. 160).

References for further reading

Guarino, A. J. 'The Biogenesis of Purine and Pyrimidine Nucleotides', *Biogenesis of Natural Compounds*, ed. P. Bernfeld (Academic Press, Oxford, 1967), pp. 45–102.

Luckner, M. 'Purine, Pteridine und Alloxazine', *Biosynthese der Alkaloide*, ed. K. Mothes and H. R. Schütte (VEB Deutscher Verlag der Wissenschaften, Berlin, 1969), pp. 568–92.

Skoog, F. and Armstrong, D. J. Cytokinins, *Ann. Rev. Plant Physiol.* **21** (1970), pp. 359–84.

Murray, A. W. The Biological Significance of Purine Salvage, *Ann. Rev. Biochem.* **40** (1971), pp. 811–26.

8.2.2 Biosynthesis of purine alkaloids and methylated aminopurines

Xanthine, which originates from xanthosine monophosphate by elimination of phosphate and ribose, is the starting material for formation of the purine alkaloids, caffeine, theobromine and theophylline, which occur in *Coffea arabica*, *Camellia sinensis*, *Theobroma cacao* and other species, and are important as the active substituents in coffee and tea. In *Coffea arabica*, xanthine is converted to theobromine and caffeine either via N^3-methylxanthine or N^7-methylxanthine (fig. 161). The methyl groups originate from methionine (cf. C.3.3).

In the case of *Coffea arabica*, caffeine, fed to the plant, is demethylated to monomethylxanthines and degraded to aliphatic compounds (cf. D.8.2.6).

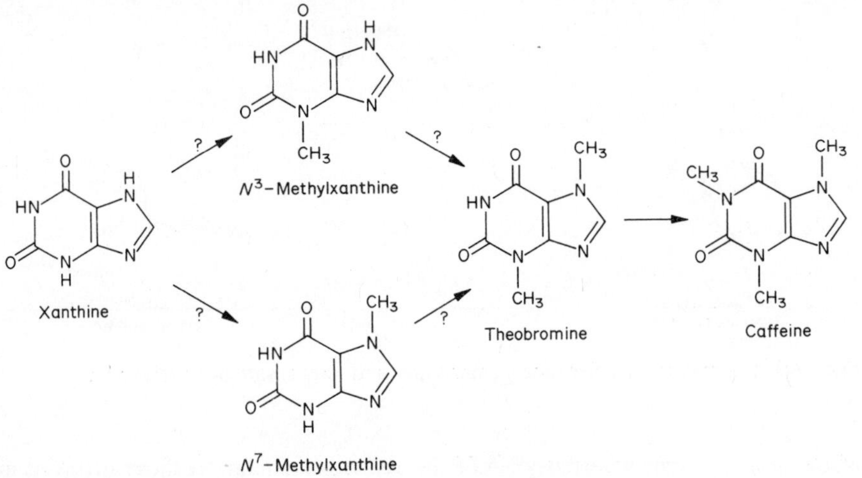

Fig. 161. Formation of theobromine and caffeine from xanthine.

The presence of methylated pyrimidines and purines (e.g. N^1-methylguanine, dimethylguanine, N^7-methylguanine, N^1-methyladenine, 2-methyladenine, dimethyladenine and N^1-methylhypoxanthine, cf. fig. 162) in soluble ribonucleic acids, especially in the tRNA species of bacteria and mammals, has been known for several years. Methylated bases are also found in ribosomal RNA and in DNA.

The function of the methylated purine and pyrimidine bases in the nucleic acid chain is unclear. Possibly, the secondary and tertiary structure of the polynucleotide chains are changed by methylation, due to the decrease in the amount of hydrogen bonding between the individual bases.

Experiments with micro-organisms show that the methylated bases are not incorporated into the nucleic acids as such, but originate by methylation of the purines and pyrimidines already polymerized in the RNA chain. Methionine acts as the donor of methyl groups both in intact cells and in cell-free extracts.

The methylated purines found in the urine of mammals and human beings (N^7-methylguanine was the first to be detected; recently, however, N^1-methyl-

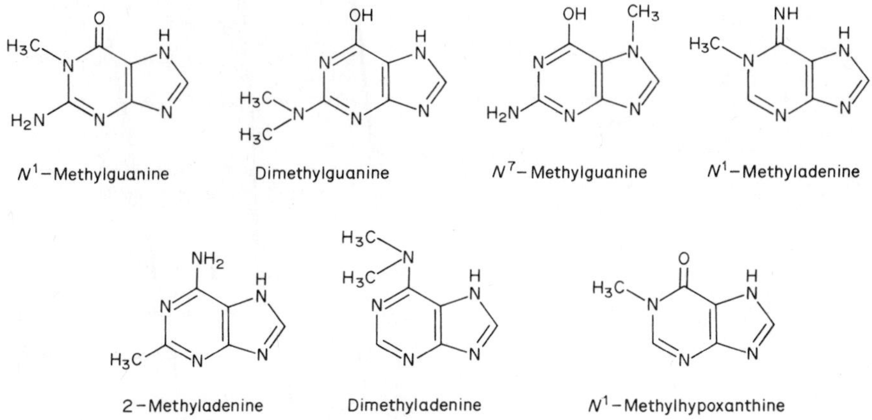

N^1–Methylguanine   Dimethylguanine N^7–Methylguanine N^1–Methyladenine

2–Methyladenine   Dimethyladenine N^1–Methylhypoxanthine

Fig. 162. Methylated purines from nucleic acids.

adenine, as well as N^1-methylhypoxanthine and dimethylguanine, have been detected after injection of methionine-$^{14}CH_3$) possibly originate from tRNA. They probably accumulate because the enzymes degrading the purine ring system are hindered in their attack by the methylation (cf. D.8.2.6).

Methylation of the freely existing aminopurines is also possible, judging from results from mammalian tissue preparations. Thus, the methylation of adenine to N^3-methyladenine by a preparation from rabbit lungs has been shown to occur in the presence of S-adenosyl-$^{14}CH_3$-methionine.

References for further reading

See D.8.2.1.

8.2.3 Formation and metabolism of pteridines

The pteridine ring system is a component of the compounds of the folic acid group which play an important role in 'one-carbon' metabolism (cf. C.3.2). They are also found in the free form as pterins (2-amino-4-hydroxypteridines) which exist in high concentrations in the wings of insects, in the eyes and skin of fishes, amphibians and reptiles, as well as in certain blue-green algae.

In the formation of the pteridine ring system (fig. 163), guanine (2-amino-6-hydroxy purine) is first converted to guanosine triphosphate. Then, catalysed by the enzyme cyclohydrolase, the ring system is hydrolytically cleaved between positions 7 and 8 (compound I) and 2,5-diamino-6-(5'-triphosphoribosyl)-amino-4-hydroxy pyrimidine is formed together with formic acid. After an Amadori rearrangement of ribose to ribulose (cf. formation of tryptophan, D.19, and of histidine, cf. D.18), ring closure takes place with the formation of an azomethine

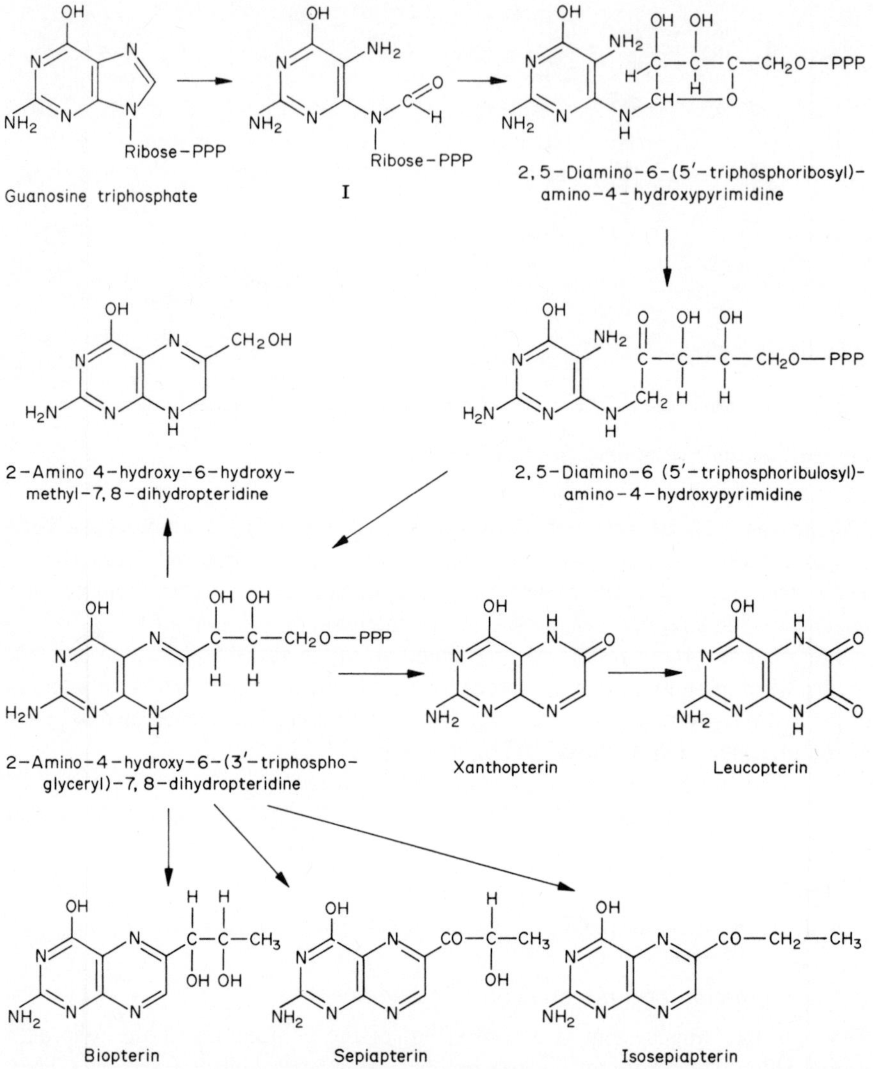

Fig. 163. Formation of various pteridine derivatives.

grouping to yield 2-amino-4-hydroxy-6-(3'-triphosphoglyceryl)-7,8-dihydropteri-dine. This substance, also termed 7,8-dihydroneopterin-3'-triphosphate, may perhaps after elimination of the phosphate groups be incorporated either into folic acid with the elimination of glycolaldehyde in an aldolase reaction via 2-amino-4-hydroxy-6-hydroxymethyl-7,8-dihydropteridine and its pyrophos-phate, or be converted to the secondary natural products biopterin, sepiapterin and isosepiapterin by isomerization and oxidoreductions. The pteridine pigments of the wings of butterflies, e.g. xanthopterin and leucopterin (fig. 163), are formed by the complete elimination of the side chain. The conversion of xanthopterin to leucopterin is possibly brought about by a xanthine oxidase (cf. C.2.2).

In the formation of tetrahydrofolic acid, 2-amino-4-hydroxy-6-hydroxy-methyl-7,8-dihydropteridine diphosphate and *p*-aminobenzoic acid are first converted to dihydropteroic acid in the presence of ATP and dihydropteroic acid synthetase. Condensation of dihydropteroic acid with glutamic acid in an ATP dependent reaction yields dihydrofolic acid. Dihydrofolic acid is reduced to tetra-hydrofolic acid in the presence of NADPH by dihydrofolic acid reductase (fig. 164).

p-Amino benzoic acid, which is necessary for the synthesis of dihydropteroic acid, may be displaced by sulphonamides from the dihydropteroic acid synthetase-substrate complex (competitive inhibition). This inhibits the formation of tetra-hydrofolic acid and as a result 'one-carbon'-metabolism comes to an end. It has been shown by quantitative experiments that 500 molecules of sulphanilamide, fifty molecules of sulphapyridine or ten molecules of sulphapyrimidine are necessary for the displacement of one molecule of *p*-aminobenzoic acid. The mechanism of action of the sulphonamides is one of the best examples of the action of drugs which is understood at the molecular level.

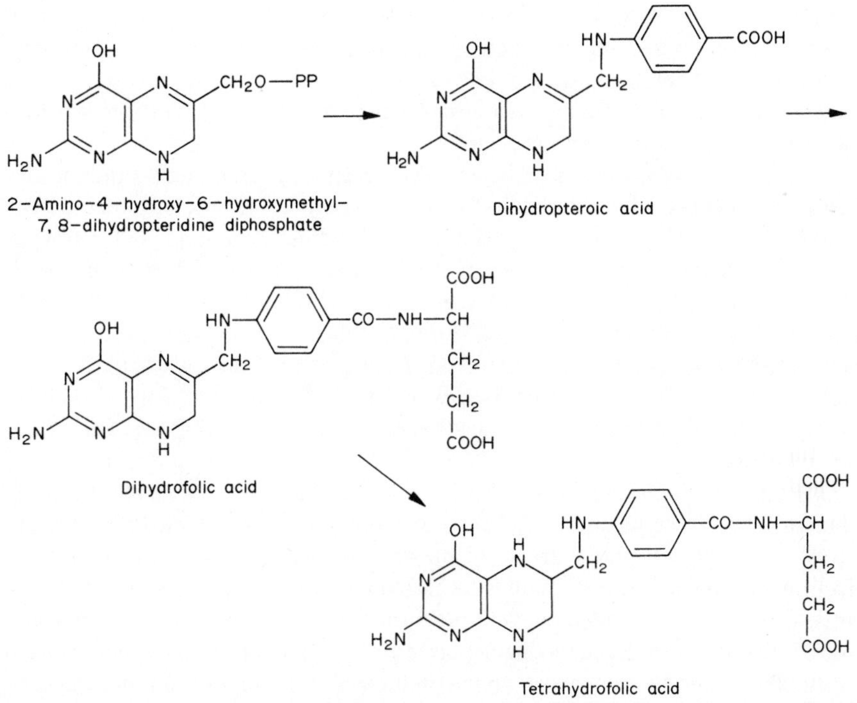

Fig. 164. Biosynthesis of tetrahydrofolic acid.

References for further reading

Luckner, M. 'Purine, Pteridine und Alloxazine', *Biosynthese der Alkaloide*, ed. K. Mothes and H. R. Schütte (VEB Deutscher Verlag der Wissenschaften, Berlin, 1969), pp. 568–92.

Blakley, R. L. *The Biochemistry of Folic Acid and Related Pteridines* (North-Holland Publishing Company, Amsterdam, 1969).

Forrest, H. S. and Van Baalen, C. Microbiology of Unconjugated Pteridines, *Ann. Rev. Microbiology* **24** (1970), pp. 91–108.

8.2.4 Formation of isoalloxazines (benzopteridines)

Riboflavin (vitamin B_2 or lactoflavin), which is particularly abundant in milk and its products, is one of the most important compounds possessing an iso-alloxazine ring system. Riboflavin is a constituent of a number of enzymes which catalyse oxidation-reduction reactions, as well as of flavin mononucleo-tide (FMN) and flavin adenine dinucleotide (FAD) which are of importance in many reactions as hydrogen acceptors, or in the reduced form as hydrogen donors (cf. C.2.2).

The formation of riboflavin has been investigated in a number of micro-organisms (e.g. *Asbya gossypii*, *Eremothecium ashbyii* and *Candida flareri*) in which the compound is formed in large quantities. It has been shown that synthesis of riboflavin is considerably increased on feeding purine derivatives.

In addition, by feeding radioactively labelled CO_2, formic acid, glycine, serine and threonine riboflavin was obtained in which the pyrimidine ring (A) was labelled in the same manner as the corresponding purine derivatives would be labelled after feeding these compounds.

It was later shown by feeding labelled adenine, guanine and xanthine that, with the exception of carbon atom 8, the entire purine ring system is incorporated into riboflavin. The actual precursor is a guanidine compound. It is, however, not clear whether it is in the form of the free base, or as the riboside or ribotide. The biosynthetic pathway proceeds via compound 1 (fig. 165), formed by the action of cyclohydrolase (cf. D.8.2.3) on guanine, and 4-ribitylamino-5-aminour-acil. In the case of the yeast, *Candida flareri*, 70% of the radioactivity of ^{14}C-labelled guanine added as precursor was found in riboflavin. The origin of the ribitol group of riboflavin is not yet clear, but probably originates from ribose by reduction (cf. D.1.4.1).

4-Ribitylamino-5-aminouracil reacts with diacetyl (cf. D.13) to form 6,7-dimethyl-8-ribityllumazine. 6,7-Dimethyl-8-ribityllumazine is directly converted to riboflavin by cell-free extracts of micro-organisms as well as by preparations from higher plants. The reaction is catalysed by the enzyme riboflavin synthetase. Investigations with purified enzyme preparations have shown that two molecules of 6,7-dimethyl-8-ribityllumazine are necessary for the formation of one molecule of riboflavin and that the entire riboflavin molecule originates from this precursor. In this reaction carbon atoms 6 and 7, together with both the attached methyl groups of one molecule, are transferred to the second molecule of 6,7-dimethyl-

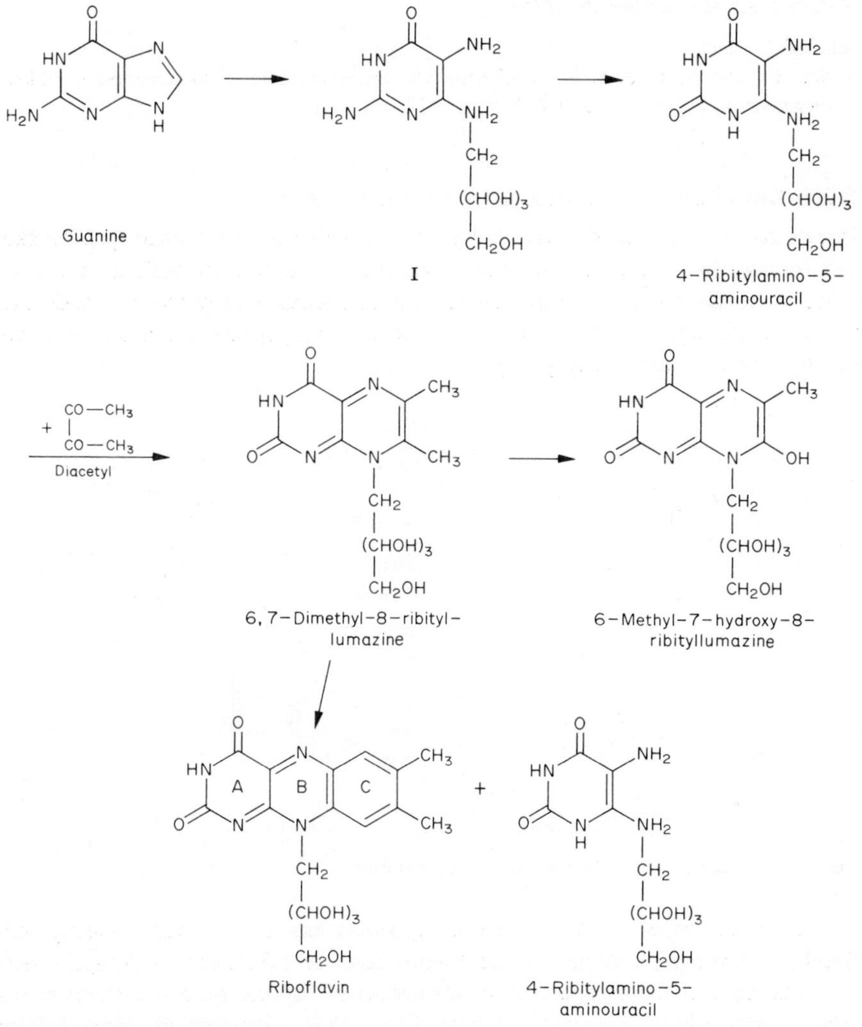

Fig. 165. Biosynthesis of riboflavin.

8-ribityllumazine to form 4-ribityl-amino-5-aminouracil and riboflavin. The transferred 'four-carbon' unit may be considered to be analogous to diacetyl; both methyl groups of dimethyllumazine are activated to such an extent that reaction with the carbonyl groups of the four-carbon unit is possible, with the elimination of water. Incubation of 6,7-dimethyl-8-ribityllumazine with extracts from *Candida flareri* or the green leaves of higher plants in the presence of oxygen leads to the formation of 6-methyl-7-hydroxy-8-ribityllumazine as well as riboflavin. 6-Methyl-7-hydroxy-8-ribityllumazine has been isolated from cultures of various micro-organisms and is not an intermediate in the biosynthesis of riboflavin.

References for further reading

See D.8.2.1 and

Schlee, D. and Zurnieden, K. Biochemie und Physiologie der Flavinogenese in Mikro-organismen, *Pharmazie* **25** (1970), pp. 651–69.

8.2.5 Conversion of purines to pyrrolopyrimidines

In addition to the real purines, compounds which may be termed 'purine like' are formed by a number of micro-organisms. Compounds such as the anti-biotics toxoflavin and fervenulin as well as nucleosides with pyrrol pyrimidine as the aglycone, e.g. tubercidin, toyocamycin and the peptide antibiotic viomycin (cf. fig. 166), belong to this group.

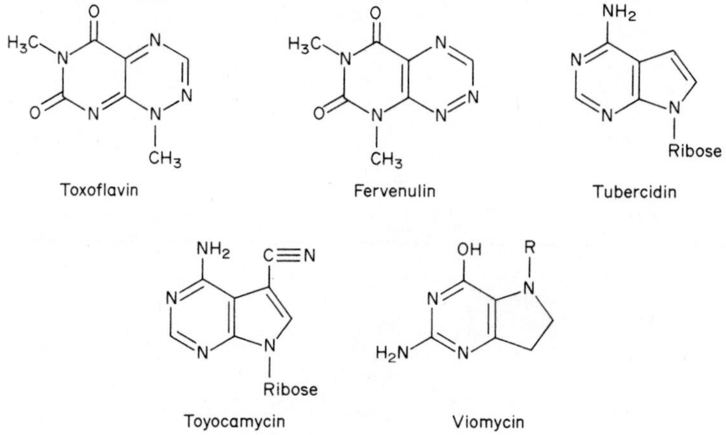

Toxoflavin   Fervenulin   Tubercidin

Toyocamycin   Viomycin

Fig. 166. Antibiotics with 'purine-like' ring systems.

The formation of most of these compounds has not yet been investigated. Recently, however, outlines of the biosynthesis of tubercidin in *Streptomyces tubercidicus* and the mechanism of formation of toyocamycin in *Streptomyces rimosus* have been reported. It was found that radioactivity derived from adenine-^{3}H and adenine-U-^{14}C was incorporated into the heterocyclic ring while radioactivity from adenosine-8-^{14}C was not incorporated. Radioactivity from ribose-1-^{14}C, glucose-1-^{14}C and glucose-2-^{14}C was incorporated into the pyrrolopyrimidine ring system as well as into the ribose moiety linked to carbon atom 9. After feeding ribose-^{14}C the incorporation rate was especially high.

These results indicate that both antibiotics originate from adenine (or adenosine), in such a way that the nitrogen atom at position 7 and the carbon atom at position 8 of the purine ring are lost and replaced by the first two carbon atoms of ribose. 6-Aminopyrrolopyrimidine is supposed to be an intermediate in the synthesis of tubercidin which in a later reaction is glycosidized by UDP-ribose (fig. 167). However, during the biosynthesis of toyocamycin a second ribose group

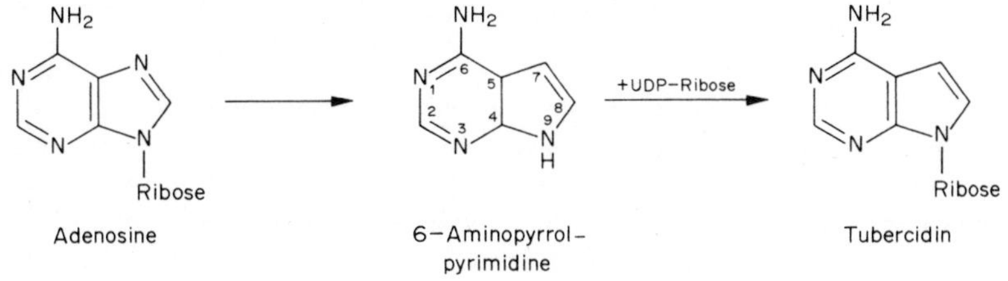

Fig. 167. Formation of tubercidin.

appears to be attached to the amino group at position 9 before the formation of the pyrrol ring.

The pathway of pyrrolopyrimidine biosynthesis is especially interesting since it involves the loss of the carbon atom at position 8 of the purine ring system (cf. D.8.2.3 and D.8.2.4). This has already been observed in the formation of pteridines and alloxazines and is possibly an indication that the other 'purine-like' compounds which occur in micro-organisms are also derived from purines.

References for further reading

Suhadolnik, R. J. *Nucleoside Antibiotics* (John Wiley & Sons, New York, 1970).
Schlee, D. Purin- und Pyrimidinantibiotica, *Biol. Rundschau* **8** (1970), pp. 317–30.

8.2.6 Degradation of purine ring system

While human beings, higher monkeys, birds, terrestrial reptiles and most insects excrete uric acid as the end product of purine metabolism, almost all other animals, as well as plants and micro-organisms, degrade the purine ring system to allantoin and allantoic acid, or via both these compounds to urea and glyoxylic acid. The reactions involved in these conversions were first studied in mammals and bacteria, and later in higher and lower plants (fig. 168).

The naturally occurring purines are first oxidized to uric acid via hypoxanthine and xanthine. This oxidation is catalysed by xanthine oxidase (cf. C.2.2) or xanthine dehydrogenase.

The formation of hydrogen bonds between the enzyme and the purine is facilitated if the nitrogen atoms are not substituted by methyl groups.

Before oxidation occurs, ATP is converted to inosine via inosine monophosphate, and then to hypoxanthine and ribose-l-phosphate by a nucleoside phosphorylase. Adenosine is similarly converted to hypoxanthine via inosine. Guanine may be directly deaminated to xanthine.

The conversion of uric acid to allantoin is catalysed by the enzyme uricase. During this reaction, molecular oxygen is reduced to hydrogen peroxide. Uricase has been detected in animals, fungi and bacteria, as well as in the seeds of many plants.

Absence of the enzyme uricase in human beings, with a resultant increased uric acid level in the tissues, leads in certain cases to symptoms of gout and chronic arthritis. The cause of these diseases is unknown, however. The use of isotopically labelled uric acid has shown that the uric acid pool of a gout patient is up to fifteen times greater than that of a healthy person; however, the uric acid level of a person who does not show symptoms of gout and chronic arthritis may also lie above the normal. Uric acid itself, therefore, does not appear to be responsible for cases of acute gout, since introduction of larger quantities of uric acid into gout patients does not cause an attack, and colchicine, which exerts a soothing influence on the attack, does not lower the uric acid level in either the blood or the tissues. Drugs which dissolve uric acid such as salicylates, facilitate a slow resorption of the deposited uric acid crystals, but are not a defence against acute attacks of gout.

The conversion of uric acid to allantoin catalysed by uricase probably proceeds via a symmetrical intermediate, since the urea molecule originating from the pyrimidine ring and that formed from the imidazole ring within the molecule of allantoin cannot be distinguished from one another. During conversion of the symmetrical compound to allantoin, carbon atom 6 of the purine ring system is eliminated as CO_2 in a nonenzymatic reaction.

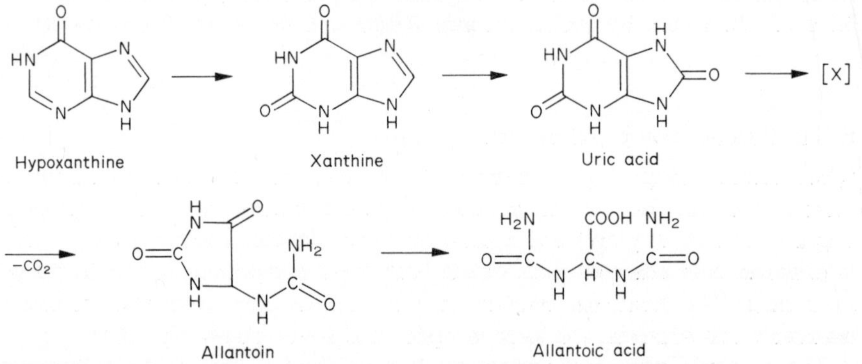

Hypoxanthine   Xanthine   Uric acid   [x]

Allantoin   Allantoic acid

Fig. 168. Degradation of purine ring system via allantoin and allantoic acid.

While, most probably, methylated purines cannot be degraded in mammals and human beings (cf. D.8.2.2), the *Pseudomonas* bacteria are capable of oxidatively degrading caffeine, theobromine, theophylline, N^7-methylxanthine and N^7-methyluric acid without prior demethylation. Methylated allantoins are probably formed. Degradation stops at the level of the methylated urea derivatives. A pathway via allantoin, allantoic acid and urea, and which possibly includes N^3-methyl xanthine and N^7-methyl xanthine, has been suggested for the degradation of caffeine by the older leaves of *Coffea arabica*. Carbon dioxide is ultimately formed from the ring carbon atoms, as well as from the methyl groups.

A number of other, but less important degradative pathways are known to exist besides the widespread pathway presented above. Thus, in humans, though the enzyme uricase is absent, about 20% of uric acid introduced from outside the body is degraded to urea. How this happens is still not clear; however, it has been shown that leucocytes synthesize alloxan from uric acid in the presence of small quantities of H_2O_2.

Extracts capable of degrading purines to ammonia, glycine, formic acid and carbon dioxide have been isolated from anaerobic *Clostridium* species (e.g. *C. cylindrosporum* and *C. acidi-urici*). Xanthine is converted to formimino glycine via 4-ureido-5-carboxyimidazole, 4-amino-5-carboxyimidazole and 4-aminoimidazole as well as possibly 4-hydroxyimidazole (fig. 169). Formimino-glycine is either further degraded to glycine, formic acid and ammonia or the formimino group is transferred to tetrahydrofolic acid to form N^5-formimino-tetrahydrofolic acid and glycine (cf. C.3.2).

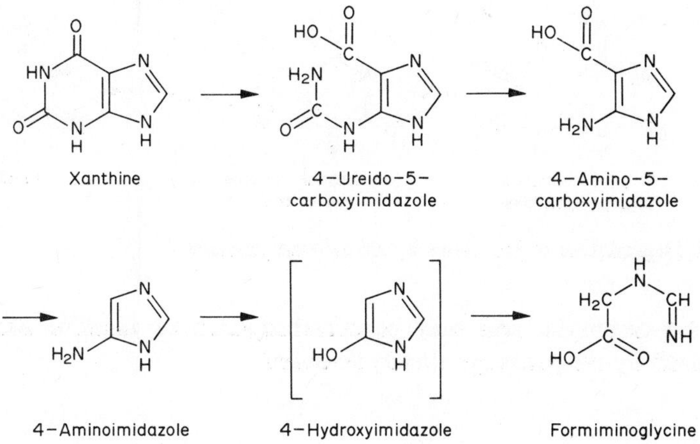

Fig. 169. Degradation of purine ring system by *Clostridium* species.

Alcaligenes faecalis, a micro-organism which is capable of growing on xantho-pterin and isoxanthopterin as the only carbon and nitrogen source, deaminates these pterins to 2,4,6-trioxohexahydropteridine or 2,4,7-trioxohexahydro-pteridine respectively and then oxidizes them to 2,4,6,7-tetraoxooctahydrop-terdine (fig. 170). The enzyme which performs this oxidation is similar to xanthine oxidase (cf. C.2.2). Other pteridines may be oxidized by the xanthine oxidase obtained from milk.

2,4,6,7-Tetraoxooctahydropteridine is rearranged to xanthine-8-carboxylic acid by the enzyme tetraoxopteridine isomerase (fig. 170). It has been shown with the help of radioactive labelled compounds that the carboxyl group of xanthine-8-carboxylic acid originates from carbon atom 7 of the pteridine ring system.

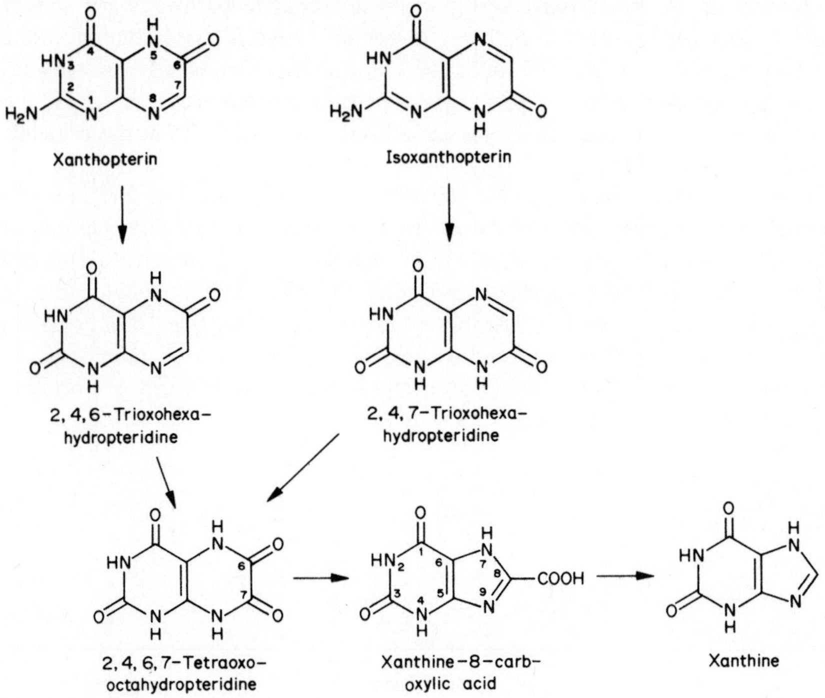

Fig. 170. Degradation of pteridines by *Alcaligenes faecalis*.

Xanthine-8-carboxylic acid may be decarboxylated to xanthine and further metabolized by the pathways already described.

References for further reading

See D.8.2.1

8.3 Formation of glycine conjugates in animals

The detoxication of acids which cannot be further used in metabolism frequently takes place in animals by formation of an acid amide like linkage with glycine. This reaction proceeds in liver and kidney. The acids must first be converted to their CoA-ester (cf. C.1.2). The conjugates produced are excreted in the urine.

The best-known conjugate is hippuric acid which is formed from benzoic acid and glycine (cf. fig. 171). Heterocyclic acids such as nicotinic acid, as well as phenylacetic acid and their derivatives also form conjugates.

Glycine conjugation is widespread in animals, but does not occur in certain birds where glycine is replaced by the amino acid ornithine (cf. D.15.4).

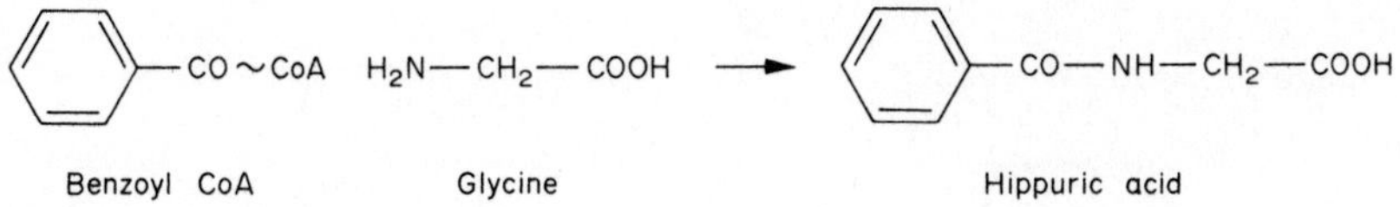

Fig. 171. Biosynthesis of hippuric acid.

References for further reading

Williams, R. T. 'The Biogenesis of Conjugation and Detoxication Products', *Biogenesis of Natural Compounds*, ed. P. Bernfeld (Pergamon Press, Oxford, 1967), pp. 589–639.

9.   Biosynthesis of secondary products from cysteine

Inorganic sulphate taken up by the cell must first be reduced to sulphide berore being used in metabolism, i.e. before it is incorporated into the sulphur-containing amino acids, cysteine and methionine. The first step is the activation of sulphate by a reaction with ATP forming adenosine-5'-phosphosulphate and 3'-phospho-adenosine-5'-phosphosulphate ('activated sulphate'). Both phosphosulphates can then be enzymically reduced to sulphite. The sulphite is either directly reduced to sulphide, which reacts with serine or O-acetyl serine to form cysteine, or is reduced to thiosulphate and then converted to cysteine via S-sulphocysteine (fig. 172).

The sulphate group of 3'-phosphoadenosine-5'-phosphosulphate may be transferred to other compounds by means of sulphotransferases.

$$R_1-\underset{\underset{O}{\|}}{\overset{\overset{OH}{|}}{P}}-O-\underset{\underset{O}{\|}}{\overset{\overset{O}{\|}}{S}}-OH + R_2-H \longrightarrow R_1-\underset{\underset{O}{\|}}{\overset{\overset{OH}{|}}{P}}-OH + R_2-\underset{\underset{O}{\|}}{\overset{\overset{O}{\|}}{S}}-OH$$

The formation of sulphate esters occurs most frequently, but reaction with amines and sulphhydryl compounds is also possible. Phenols especially are converted to the sulphate ester in animals before their elimination by the kidneys (cf. the formula for urinary indican, D.19.6). Corresponding phenol derivatives are also found in plants (e.g. the flavonol sulphate ester persicarin in *Polygonum thunbergii*). Sulphate esters of carbohydrates are synthesized by micro-organisms and higher plants as well as by animals.

References for further reading

Meister, A. *Biochemistry of the Amino Acids*, Vol. II (Academic Press, New York, 1965).

Williams, R. T. 'The Biogenesis of Conjugation and Detoxication Products', *Biogenesis of Natural Compounds*, ed. P. Bernfeld (Pergamon Press, Oxford, 1967), pp. 589–639.

Roy, A. B. and Trudinger, P. A. *The Biochemistry of Inorganic Compounds of Sulphur* (University Press, Cambridge, 1970).

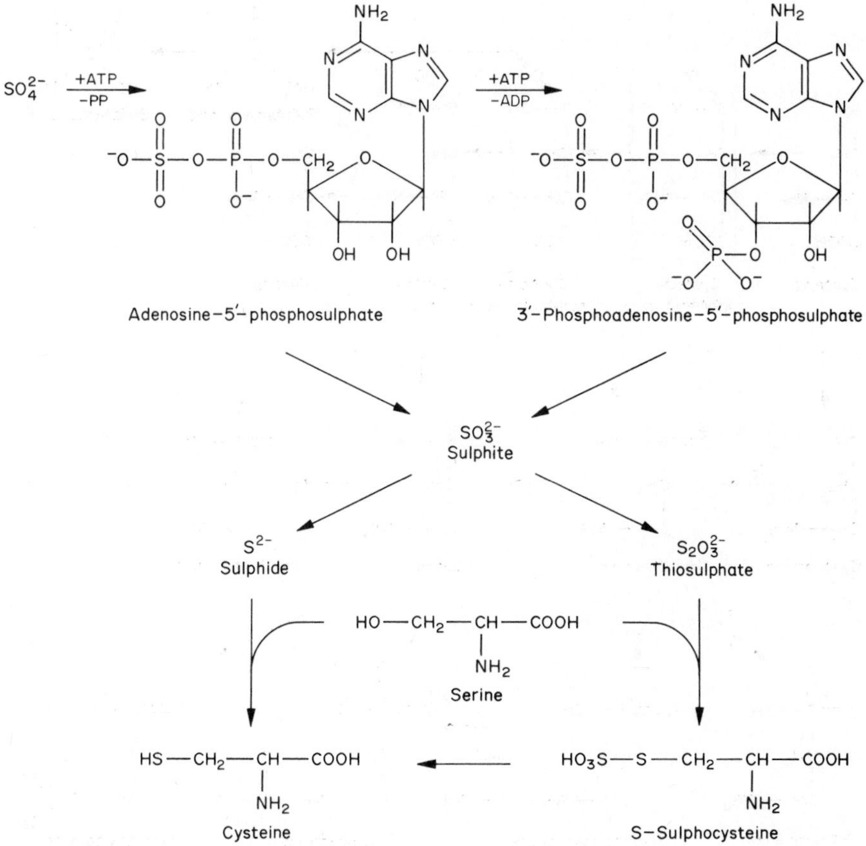

Fig. 172. Formation of cysteine.

9.1 Oxidation and decarboxylation of cysteine

Oxidized derivatives of the amino acid cysteine occur in a number of living organisms but especially in animals and micro-organisms. In the formation of these derivatives cysteine is oxidized to cysteic acid via cysteine sulphenic acid (cysteine sulphoxide) and cysteine sulphinic acid. Cysteine sulphenic acid does not occur in the free form and is probably enzyme bound. Cysteine sulphinic acid may be degraded to alanine and sulphurous acid by aspartate-β-decarboxylase. Sulphurous acid is comparatively easily oxidized to sulphuric acid (fig. 173).

On oxidation cysteamine yields taurine and hypotaurine. Both these compounds occur in large quantities in many tissues and in the urine of mammals and human beings. They may also arise by decarboxylation of cysteine sulphinic

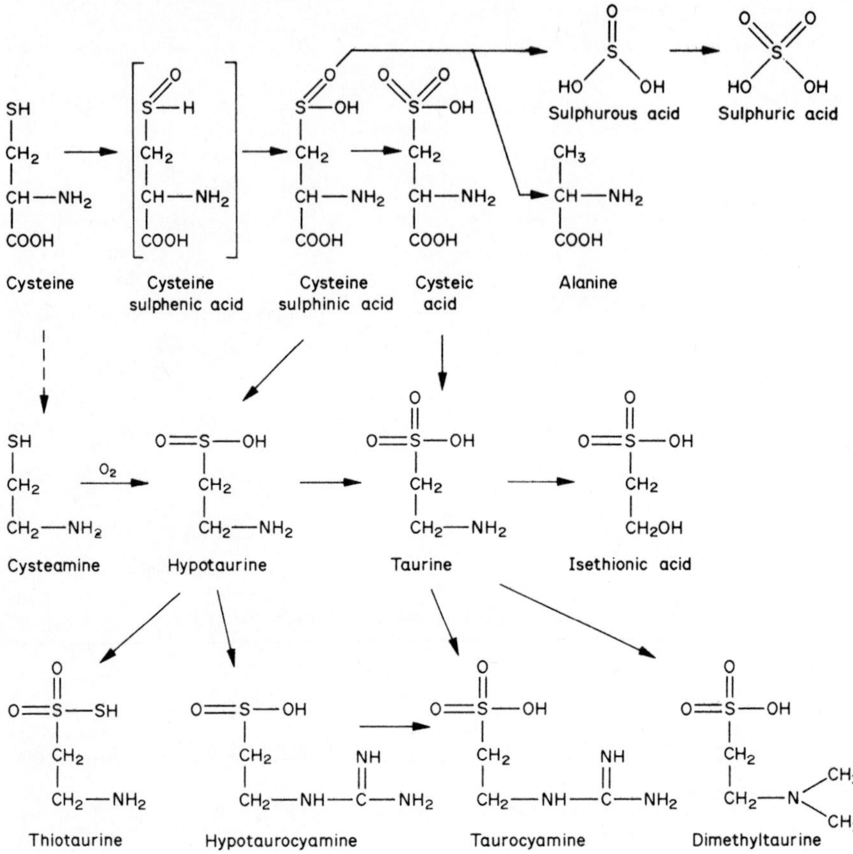

Fig. 173. Oxidation products and other metabolites from cysteine and cysteamine.

acid and cysteic acid. Both atoms of an oxygen molecule are introduced into cysteamine by a dioxygenase during the formation of hypotaurine.

The nerve fibres of cuttle fish as well as the heart muscles of mammals contain, besides taurine, a large amount of isethionic acid. This latter compound is derived from taurine by deamination. Taurine may be further converted to dimethyltaurine and taurocyamine and hypotaurine to thiotaurine and hypotaurocyamine. Hypotaurocyamine is easily oxidized to taurocyamine (fig. 173).

Cysteamine is a constituent of coenzyme A. In the formation of coenzyme A cysteine first condenses with 4'-phosphopantothenic acid to form 4-phosphopantothenylcysteine, which is then decarboxylated to 4'-phosphopantetheine. 4'-Phosphopantetheine is then converted to coenzyme A, via dephospho-coenzyme A, in two ATP dependent reactions (fig. 174).

Cysteamine is formed during the degradation of coenzyme A; other pathways for the formation of the amine cannot be ruled out, however (e.g. the direct decarboxylation of cysteine, cf. D.7.3).

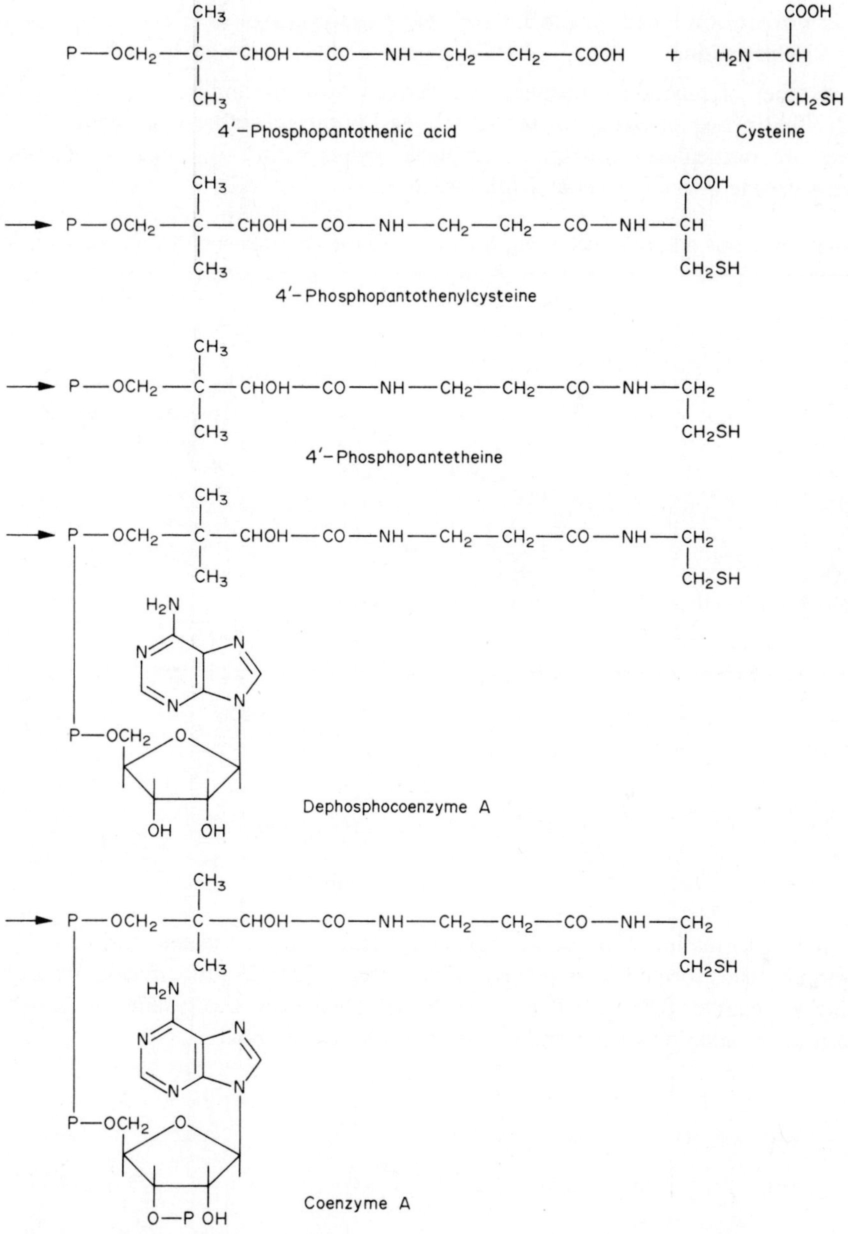

Fig. 174. Biosynthesis of coenzyme A.

References for further reading

See D.9.

9.2 Formation and metabolism of *S*-alkylcysteine derivatives and sulphoxides

A number of secondary products are derived from the amino acid cysteine by substitution and oxidation of the sulphhydryl group (cf. table 16). These derivatives are particularly widespread in higher plants within the families Liliaceae, Brassicaceae (Crucifereae) and Mimosaceae.

Table 16. Some naturally occurring *S*-alkylcysteine derivatives and cysteine sulphoxides

R	$\begin{array}{c} CH_2-CH-COOH \\ \mid \quad\quad \mid \\ R-S \quad NH_2 \end{array}$	$\begin{array}{c} CH_2-CH-COOH \\ \mid \quad\quad \mid \\ R-S_{\diagdown O} \quad NH_2 \end{array}$
H—	Cysteine	Cysteine sulphoxide (cysteine sulphenic acid)
CH_3—	*S*-Methylcysteine	Methylalliin
$CH_3-CH_2-CH_2-$	*S*-Propylcysteine	Propylalliin
$CH_2=CH-CH_2-$	*S*-Allylcysteine	Allylalliin (Alliin)
$CH_3-CH=CH-$	*S*-Δ^1-Propenylcysteine	Δ^1-Propenylalliin
$HOOC-CH_2-CH_2-$	*S*-2-Carboxyethylcysteine	—
$\begin{array}{c} HOOC-CH-CH_2- \\ \mid \\ CH_3 \end{array}$	*S*-2-Carboxyisopropylcysteine	—

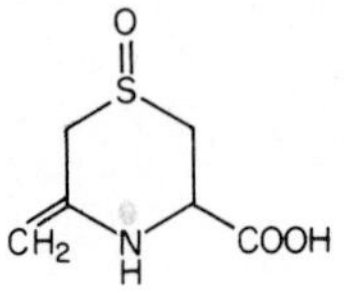

Cycloalliin 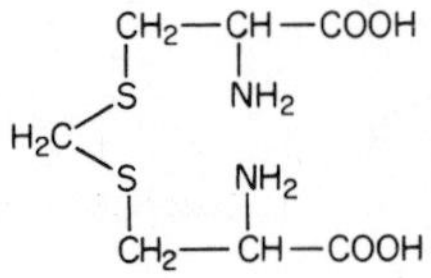   Dyenkolic acid

In the formation of these substances cysteine is first alkylated, and the alkylcysteines thus formed are oxidized to alliins (fig. 175). Δ^1-Propenylalliin is probably a precursor for cycloalliin, found in *Allium* species. Cycloalliin seems to be formed by addition of the amino group to the double bond.

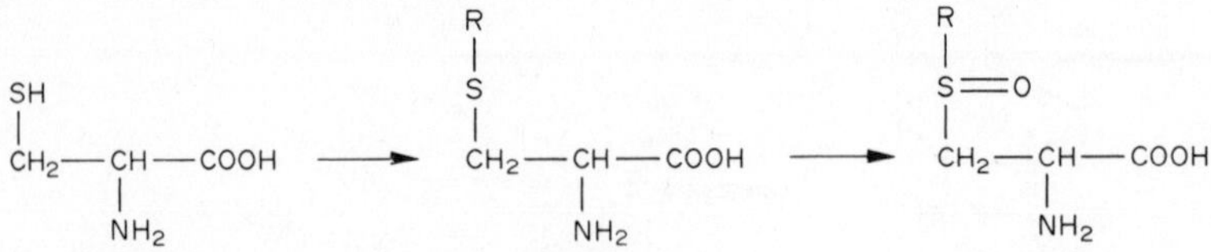

Fig. 175. Formation of alkylcysteines and alliins.

Alliins are degraded to an alkyl sulphenic acid and α-aminoacrylic acid by the action of the enzyme alliinase, which up until now has been found only in the genera *Allium* and *Nothoscordum*. Allylsulphenic acid is formed from allylalliin and Δ^1-propenylsulphenic acid is formed from Δ^1-propenylalliin by this reaction. Δ^1-Propenylsulphenic acid is probably the tear-producing substance which is liberated on crushing the cells of onion, *Allium cepa*.

α-Aminoacrylic acid decomposes spontaneously into pyruvic acid and ammonia. The sulphenic acids condense easily to dimeric compounds of the allicin type which on steam distillation give up oxygen and change to disulphides (fig. 176). Compounds of the allicin type which possess considerable antibiotic activity, as well as the disulphides, have mostly a strong unpleasant smell.

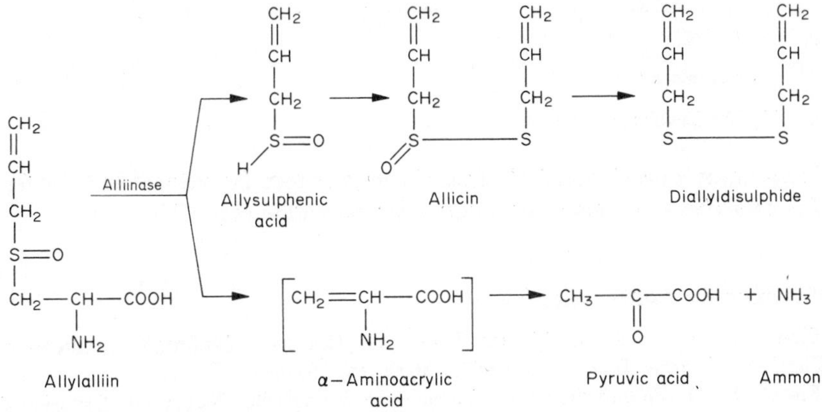

Fig. 176. Degradation of allylalliin by alliinase.

References for further reading

Virtanen, A. I. Organische Schwefelverbindungen in Gemüse- und Futterpflanzen, *Angew. Chem.* **74** (1962), pp. 374–82.

Hegnauer, R. *Chemotaxonomie der Pflanzen*, Vol. II (Birkhäuser Verlag, Basel, 1963).

Meister, A. *Biochemistry of the Amino Acids*, Vol. I (Academic Press, New York, 1965).

Kjaer, A. 'The Distribution of Sulphur Compounds', *Comparative Phytochemistry*, ed. T. Swain (Academic Press, London, 1966), pp. 187–94.

9.3 Luciferin and luciferase

Certain micro-organisms, a number of insects (e.g. the glow worm) and a few other animals possess organs which are capable of emitting light. The substance involved in this process is D-luciferin, which contains a D-cysteine unit in the molecule. The biosynthesis of luciferin has not yet been elucidated.

Luciferin reacts with ATP in the presence of luciferase to yield luciferyl-AMP, with the elimination of pyrophosphate. This process resembles the activation of other acids by ATP (cf. D.2.1). Activated luciferin then reacts with oxygen to form an intermediate, whose structure is not yet known. On emitting light the intermediate is converted to dehydroluciferin. For every molecule of activated luciferin conve:ted to dehydroluciferin, one molecule of oxygen is used and one

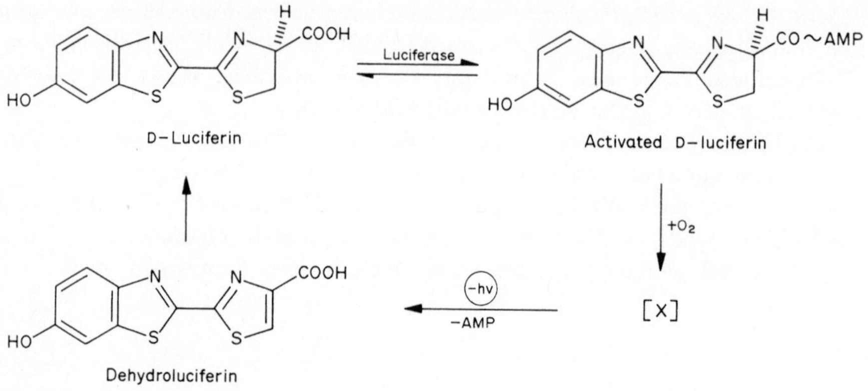

Fig. 177. The Luciferin cycle.

light quantum is produced. Dehydroluciferin may then be reduced to D-luciferin and is available for renewed oxidation after activation (fig. 177).

References for further reading

McElroy, W. D., DeLuca, M. and Travis, G. Molecular Uniformity in Biological Catalysis, *Science* (*Washington*) **157** (1967), pp. 150–60.

McElroy, W. D. Biolumineszenz—Chemie und biologische Bedeutung, *Umschau* **69** (1969), pp. 472–4.

9.4 Biosynthesis of biotin

Biotin is the coenzyme of carboxylases and transcarboxylases (cf. C.3.1). The feeding of radioactive-labelled precursors to *Achromobacter*, followed by degradation of the radioactive biotin formed has shown that this compound probably originates from cysteine, pimelyl CoA and carbamyl phosphate (fig. 178). Most recent experiments indicate that the entire pimelic acid molecule, including both the carboxyl groups, serves as the precursor of biotin. Desthiobiotin which occurs in various micro-organisms is also converted to biotin by yeast and *Escherichia coli*. In this transformation methionine and methionine sulphoxide as well as *S*-adenosyl methionine are particularly suitable as sources of sulphur.

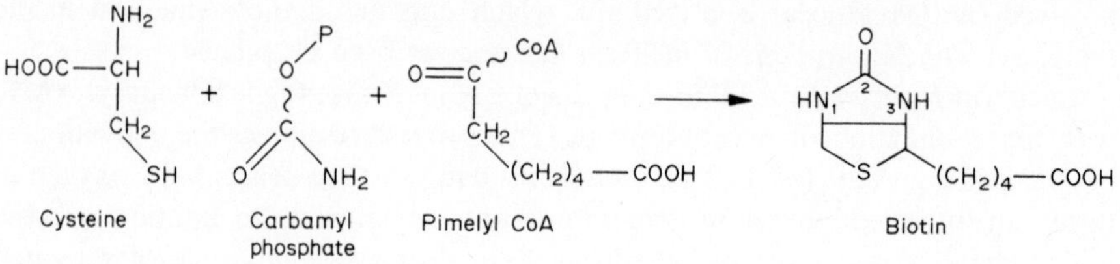

Fig. 178. Formation of biotin.

Incorporation of ^{35}S from thioglycollic acid has also been observed. It seems that alanine (in place of cysteine) reacts with pimelyl CoA and carbamyl phosphate in the biosynthesis of desthiobiotin.

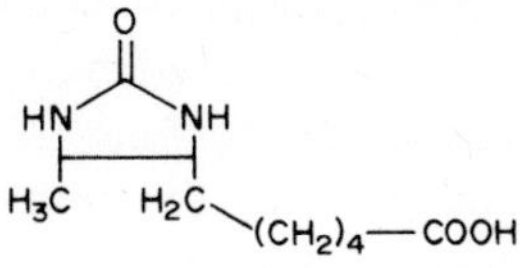

Fig. 179. Desthiobiotin.

References for further reading

Luckner, M. Über neue Arbeiten zur Biosynthese *N*-heterocyclischer Verbindungen, 6. Teil: Bildung und Abbau von Verbindungen mit Imidazol und Tetrahydroimidazolringsystem. *Pharmazie* **22** (1967), pp. 65–75.

9.5 Formation of premercapturic acids and mercapturic acids

Mercapturic acids are derivatives of *N*-acetyl-L-cysteine and are synthesized by animals after feeding aromatic hydrocarbons or aliphatic or aromatic

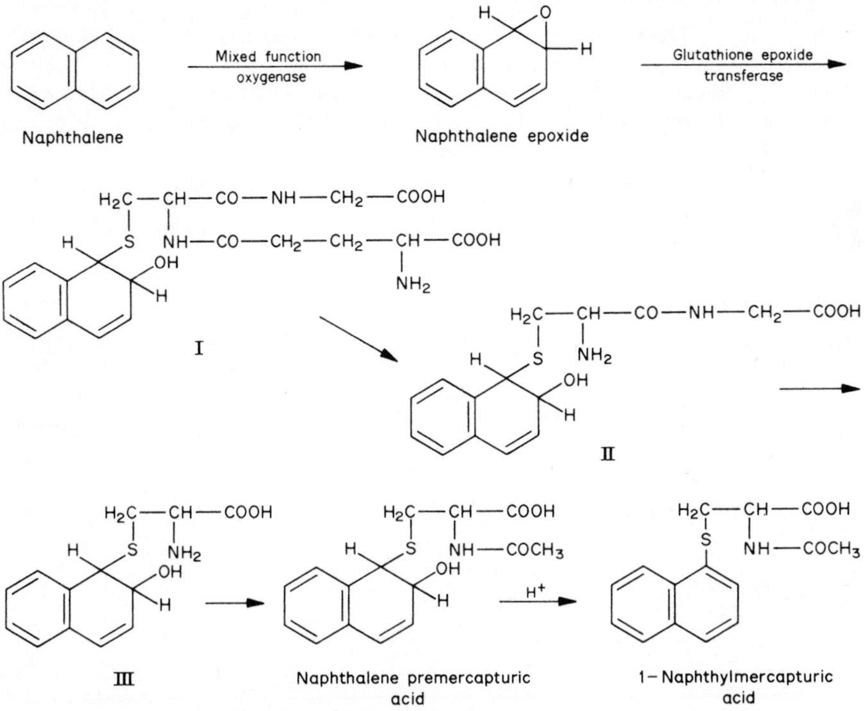

Fig. 180. Formation of 1-naphthyl-mercapturic acid.

hydrocarbons substituted by chlorine atoms or nitro groups. They are involved in the detoxication of foreign compounds.

Mercapturic acids are formed by two pathways. The first pathway operates in the case of unsubstituted aromatic hydrocarbons, such as naphthalene. This is first converted to an epoxide by a mixed function oxygenase contained in the liver microsomes (cf. C.2.6.5). The epoxide grouping is then opened by attack of the sulphhydryl group of glutathione with the formation of compound I. The glutamyl group is then eliminated through a transpeptidization reaction (compound II), and the glycine residue is hydrolytically eliminated by a peptidase (compound III). Acetylation of the cysteine group forms 1-naphthyl premercapturic acid which is excreted in the urine. Premercapturic acids may be converted to mercapturic acids *in vitro* by treatment with acids, when the elements of water are eliminated (fig. 180).

In the case of aromatic compounds substituted by halogen or nitro groups (e.g. chlorobenzene or *p*-fluoronitrobenzene), glutamine-*S*-aryl transferase catalyses the addition of glutathione with the elimination of the substitute groups and maintenance of the aromatic character. The corresponding mercapturic acid is formed *in vivo* through this pathway.

References for further reading

Williams, R. T. 'The Biogenesis of Conjugation and Detoxication Products', *Biogenesis of Natural Compounds*, ed. P. Bernfeld (Pergamon Press, Oxford, 1967), pp. 589–639.

Wood, J. L. 'Biochemistry of Mercapturic Acid Formation', *Metabolic Conjugation and Metabolic Hydrolysis*, ed. W. H. Fishman (Academic Press, New York, 1970), pp. 261–99.

10. Formation of secondary products from methionine

The amino acid L-methionine originates from L-homoserine which in turn is synthesized from aspartic acid in a reductive process via aspartic acid semialdehyde.

L-Cysteine is usually the donor of the —SH group. L-Homoserine and L-cysteine condense to form cystathionine with the elimination of water. Cystathionine is then hydrolytically decomposed to L-homocysteine and L-serine (fig. 181).

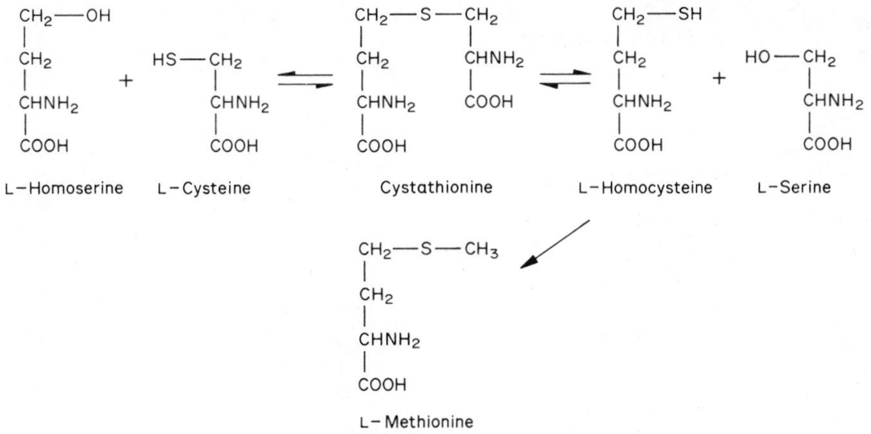

```
CH₂——OH                          CH₂——S——CH₂                    CH₂——SH
 |                                |         |                    |
CH₂         HS——CH₂              CH₂       CHNH₂                 CH₂           HO——CH₂
 |     +     |          ⇌        |         |          ⇌         |     +        |
CHNH₂       CHNH₂                CHNH₂     COOH                  CHNH₂         CHNH₂
 |           |                    |                              |             |
COOH        COOH                 COOH                           COOH          COOH

L–Homoserine  L–Cysteine              Cystathionine           L–Homocysteine   L–Serine


                                 CH₂——S——CH₃
                                  |
                                 CH₂
                                  |
                                 CHNH₂
                                  |
                                 COOH

                                 L–Methionine
```

Fig. 181. Biosynthesis of methionine.

Esters of homoserine (O-oxalylhomoserine, O-acetyl- or O-succinylhomoserine, cf. D.7.2) are important intermediates of this reaction in certain microorganisms and higher plants.

Methionine may also be formed from O-acetylhomoserine and sulphide in spinach, *Spinacia oleracea*, as shown below:

$$S^{2-} + O\text{-Acetylhomoserine} \longrightarrow \text{Homocysteine} + \text{Acetate}$$

N^5-tetrahydrofolic acid acts as the 'one-carbon' donor in the methylation of L-homocysteine to L-methionine. Vitamin B_{12} is also required (cf. D.8.2.1).

References for further reading

Meister, A. *Biochemistry of the Amino Acids*, Vol. II (Academic Press, New York, 1965).

10.1 Sulphonium compounds as secondary natural products

The most important sulphonium compound in metabolism is 'activated methionine' (*S*-adenosylmethionine) which functions as the donor of methyl groups in transmethylation reactions (cf. C.3.3). Various secondary natural products are derived from this compound.

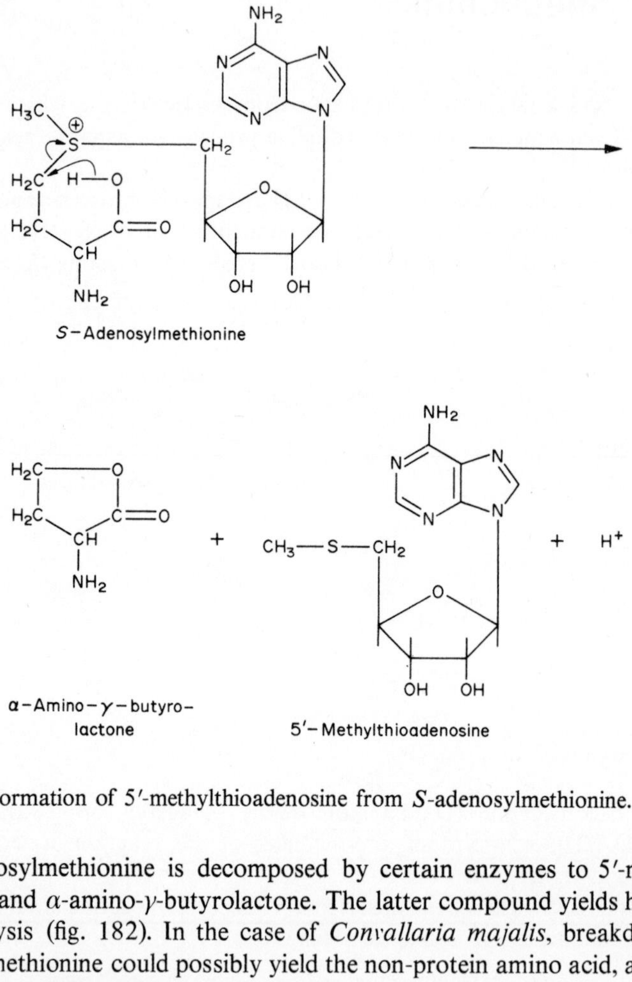

Fig. 182. Formation of 5'-methylthioadenosine from *S*-adenosylmethionine.

S-Adenosylmethionine is decomposed by certain enzymes to 5'-methylthioadenosine and α-amino-γ-butyrolactone. The latter compound yields homoserine on hydrolysis (fig. 182). In the case of *Convallaria majalis*, breakdown of *S*-adenosyl methionine could possibly yield the non-protein amino acid, azetidine-2-carboxylic acid (fig. 183).

5'-Methyladenosine which is found in a large number of micro-organisms, is also formed during the synthesis of spermidine (fig. 184). *S*-Adenosylmethionine is decarboxylated, and the propylamine group of *S*-adenosyl-*S*-methylmercaptopropylamine so formed is transferred to putrescine.

Dimethylpropiothetin occurs in various algae as well as in animals and α-aminodimethyl-γ-butyrothetin (*S*-methylmethionine) occurs in higher plants

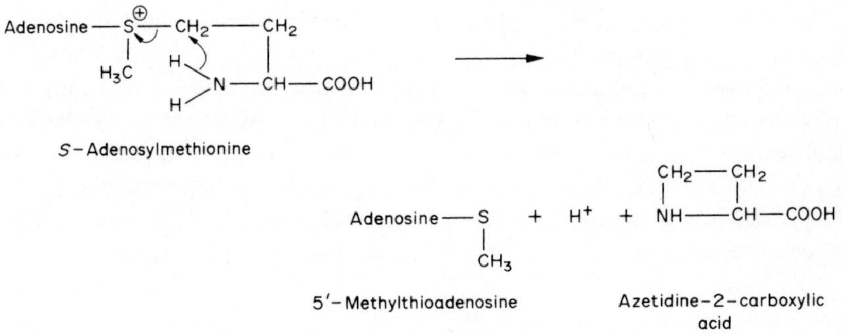

Adenosine—S⁺—CH₂—CH₂
      H₃C   H
         N—CH—COOH
      H

S–Adenosylmethionine

$$\longrightarrow$$

Adenosine—S   +   H⁺   + $\overset{\displaystyle CH_2 — CH_2}{\underset{\displaystyle NH——CH—COOH}{|\qquad\quad|}}$
          |
          CH₃

5′–Methylthioadenosine   Azetidine–2–carboxylic
                                                  acid

Fig. 183. Possible mechanism for the formation of azetidine-2-carboxylic acid.

H₃C   Adenosine
    ⊕
    S
    |
    CH₂
    |
    CH₂
    |
    CHNH₂
    |
    COOH

S–Adenosylmethionine

$$\longrightarrow$$

H₃C   Adenosine
    ⊕
    S
    |
    CH₂
    |
    CH₂
    |
    CH₂NH₂

+

              H
              |
    H—N—(CH₂)₄—NH₂

                         Putrescine

S–Adenosyl–S–methyl–
  mercaptopropylamine

$$\downarrow$$

H₃C   Adenosine
    S

5′–Methylthioadenosine

+

CH₂—NH—(CH₂)₄—NH₂   +   H⁺
|
CH₂
|
CH₂NH₂   Spermidine

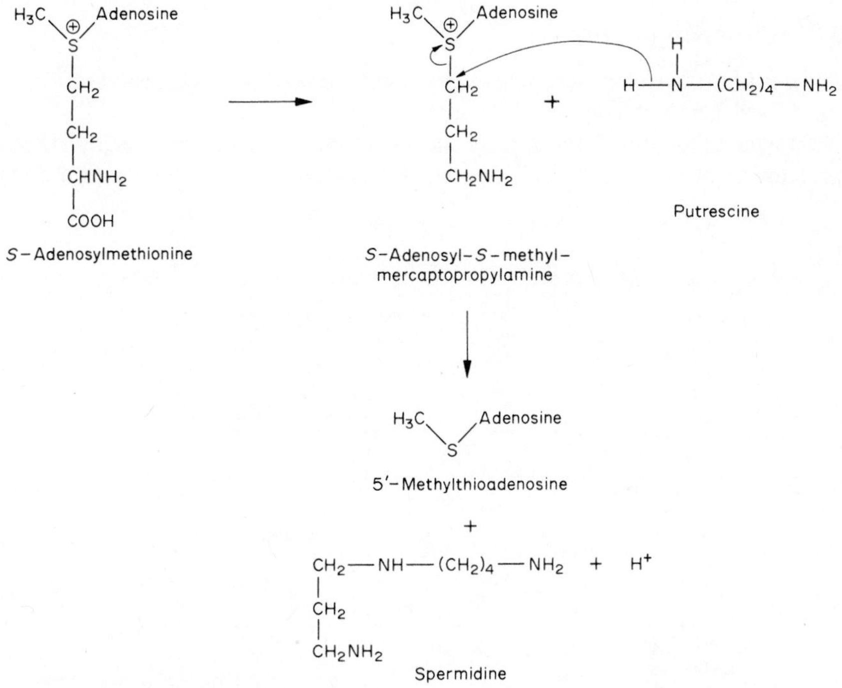

Fig. 184. Formation of spermidine.

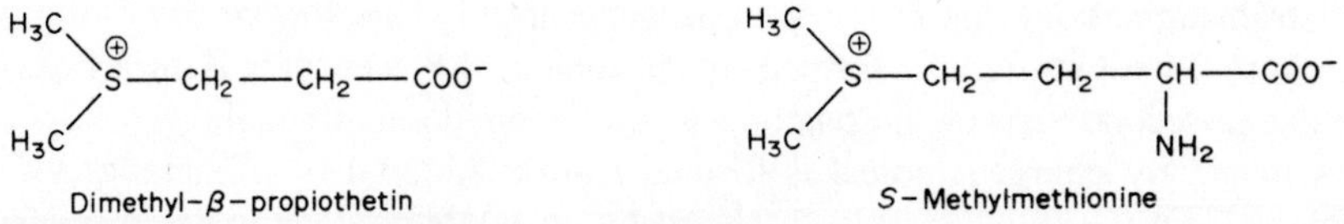

H₃C
    ⊕
    S—CH₂—CH₂—COO⁻
H₃C

Dimethyl–β–propiothetin

H₃C
    ⊕
    S—CH₂—CH₂—CH—COO⁻
H₃C |
                  NH₂

S–Methylmethionine

Fig. 185. Dimethyl-β-propiothetin. S-Methylmethionine.

(fig. 185). Both these compounds originate from methionine and may act as donors of methyl groups.

An enzyme from the sea algae, *Polysiphonia lanulosa*, decomposes dimethyl-propiothetin to dimethyl sulphide, acrylic acid and a hydrogen ion. In bacteria (e.g. *Pseudomonas* strains) and in patients suffering from liver diseases degradation of methionine is possible with the formation of methylmercaptan and dimethyl disulphide. Methylmercaptan is probably responsible for the condition of Foetor hepaticus which is apparent in a number of liver diseases.

References for further reading

See D.10.

10.2 Formation of ethylene

Ethylene is a physiologically active gas which accelerates the ripening of fruits, and is used commercially for this purpose.

Ethylene originates from carbon atoms 3 and 4 of methionine (fig. 185a). Degradation of this amino acid is initiated by means of a transaminase (cf. C.4).

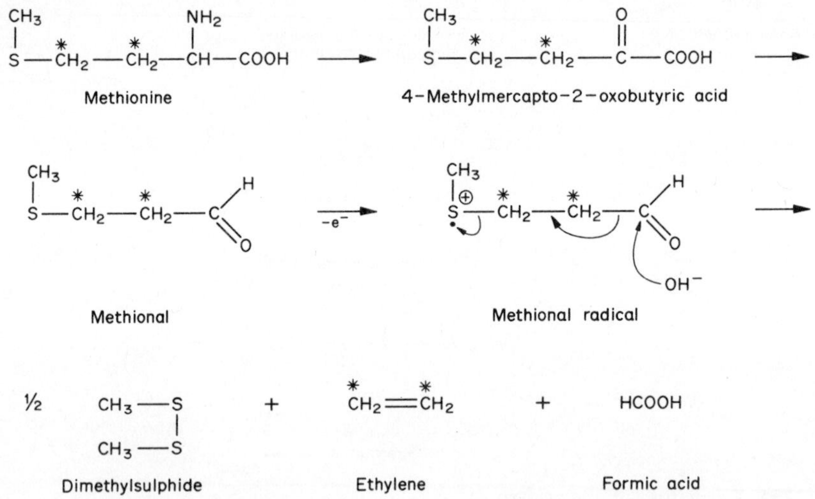

Fig. 185a. Formation of ethylene.

The transamination step is the limiting stage in ethylene formation. 4-Methyl-mercapto-2-oxobutyric acid formed by the action of this enzyme is decarboxylated in the second step of the reaction to the aldehyde, methional.

A positively charged radical is formed from methional by elimination of an electron from the sulphur atom catalysed by an FMN-containing flavin enzyme. Oxygen acts as the acceptor of the eliminated electron and hydrogen peroxide is formed (cf. C.2.2). The radical decomposes in a concerted reaction on addition of an hydroxyl ion with formation of formic acid, ethylene and a methyl-sulphydryl radical which stabilizes to form dimethyldisulphide.

The synthesis of ethylene which is occasionally observed from unsaturated fatty acids, β-alanine, acrylic acid etc., appears to be of no significance in intact living organisms, and is considered to be an artefact.

References for further reading

Mapson, L. W., March, J. F. and Wardale, D. A. Biosynthesis of Ethylene, *Biochem. J.* **115** (1969), pp. 653–61.

Yang, S. F. Ethylene Evolution from 2-Chloroethylphosphonic Acid, *Plant Physiol.* **44** (1969), pp. 1203–4.

Mapson, L. W. Biosynthese von Äthylen und der Reifeprozess von Früchten, *Endeavour* **29** (1970), pp. 29–33.

10.3 Formation and reactions of thiamine

The vitamin thiamine as its pyrophosphate is the coenzyme for transketolases and a number of α-keto acid decarboxylases (cf. C.3.4). It consists of one sub-stituted pyrimidine ring and a substituted thiazole ring linked to each other by a —CH_2— group.

The feeding of ^{35}S, $^{14}CH_3$-methionine has indicated that the whole molecule of this amino acid is incorporated into the thiazole moiety of thiamine. Alanine is possibly the other precursor. During the biosynthesis, 4-methyl-5-alanylthiazole

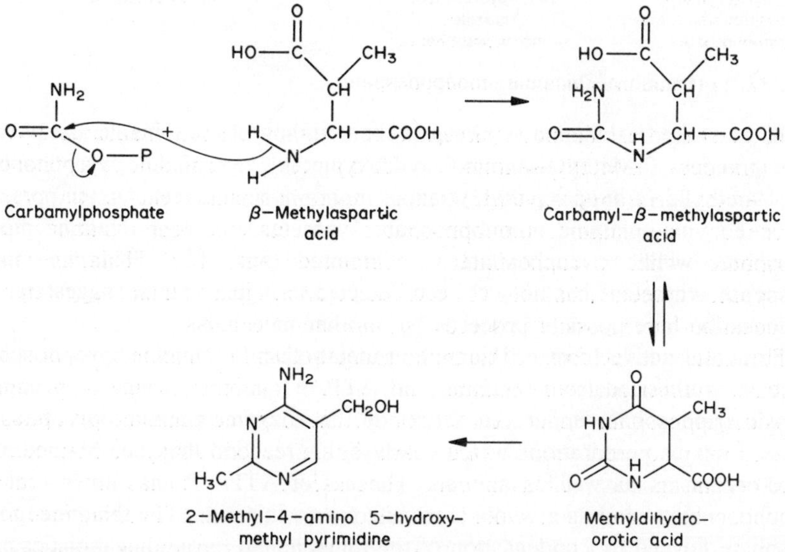

Carbamylphosphate

β—Methylaspartic acid

Carbamyl—β—methylaspartic acid

2—Methyl—4—amino 5—hydroxy-methyl pyrimidine

Methyldihydro-orotic acid

Fig. 186. Possible way for formation of 2-methyl-4-amino-5-hydroxymethyl pyrimidine.

is first formed, and is then converted to 4-methyl-5-(β-hydroxyethyl)-thiazole monophosphate via 4-methyl-5-(β-hydroxyethyl)-thiazole (fig. 187).

There is considerable uncertainty about the formation of the pyrimidine part. However, it probably originates from β-methylaspartic acid by condensation with carbamyl phosphate as described for the biosynthesis of orotic acid (cf. D.14.1). The methyldihydroorotic acid built up in this way is transformed to 2-methyl-4-amino-5-hydroxymethyl pyrimidine which is later converted by ATP to the mono- and then to the pyrophosphate.

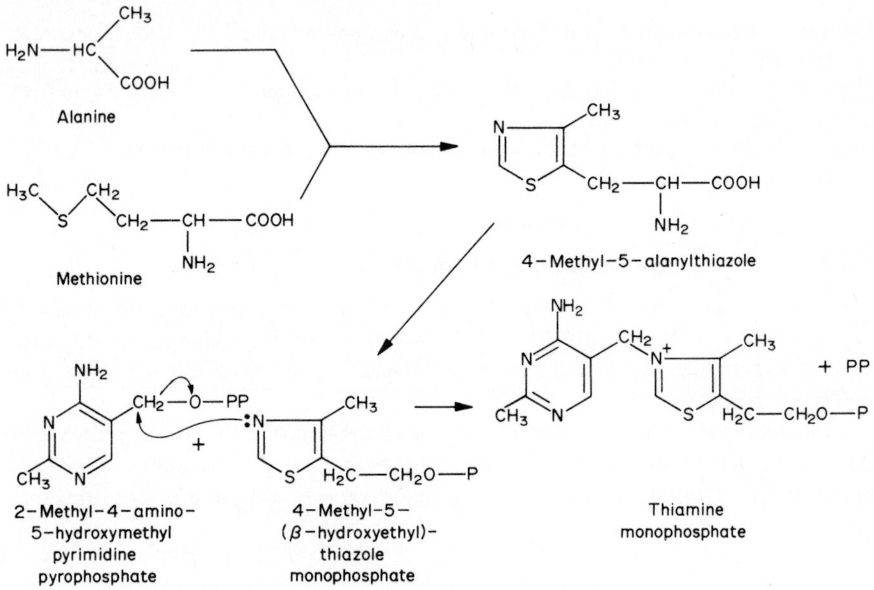

Fig. 187. Formation of thiamine monophosphate.

The last step in thiamine synthesis is the coupling of the thiazole and pyrimidine moieties. 2-Methyl-4-amino-5-hydroxymethyl pyrimidine pyrophosphate and 4-methyl-5-(β-hydroxyethyl)-thiazole monophosphate react in the presence of the enzyme thiamine monophosphate synthetase to form thiamine monophosphate, while pyrophosphate is eliminated (fig. 187). Thiamine monophosphate synthetase has not yet been detected in higher plants, suggesting that condensation here possibly proceeds by another mechanism.

The actual active form of thiamine in metabolism is thiamine pyrophosphate which is synthesized from thiamine and ATP or another purine or pyrimidine riboside triphosphate under the action of the enzyme thiamine pyrophosphokinase. Enzyme preparations which catalyse this reaction may be obtained from micro-organisms as well as animals. The use of ATP-^{32}P has shown that the pyrophosphate group as a whole is transferred to thiamine. The thiamine monophosphate formed by condensation of the thiazole and pyrimidine moieties must therefore be converted to thiamine and phosphate before reaction with ATP.

Diphosphothiamine disulphide isolated from yeast is probably formed from the thiol form of thiamine pyrophosphate. In aqueous solution this remains in equilibrium with the ammonium form (fig. 188). It is still not clear whether diphosphothiamine disulphide possesses any physiological importance.

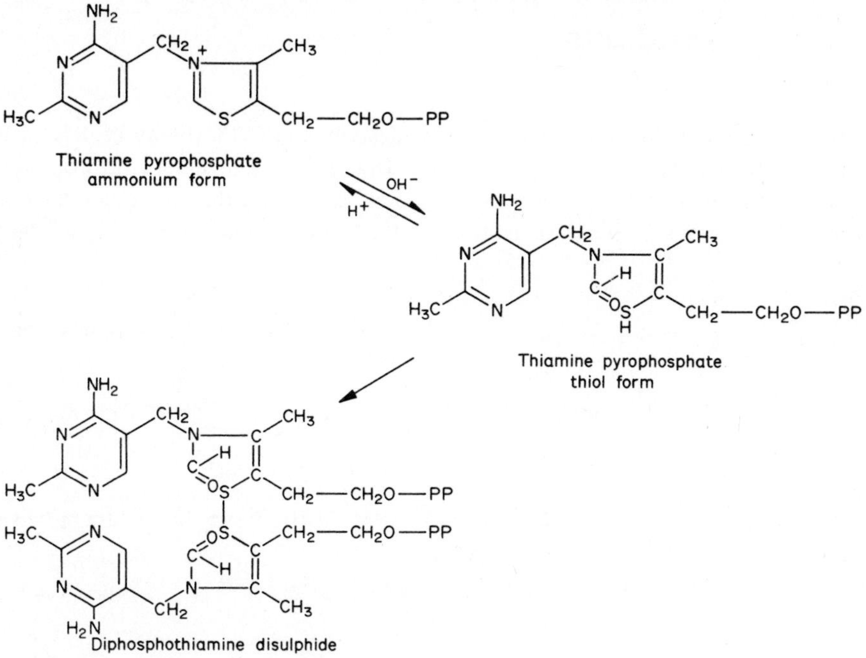

Fig. 188. Formation of diphosphothiamine disulphide.

References for further reading

Fragner, J. *Vitamine, Chemie und Biochemie*, Vol. II (VEB Gustav Fischer Verlag, Jena, 1965).

Luckner, M. and Wasternack, C. Über neue Arbeiten zur Biosynthese *N*-heterocyclischer Verbindungen, 7. Teil: Bildung und Abbau von Verbindungen mit Pyrimidinring-system, *Pharmazie* **22** (1967), pp. 181–98.

11. Valine as precursor of secondary natural products

The formation of the amino acid valine is described in D.13. It may be degraded to methylmalonyl CoA as outlined in fig. 189. This is a precursor of propionyl CoA and propionic acid as well as of succinyl CoA, and thus of compounds of primary metabolism (cf. D.3). Isobutyryl CoA, through which the degradation

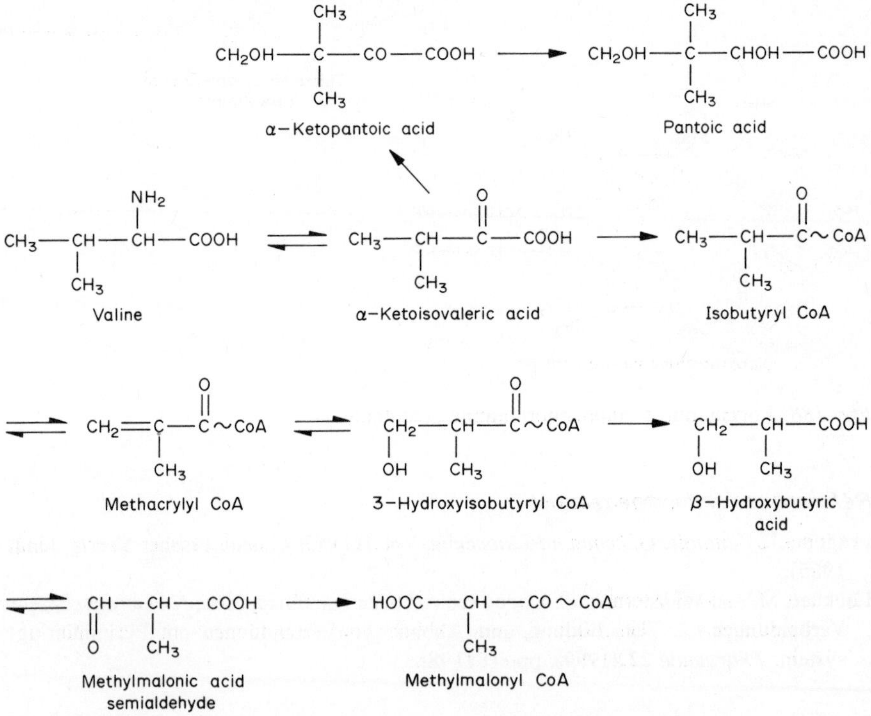

Fig. 189. Biosynthesis of pantoic acid and methylmalonyl CoA from valine.

proceeds, is the starter molecule in the synthesis of certain fatty acids (cf. D.2.2.1). The free acid formed by hydrolysis of methacrylyl CoA is found in high concentration in the defence secretions of certain insects (cf. fig. 1). Pantoic acid is formed by reaction of α-ketoisovaleric acid with 'active formaldehyde' (cf. C.3.2) via α-ketopantoic acid (fig. 189). Pantoic acid is a constituent of coenzyme

A (cf. D.2.1). Valine is also one of the precursors of the antibiotic magnamycin (cf. D.3.1) and of certain necic acids (cf. D.15.2) as well as of erythroskyrin, a pigment from *Penicillium islandicum*.

Erythroskyrin is formed from one molecule of valine and ten molecules of acetate or malonate (fig. 190). Intermediates of the biosynthesis are not yet known. However, the polyketide I is probably formed first from the activated acids. Then the activated carboxyl group of the polyketide is substituted by the amino grouping of valine and the activated carboxyl group of this compound by

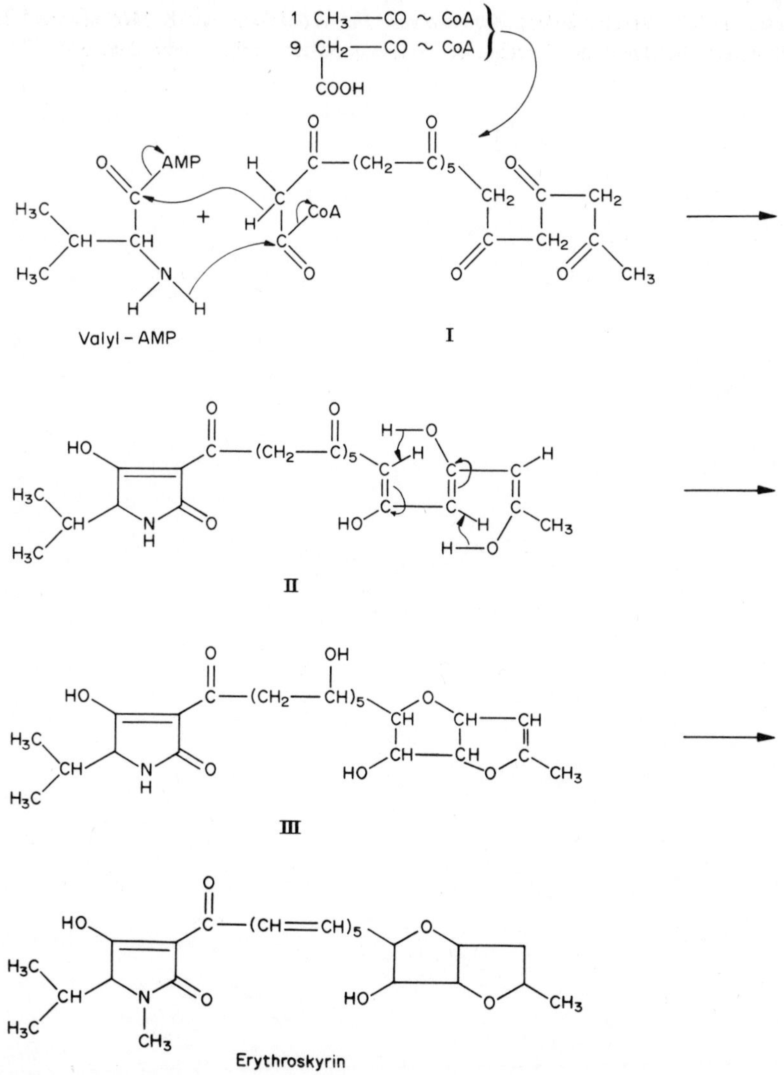

Fig. 190. Formation of erythroskyrin.

the α-carbon atom of the β-keto acid with the elimination of CoA and AMP respectively (cf. the mechanism of formation of tenuazonic acid, D.13.2). The further biosynthesis probably proceeds via compounds II and III.

References for further reading

Cheldelin, V. H. and Baich, A. The Biosynthesis of the Water-Soluble Vitamins, *Biogenesis of Natural Compounds*, ed. P. Bernfeld (Pergamon Press, Oxford, 1967), pp. 679–742.

Schütte, H. R. 'Verschiedenes', *Biosynthese der Alkaloide*, ed. K. Mothes and H. R. Schütte (VEB Deutscher Verlag der Wissenschaften, Berlin, 1969), pp. 645–77.

12. Leucine as precursor of secondary natural products

Leucine is synthesized from the amino acid valine, which is shorter by one —CH_2— group, as outlined in fig. 191 (cf. D.13 for formation of valine). Elongation of the carbon chain takes place by means of enzymes of the α-keto acid elongation-complex. This sequence of reactions, which closely resembles the conversion of oxalacetic acid to α-ketoglutaric acid in the tricarboxylic acid cycle (cf. D.4), begins with the nucleophilic substitution of the α-keto acid by a molecule of acetyl CoA. A β-hydroxy-β-carboxy carboxylic acid is thus formed, which structurally resembles citric acid. This compound is then rearranged to an α-hydroxy-β-carboxy carboxylic acid of the isocitric acid type by the elimination and renewed addition of water. An α-keto-β-carboxy carboxylic acid is formed from this by dehydrogenation which is similar to oxalosuccinic acid and changes by decarboxylation to the higher homologue of the starting α-keto acid. Since α-amino acids by transaminases (cf. D.4) are reversibly convertible to α-keto acids the carbon skeleton of these compounds can be elongated by a CH_2-group during the described cycle of reactions.

α-Ketoisovaleric acid originating from valine by transamination (cf. C.4) is converted in this way via β-carboxy-β-hydroxyisocaproic acid, α-isopropylmalic

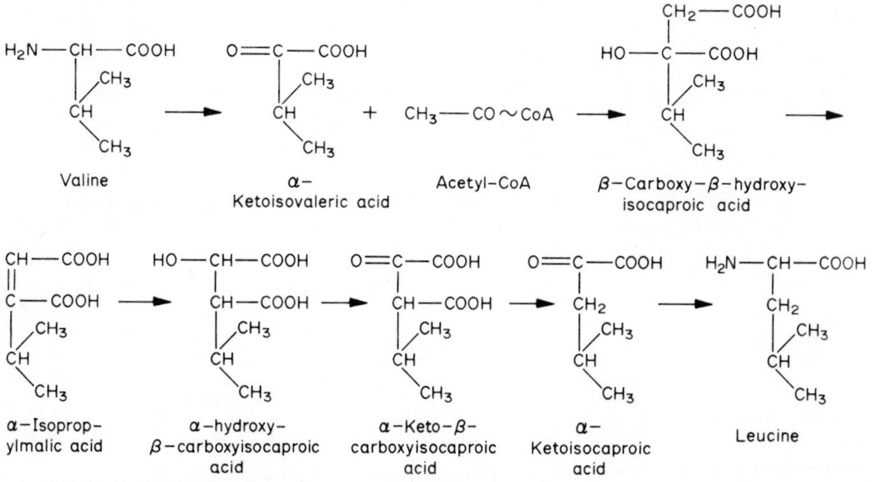

Fig. 191. Formation of leucine.

acid, α-hydroxy-β-carboxyisocaproic acid and α-keto-β-carboxyisocaproic acid to α-ketoisocaproic acid which is longer by one —CH_2— group. The latter compound is converted by transamination to leucine, a valine homologue (fig. 191).

α-Keto acid elongation-complexes also participate in the formation of lysine via α-ketoadipic acid (cf. D.16) and probably in the biosynthesis of a number of secondary amino acids which are precursors of mustard oil glucosides and other nitrogen-containing natural products (cf. D.7.6 and D.17.2).

References for further reading

See D.10.

12.1 Formation of isopentenyl pyrophosphate from leucine

β-Hydroxy-β-methylglutaryl CoA, an intermediate in the synthesis of isoprenoid compounds, may also originate from leucine, as well as from acetyl CoA, as described in D.5.1. The α-keto acid formed from this amino acid by transamination is first converted by oxidative decarboxylation to isovaleryl CoA, which may be further dehydrogenated to β-methylcrotonyl CoA (senecioyl CoA) (cf. the degradation of isoleucine and valine, D.13.1 and D.11). β-Methylglutaconyl CoA

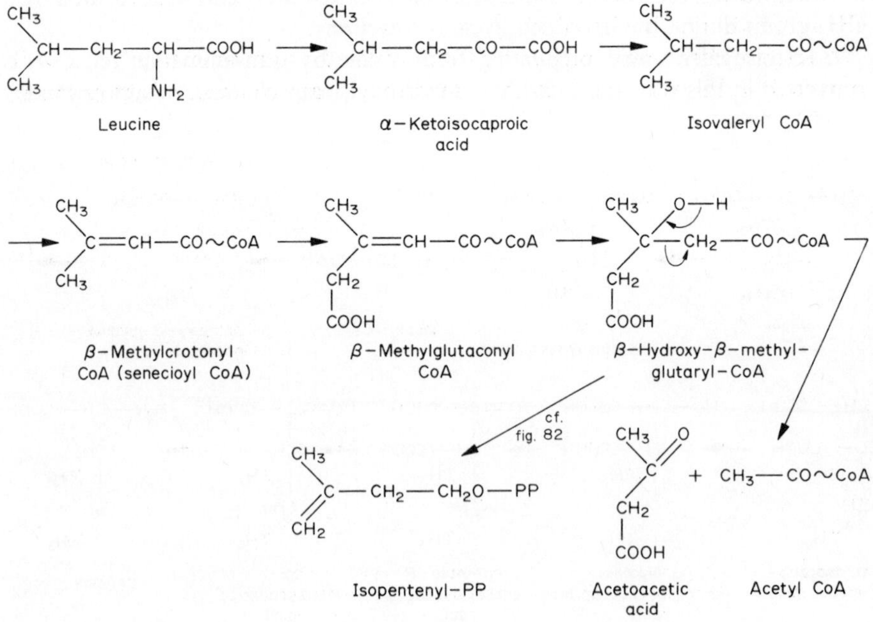

Fig. 192. Conversion of leucine to isopentenyl pyrophosphate.

is formed from the latter compound in a carboxylation reaction catalysed by methylcrotonyl-CoA-carboxylase, a biotin enzyme (cf. C.3.1) and is then converted to β-hydroxy-β-methylglutaryl CoA by the addition of water to the double bond (fig. 192).

β-Hydroxy-β-methylglutaryl CoA is normally degraded to acetoacetic acid and acetyl CoA and thus the carbon skeleton of leucine is converted to compounds which may undergo further reactions in primary metabolism (fig. 192). β-Hydroxy-β-methylglutaryl CoA may, however, be converted to isopentenyl pyrophosphate via mevalonic acid as described in D.5.1.

References for further reading

See D.10.

13. Formation of secondary products from isoleucine

The amino acid isoleucine originates from α-ketobutyric acid and 'active acetaldehyde' (cf. C.3.4) as outlined in fig. 193. Both these compounds condense to form α-aceto-α-hydroxybutyric acid. Then a shift of the acetyl side chain from the α- to the β-carbon atom takes place, forming α,β-dihydroxy-β-methylvaleric acid. This is converted to L-isoleucine via α-keto-β-methylvaleric acid by dehydration and transamination. The amino acid L-valine is formed from α-ketopropionic acid (pyruvic acid) through similar reaction steps. α-Acetolactic acid can be converted to acetoin and diacetyl by decarboxylation and oxidation. The latter compound is a precursor of riboflavin (cf. D.8.2.4).

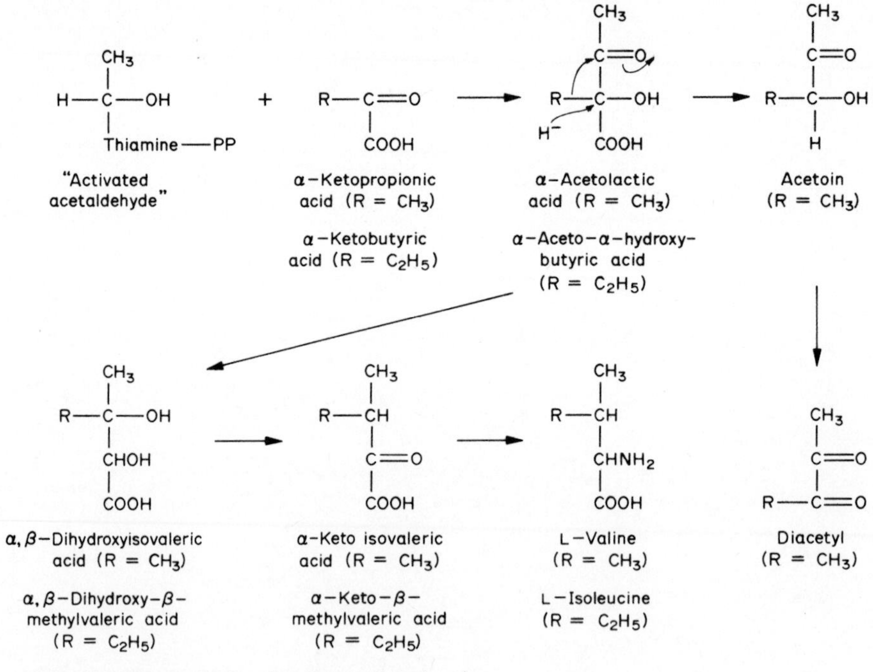

Fig. 193. Biosynthesis of valine and isoleucine.

References for further reading

See D.10.

13.1 Formation of tiglic acid and α-methylbutyric acid

Tiglic acid and α-methylbutyric acid occur as constituents of certain tropane alkaloids (cf. D.15.1). Both these compounds originate from L-isoleucine in higher plants and animals. Since isoleucine-2-^{14}C, but not isoleucine-1-^{14}C, gives rise to tiglic acid-[14]COOH, the biosynthetic pathway is probably the same as the one outlined in fig. 192. L-Isoleucine is probably converted first to the corresponding α-keto acid and then oxidatively decarboxylated to α-methylbutyryl CoA. Tiglyl CoA is derived from the latter compound by dehydrogenation (fig. 194).

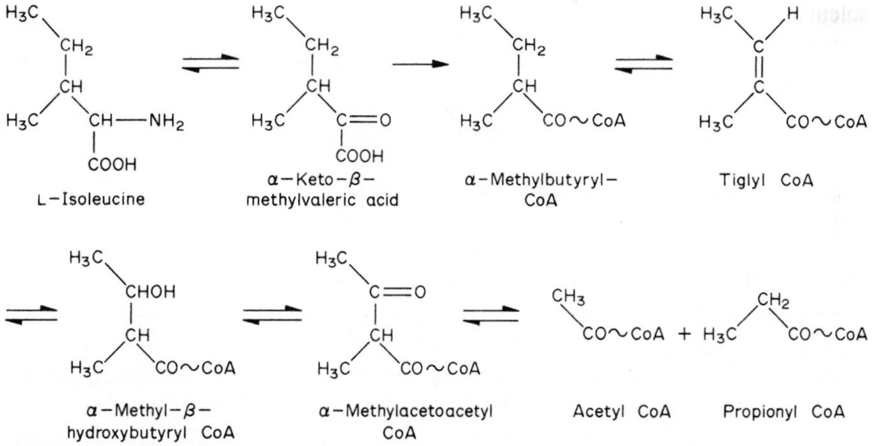

Fig. 194. Formation and degradation of tiglyl- and α-methylbutyryl CoA.

Usually tiglic acid is further degraded to acetyl CoA and propionyl CoA as outlined in fig. 194. Since, apart from oxidative decarboxylation, all reactions of this degradative pathway are reversible, tiglic acid and methylbutyric acid may be synthesized from acetate and propionate, e.g. as in *Ascaris lumbricoides.*

References for further reading

See D.10.

13.2 Biosynthesis of tenuazonic acid

Tenuazonic acid is a pyrrol derivative which is synthesized by the fungus *Alternaria tenuis*. If this organism is grown in the presence of acetate-1-^{14}C the major portion of the radioactivity incorporated into the alkaloid is found in carbon atoms 2 and 6. Radioactivity is also incorporated to a small extent into carbon atoms 4 and 10. It was found that the corresponding carbon atoms 1 and 5 of isoleucine were similarly labelled when the amino acid was isolated after feeding acetate-1-^{14}C. It is thus probable that tenuazonic acid is formed from isoleucyl-AMP, and acetoacetyl CoA by acylation of the amino group and nucleophilic substitution of the activated carboxyl group of the amino acid by the CH_2—group of the acetoacetate (cf. formation of erythroskyrin, D.11) (fig. 195).

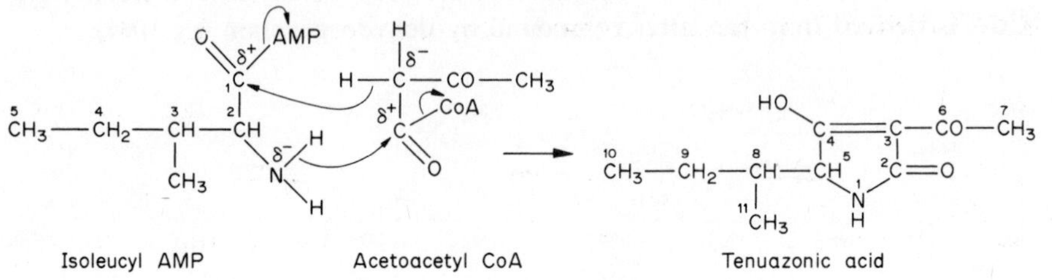

Fig. 195. Formation of tenuazonic acid.

References for further reading

Birkinshaw, J. H. and Stickings, C. E. Nitrogen-Containing Metabolites of Fungi, *Fortschr. Chem. org. Naturstoffe* **20** (1962), pp. 1–40.

Schütte, H. R. 'Verschiedenes', *Biosynthese der Alkaloide*, ed. K. Mothes and H. R. Schütte (VEB Deutscher Verlag der Wissenschaften, Berlin, 1969), pp. 645–77.

14.   Biosynthesis of secondary products from aspartic acid

L-Aspartic acid, a dicarboxy monoamino acid, and its γ-amide, L-asparagine, occur in all organisms. L-Aspartic acid is formed by transamination from oxalacetic acid which originates from the tricarboxylic acid cycle (cf. D.4). In addition, L-aspartic acid remains in equilibrium with fumaric acid by a reversible reaction, catalysed by the enzyme aspartase (fig. 196). L-Aspartic acid may be aminated to L-asparagine with ammonia in an ATP dependent reaction. L-Aspartyl phosphate is supposed to be the intermediate.

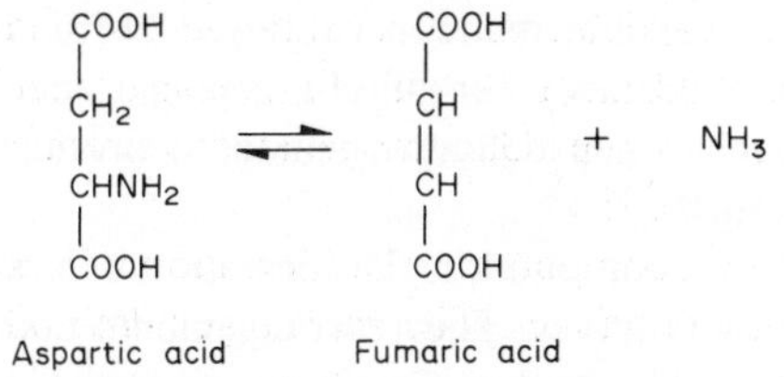

$$
\begin{array}{ccc}
\text{COOH} & & \text{COOH} \\
| & & | \\
\text{CH}_2 & & \text{CH} \\
| & \rightleftharpoons & \| \quad + \quad \text{NH}_3\\
\text{CHNH}_2 & & \text{CH} \\
| & & | \\
\text{COOH} & & \text{COOH} \\
\text{Aspartic acid} & & \text{Fumaric acid}
\end{array}
$$

Fig. 196. Degradation of aspartic acid to fumaric acid and ammonia catalysed by aspartase.

L-Aspartic acid is the starting molecule for the biosynthesis of certain pyrimidine compounds and pyridine derivatives. The nitrogen of the amino group is retained in these heterocyclic compounds. The amino acid may be converted by decarboxylation (cf. D.7.3) and oxidation to β-nitropropionic acid.

References for further reading

See D.10.

14.1 Biosynthesis and degradation of orotic acid and orotic acid derivatives

Compounds which are derived from orotic acid, e.g. uracil, thymine, cytosine and 5-methylcytosine (fig. 197), are of vital importance to all organisms as constituents of the nucleic acids. In addition, they play an important role in the metabolism of sugars (cf. D.1.1.1), lecithins and cephalins (cf. D.2.2.7) and are constituents of compounds of similar structure to ATP (cf. C.1.1).

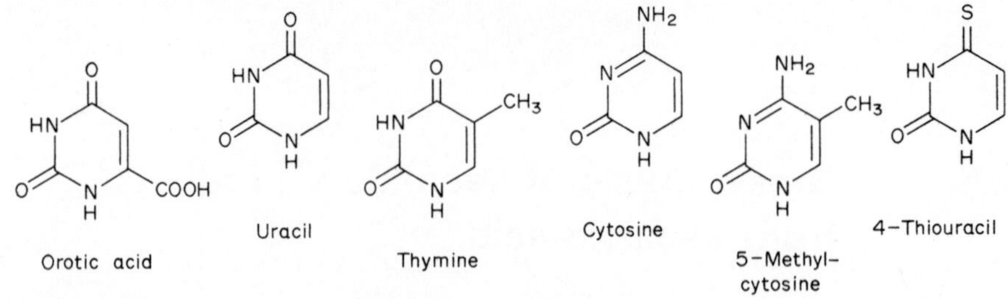

Fig. 197. Widely occurring pyrimidine derivatives

The biosynthetic pathway of these compounds leads from carbamyl phosphate and aspartic acid via orotic acid, and is found in animals, higher plants and micro-organisms. The details of the pathway were worked out in recent years with the aid of isotopically labelled compounds and mutants, as well as by the isolation and characterization of individual enzymes.

Carbamyl phosphate is first transferred to aspartic acid in an irreversible reaction catalysed by the enzyme aspartate carbamyl transferase to form carbamyl-L-aspartic acid. In a reversible reaction catalysed by dihydroorotase (4,5-L-di-hydro-orotate amino hydratase) carbamyl-L-aspartic acid is then converted to dihyroortic acid, which is then dehydrogenated to orotic acid by dihydroorotic acid dehydrogenase (fig. 198).

Orotic acid is the key compound in the formation of uracil, thymine, cytosine and 5-methylcytosine derivatives. The rearrangements take place usually at the level of the ribosides and ribotides. Orotidine monophosphate is, for example, decarboxylated to uridine monophosphate by orotidyl decarboxylase (fig. 198).

Other uracil derivatives originating from uridine monophosphate are precursors for cytosine compounds (e.g. uridine triphosphate reacts with the amide nitrogen

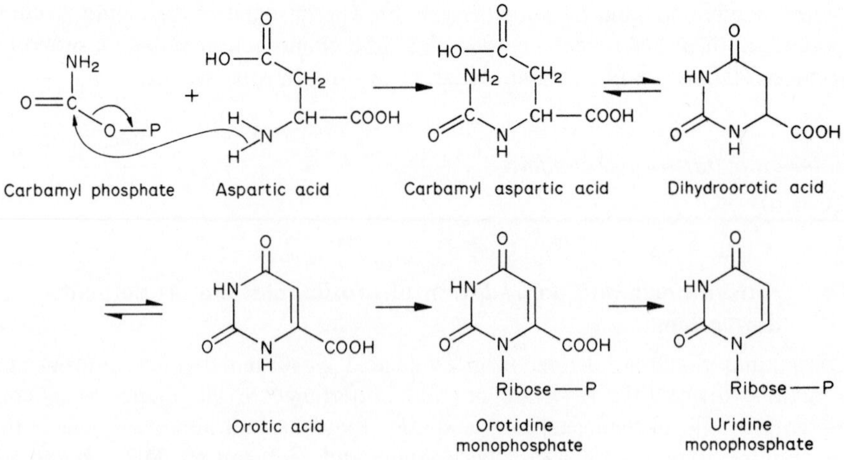

Fig. 198. Formation of orotic acid and uridine monophosphate.

of glutamine to form cytidine triphosphate), as well as for 5-methyluracil (thymine) (here, deoxyuridine monophosphate is methylated to thymidine mono-phosphate by N^5-methyltetrahydrofolic acid (cf. C.3.2).* The corresponding 5-methyl- and 5-hydroxymethylcytosine derivatives are formed from the cytosine derivatives in a similar manner. Pseudouridine, a C-glycosyl compound in which ribose is linked to uracil at position 5 by a carbon–carbon bond, is synthesized from either cytidine or uridine.

The sulphhydryl group of cysteine is the precursor for the thio group of 4-thiouracil (fig. 197), which occurs as a rare base in the tRNA of various organisms. The thiol transferase which catalyses this transfer contains pyridoxal phosphate as coenzyme.

The most important metabolic pathway for the degradation of orotic acid derivatives starts from uracil and thymine. Cytosine and 5-methylcytosine, the pyrimidine derivatives carrying amino groups, must therefore be first deaminated.

Uracil and thymine are first reduced to the corresponding dihydro compounds where, depending on the enzyme involved, either NADH or NADPH acts as coenzyme. Dihydrouracil and dihydrothymine are then converted to carbamyl-β-alanine and carbamyl-β-aminoisobutyric acid respectively by the enzyme dihydropyrimidine hydrase. The latter compounds are further degraded by hydro-lases to carbamic acid (which decomposes spontaneously to ammonia and CO_2) and β-alanine or β-aminoisobutyric acid (fig. 199).

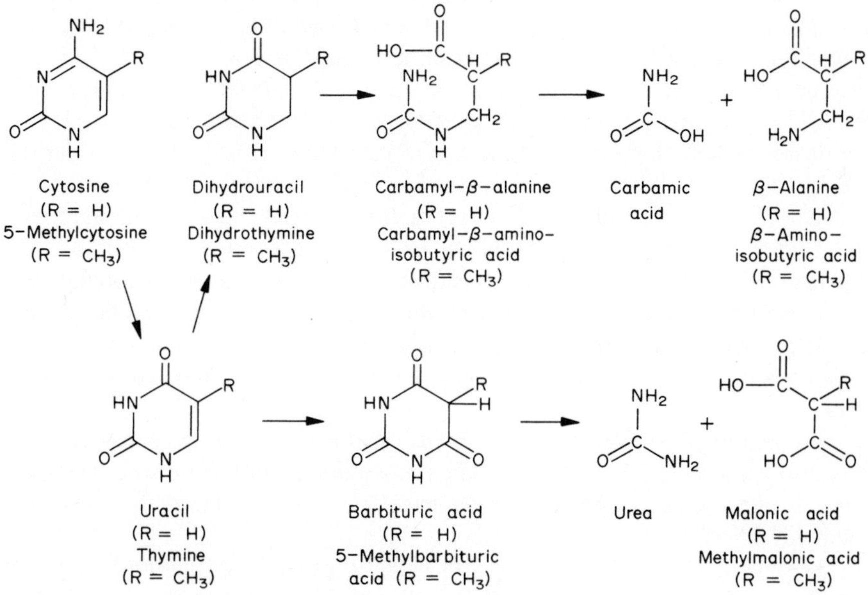

Fig. 199. Degradation of pyrimidine derivatives.

* The formation of thymine by transmethylation with S-adenosylmethionine occurs in preformed tRNA and DNA chains (cf. D.8.2.2).

Uracil and thymine may be oxidized to barbituric acid and 5-methylbarbituric acid by certain bacteria. Both these compounds are then hydrolysed by the enzyme barbiturase (barbituric acid amino hydrolase), to form urea and malonic acid or methylmalonic acid (fig. 199).

References for further reading

Luckner, M. and Wasternack, C. Über neue Arbeiten zur Biosynthese *N*-heterocyclischer Verbindungen, 7. Teil: Bildung und Abbau von Verbindungen mit Pyrimidinring-system, *Pharmazie* **22** (1967), pp. 181–98.

Schlee, D. Thiolierung von Nucleinsäuren, *Pharmazie* **24** (1969), pp. 241–4.

14.2 Formation of nicotinic acid and the alkaloids derived from nicotinic acid

In the case of most bacteria (with the exception of *Xanthomonas* species) and higher plants, pyridine carboxylic acids (e.g. quinolinic acid and nicotinic acid) are not formed from tryptophan by the metabolic pathway described in D.19.5.3 but are synthesized from aspartic acid and a three-carbon unit derived from glycerol (fig. 200). Glyceraldehyde phosphate probably condenses first with the amino group of aspartic acid to form the Schiff base I which is then converted to quinolinic acid probably via intermediates II and III by the elimination of phosphate and a proton, dehydration, and dehydrogenation as outlined in fig. 200.

Nicotinic acid does not originate from quinolinic acid by decarboxylation as was once assumed, but is synthesized in the so-called pyridine nucleotide cycle (fig. 200). Quinolinic acid is first converted to nicotinic acid mononucleotide by condensation with 5-phosphoribosyl-1-pyrophosphate and elimination of carbon dioxide. Quinolinic acid mononucleotide is possibly formed as an intermediate.

Nicotinic acid mononucleotide changes either directly to nicotinic acid with the elimination of ribosyl-5-phosphate or is first converted to nicotinic acid adenine dinucleotide in an ATP dependent reaction. The latter compound may be converted to nicotinamide adenine dinucleotide (NAD^+) by reaction with the amide nitrogen of glutamine. NAD^+ is degraded in an enzyme catalysed hydrolysis with the formation of nicotinamide, which is then deaminated to nicotinic acid.

This metabolic pathway has been intensively investigated in mycobacteria, but has also been identified in higher plants and animals. In most of the mycobacterial strains, some of the synthesized nicotinic acid is reduced to 3-hydroxymethylpyridine.

The degradation of nicotinic acid in bacteria begins with the introduction of a hydroxyl group at position 6. 6-Hydroxynicotinic acid is then either oxidatively decarboxylated and degraded via 2,5-dihydroxypyridine to maleic acid, ammonia and formic acid, or via 1,4,5,6-tetrahydro-6-oxonicotinic acid, which is then converted to α-methyleneglutaric acid with the loss of ammonia, α-Methylene-

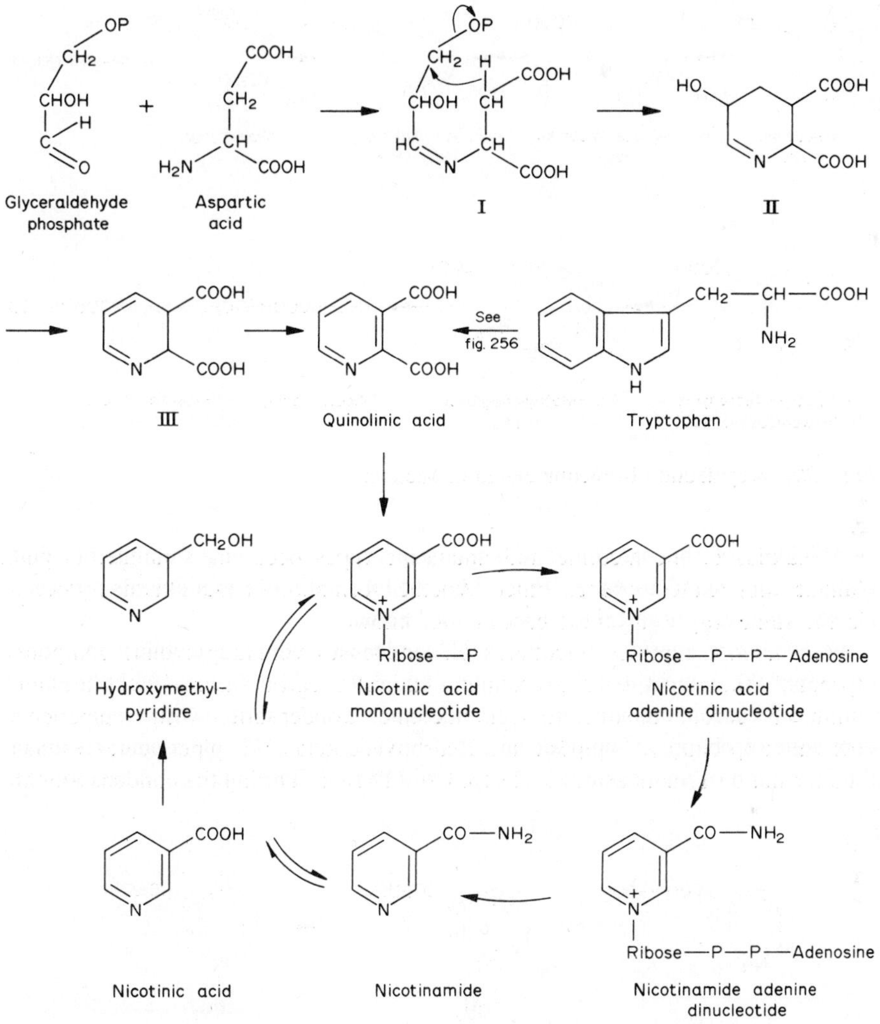

Fig. 200. Biosynthesis of quinolinic acid and the pyridine nucleotide cycle.

glutaric acid may be degraded further to acetic acid, propionic acid and CO_2 (fig. 201).

In higher plants nicotinic acid is the precursor of the betaine trigonelline (cf. D.7.4), as well as of the alkaloids nicotine, anabasine and ricinine.

The cyano group of ricinine originates from the acid amide group of nicotinamide. N-Methyl-3-cyanopyridine, which is probably formed first, is oxidized by a specific enzyme isolated from *Ricinus communis* to the 4-pyridone, which is then converted by methylation and further oxygenation to ricinine (fig. 202). The possibility that some biosynthesis of ricinine takes place from nicotinic acid nucleotide cannot be ruled out, however.

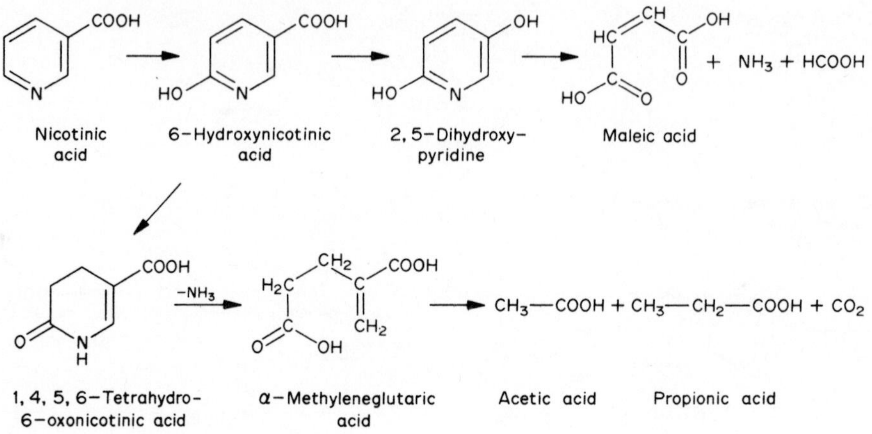
COOH
Nicotinic acid
6-Hydroxynicotinic acid
2,5-Dihydroxy-pyridine
Maleic acid
+ NH3 + HCOOH
-NH3
1,4,5,6-Tetrahydro-6-oxonicotinic acid
α-Methyleneglutaric acid
CH3—COOH + CH3—CH2—COOH + CO2
Acetic acid
Propionic acid

Fig. 201. Degradation of nicotinic acid in bacteria.

Alkaloids of the nicotine and anabasine types occur in a number of plant families not related to each other. Whether formation of the alkaloids proceeds in the same way in all these cases is not known.

In *Nicotiana* species, nicotinic acid condenses with a pyrroline compound (probably the *N*-methyl-Δ^1-pyrrolinium cation) at carbon atom 3, with the elimination of carbon dioxide, to form nicotine. Condensation with a piperideine derivative (probably Δ^1-piperideine-2-carboxylic acid or Δ^1-piperideine) results in the formation of anabasine (cf. D.15.3 and D.16.4). During the condensation the

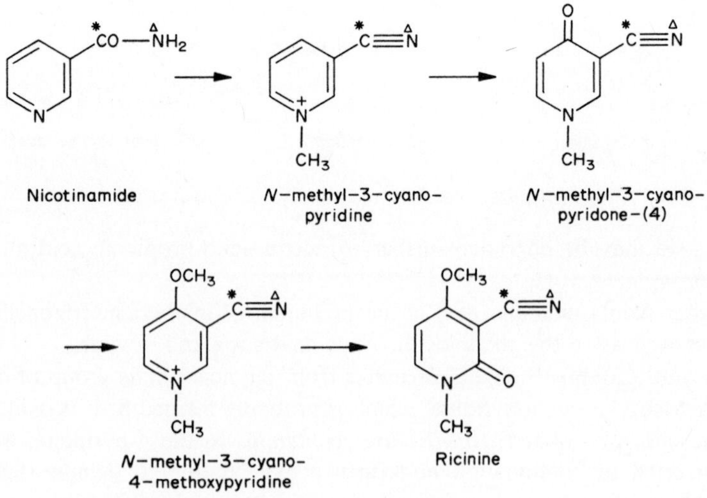
Nicotinamide
N-methyl-3-cyano-pyridine
N-methyl-3-cyano-pyridone-(4)
N-methyl-3-cyano-4-methoxypyridine
Ricinine

Fig. 202. Possible biosynthetic pathway of ricinine.

hydrogen atom at position 6 of the nicotinic acid is becoming loose. The pathway shown in fig. 203 is suggested in which the quinoid pyridine derivatives I and II may be formed as intermediates.

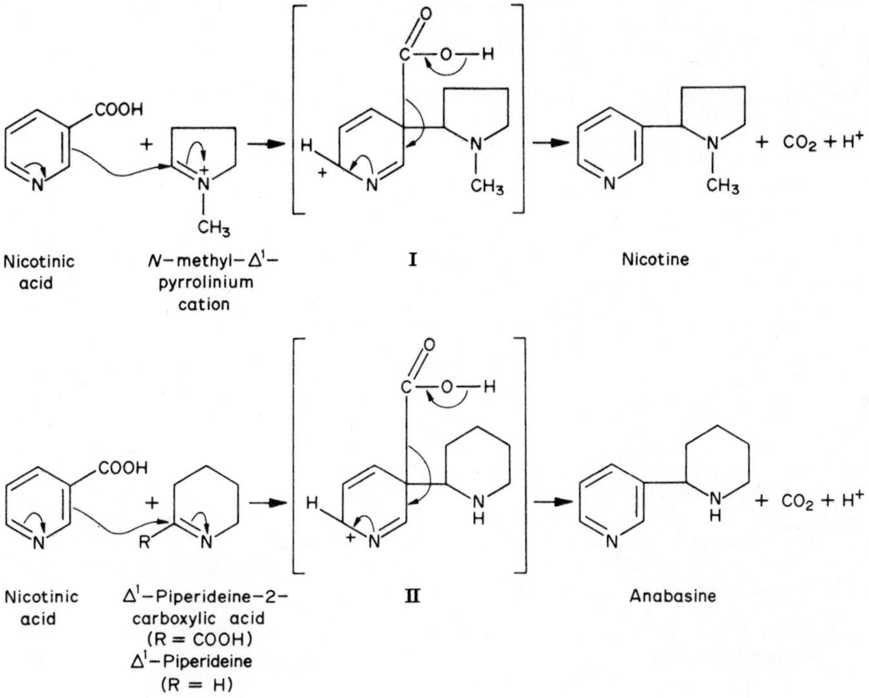

Fig. 203. Formation of nicotine and anabasine.

Nornicotine is formed during the degradation of nicotine which occurs when tobacco leaves are dried, or subjected during fermentation to bacterial attack. Several rearrangement products of nicotine which still contain the intact ring system (such as cotinine) as well as degradation products such a pyridyl butyric acid and nicotinic acid are also formed.

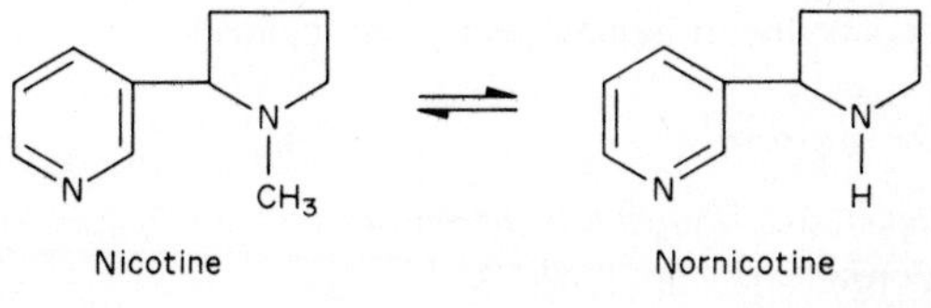

Fig. 204. Interconversion of nicotine and nornicotine.

The degradation of nicotine in humans and animals proceeds as shown in fig. 205. In bacteria adapted to nicotine, such as those obtained from soil samples taken from tobacco fields, the reactions outlined in fig. 206 take place.

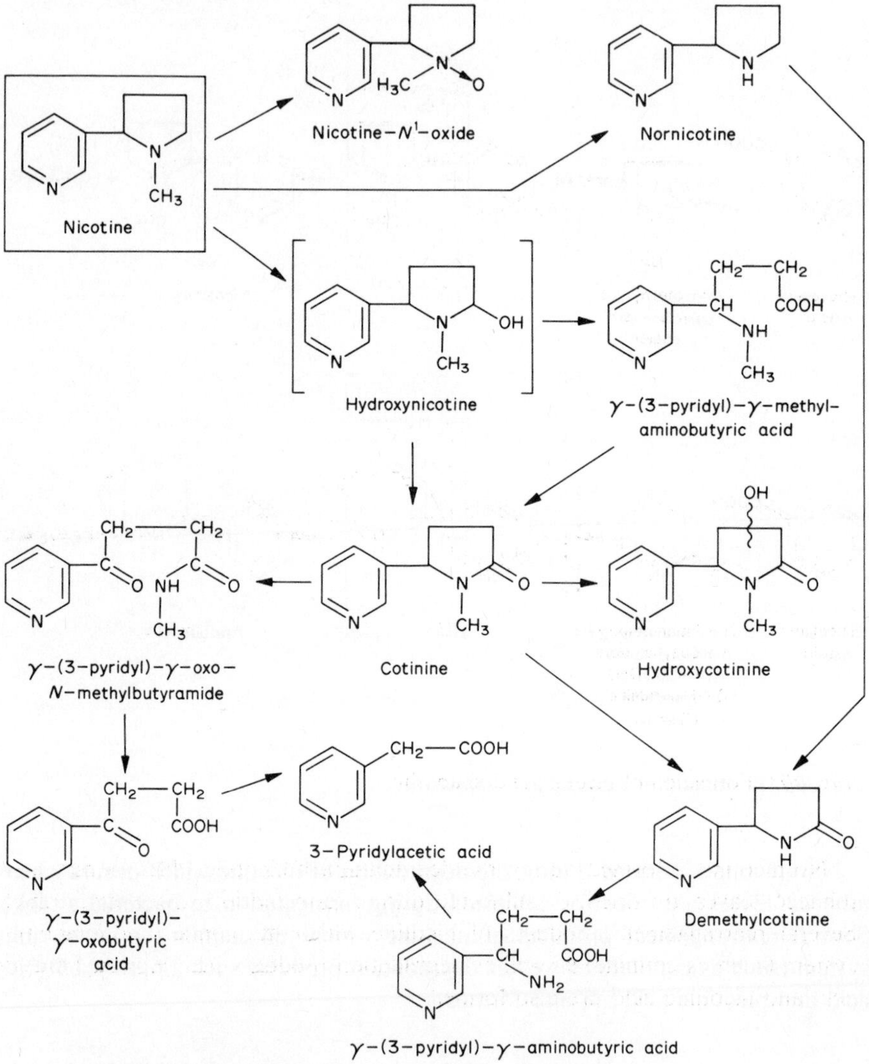

Fig. 205. Degradation of nicotine in human beings and animals.

References for further reading

Leete, E. 'Alkaloid Biogenesis', *Biogenesis of Natural Compounds*, ed. P. Bernfeld (Pergamon Press, Oxford, 1967), pp. 953–1023.

Gross, D. 'Pyridinalkaloide', *Biosynthese der Alkaloide*, ed. K. Mothes and H. R. Schütte (VEB Deutscher Verlag der Wissenschaften, Berlin, 1969), pp. 215–74.

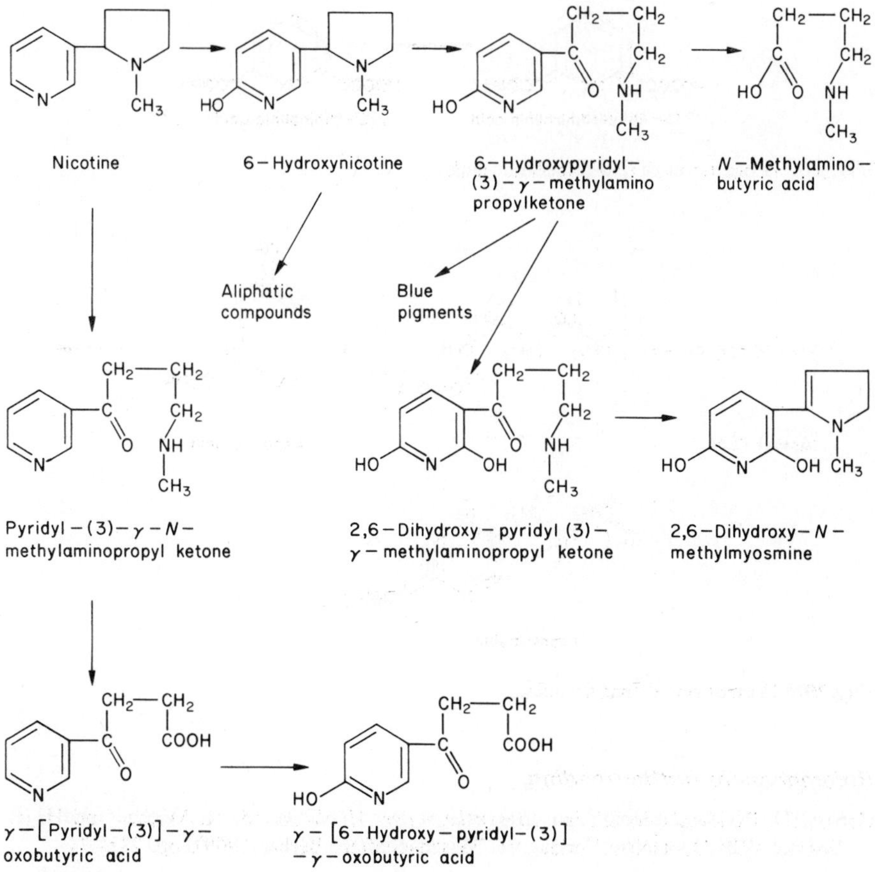

Nicotine   6 – Hydroxynicotine   6 – Hydroxypyridyl – (3) – γ – methylamino propylketone   N – Methylamino – butyric acid

Aliphatic compounds   Blue pigments

Pyridyl – (3) – γ – N – methylaminopropyl ketone   2,6 – Dihydroxy – pyridyl (3) – γ – methylaminopropyl ketone   2,6 – Dihydroxy – N – methylmyosmine

γ – [Pyridyl – (3)] – γ – oxobutyric acid   γ – [6 – Hydroxy – pyridyl – (3)] – γ – oxobutyric acid

Fig. 206. Degradation of nicotine in bacteria.

14.3 Formation of 2,6-dipicolinic acid and fusaric acid

While carbon atom 4 of aspartic acid is lost during the formation of nicotinic acid (cf. D.14.2), it is retained during the formation of other pyridine compounds.

2,6-Dipicolinic acid, which occurs in large quantities together with α,ϵ-di-aminopimelic acid in *Bacillus cereus* and *Penicillium citro-viride*, belongs to this group of substances. 2,6-Dipicolinic acid is probably formed from 2,3-dihydrodi-picolinic acid, an intermediate in the biosynthesis of the amino acid lysine (cf. D.16) (fig. 207).

Fusaric acid occurs in various *Fusarium* species. The specific incorporation rate obtained by feeding acetate and aspartic acid to *Fusarium oxysporum* suggests the biosynthetic pathway outlined in fig. 208. However, aspartic acid β-semialdehyde may possibly also be a precursor.

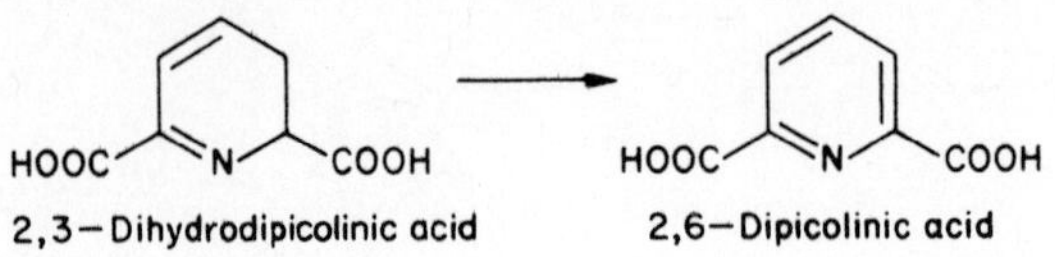

Fig. 207. Formation of 2,6-dipicolinic acid.

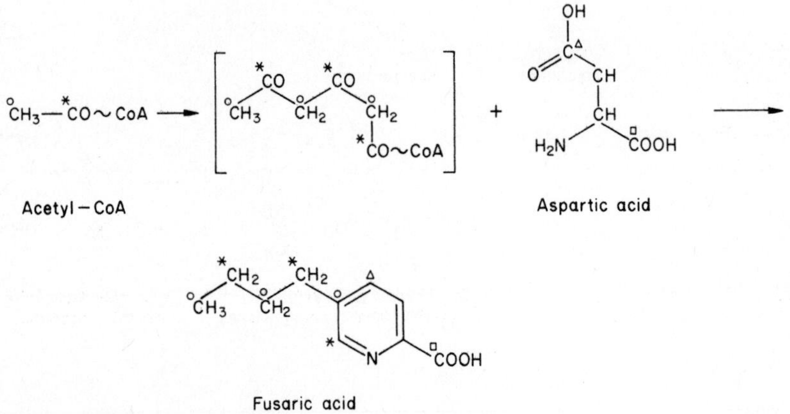

Fig. 208. Formation of fusaric acid.

References for further reading

Gross, D. 'Pyridinalkaloide', in: *Biosynthese der Alkaloide*, ed. K. Mothes and H. R. Schütte (VEB Deutscher Verlag der Wissenschaften, Berlin, 1969), pp. 215–74.

15. Compounds of the glutamic acid-proline-ornithine-group as precursors of secondary natural products

A number of amino acids with five carbon atoms which may be converted easily and reversibly into one another are the precursors of alkaloids with nitrogen-containing five-membered rings. Glutamic acid, proline and ornithine belong to these amino acids.

The relationships which exist between these compounds are shown in fig. 209. Glutamic acid may be reduced directly in a few organisms to glutamic acid-γ-semialdehyde and this compound may be aminated to ornithine. In others α-N-acetylglutamic acid is formed first and is then converted to ornithine via α-N-acetylglutamic acid-γ-semialdehyde and α-N-acetylornithine. Glutamic acid-γ-semialdehyde is in equilibrium with the amino acid proline via Δ^1-pyrroline-5-carboxylic acid. By reaction with carbamyl phosphate, ornithine is converted to citrulline, and this amino acid reacts with ammonia to form arginine (fig. 209, cf. D.17).

Putrescine and agmatine are formed from ornithine and arginine respectively by decarboxylation (cf. D.7.3). Agmatine may be converted via carbamyl-putrescine to putrescine so that the latter compound may also arise (e.g. in higher plants) from ornithine via arginine. γ-Aminobutyraldehyde is formed from putrescine by transamination (cf. C.4) or oxidative deamination (cf. C.2.1.2), and may cyclize spontaneously to Δ^1-pyrroline. γ-Aminobutyraldehyde may also be formed by decarboxylation from α-keto-δ-aminovaleric acid which is formed from ornithine by transamination. α-Keto-δ-aminovaleric acid cyclizes spontaneously to Δ^1-pyrroline-2-carboxylic acid (fig. 209). N-Acetylornithine as well as proline, citrulline and arginine are used as storage products for bound nitrogen in certain higher plants.

See D.10.1 for the formation of spermidine from S-adenosyl-S-methyl-mercaptopropylamine and putrescine.

References for further reading.

Meister, A. *Biochemistry of the Amino Acids*, Vol. II (Academic Press, New York, 1965).

Schütte, H. R. 'Einfache Amine', *Biosynthese der Alkaloide*, ed. K. Mothes and H. R. Schütte (VEB Deutscher Verlag der Wissenschaften, Berlin, 1969), pp. 168–82.

Reinbothe, H. 'Metabolismus der in die Alkaloidbiogenese eintretenden Aminosäuren', *Biosynthese der Alkaloide*, ed. K. Mothes and H. R. Schütte (VEB Deutscher Verlag der Wissenschaften, Berlin, 1969), pp. 40–100.

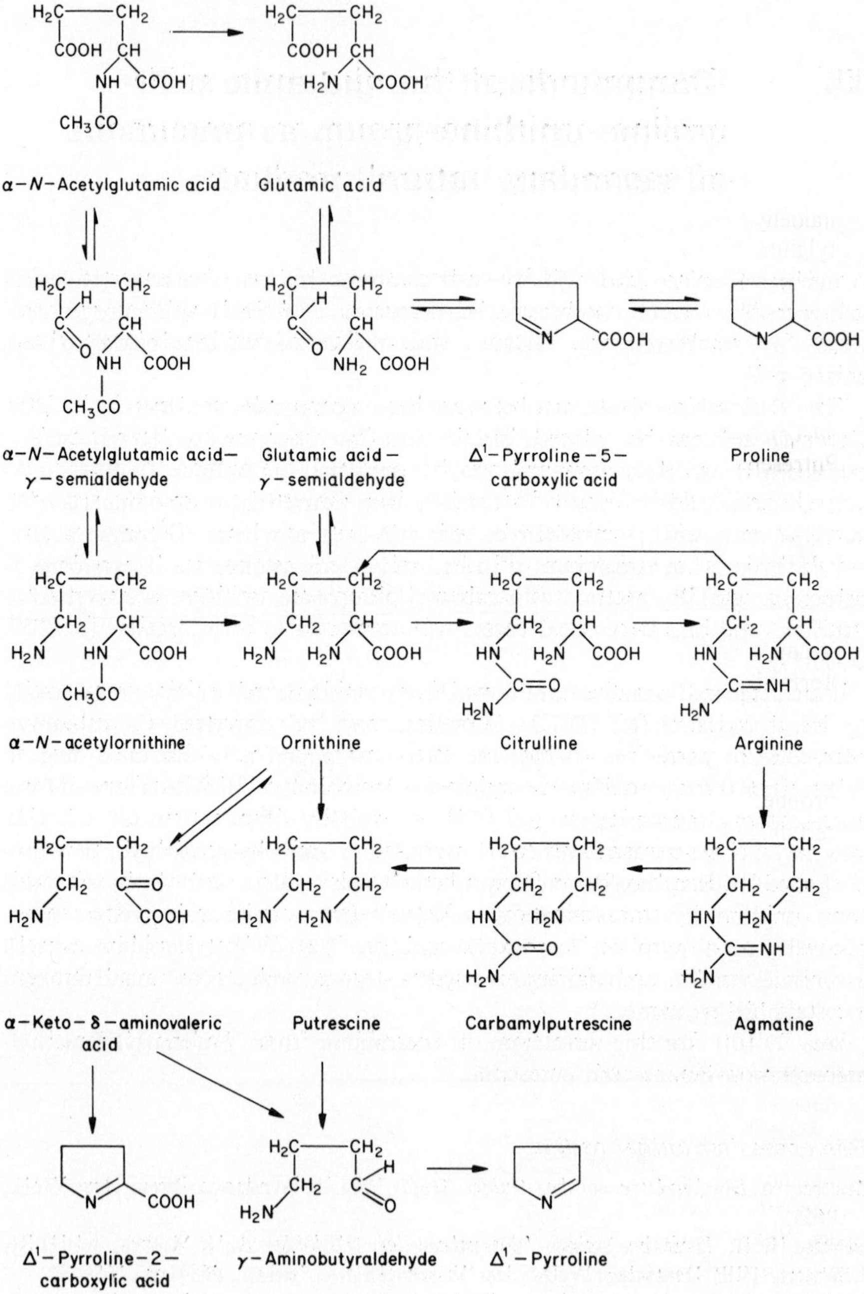

Fig. 209. Metabolic reactions of the amino acids of the glutamic acid-proline-orthinine-group.

15.1 Biosynthesis of tropane alkaloids

Tropane alkaloids have as yet been found only in higher plants. In most of these alkaloids the ring system of tropane is substituted by an oxygen function at position 3. If it is a hydroxyl group then two geometrical isomers are formed which are known as pseudotropine and tropine. Other important basic skeletons of the alkaloids are scopine and ecgonine (fig. 210).

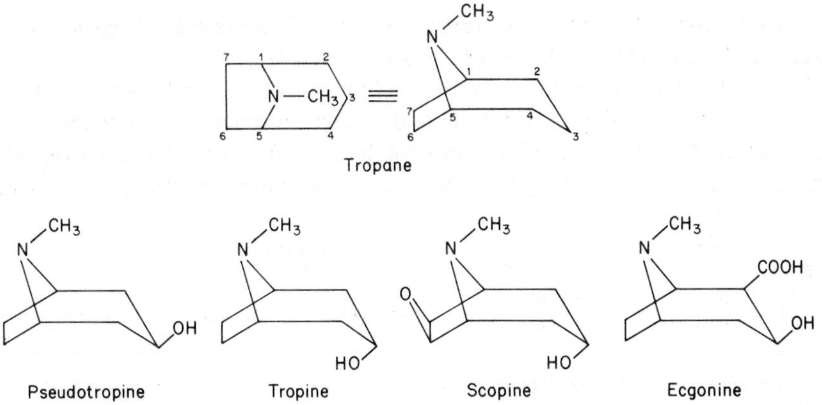

Tropane

Pseudotropine   Tropine   Scopine   Ecgonine

Fig. 210. Frequently occurring alcohols derived from tropane.

Most of the tropane alkaloids possess an ester linkage with an acid at carbon atom 3. The most frequently occurring acid components are tropic acid, atropic acid and phenyl glyceric acid (cf. D.20.3) as well as benzoic acid (cf. D.20.6), tiglic acid and methylbutyric acid (cf. D.13.1). The alkaloid hyoscyamine, widespread in the Solanacaeae, as well. as other species, is the ester of tropine and L-tropic acid, L-scopolamine is the ester of scopine and L-tropic acid, and the alkaloid cocaine, occurring in *Erythroxylon coca*, is the ester of methylecgonine and benzoic acid. L-Tropic acid racemizes easily, thus changing L-hyoscyamine to D,L-hyoscyamine (= atropine).

Reciprocal graftings using alkaloid-abundant and alkaloid-free Solanaceae (e.g. *Datura* and *Lycopersicon*) showed that the major portion of the tropane alkaloids are synthesized in the roots but accumulate in the shoot, and that they may undergo secondary changes during transport.

The tropane moiety of the alkaloid originates from a member of the glutamine-proline-ornithine-group, and acetate. Ornithine is incorporated in such a way that carbon atom 1 of tropine originates from the α-carbon atom of this amino acid. A symmetrical intermediate cannot therefore be formed in this process. α-Keto-δ-aminovaleric acid (cf. D.15.) therefore is probably formed as an intermediate substance.

Labelled putrescine is incorporated into tropine to almost the same extent as ornithine. After feeding putrescine-1,4-^{14}C about 50% of the radioactivity were located at carbon atom 1 of tropine. Competition experiments also indicate that the diamine actually is a precursor. Thus the enzyme system synthesizing tropine appears to be comparatively unspecific and unable to differentiate between putrescine and ornithine. Since methylputrescine-^{15}N,^{14}CH$_3$ is incorporated into hyoscyamine and scopolamine without change in the isotopic ratio and with higher specific incorporation rate, it may be assumed that as in the case of nicotine (cf. D.15.3) methylation of the nitrogen atom takes place before cyclization.

Carbon atoms 2, 3 and 4 of tropane originate from acetate. Acetoacetate or acetoacetyl CoA appears, however, to be the actual precursor.

There are two possibilities for the mechanism of the condensation between the pyrroline derivative originating either from ornithine or putrescine and acetoacetate. On the one hand it is possible that a double condensation takes place immediately, to form dehydroecgonine as well as tropinone (the point at which

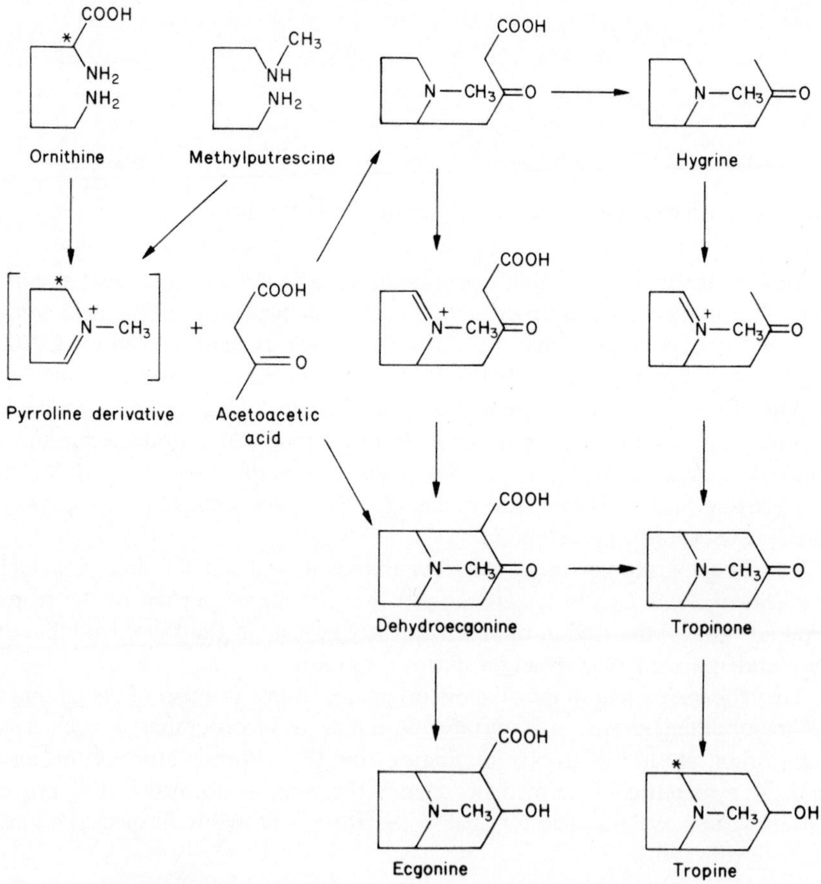

Fig. 211. Possible pathways for the formation of ergonine and tropine.

the carboxyl group of acetoacetate is eliminated during the formation of tropinone is unknown). Ecgonine and tropine may be formed from these ketones by reduction (fig. 211).

It is also possible, however, that linkage of acetoacetate occurs in two steps and monocyclic bases of the hygrine type are formed first. After dehydrogenation to immonium compounds they might then be precursors of ecgonine and tropine (fig. 211).

The esterification of the tropane moiety with acids is catalysed by the enzyme tropine esterase. Tropine esterase has been detected in the roots of various Solanaceae. The enzyme also occurs in many micro-organisms, and also in a number of mammals where it will degrade ingested tropane alkaloids into the acid and the tropane moiety.

In *Datura ferox* the enzymes necessary for the conversion of L-hyoscyamine to L-scopolamine are located in the stem tissues through which the hyoscyamine formed in the roots passes *en route* to the shoot. During scopolamine formation, 6-hydroxyhyoscyamine is formed as an intermediate (an accompanying alkaloid in many *Datura* species) (fig. 212).

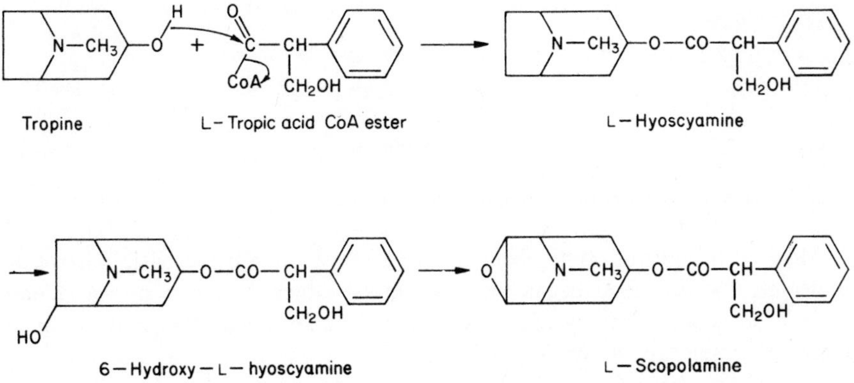

Fig. 212. Formation of L-hyoscyamine and L-scopolamine from tropine and L-tropic acid.

References for further reading

Luckner, M Über neue Arbeiten zur Biosynthese N-heterocyclischer Verbindungen, 4. Teil: Die Verbindungen mit Pyridin-, Piperidin-, Chinolizidin-, Pyrrolizidin- und Tropanringsystem, *Pharmazie* **19** (1964), pp. 1–14.

Leete, E. 'Alkaloid Biogenesis', *Biogenesis of Natural Compounds*, ed. P. Bernfeld (Pergamon Press, Oxford, 1967), pp. 953–1023.

Liebisch, H. W. 'Tropanalkaloide und Pyrrolidinbasen', *Biosynthese der Alkaloide*, ed. K. Mothes and H. R. Schütte (VEB Deutscher Verlag der Wissenschaften, Berlin, 1969), pp. 103–214.

15.2 Formation of pyrrolizidine alkaloids

Alkaloids possessing a pyrrolizidine ring are characteristic constituents of the genera *Senecio, Erechtites* and *Petasites* (Compositae). They also occur in other Compositae as well as in the Papilionaceae and Borraginaceae.

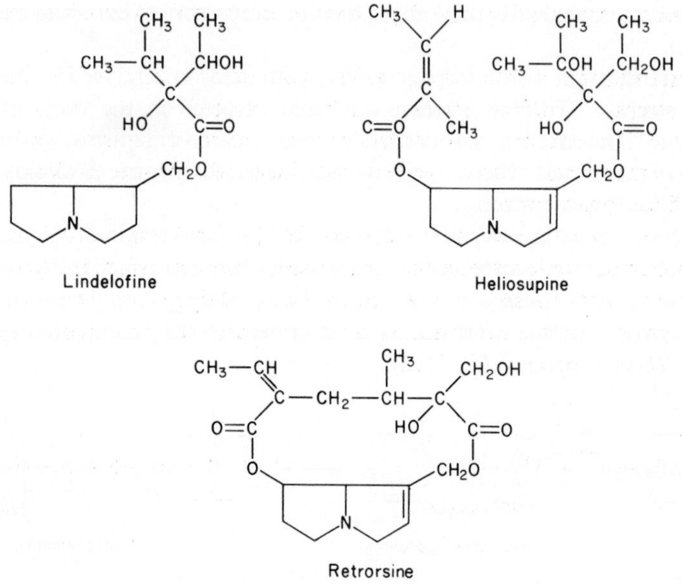

Lindelofine

Heliosupine

Retrorsine

Fig. 213. Naturally occurring pyrrolizidine alkaloids.

Most of the pyrrolizidine alkaloids are esters in which the basic part of the molecule, the so-called necine base, is linked either with one or two mono-carboxylic acids or with a dicarboxylic acid, the so-called necic acids. Some of the esters found are shown in fig. 213.

The necine bases originate from compounds of the glutamic acid-proline-ornithine-group, putrescine probably being the actual precursor (fig. 214). The biosynthesis appears to be analogous to the biosynthesis of the quinolizidine alkaloids from cadaverine (cf. D.16.3), where transaminations (cf. C.4) or oxidative deaminations (cf. C.2.1.2) play an important role. The aldehyde I may first be formed, via bisaminobutylamine and bisbutanalamine, which is then reduced in the next step to the primary alcohol II. According to its configuration this alcohol is known as isoretronecanol, lindelofidine, trachelanthamidine or laburnine. The other necine bases are formed from these by secondary changes. Introduction of a hydroxyl group at position 7 is frequently observed (compound III).

The necic acids known at present are formed at least partly from branched amino acids. Thus the carbon skeleton of isoleucine, with the exception of the carboxyl group, is incorporated as a whole into seneciphyllic acid (fig. 215). The carbon atom marked '*' originates from a one-carbon unit. High incorporation

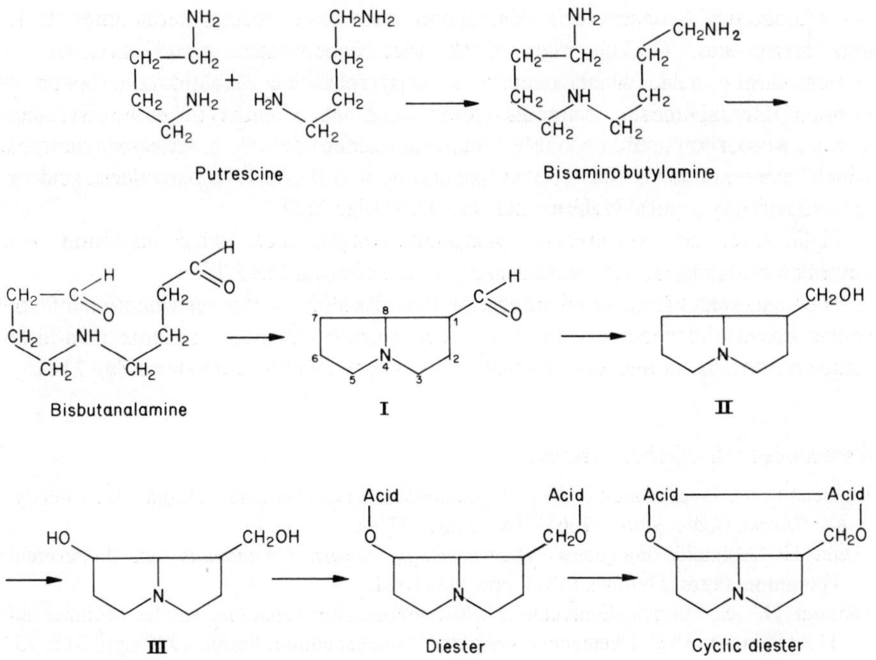

Fig. 214. Formation of necine bases.

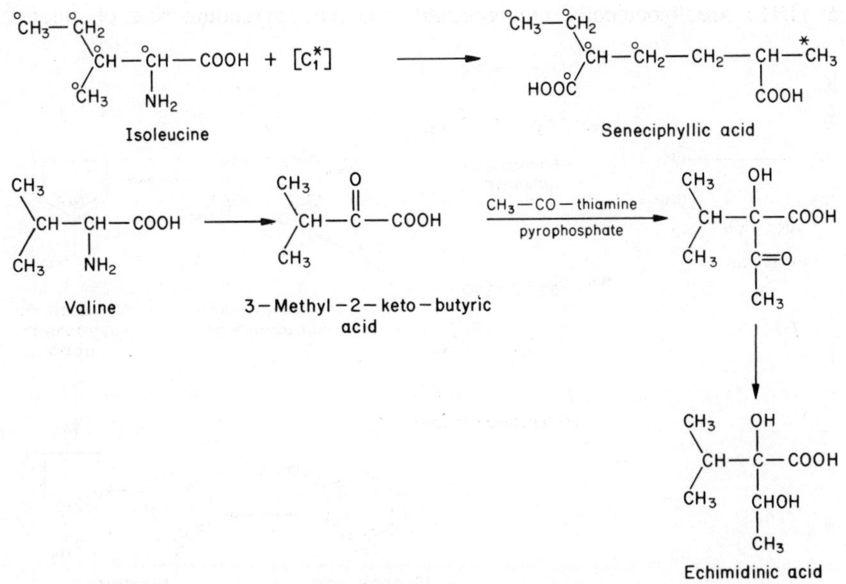

Fig. 215. Biosynthesis of seneciphyllic acid and echimidinic acid.

of radioactivity is observed in this carbon atom after feeding methionine-$^{14}CH_3$ and formic acid-^{14}C. The origin of the other carbon atoms is still unknown.

Echimidinic acid, which occurs as a pyrrolizidine alkaloid component of various Borraginaceae, originates from valine via 3-methyl-2-ketobutyric acid and a two-carbon unit, probably 'active acetaldehyde', by a series of reactions which corresponds to that of the formation of α,β-dihydroxyisovaleric acid or α,β-dihydroxy-β-methylvaleric acid (cf. D.13) (fig. 215).

Tiglic acid and its isomeric compound, angelic acid, which are found in a number of alkaloids, are synthesized as described in D.13.1.

The last step in the biosynthesis of the alkaloids is the esterification of the necine bases and necine acids. The cyclic diesters possibly originate at a later stage by the intramolecular coupling of two pre-existing half-esters (fig. 214).

References for further reading

Warren, F. L. Biosynthesis of Pyrrolizidine Alkaloids, *Abh. dtsch. Akad. Wiss. Berlin, Kl. Chem., Geol., Biol.* (1966), No. 3, pp. 571–6.

Leete, E. 'Alkaloid Biogenesis', *Biogenesis of Natural Compounds*, ed. P. Bernfeld (Pergamon Press, Oxford, 1967), pp. 953–1023.

Schütte, H. R. 'Pyrrolizidinalkaloide', *Biosynthese der Alkaloide*, ed. K. Mothes and H. R. Schütte (VEB Deutscher Verlag der Wissenschaften, Berlin, 1969), pp. 312–23.

15.3 Formation of the pyrrolidine ring of nicotine

Ornithine, glutamic acid, proline and other biogenetically related compounds (cf. D.15) are specifically incorporated into the pyrrolidine ring of nicotine.

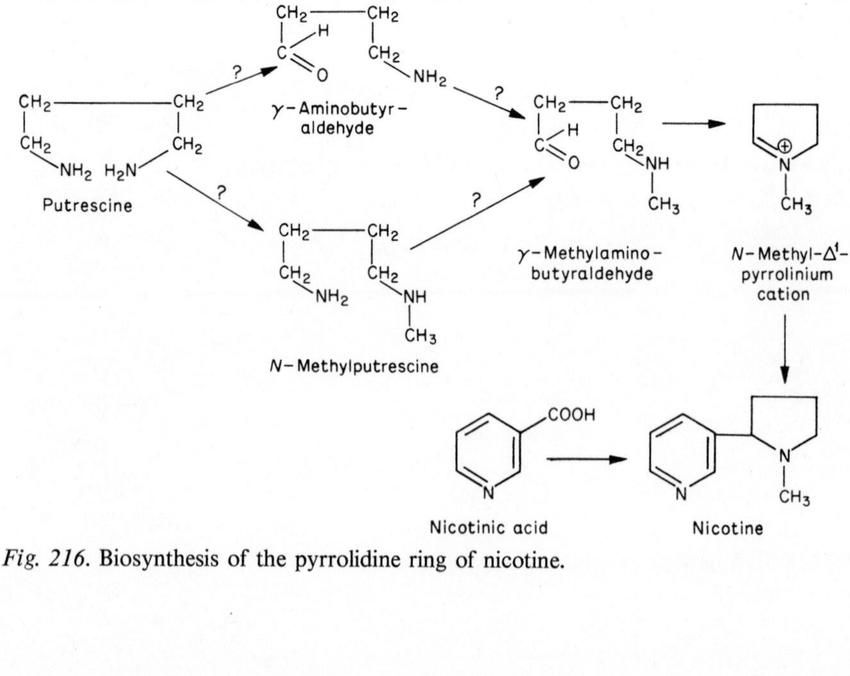

Fig. 216. Biosynthesis of the pyrrolidine ring of nicotine.

Since after feeding ornithine-2-^{14}C and ornithine-5-^{14}C the radioactivity is randomized between the positions 2 and 5 of the pyrrolidine ring, a symmetrical intermediate of the putrescine or Δ^1-pyrroline type must be formed in the process.

During the formation of nicotine in tobacco, the amino acid ornithine is most probably converted to putrescine via citrulline, arginine, agmatine and carbamyl-putrescine (fig. 209). Putrescine is then methylated and oxidized to γ-methyl aminobutyraldehyde by an amine oxidase (cf. C.2.1.2). This aldehyde cyclizes spontaneously to an N-methyl-Δ^1-pyrrolinium cation which is probably the actual precursor, and which reacts with carbon atom 3 of nicotinic acid (cf. D.14.2) with the formation of nicotine (fig. 216). It is, however, possible that the deamination product of putrescine, γ-aminobutyraldehyde, is methylated, rather than putrescine itself.

References for further reading

Leete, E. 'Alkaloid Biogenesis', *Biogenesis of Natural Compounds*, ed. P. Bernfeld (Pergamon Press, Oxford, 1967), pp. 953–1023.

Gross, D. 'Pyridinalkaloide', *Biosynthese der Alkaloide*, ed. K. Mothes and H. R. Schütte (VEB Deutscher Verlag der Wissenschaften, Berlin, 1969), pp. 215–74.

15.4 Formation of ornithine and glutamine conjugates in animals

While in most animals, potentially harmful organic acids are removed by excretion in the urine after 'conjugation' with glycine (cf. D.8.3), conjugates with ornithine are formed in certain birds and reptiles. Thus ornithuric acid (N,N'-dibenzoylornithine) discovered by Jaffé in 1877 is formed from benzoyl CoA

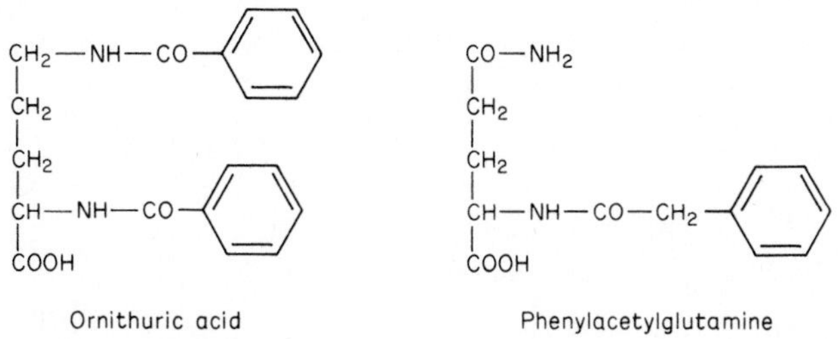

Ornithuric acid   Phenylacetylglutamine

Fig. 217. The structural formulae of ornithuric acid and phenylacetylglutamine.

and ornithine. Similar compounds originate from substituted benzoic acids, heterocyclic aromatic acids (e.g. nicotinic acid) as well as from phenylacetic acid and its derivatives.

Detoxication of phenylacetic acid (formed during the degradation of phenyl-alanine, cf. D.20.6) and indolyl acetic acid (formed from tryptophan, cf. D.19.6) occurs by linkage of the CoA-esters of these acids to the α-amino group of glutamine in the liver and kidneys of human beings and some anthropoid monkeys (e.g. chimpanzees) (fig. 217).

References for further reading

Williams, R. T. 'The Biogenesis of Conjugation and Detoxication Products', *Biogenesis of Natural Compounds*, ed. P. Bernfeld (Pergamon Press, Oxford, 1967), pp. 589–639.

16. Synthesis of secondary metabolites from lysine

The amino acid lysine is formed through two independent pathways. In certain fungi and algae it is formed from acetate and α-ketoglutaric acid via α-aminoadipic acid, and in the case of bacteria and a few algae from pyruvate and aspartic acid semialdehyde via α,ϵ-diaminopimelic acid. Both these metabolic pathways probably occur in higher plants.

(a) Diaminopimelic acid pathway

Pyruvic acid and aspartic acid semialdehyde condense to form the unsaturated -α-amino-ϵ-ketopimelic acid, I. This does not, however, occur in the free form, but cyclizes spontaneously to 2,3-dihydrodipicolinic acid. Δ^1-Piperideine-2,6-dicarboxylic acid originates from the latter compound by reduction, and by reaction of the corresponding open-chain compound α-amino-ϵ-ketopimelic acid, II, with succinyl CoA, N-succinyl-ϵ-keto-L-α-aminopimelic acid is formed (this compound is prevented from further cyclization by substitution of the amino group). L-α,ϵ-Diaminopimelic acid is formed from N-succinyl-ϵ-keto-L-α-aminopimelic acid by transamination, followed by elimination of the succinyl residue. L-Diaminopimelic acid is then converted to meso-α,ϵ-diaminopimelic acid (cf. D.7.1) by racemization at the ϵ-amino group. The latter compound yields L-lysine on decarboxylation (fig. 218).

(b) Aminoadipic acid pathway

The enzymes of the α-keto acid elongation-complex (cf. D.12) play an important role in the formation of lysine from α-ketoglutaric acid and acetyl CoA. α-Ketoadipic acid, which may be converted to α-aminoadipic acid by transamination (cf. C.3), is formed from α-ketoglutaric acid via homocitric acid, homo-cis-aconitic acid, homoisocitric acid and oxaloglutaric acid (fig. 219).

During conversion to lysine α-aminoadipic acid is first reduced to the corresponding semialdehyde, which either may then be converted directly to L-lysine by transamination (cf. C.4), or reacts with glutamic acid to form the Schiff base III. Saccharopine is formed from this compound by reduction, and may then be degraded to L-lysine and α-ketoglutaric acid by dehydrogenation (fig. 219). The

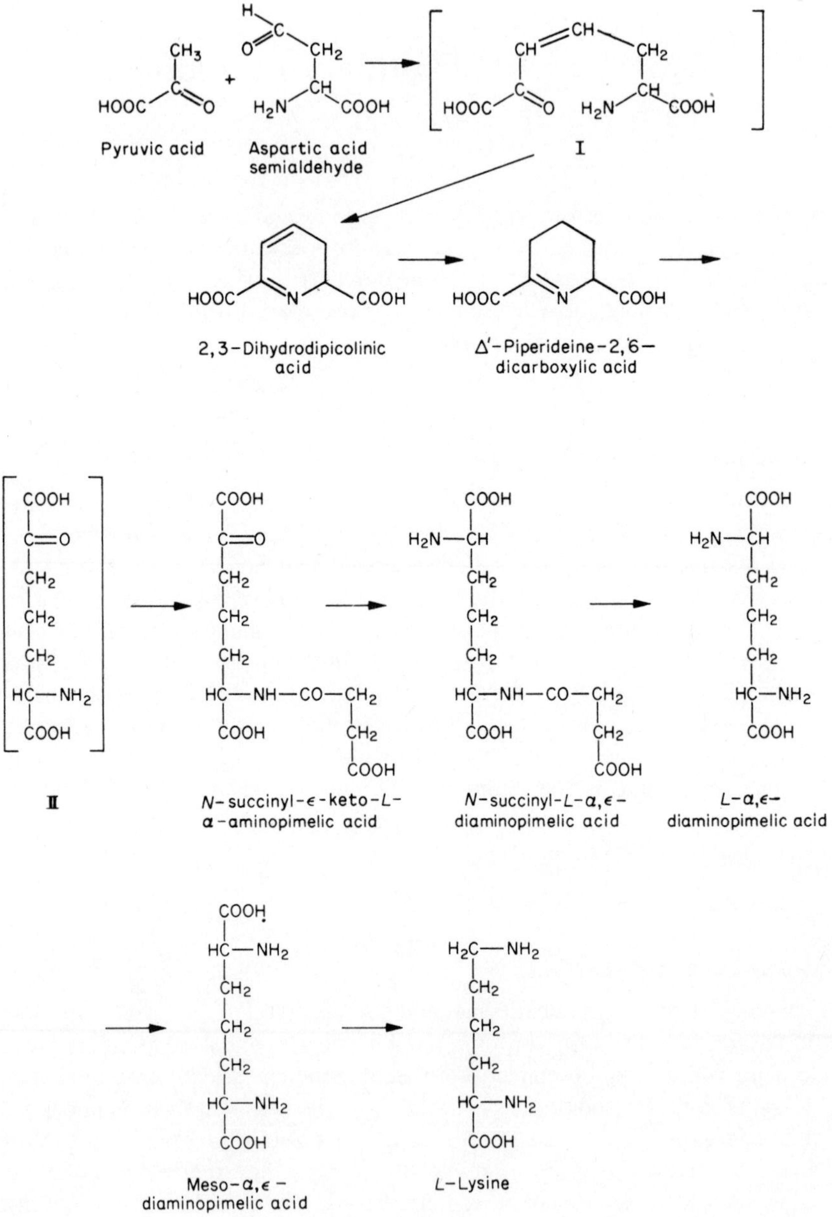

Fig. 218. Formation of lysine by the diaminopimelic acid pathway.

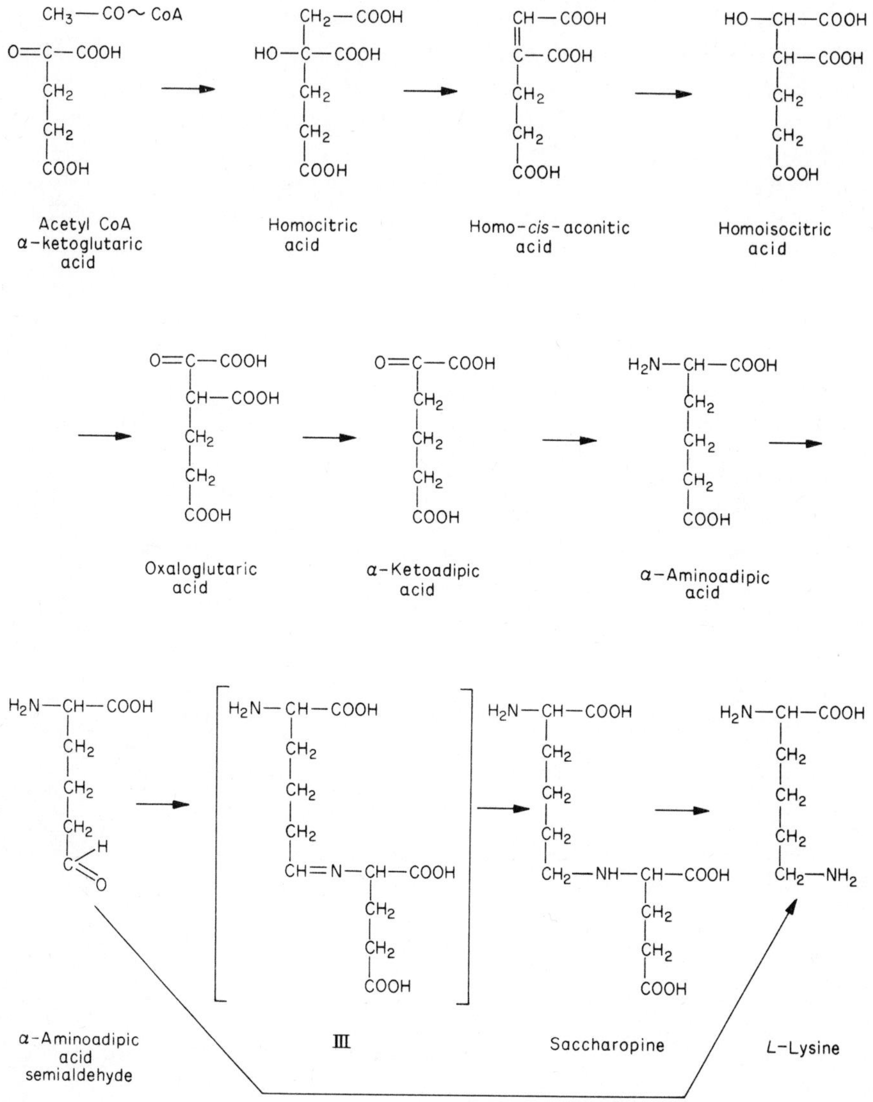

Fig. 219. Formation of lysine by the α-aminoadipic acid pathway.

formation of L-lysine from α-aminoadipic acid corresponds to the conversion of glutamic acid to ornithine (cf. D.15).

References for further reading

See D.10.

16.1 Biosynthesis of pipecolic acid and pipecolic acid derivatives

Pipecolic acid is widespread in higher plants, animals and micro-organisms. This imino acid is frequently accompanied by derivatives substituted at positions 4, 5 and 6. The biosynthesis of these compounds is closely connected with the synthesis and metabolism of lysine (fig. 220).

In animals the degradation of lysine begins with the elimination of the amino group at the α-position with the formation of α-keto-ϵ-aminocaproic acid. This acid cyclizes spontaneously to Δ^1-piperideine-2-carboxylic acid which may then be stereospecifically reduced to pipecolic acid in a further reaction (fig. 220).

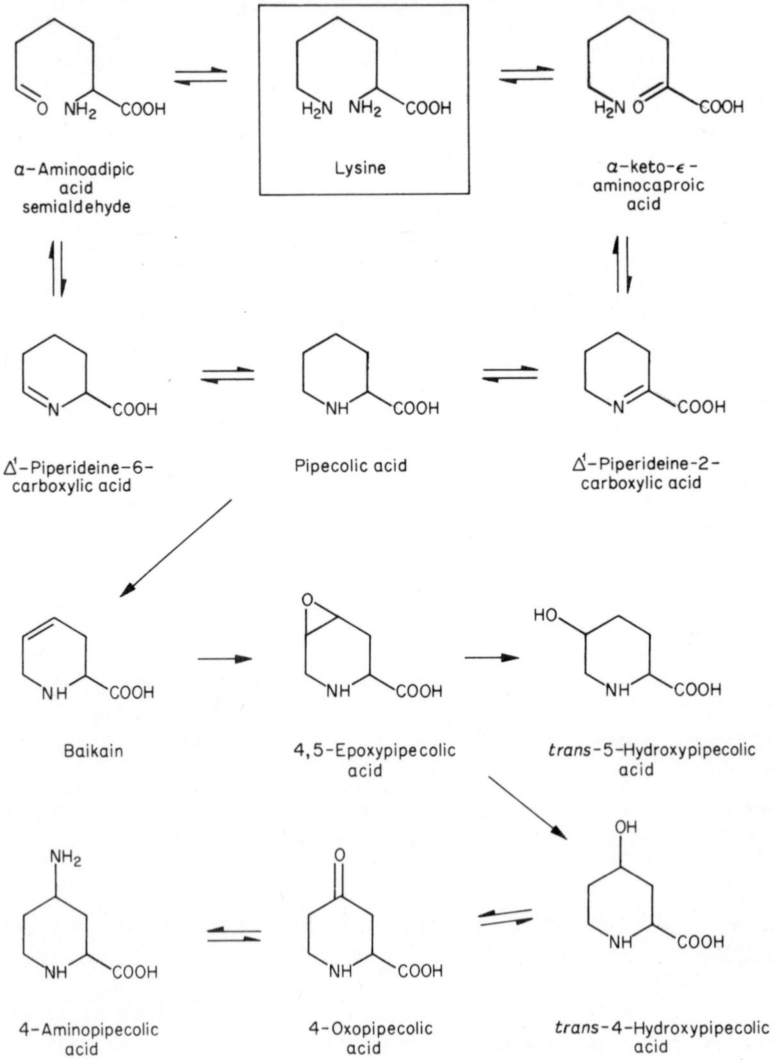

Fig. 220. Formation of pipecolic acid and pipecolic acid derivatives.

Pipecolic acid is partly excreted in the urine and partly oxidized to α-amino-adipic acid via α-aminoadipic acid semialdehyde; this may be further degraded via α-ketoadipic acid and glutaryl CoA.

In higher plants (e.g. *Phaseolus vulgaris*), on the other hand, the formation of pipecolic acid takes place by the elimination of the ϵ-amino group via α-amino-adipic acid semialdehyde and Δ^1-piperideine-6-carboxylic acid. Pipecolic acid is probably an intermediate in lysine synthesis in *Euglena gracilis* and *Aspergillus nidulans*. Here the pathway α-aminoadipic acid $\rightarrow$ α-aminoadipic acid semi-aldehyde $\rightarrow$ pipecolic acid $\rightarrow$ lysine is probable.

Strophanthus scandens, a species of Acacia, and a number of other plants, can convert pipecolic acid to *trans*-4-hydroxypipecolic acid. The simultaneous occurrence of these compounds with baikain and *trans*-5-hydroxypipecolic acid indicates that the formation of the hydroxylated pipecolic acids is catalysed by the action of a mixed function oxygenase on baikain, with the formation of 4,5-epoxypipecolic acid as shown in fig. 220.

In *Strophanthus scandens*, 4-hydroxypipecolic acid is converted to 4-amino-pipecolic acid. This reaction is reversible and probably proceeds via 4-oxo-pipecolic acid (fig. 220).

References for further reading

Liebisch, H. W. 'Piperidinalkaloide', *Biosynthese der Alkaloide*, ed. K. Mothes and H. R. Schütte (VEB Deutscher Verlag der Wissenschaften, Berlin, 1969), pp. 275–311.

Gross, D. Vorkommen, Struktur und Biosynthese natürlicher Piperidinverbindungen, *Fortschritte Chem. org. Naturstoffe* **29** (1971), pp. 1–59.

16.2 Formation of Punica-, Sedum-, and Lobelia alkaloids

Piperidine alkaloids in which the piperidine ring is substituted by side chains at position 2, or positions 2 and 6, have been isolated from *Punica granatum*, a large number of species of *Sedum* and *Lobelia*, as well as from other higher plants. Frequently these compounds are accompanied by pyrrolidine alkaloids possessing a similar structure.

Experiments on *Punica granatum*, *Sedum acre* and *Lobelia cardinalis* have shown that lysine and cadaverine are precursors of the heterocyclic ring of these alkaloids (fig. 221 and 222). It may be assumed, by analogy with the formation of compounds of the hygrine type, and the tropane alkaloids, that the remaining carbon atoms in isopelletierine and pseudopelletierine originate from acetate (cf. D.15.1) (fig. 221).

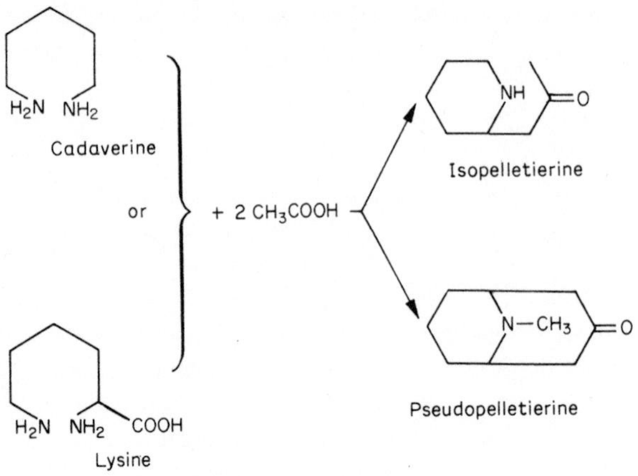

Fig. 221. Formation of isopelletierine and pseudopelletierine.

The phenylethyl side chain is derived from phenylpropane units. Phenylalanine is incorporated into sedamine as well as lobinaline with the loss of the carboxyl group. In the formation of sedamine, the side chain is linked to carbon atom 2 of the precursor lysine (fig. 222).

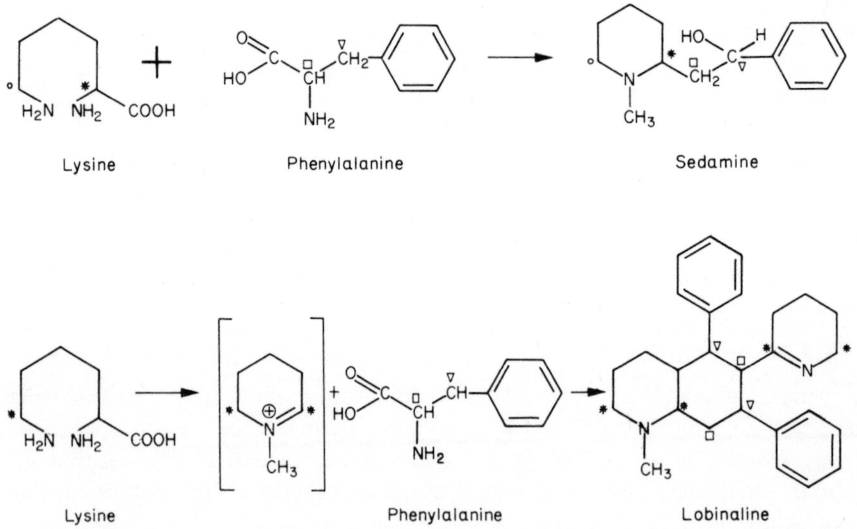

Fig. 222. Biosynthesis of sedamine and lobinaline.

On the other hand, during the formation of lobinaline a symmetrical intermediate is formed (fig. 222). The linkage of the second side chain at position 6 of alkaloids of the lobeline type probably takes place after the 1,6-dehydrogenation of a monosubstituted piperidine to a quarternary base.

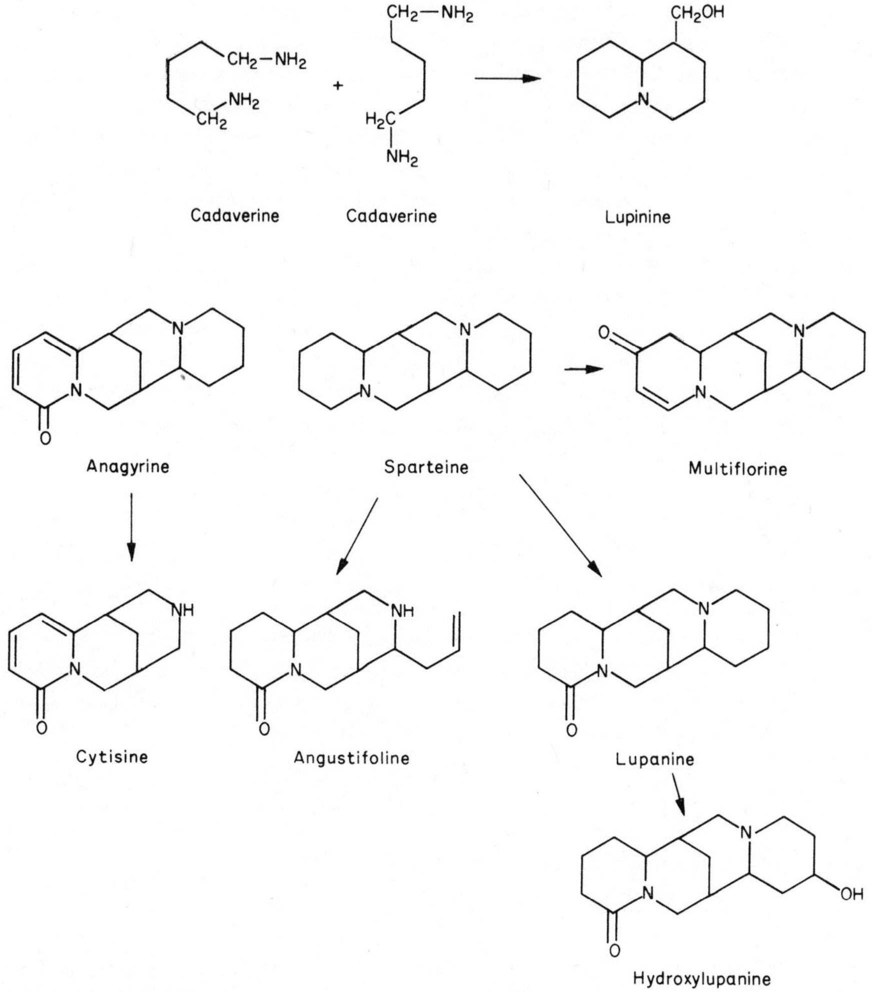

Fig. 223. Formation of various quinolizidine alkaloids.

References for further reading

Gross, D. Vorkommen, Struktur und Biosynthese natürlicher Piperidinverbindungen, *Fortschritte Chem. org. Naturstoffe* **29** (1971), pp. 1–59.

16.3 Biosynthesis of quinolizidine alkaloids

The quinolizidine alkaloids are derived from norlupinan and are especially widespread in the Papilionaceae (e.g. species of *Lupinus* and *Sarothamnus*), but also occur in the Chenopodiaceae, Ranunculaceae, Berberidaceae, Papaveraceae and Solanaceae.

The compounds originate from lysine. This amino acid is probably decarboxylated to cadaverine (cf. D.7.3). Further intermediates have, however, not been detected so that it is assumed that the biosynthesis proceeds on a multi-enzyme complex from which only the synthesized alkaloid is removed.

The alkaloid lupinine found in various species of *Lupinus* is derived from two molecules of cadaverine. Anagyrine and sparteine are derived from three molecules. The latter compound may be oxidized to multiflorine and lupanine, and is also a precursor of angustifoline, which is shorter by one carbon atom. The formation of angustifoline from sparteine probably does not proceed via lupanine. Hydroxylupanine is derived from lupanine. This alkaloid occurs esterified with *trans-* and *cis*-cinnamic acid (cf. D.20.4.1), veratric acid or benzoic acid (cf. D.20.6), and tiglic acid (cf. D.13.1) in high concentrations in various species of *Lupinus*. Cytisine, an alkaloid found in *Laburnum anagyroides*, is derived from anagyrine (fig. 223). Nothing is known about the mechanisms of the conversion reactions.

References for further reading

Leete, E. 'Alkaloid Biogenesis', *Biogenesis of Natural Compounds,* ed. P. Bernfeld (Pergamon Press, Oxford, 1967), pp. 953–1023.

Schütte, H. R. 'Chinolizidinalkaloide', *Biosynthese der Alkaloide,* ed. K. Mothes and H. R. Schütte (VEB Deutscher Verlag der Wissenschaften, Berlin, 1969), pp. 324–43.

Schütte, H. R., Seelig, G. and Knöfel, D. Zur Biosynthese des Angustifolins in Lupinus angustifolius, *Z. Pflanzenphysiologie* **63** (1970), p. 393.

16.4 Formation of the piperidine ring of anabasine

The piperidine ring of the alkaloid anabasine, which occurs in species of *Anabasis* and *Nicotiana*, is derived from lysine. The incorporation of the amino acid takes place in such a manner that the ω-amino group is retained and carbon atom 2′ of anabasine is formed from the α-carbon atom. An asymmetric intermediate, probably Δ^1-piperideine-2-carboxylic acid, may be formed during this conversion. This then reacts with carbon atom 3 of nicotinic acid (cf. D.14.2) (fig. 224).

Feeding of the diamine, cadaverine, which may be formed from lysine by decarboxylation (cf. D.7.1), to species of *Nicotiana* also results in incorporation of label. Carbon atom 1 is incorporated into positions 2′ and 6′ of the alkaloid, so it may be assumed that a semialdehyde is formed first from the diamine by the action of an amine oxidase (cf. C.2.1.1). This then spontaneously cyclizes to a symmetrical Δ^1-piperideine.

A further biosynthetic pathway for formation of anabasine has been found in *in vitro* experiments with extracts from the seedings of peas and lupins. Here the whole molecule of the alkaloid originates from cadaverine. Δ^1-Piperideine is formed first from cadaverine under cellular conditions, and then dimerizes to

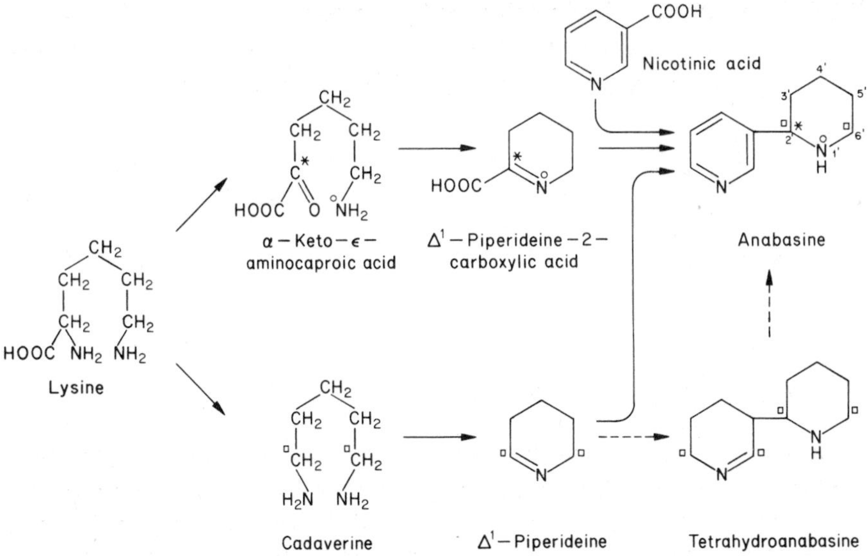

Fig. 224. Formation of anabasine.

tetrahydroanabasine. This compound is then dehydrogenated to anabasine by a low molecular weight co-factor of which little is known (fig. 224).

This biosynthetic pathway for anabasine probably does not occur in intact plants (pea and lupin plants do not normally contain anabasine). The tetrahydroanabasine alkaloids such as ammodendrine (fig. 225) found in *Ammodendron conollyi* possibly originate in this way.

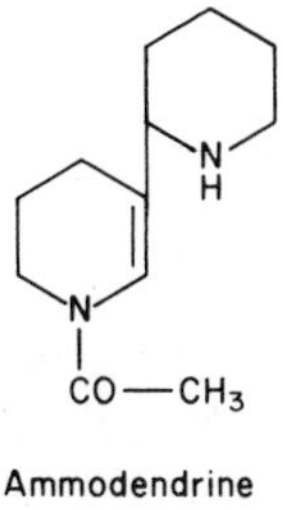

Ammodendrine

Fig. 225. Ammodendrine.

References for further reading

Gross, D. 'Pyridinalkaloide', *Biosynthese der Alkaloide*, ed. K. Mothes and H. R. Schütte (VEB Deutscher Verlag der Wissenschaften, Berlin, 1969), pp. 215–74.

Gross, D. Vorkommen, Struktur und Biosynthese natürlicher Piperidinverbindungen, *Fortschritte Chem. org. Naturstoffe* **29** (1971), pp. 1–59.

16.5 Biosynthesis of mimosine and desmosine

The amino acid mimosine has been found in *Leucaena glauca* and *Mimosa pudica* together with pipecolic acid and 5-hydroxypipecolic acid.

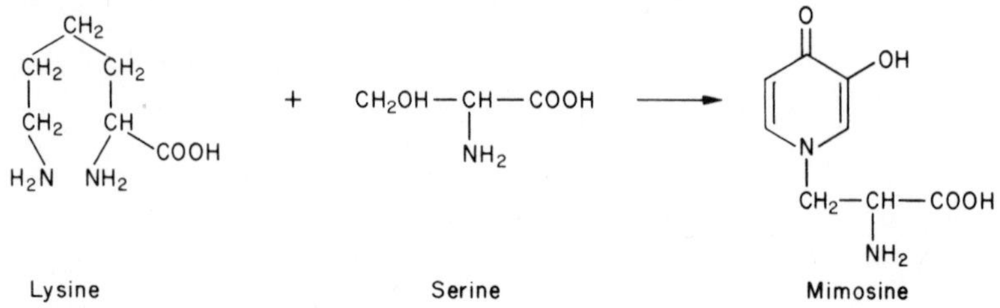

Fig. 226. Biosynthesis of mimosine.

Mimosine is derived from the amino acids lysine and serine (fig. 226). Intermediates of the biosynthetic pathway have not yet been found. δ-Hydroxylysine is not incorporated into mimosine, a result which suggests that introduction of the hydroxyl group into the mimosine molecule occurs at a later stage of the synthesis. It is not yet clear whether the ring nitrogen atom is derived from the α- or the ω-amino group of lysine.

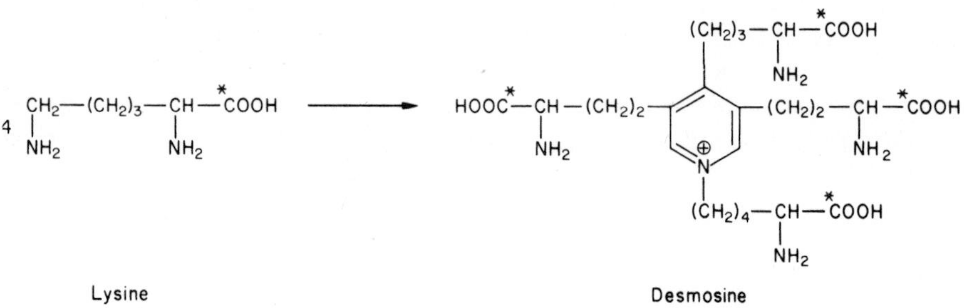

Fig. 227. Formation of desmosine.

The amino acid desmosine is a constituent of the animal protein elastin. Feeding experiments using rats and embryonic chicken aorta have shown that this compound originates from four molecules of lysine (fig. 227). α-Amino adipic acid semialdehyde was found to be an intermediate.

References for further reading

Gross, D. 'Pyridinalkaloide', *Biosynthese der Alkaloide*, ed. K. Mothes and H. R. Schütte (VEB Deutscher Verlag der Wissenschaften, Berlin, 1969), pp. 215–74.

17. Arginine as precursor of secondary natural products

Arginine is an intermediate of the urea cycle (Krebs-Henseleit cycle) (fig. 228) and originates from ornithine and carbamyl phosphate via citrulline and argininosuccinate. The enzymes of the urea cycle have been found in micro-organisms as well as in higher plants and animals. An actual cycle in which urea is produced

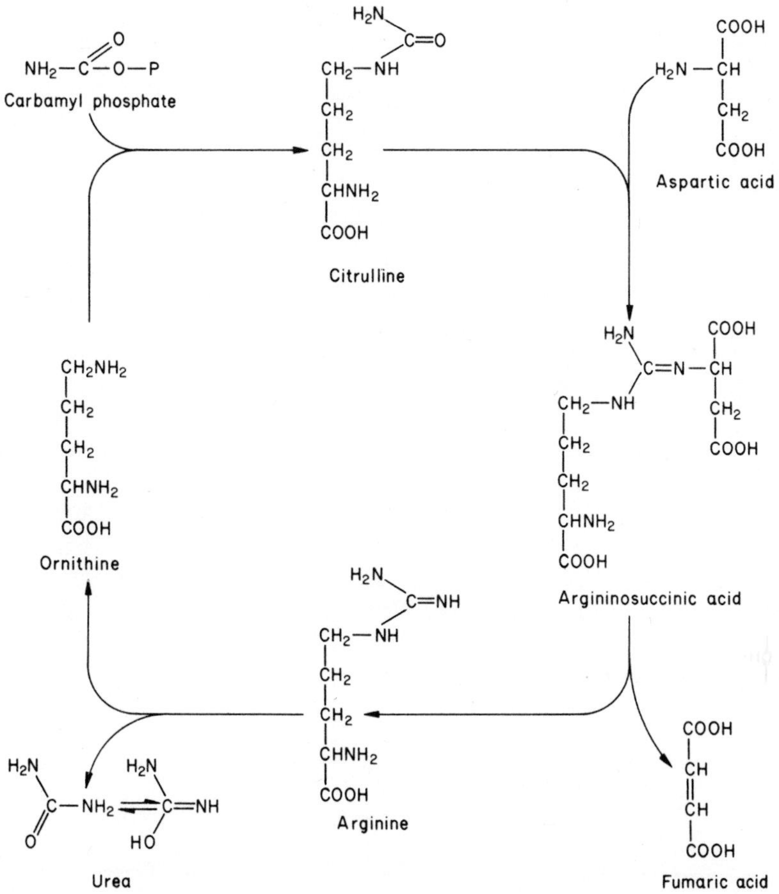

Fig. 228. The urea cycle.

in large quantities appears to proceed only in the case of animals. Urea serves as the principal end product of nitrogen metabolism and is excreted in the urine.

References for further reading

See D.10.

17.1 Formation of secondary guanidine compounds

Secondary products with the guanidine grouping occur comparatively rarely, but have been found in certain animals as well as in micro-organisms and higher plants

Octopine, found in species of *Octopus*, originates from arginine and pyruvate via the Schiff base I (fig. 229). The alanine unit formed from it by reduction possesses the D-configuration.

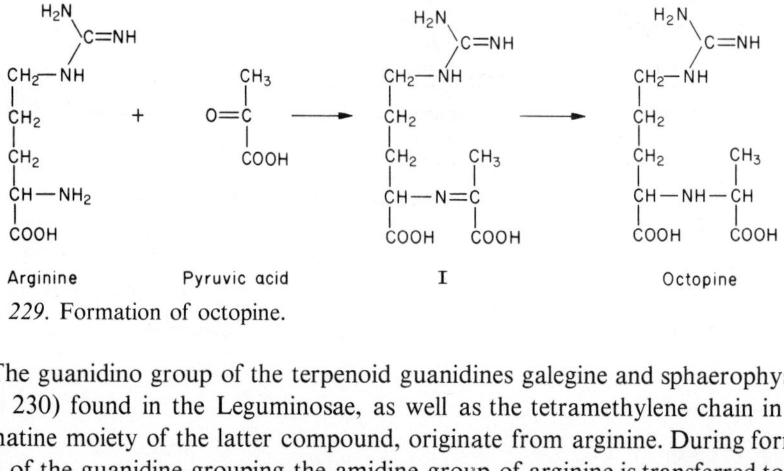

Arginine Pyruvic acid I Octopine

Fig. 229. Formation of octopine.

The guanidino group of the terpenoid guanidines galegine and sphaerophysine (fig. 230) found in the Leguminosae, as well as the tetramethylene chain in the agmatine moiety of the latter compound, originate from arginine. During formation of the guanidine grouping the amidine group of arginine is transferred to the amine, which is the precursor, by the action of an amidino transferase, with the formation of ornithine. Nothing definite is known about the biosynthesis of the isoprenoid portion of both these alkaloids; the results, however, indicate that biosynthesis does not proceed via mevalonic acid.

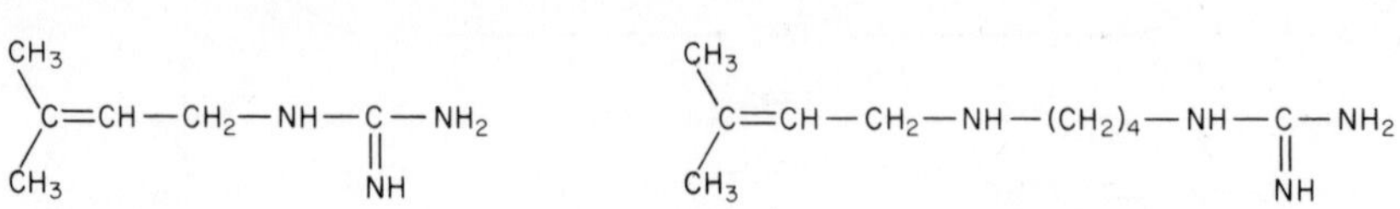

Galegine Sphaerophysine

Fig. 230. The structural formulae of the terpenoid guanidines, galegine and sphaerophysine.

References for further reading

Meister, A. *Biochemistry of the Amino Acids*, Vol. II (Academic Press, New York, 1965).

Reuter, G. Die Biosynthese terpenoider Guanidine, *Abh. dtsch. Akad. Wiss. Berlin, Kl. Chem., Geol., Biol.* (1966), No. 3, pp. 617–21.

17.2 Conversion of homoarginine to lathyrine

Lathyrine, an amino acid possessing a pyrimidine ring, has recently been found in the seeds of *Lathyrus tingitanus*. The non-protein amino acids, homoarginine and γ-hydroxyhomoarginine, also occur.

By feeding radioactively labelled homoarginine it has been shown that this compound is first hydroxylated at the γ-position and then incorporated into lathyrine. This pathway is outlined in fig. 231.

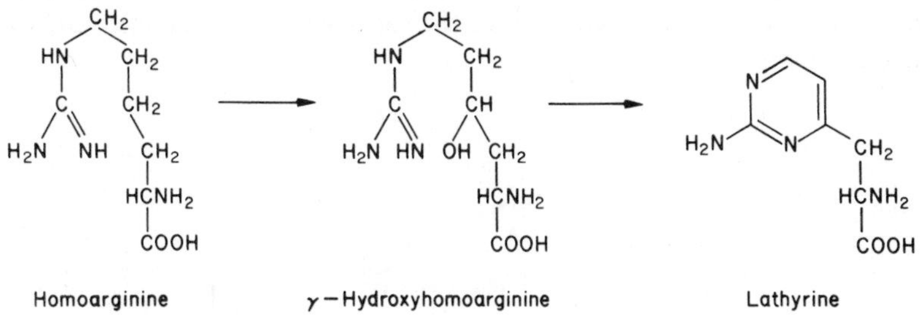

Homoarginine γ — Hydroxyhomoarginine   Lathyrine

Fig. 231. Formation of lathyrine.

The mode of biosynthesis of homoarginine is unknown. Synthesis from arginine by enzymes of an α-keto acid elongation-complex is possible (cf. D.12).

References for further reading

Meister, A. *Biochemistry of the Amino Acids*, Vol. I (Academic Press, New York, 1965).

Luckner, M. and Wasternack, C. Über neue Arbeiten zur Biosynthese *N*-heterocyclischer Verbindungen, 7. Teil: Bildung und Abbau von Verbindungen mit Pyrimidinring-system, *Pharmazie* **22** (1967), pp. 181–98.

18. Biosynthesis of secondary natural products from histidine

Histidine and the tissue hormone histamine, derived from this amino acid by decarboxylation (cf. D.7.3.), are the most widespread imidazole derivatives in nature. No information is available on the biosynthesis of alkaloids possessing an imidazole ring, e.g. pilocarpine, pilosine, spinacin and zapotidin (fig. 232). However, their structural relationship with histidine and the simultaneous occurrence of the imidazole alkaloids, glochidine and glochicidine, with N-4'-oxodecanoyl histamine in species of *Glochidion* make the formation of these compounds from histidine probable.

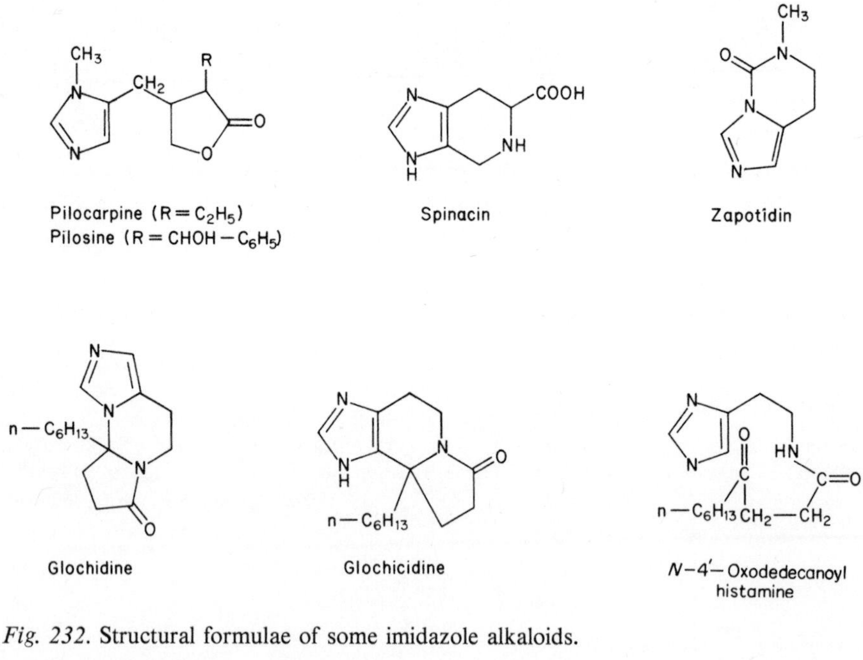

Pilocarpine (R = C$_2$H$_5$)
Pilosine (R = CHOH — C$_6$H$_5$)

Spinacin

Zapotidin

Glochidine

Glochicidine

N-4'-Oxodecanoyl histamine

Fig. 232. Structural formulae of some imidazole alkaloids.

The biosynthesis of histidine, which is very closely connected with purine metabolism, has been clarified to a great extent in recent years. ATP and 5-phosphoribosyl-1-pyrophosphate first condense to form 1-(5'-phosphoribosyl)-ATP and the ring system is then cleaved between position 1 and 6. 1-(5'-Phosphoribosyl)-4-carboxamido-5-N-(N'-5''-phosphoribosyl)-formamidinoimidazole

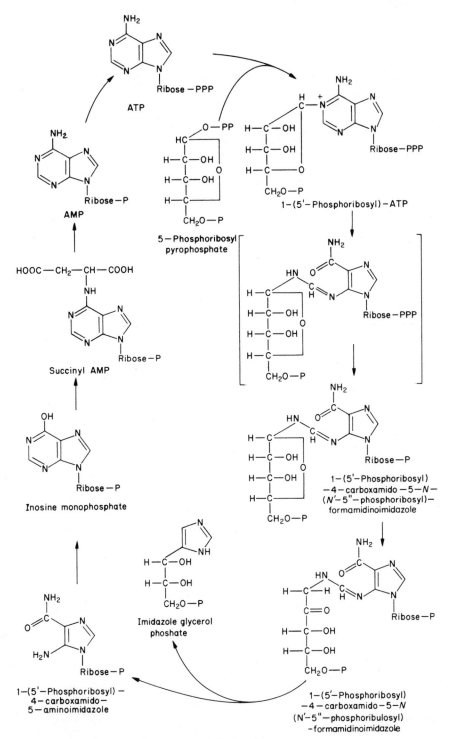

Fig. 233. Formation of imidazole glycerol phosphate.

which is thus formed via an unidentified intermediate change to 1-(5'-phosphori-bosyl)-4-carboxamido-5-N-(5''-phosphoribulosyl)-formamidinoimidazole by an Amadori rearrangement. This latter compound then reacts with glutamine to form D-erythro-imidazole glycerol phosphate and 1-(5'-phosphoribosyl)-4-carbo-xamido-5-aminoimidazole via an unknown intermediate. 1-(5'-phosphoribosyl)-4-carboxamido-5-aminoimidazole may be converted to ATP via inosine mono-phosphate, succinyl adenosine monophosphate and AMP (fig. 233) (cf. D.8.2.1).

The conversion of D-erythro-imidazole glycerol phosphate to histidine has been particularly investigated in micro-organisms. It was found that imidazole glycerol phosphate is first converted to imidazole acetol phosphate (imidazole hydroxy acetone phosphate) by the enzyme imidazole glycerol phosphate de-hydrase. This involves the shift of a hydroxy group from the β- to the α-carbon atom and the elimination of the elements of water. This compound is then con-verted to L-histidinol phosphate by a transaminase. After elimination of the phosphate group of L-histidinol phosphate by the enzyme L-histidinol phosphate phosphatase, the L-histidinol thus formed is oxidized to L-histidine. Oxidation takes place in the presence of L-histidinol dehydrogenase and NAD^+ and probably proceeds via L-histidinal (fig. 234).

A number of secondary natural products originate from the amino acid histidine. Methylation of the imidazole ring has been detected in various organ-isms. After feeding D,L-histidine-[α-^{14}C] to rats, radioactively labelled N^1-methylhistidine was found in the urine. Histamine is oxidized to N^1-methylimi-dazole acetic acid via N^1-methylhistamine and N^1-methylimidazole acetaldehyde

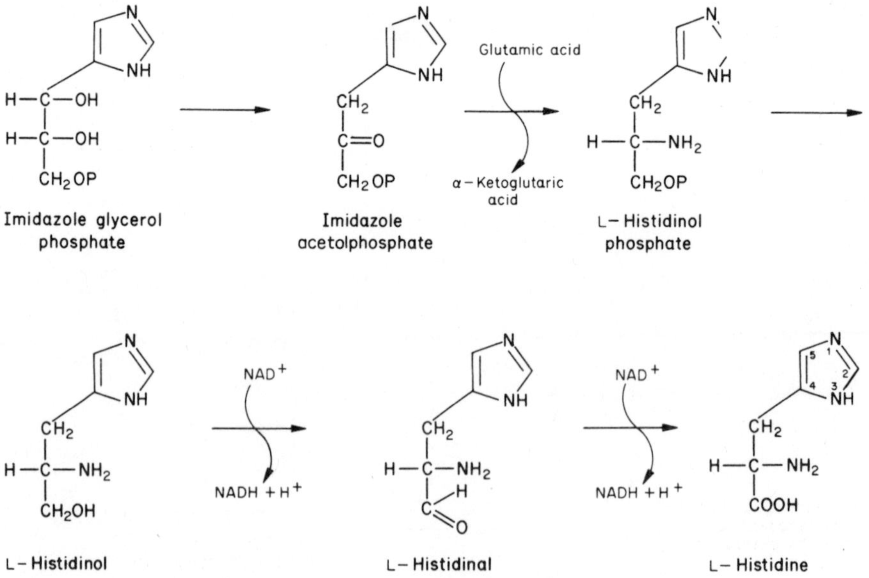

Fig. 234. Conversion of imidazole glycerol phosphate to histidine.

by human placental tissue. Introduction of histamine into mammals results in the presence of N^3-methylhistamine and other metabolites in the urine.

The tissue of molluscs can convert histamine to imidazole pyruvic acid, imidazole acetic acid, imidazole methanol and other substances by transamination or oxidative deamination, oxidation and decarboxylation. The administration of histidine to mammals resulted in the appearance of imidazole pyruvic acid, imidazole lactic acid, imidazole propionic acid and imidazole ethanol in the urine (cf. fig. 235).

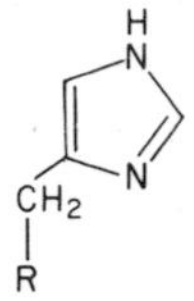

Imidazole pyruvic acid $(R = CO - COOH)$
Imidazole lactic acid $(R = CHOH - COOH)$
Imidazole propionic acid $(R = CH_2 - COOH)$
Imidazole acetic acid $(R = COOH)$
Imidazole ethanol $(R = CH_2OH)$
Imidazole methanol $(R = OH)$

Fig. 235. The structural formulae of some imidazole derivatives.

The sulphur-containing imidazole derivative, ergothioneine, has been detected in a number of plant organisms and animal tissues after its initial discovery in *Claviceps purpurea*. Animals are not capable of synthesizing this compound but accumulate it, however, after ingestion. The formation of ergothioneine starts from histidine both in ergot as well as in *Neurospora crassa* and proceeds via the betaine, hercynine (cf. D.7.4), into which sulphur is introduced from cysteine. Degradation of ergothioneine in the bacterium *Alcaligenes faecalis* leads to the formation of trimethylamine and thiolurocanic acid which may then be converted to aliphatic compounds (fig. 236).

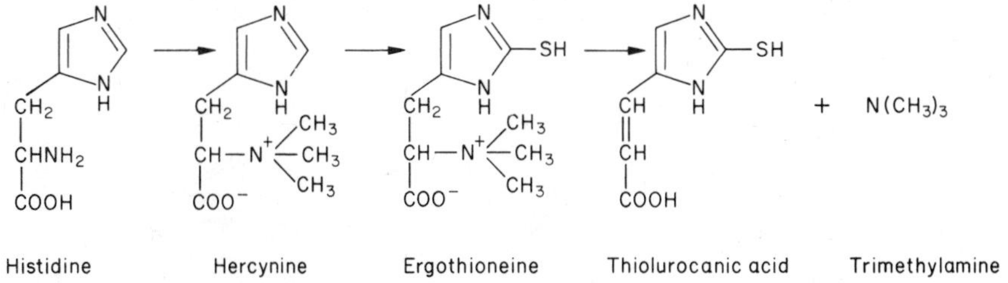

Histidine   Hercynine   Ergothioneine   Thiolurocanic acid   Trimethylamine

Fig. 236. Biosynthesis and degradation of ergothioneine.

Phosphorylated derivatives of histidine (e.g. N^3-phosphohistidine) are possibly intermediates in oxidative phosphorylation.

One of the most important degradative pathways of the imidazole ring leads from histidine via urocanic acid and imidazolone propionic acid to glutamic acid derivatives. The formation of the necessary enzymes is induced by histidine in e.g. *Aerobacter aerogenes*.

Urocanic acid is a normal constituent of human urine and usually occurs there in very minute quantities, but under certain conditions may appear in large amounts. The formation of urocanic acid catalysed by histidine deaminase (= histidase) is analogous to the conversion of aspartic acid to fumaric acid (cf. D.14) or of phenylalanine or tyrosine to cinnamic acid and *p*-coumaric acid (cf. D.20.4.1). The enzyme catalyses a transelimination in which both the NH₂-group, and a hydrogen atom in the β-position is lost.

In mammals histidine deaminase is present in especially large quantities in the

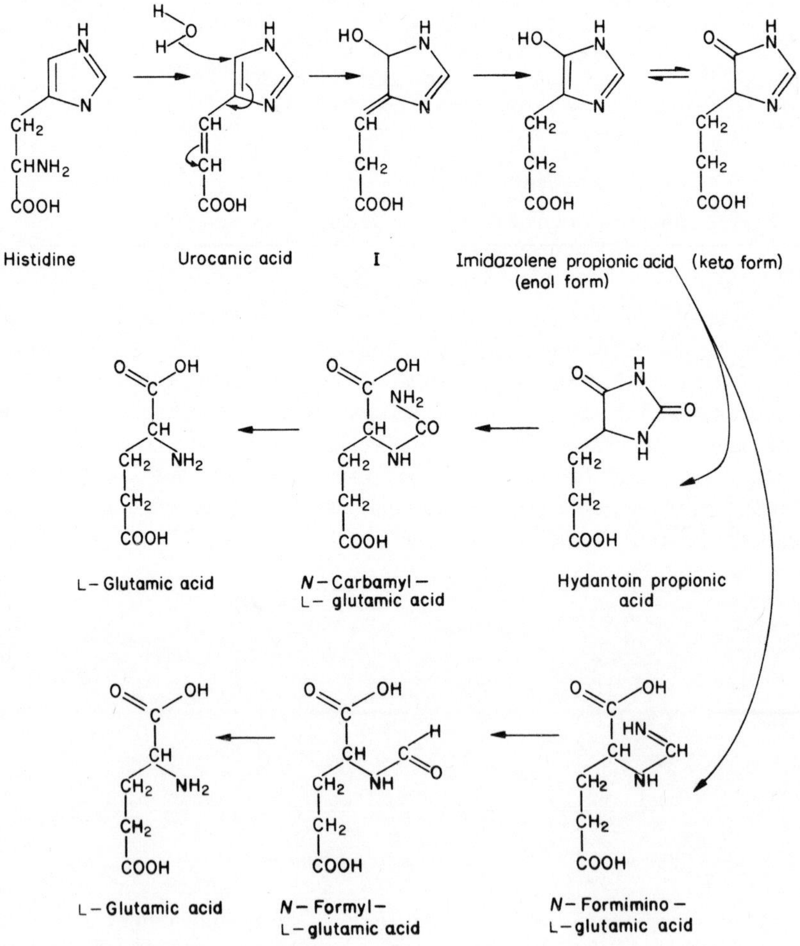

Fig. 237. Degradation of histidine.

liver and in the epidermal layer of the skin. The enzyme is absent in patients with histidinaemia.

A transaminase reaction (cf. C.4) catalysed by histidine transaminase, in which the α-amino group of histidine is transferred to α-keto acids and imidazole propionic acid is formed, competes in many tissues with the formation of urocanic acid from histidine by histidine deaminase. The activity of the transaminase is greater than that of the deaminase, for example, in the liver of rats.

Urocanic acid is converted by hydration (compound I) and rearrangement to imidazolone propionic acid which may be further degraded via either N-formimino-L-glutamic acid or hydantoin propionic acid.

N-Formimino-L-glutamic acid is degraded in bacteria via N-formyl-L-glutamic acid to formic acid, ammonia and glutamic acid. Mammals are capable of directly degrading N-formimino glutamic acid to glutamic acid in the presence of tetrahydrofolic acid, N^5-formiminotetrahydrofolic acid (cf. C.3.2) being formed in the process. Rats which suffer from folic acid deficiency excrete large amounts of N-formimino-L-glutamic acid in the urine. Degradation of N-formimino-L-glutamic acid to glutamic acid and formamide appears to occur in *Clostridium tetanomorphum* and *Aerobacter aerogenes*.

Certain bacteria are capable of converting imidazolone propionic acid to hydantoin propionic acid. The enzymes which catalyse this reaction appear to have the character of peroxidases (cf. C.2.4). Imidazolone propionic acid is also converted to hydantoin propionic acid by a xanthine oxidase obtained from

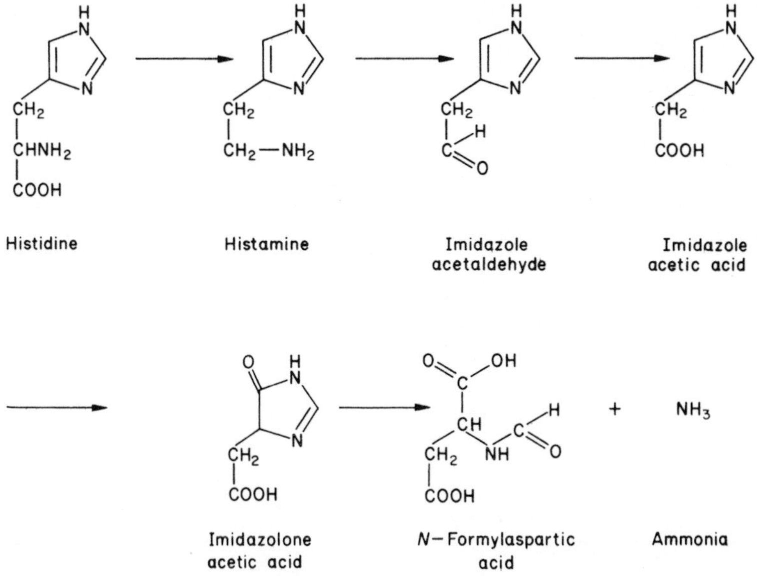

Fig. 238. Degradation of histamine.

milk (cf. C.2.2). Hydantoin propionic acid may be hydrolytically degraded to N-carbamyl-L-glutamic acid and via this compound to glutamic acid (fig. 237).

Histamine formed by the decarboxylation of histidine is converted to imidazole acetic acid via imidazole acetaldehyde (fig. 238). The amine is first converted to an aldehyde by an amine oxidase (cf. C.2.1.2). Its further oxidation to imidazole acetic acid is catalysed by xanthine oxidases (cf. C.2.2) or aldehyde dehydrogenases in the presence of NAD^+ (cf. C.2.1). Imidazole acetic acid is excreted in the urine of mammals, either in the free form or as the riboside.

Certain strains of *Pseudomonas* convert imidazole acetic acid to imidazolone acetic acid by mixed function oxidation (cf. C.2.6). They are also capable of degrading this compound hydrolytically to formyl aspartic acid and ammonia. The formyl aspartic acid is converted to aspartic acid.

References for further reading

Meister, A. *Biochemistry of the Amino Acids*, Vol. II (Academic Press, New York, 1965).

Luckner, M. Über neue Arbeiten zur Biosynthese N-heterocyclischer Verbindungen, 6. Teil: Bildung und Abbau von Verbindungen mit Imidazol- und Tetrahydroimidazolringsystem, *Pharmazie* 22 (1967), pp. 65–75.

Luckner, M. 'Imidazole', *Biosynthese der Alkaloide*, ed. K. Mothes and H. R. Schütte (VEB Deutscher Verlag der Wissenschaften, Berlin, 1969), pp. 593–6.

Seiler, N., Demisch, L. and Schneider, H. Biochemie und Funktion von biogenen Aminen in Zentralnervensystem, *Angew. Chem.* 83 (1971), pp. 53–69.

19. Formation of secondary natural products from tryptophan

The amino acid tryptophan is derived from anthranilic acid, 5-phosphoribosyl-1-pyrophosphate and serine as outlined in fig. 239. The formation of a riboside with elimination of pyrophosphate takes place first. N-(5'-phosphoribulosyl)-anthranilic acid is formed from this in an Amadori rearrangement.

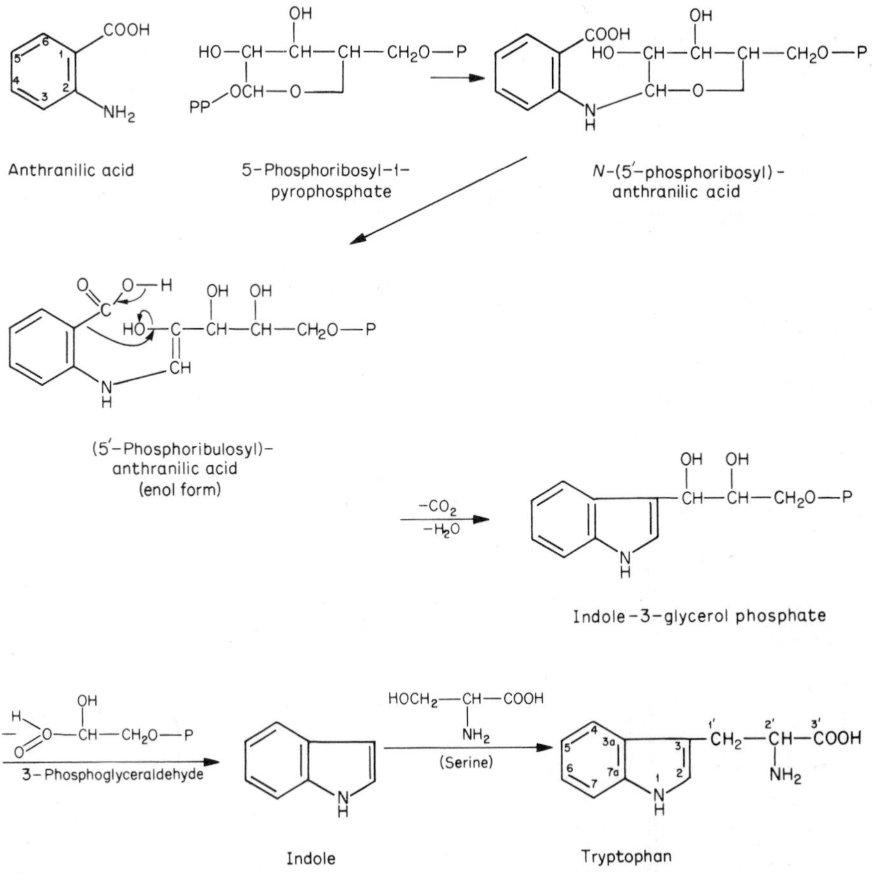

Fig. 239. Formation of tryptophan.

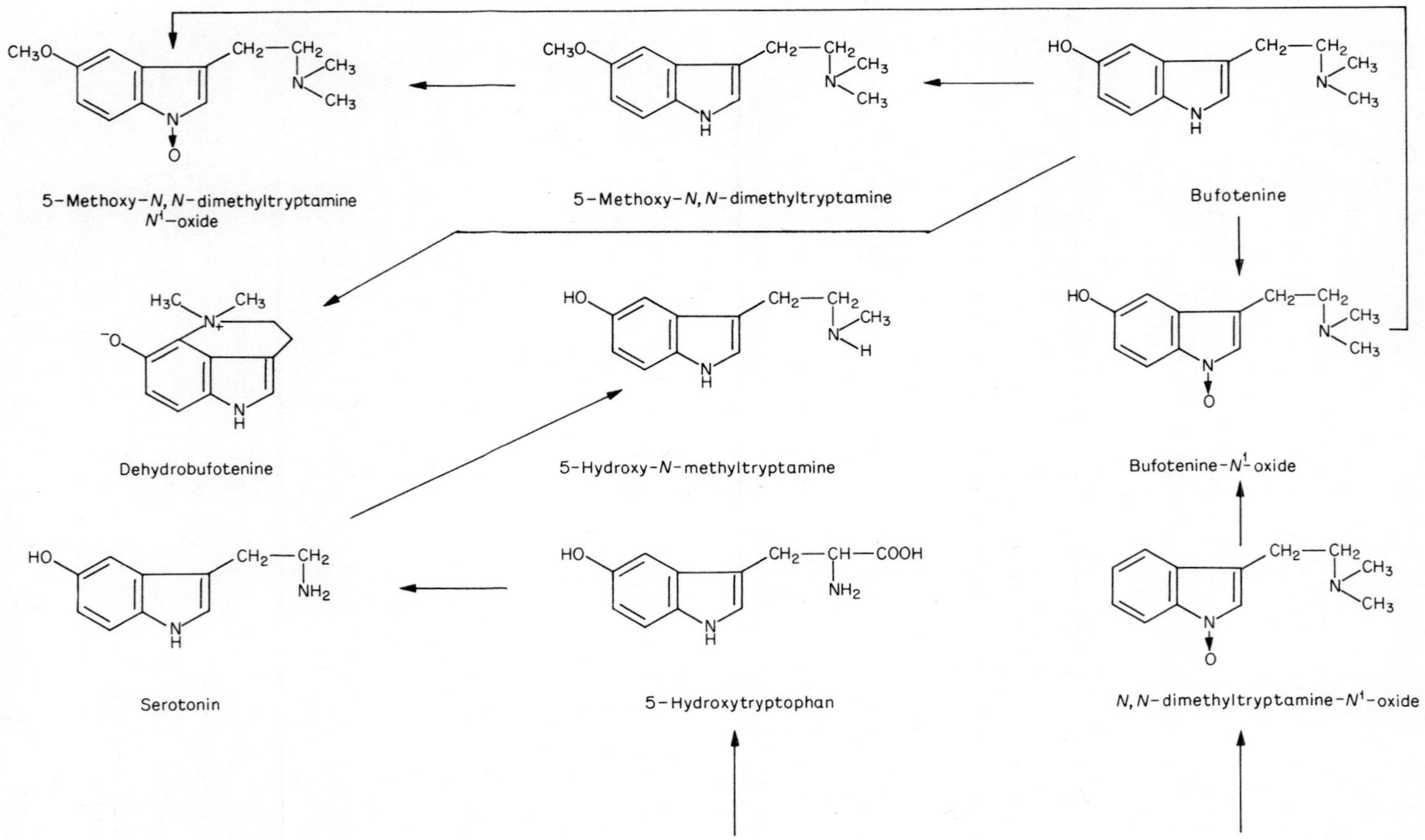

Bufotenine

Bufotenine-N^1-oxide

N,N-dimethyltryptamine-N^1-oxide

5-Methoxy-N,N-dimethyltryptamine

5-Hydroxy-N-methyltryptamine

5-Hydroxytryptophan

5-Methoxy-N,N-dimethyltryptamine N^1-oxide

Dehydrobufotenine

Serotonin

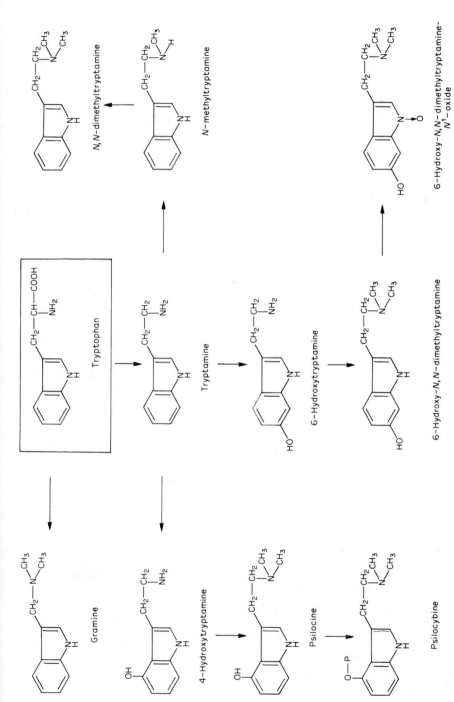

Fig. 240. Biosynthetic pathways of naturally occurring indole alkylamines.

By means of a coupled reaction the carboxyl group is then cleaved and the pyrrol ring is linked to the former carbon atom 1 of the anthranilic acid molecule. The enzyme catalysing this cyclization causes elimination of a hydroxyl anion from carbon atom 2 of the enol form of ribulose. Spontaneous ring closure then occurs with subsequent elimination of carbon dioxide and a proton, probably as described under D.6.4.2 for the synthesis of viridicatine.

Indole-3-glycerol phosphate formed in this manner is broken down in the next reaction step to indole and 3-phosphoglyceraldehyde. Indole then reacts with serine in the presence of the enzyme tryptophan synthetase, with the elimination of water and formation of tryptophan.

Both indole and tryptophan are precursors of a large number of secondary natural products.

References for further reading

See D.10.

19.1 Biosynthesis of indole alkylamines

Tryptophan is converted to a large number of indole alkylamines by animals, e.g. toads (*Bufo* sp.), as well as by plants and micro-organisms by decarboxylation (cf. C.4), hydroxylation (cf. C.2.6.5), methylation at the α-amino nitrogen and at oxygen atoms introduced into the molecule (cf. C.3.3) as well as by oxygenation of the indole nitrogen (cf. C.2.6.2). The most important compounds of this group are illustrated in fig. 240 and are arranged according to their possible or proved biosynthetic pathway. The formation of 5-hydroxytryptophan from tryptophan by a mixed function oxygenation which is accompanied by an NIH-shift has been particularly well investigated (cf. C.2.6.5).

It was found during studies on the biosynthesis of gramine, which occurs in large quantities in the seedlings of *Hordeum vulgare*, barley, that the indole ring and the methylene group of the side chain (other indole alkylamines have the original ethyl side chain instead of this grouping) are derived from tryptophan. On feeding tryptophan-(β-^{14}C, β-T) the same ^{14}C/T ratio was observed in gramine as in tryptophan. This means that the entire methylene grouping of the amino acid is incorporated into the amine. Since the amino group of tryptophan is also incorporated into gramine, a shift of this group from the α- to the β-carbon atom must take place. Pyridoxal phosphate is of importance in these reactions. The actual reaction mechanism is still unknown, however. 3-Aminomethylindole that is first formed is methylated to gramine in two steps by transmethylases (cf. C.3) (fig. 241).

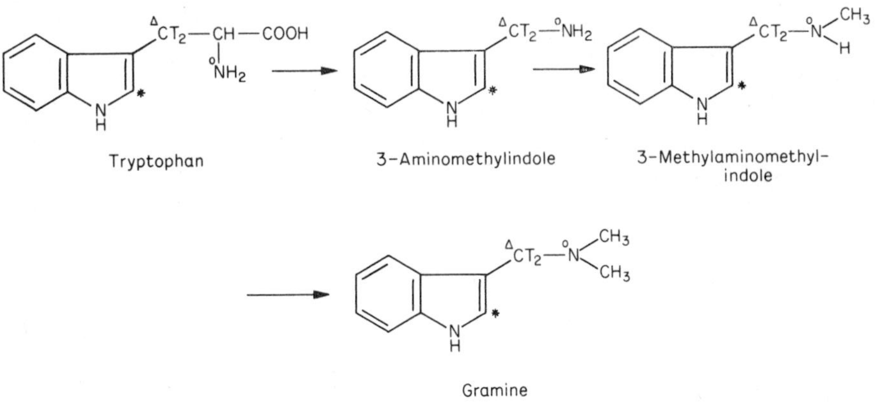

Fig. 241. Biosynthesis of gramine.

References for further reading

Gross, D. 'Indolalkylamine', *Biosynthese der Alkaloide*, ed. K. Mothes and H. R. Schütte (VEB Deutscher Verlag der Wissenschaften, Berlin, 1969), pp. 439–58.

Seiler, N., Demisch, L. and Schneider, H. Biochemie und Funktion von biogenen Aminen in Zentralnervensystem, *Angew. Chem.* **83** (1971), pp. 53–69.

19.2 Biosynthesis of ergoline alkaloids

The most important source of ergoline alkaloids are the sclerotia of various species of Ascomycetes of the genus *Claviceps*. These fungi are parasitic on rye and a number of other wild grasses. In recent years ergoline derivatives have also been found in *Aspergillus*, *Rhizopus* and species of *Penicillium* as well as in the Convolvulaceae.

The sclerotia of *Claviceps purpurea*, known as ergot, are the starting material for the industrial production of those therapeutically important ergoline alkaloids in which lysergic acid (8-carboxy-6-methyl-9,10-dehydroergoline) is joined in peptide linkage with certain amino acids (cf. the formula of ergotamine, fig. 242). Saprophytic cultures of strains of *Claviceps* are of importance in the production of simple lysergic acid derivatives such as lysergic acid ethanolamide (fig. 242). The so-called clavine alkaloids occur together with compounds derived from lysergic acid. The ergoline ring system in them is substituted not by a carboxyl group in position 8 but by a methyl, hydroxymethyl or an aldehyde group.

In fungi, and probably also in higher plants, the ergoline alkaloids originate from tryptophan and isopentenyl pyrophosphate. With the exception of the carboxyl group, L-tryptophan is incorporated *in toto*.

It is thought that an isopentenyl cation originating from isopentenyl pyrophosphate substitutes tryptophan in an electrophilic reaction at position 4, thus forming 4-dimethylallyltryptophan as the first intermediate of the biosynthetic

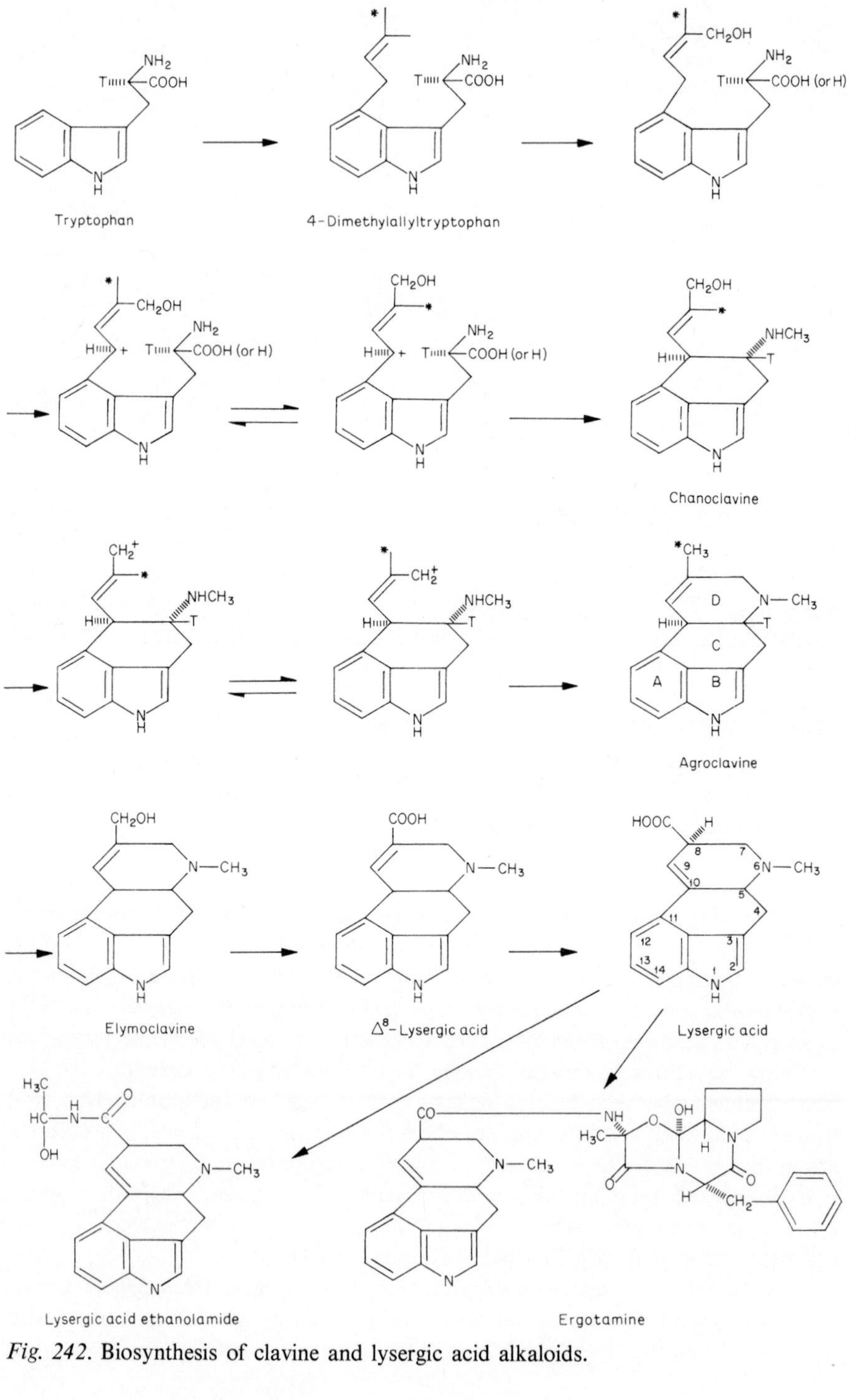

Tryptophan   4-Dimethylallyltryptophan

Chanoclavine

Agroclavine

Elymoclavine Δ^8-Lysergic acid   Lysergic acid

Lysergic acid ethanolamide   Ergotamine

Fig. 242. Biosynthesis of clavine and lysergic acid alkaloids.

chain. In further reactions the alanine residue is decarboxylated, the isopentenyl side chain hydroxylated, the amino group methylated and a new C–C bond is formed. Chanoclavine is formed through this series of reactions.

It is not clear when elimination of the carboxyl group of tryptophan takes place. 4-Dimethylallyl tryptamine is incorporated into clavines at a slower rate than the corresponding tryptophan derivative. By feeding $4R$- and $2S$-$2^{14}C$,4^3-H-mevalonic acid and determining the $^3H/^{14}C$ ratio in the alkaloids it was found that the S-hydrogen of mevalonic acid is lost during the biosynthesis while the R-hydrogen is largely retained. It may be concluded therefore that the conversion of mevalonic acid to dimethylallyl pyrophosphate proceeds by the pathway described in D.5.2. and that carbon atom 2 of mevalonic acid becomes the *trans* methyl group. A *cis-trans* isomerization takes place in the formation of chano-clavine since the group which corresponds to carbon atom 2 of mevalonic acid is no longer *trans*, but *cis*. *Cis-trans* isomerization is also involved in the conversion of chanoclavine to agroclavine, since the grouping referred to now becomes *trans* again (fig. 242).

The oxygen of the hydroxymethyl group of chanoclavine is introduced into the molecule by a mixed function oxygenase.

All other naturally occurring clavines and lysergic acid derivatives probably originate from chanoclavine. Nothing is, however, yet certain about the exact mode of synthesis of lysergic acid. It is likely that in certain *Claviceps paspali* strains the sequence of reaction agroclavine→elymoclavine→Δ^8-lysergic acid→lysergic acid exists (fig. 242). The conversion of agroclavine to elymoclavine is catalysed by a mixed function oxygenase (cf. C.2.6.4.).

The mode of synthesis of peptides of lysergic acid which occur in a number of alkaloids is still not clear. It has, however, been shown that the ethanolamine unit of lysergic acid ethanolamide originates from alanine. The nitrogen as well as carbon atoms 2 and 3 of this compound are incorporated into the amine portion of the alkaloid.

References for further reading

Gröger, D., Fortschritte der Chemie und Biochemie der Mutterkornalkaloide, *Fortschritte chem. Forsch.* **6** (1966), pp. 159–94.

Plieninger, H. Die Biosynthese der Mutterkornalkaloide, *Abh. dtsch. Akad. Wiss., Kl. Chem., Geol., Biol.* (1966), pp. 387–92.

Gröger, D. 'Ergolinalkaloide', *Biosynthese der Alkaloide*, ed. K. Mothes and H. R. Schütte (VEB Deutscher Verlag der Wissenschaften, Berlin, 1969), pp. 486–509.

Leete, E. 'Alkaloid Biogenesis', *Biogenesis of Natural Compounds*, ed. P. Bernfeld (Pergamon Press, Oxford, 1967), pp. 953–1023.

19.3 Formation of β-carboline alkaloids and related compounds

Carboline derivatives and compounds derived from them represent the biggest group of alkaloids known. Apocynaceae, Loganiaceae and Rubiaceae are rich in indole alkaloids of this group.

In all alkaloids of this type investigated up until the present, tryptophan with the exception of its carboxyl group is incorporated *in toto*. Two groups of alkaloids may be distinguished on the basis of the origin of the additional carbon atoms:

(a) Alkaloids of the harman group

It was supposed for a long time that harman alkaloids originate from tryptamine and acetaldehyde by a classical Mannich condensation (cf. D.7,7), but synthesis via *N*-acetyltryptamine, as outlined in fig. 243, was shown recently in *Passiflora edulis*. Tryptophan is first decarboxylated to tryptamine, the tryptamine acetylated and the *N*-acetyltryptamine thus formed is converted to harmalan with the elimination of water. Harmalan may either be oxidized to harman by dehydrogenases or reduced to tetrahydroharman. *N*-Acetyltryptophan is not incorporated into harman.

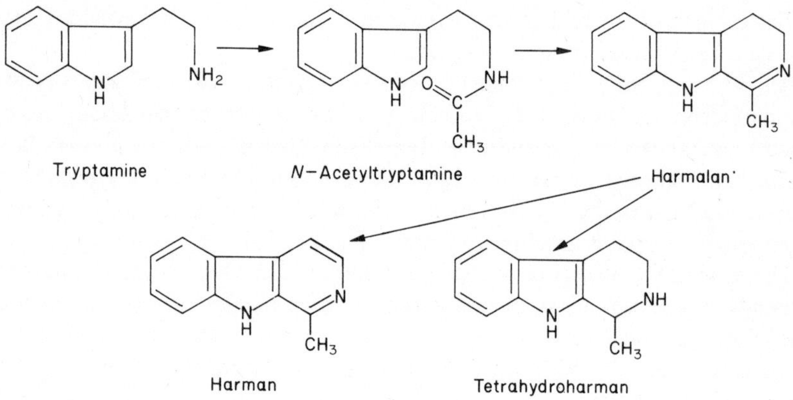

Tryptamine   *N*—Acetyltryptamine   Harmalan

Harman   Tetrahydroharman

Fig. 243. Biosynthesis of harman and tetrahydroharman.

(b) Iridoid indole alkaloids

A large number of indole alkaloids containing nine or ten additional carbon atoms in the molecule are derived from tryptamine and secologanin (cf. D.5.5). According to the structure of the iridoid portion, these alkaloids may be differentiated into those in which the carbon skeleton of secologanin is present in an unchanged form (secologanin skeleton and the Corynanthe-Strychnos skeleton derived from it by additional ring closure, cf. fig. 245), and those in which the rearrangements outlined in fig. 244 have occurred (Aspidosperma skeleton and Iboga skeleton, cf. fig. 246). The alkaloids whose iridoid moiety consists of only nine carbon atoms have lost the carbon atom corresponding to the carboxyl group of secologanin, which is separated by a dotted line in fig. 244.

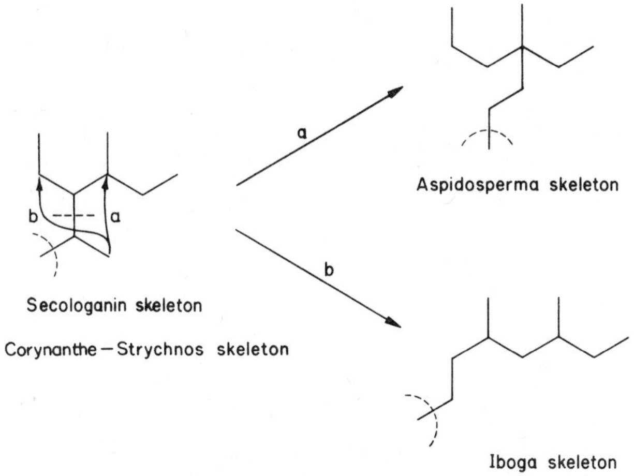

Fig. 244. Structure of the carbon skeleton of the iridoid alkaloids.

The formation of iridoid indole alkaloids (cf. figs. 245 and 246) has been especially intensively investigated in the case of *Vinca rosea*. The reaction which initiates alkaloid biosynthesis in this case appears to be similar to a Mannich condensation of tryptamine and secologanin (cf. D.7.7), resulting in the formation of vincoside. As established by feeding experiments, vincoside is the precursor for all other alkaloids and this pathway appears to proceed via ajmalicine, a compound possessing the Corynanthe-Strychnos skeleton. In the formation of ajmalicine, the second aldehyde group which is glucosylated in vincoside reacts with the nitrogen.atom of the amino grouping. The conversion (cf. fig. 245) must therefore be initiated by elimination of the glucose group (compound I) and the opening of the heterocyclic ring (compound II). Formation of a Schiff base might yield compound III, which by shift of a double bond and elimination of a proton might change to ajmalicine by reduction via compound IV. In agreement with this, only the tritium atom (H*) at position 7 of 2,7-^{3}H-secologanin is retained in ajmalicine, while the corresponding one at position 2 is replaced by a hydrogen atom.

Incorporation of radioactivity into compounds possessing the Aspidosperma skeleton, as well as into those possessing the Iboga skeleton has been shown after feeding ajmalicine to germinating seeds of *Vinca rosea*. In similar experiments the conversion of stemmadenine (Corynanthe-Strychnos skeleton) into tabersonine and vindoline (Aspidosperma skeleton) as well as into catharanthine (Iboga skeleton) were also observed.

These and similar findings can possibly be explained as follows (fig. 246): Opening of the ether ring in the molecule of ajmalicine gives compound V which is converted to compound VI by oxygenation. Elimination of water may lead to the 2-oxoindole derivative VII which by ring closure and renewed elimination of

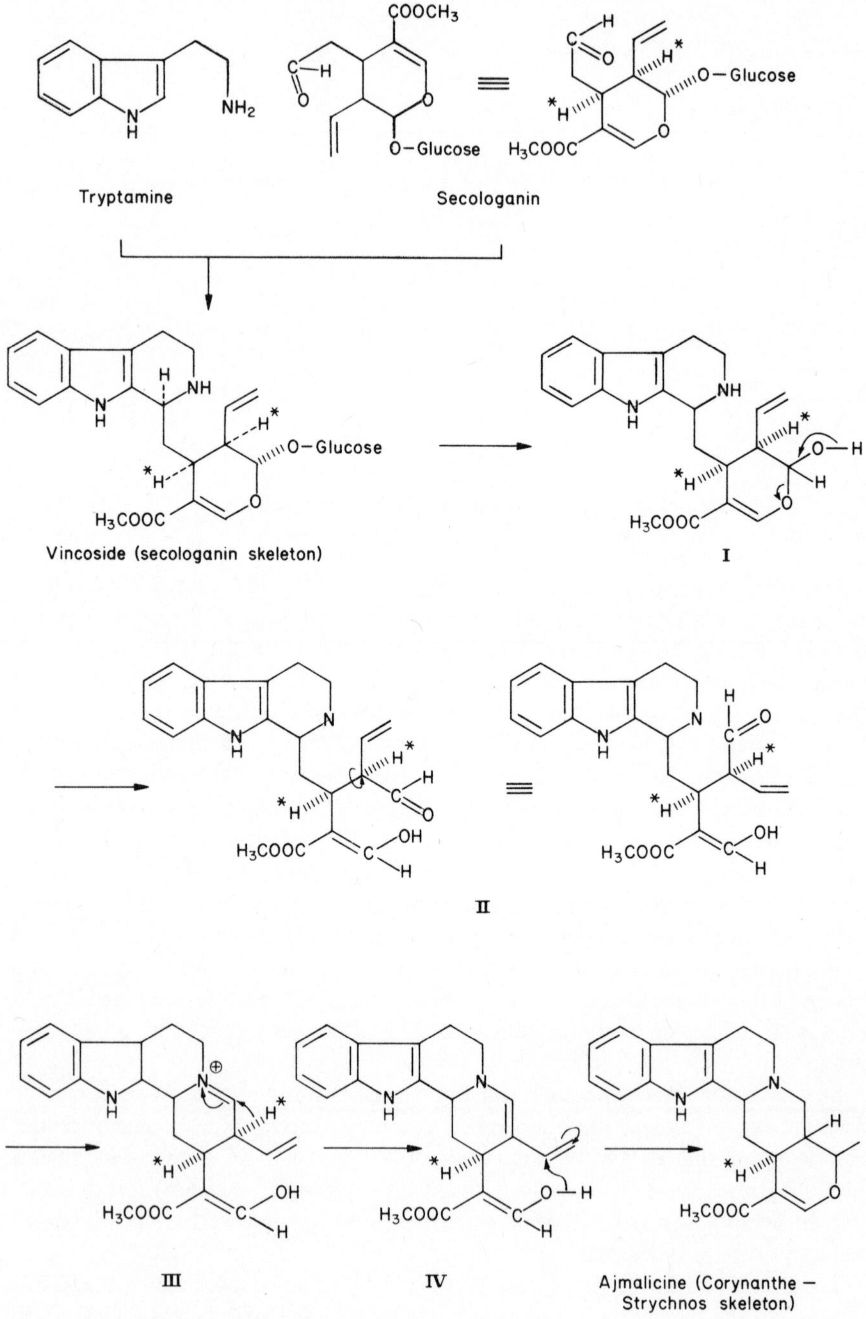

Fig. 245. Formation of iridoid carboline alkaloids possessing secologanin and Corynanthe-Strychnos skeletons.

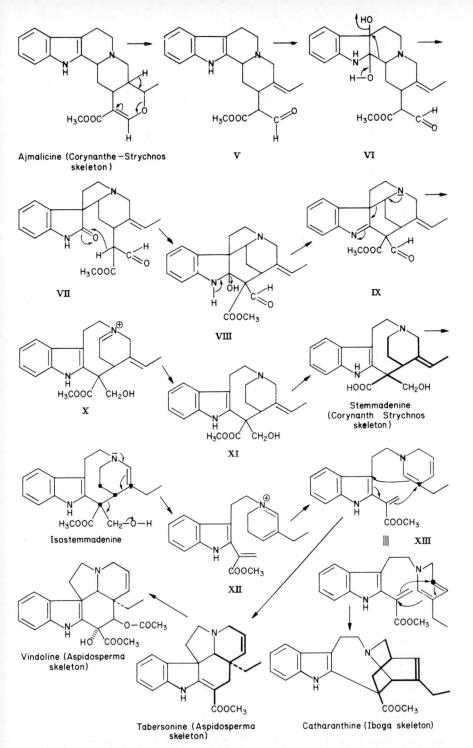

Fig. 246. Conversion of carboline alkaloids possessing a Corynanthe-Strychnos skeleton to those with an Aspidosperma and Iboga skeleton.

water may change to compound IX via compound VIII. Conversion of compound IX to X, followed by reduction, would lead to compound XI which might be a direct precursor of stemmadenine.

The conversion of the Corynanthe-Strychnos skeleton present in stemmadenine, to the Aspidosperma or the Iboga type skeleton, requires cleavage of the bond between the two carbon atoms marked '•' and the formation of a new bond at the o-position. In the first case the bond is formed with the carbon atom marked '▼' and in the second with the one marked '■'. The cleavage of the bond might occur at isostemmadenine, which originates from stemmadenine by the shift of a double bond. The alkaloid XIII formed from XII might be suitable as the direct precursor of tabersonine (Aspidosperma skeleton) as well as of catharanthine (Iboga skeleton).

References for further reading

Leete, E. 'Alkaloid Biogenesis', *Biogenesis of Natural Compounds*, ed. P. Bernfeld (Pergamon Press, Oxford, 1967), pp. 953–1023.

Gröger, D. 'Indolalkaloide', *Biosynthese der Alkaloide*, ed. K. Mothes and H. R. Schütte (VEB Deutscher Verlag der Wissenschaften, Berlin, 1969), pp. 486–509.

Leete, E. Alkaloid Biosynthesis, *Adv. in Enzymology* **32** (1969), pp. 373–422.

Gross, D. Die Biosynthese iridoider Naturstoffe, *Fortschritte der Botanik* **32** (1970), pp. 93–108.

Battersby, A. R. 'Biosynthesis of Terpenoid Alkaloids', *Natural Substances Derived Biologically from Mevalonic Acid*, ed. T. W. Goodwin, (Academic Press, London, 1970), pp. 157–68.

19.4 Secondary products originating via indolenine derivatives

19.4.1 Formation of Calycanthus alkaloids

Indole compounds such as chimonanthine and the tetrahydroquinoline, calycanthine, represent a further type of alkaloid derived from tryptophan. Calycanthine is the major alkaloid of the Calycanthaceae, a small family of higher plants which are widespread in North America and Asia.

Feeding of tryptophan-2'-^{14}C gave chimonanthine with the radioactivity in the carbon atoms marked '*'. Radioactivity was also incorporated into calycanthine. It is assumed that tryptophan is first converted to N'-methyltryptamine. This compound might then be changed to an indolenine radical which subsequently dimerizes to form compound I. The dimeric indole alkaloid of the chimonanthine type may be formed from compound I by simple ring closure whereas calycanthine may originate after hydrolytic opening of both indole rings and renewed ring closure. In the latter case the carbonyl groups formed by cleavage of the indole rings must react both with the aromatic amino groups as well as with the N-methyl groups of the other half of the molecule (fig. 247).

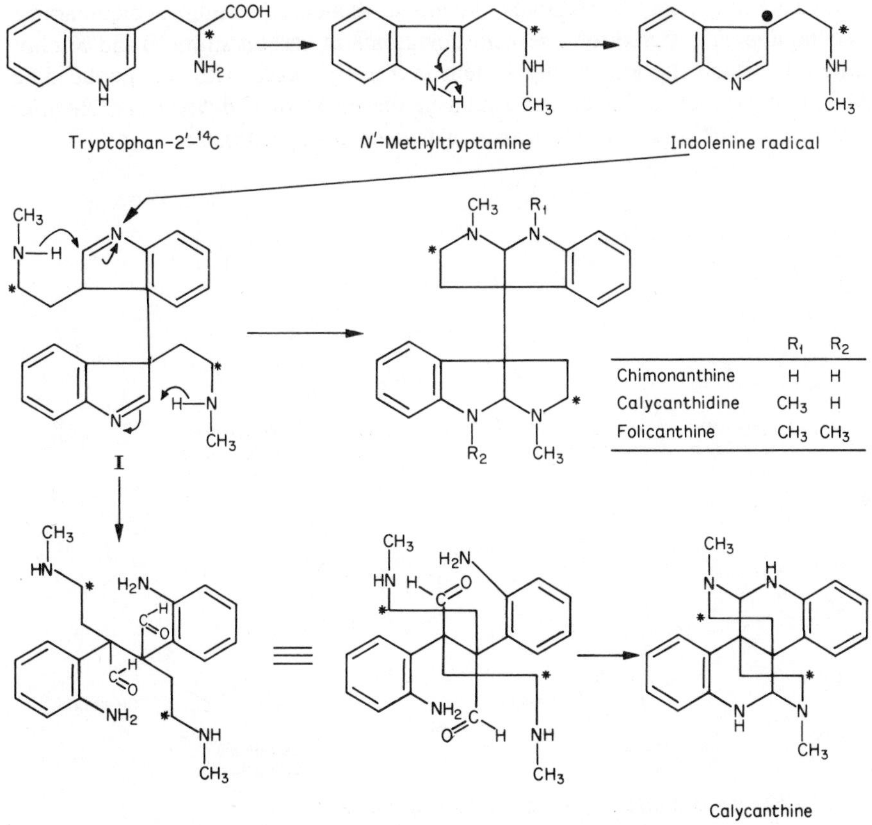

Tryptophan-2'-¹⁴C   N'-Methyltryptamine   Indolenine radical

	R_1	R_2
Chimonanthine	H	H
Calycanthidine	CH₃	H
Folicanthine	CH₃	CH₃

Calycanthine

Fig. 247. Possible pathway for the formation of Calycanthaceae alkaloids.

References for further reading

Leete, E. 'Alkaloid Biogenesis', *Biogenesis of Natural Compounds*, ed. P. Bernfeld (Pergamon Press, Oxford, 1967), pp. 953–1023.

Luckner, M. 'Chinoline', *Biosynthese der Alkaloide*, ed. K. Mothes and H. R. Schütte (VEB Deutscher Verlag der Wissenschaften, Berlin, 1969), pp. 510–50.

19.4.2 Biosynthesis of Cinchona alkaloids

Cinchona alkaloids, which occur particularly in the bark of the Rubiaceae (genus Cinchona), are one of the most chemically interesting, and economically important groups of quinoline compounds. In all these alkaloids the quinoline ring system is linked at position 4 via a secondary alcoholic group to the so-called quinuclidine nucleus, a system of two hydrogenated heterocyclic rings with a common nitrogen atom (fig. 248).

Besides these quinoline alkaloids, the indole alkaloids quinamine, epiquinamine (=conquinamine, stereoisomeric with quinamine at carbon atom 8′) and cinchonamine in which the indole ring system is directly linked with the quinuclidine nucleus at position 2, have been found in the barks of *Cinchona* and *Remijia* species, but particularly in the leaves of *Cinchona* (fig. 248).

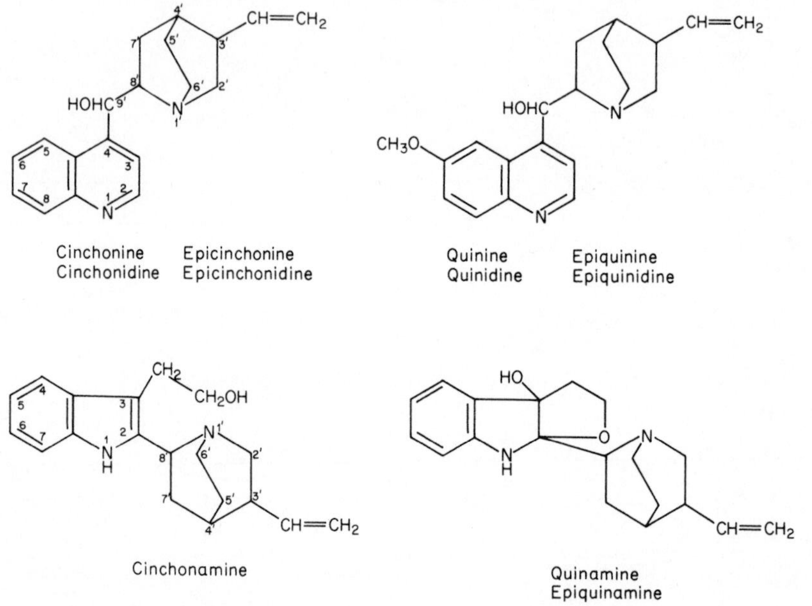

Cinchonine	Epicinchonine
Cinchonidine	Epicinchonidine

Quinine	Epiquinine
Quinidine	Epiquinidine

Cinchonamine

Quinamine
Epiquinamine

Fig. 248. Cinchona alkaloids with quinoline and indole nuclei.

A biogenetic relationship for both these groups of alkaloids has been assumed for a long time due to their simultaneous occurrence and their structural similarity. This is in agreement with the finding that in *Cinchona succirubra* the major portion of these alkaloids is synthesized in the leaves and transported from there to the bark where they are stored. While, however, indole alkaloids are found preponderantly in the leaves, quinoline derivatives are almost exclusively located in the bark, so that conversion of the former to the latter must be assumed.

Investigations with isotopically labelled compounds have shown that the quinoline ring of the alkaloids originates from tryptophan and that the quinuclidine nucleus is formed from an iridoid compound. The actual precursor appears to be compound V (fig. 249) which is derived from ajmalicine by opening of the O-heterocyclic ring (cf. D.19.3, fig. 246). By a shift of the double bond and reduction it may be changed to intermediate II which possibly changes to cinchonamine via the pyrophosphate III and the cation IV, which is the first compound of the reaction sequence to contain a quinuclidine nucleus.

In agreement with model reactions it is further supposed that aldehyde V is formed from chinchonamine after hydroxylation and oxidation. After opening of

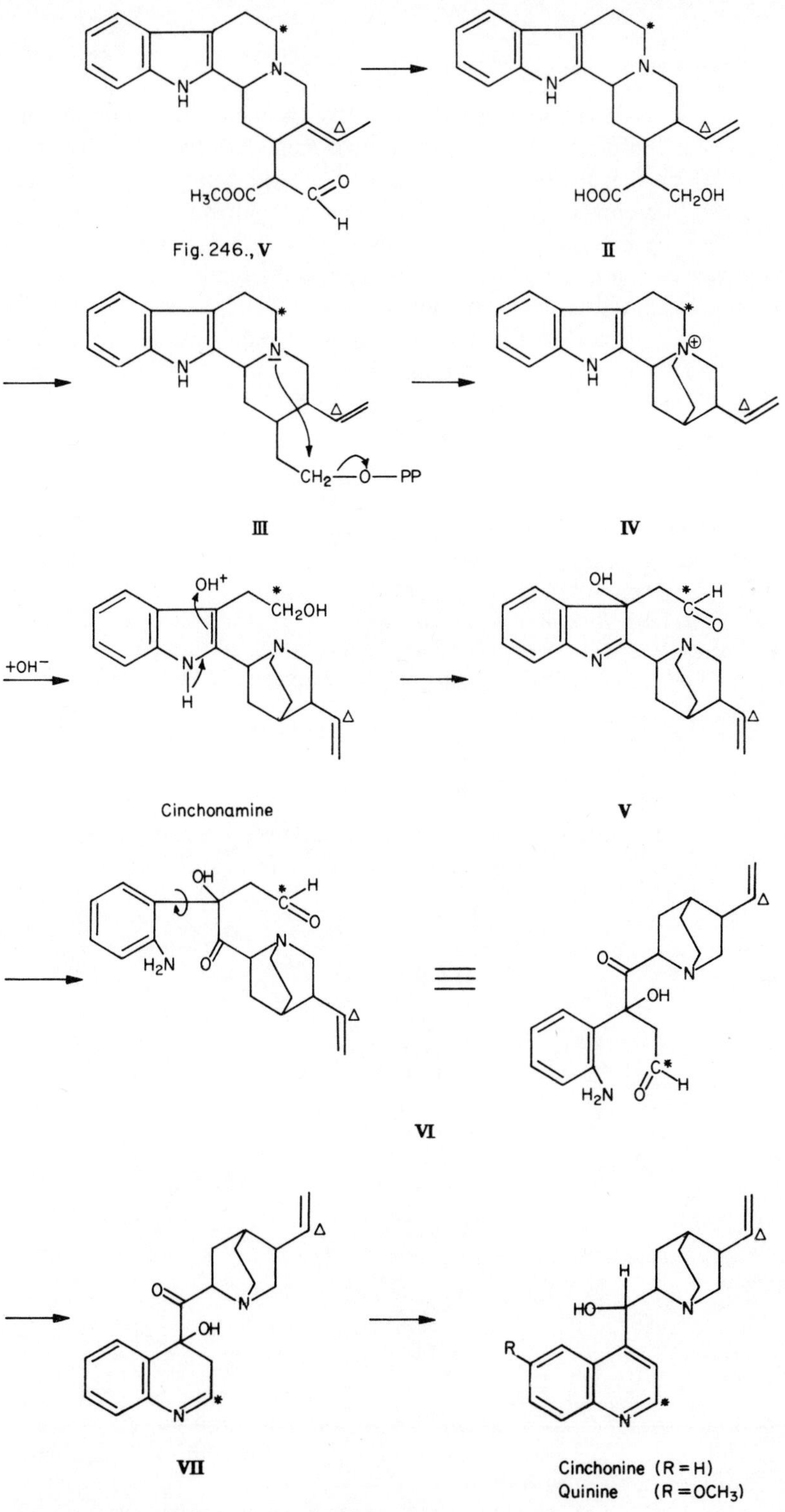

Fig. 246., V

II

III

IV

Cinchonamine

V

VI

VII

Cinchonine (R = H)
Quinine   (R = OCH₃)

Fig. 249. Biosynthesis of cinchona alkaloids.

the indole ring of this compound (compound VI) and reaction of the primary amino group with the aldehyde group, the dihydroquinoline derivative VII may be formed in which the quinuclidine nucleus is linked at position 4 to dihydroquinoline via a keto group. The alkaloids cinchonine and quinine are easily obtainable from VI by elimination of water and reduction of the carbonyl group. In the case of quinine it is not yet certain at what stage in the synthesis hydroxylation and methylation at position 6 of the quinoline ring take place.

This reaction sequence is in good agreement with feeding experiments which showed that the radioactive carbon atom of tryptophan-2'-^{14}C is located at position 2 of the quinoline ring, while that of geraniol-3-^{14}C is found in the vinyl side chain of the quinuclidine nucleus (fig. 249).

References for further reading

Leete, E. 'Alkaloid Biogenesis', *Biogenesis of Natural Compounds*, ed. P. Bernfeld (Pergamon Press, Oxford, 1967), pp. 953–1023.
Luckner, M. 'Chinoline', *Biosynthese der Alkaloide*, ed. K. Mothes and H. R. Schütte (VEB Deutscher Verlag der Wissenschaften, Berlin, 1969), pp. 510–50.
Gross, D. Die Biosynthese iridoider Naturstoffe, *Fortschritte der Botanik* **32** (1970), pp. 93–108.

19.4.3 Biosynthesis of pyrrolnitrins

Pyrrolnitrin is an antibiotic produced by some *Pseudomonas* strains. It is usually accompanied by small quantities of isopyrrolnitrin and oxypyrrolnitrin.

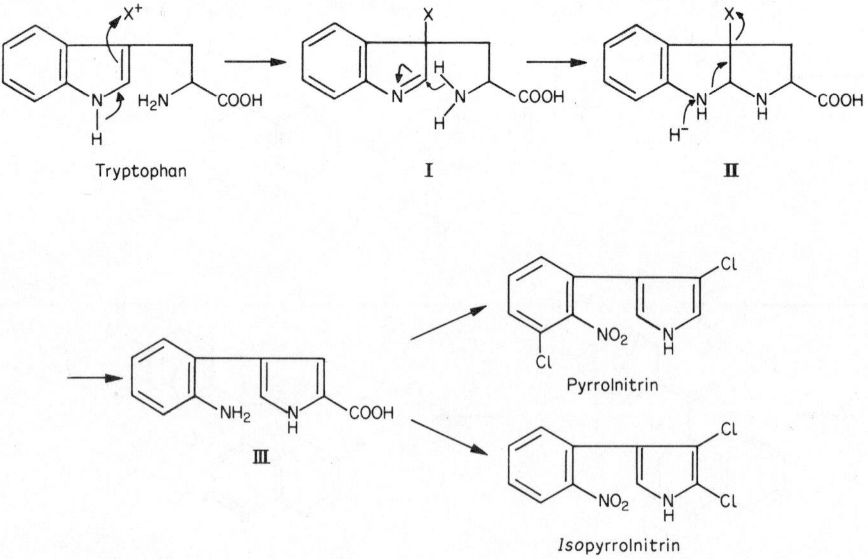

Fig. 250. Formation of pyrrolnitrins.

Compounds of the pyrrolnitrin group originate from tryptophan. It is supposed that by electrophilic attack at carbon atom 3 of the indole ring, shift of a double bond and elimination of a proton from the NH-group, the indolenine compound I is first formed, which then cyclizes to compound II (X can be a hydroxyl group, for example). By reductive decomposition of the original heterocyclic ring, compound III results, which on oxidation of the amino group and addition of chlorine (cf. C.2.4.2) is converted to compounds of the pyrrolnitrin type (fig. 250).

References for further reading

Gorman, M. and Lively, D. H. 'Pyrrolnitrin: A New Mode of Tryptophan Metabolism', *Antibiotics*, Vol. II, ed. D. Gottlieb and P. D. Shaw (Springer Verlag, Berlin, 1967), pp. 433–8.

19.5 Secondary products originating from the intermediates of tryptophan degradation by the pyrrolase pathway

Tryptophan pyrrolase is a dioxygenase (cf. C.2.5) which is capable of oxidatively cleaving the indole ring system of tryptophan between positions 2 and 3

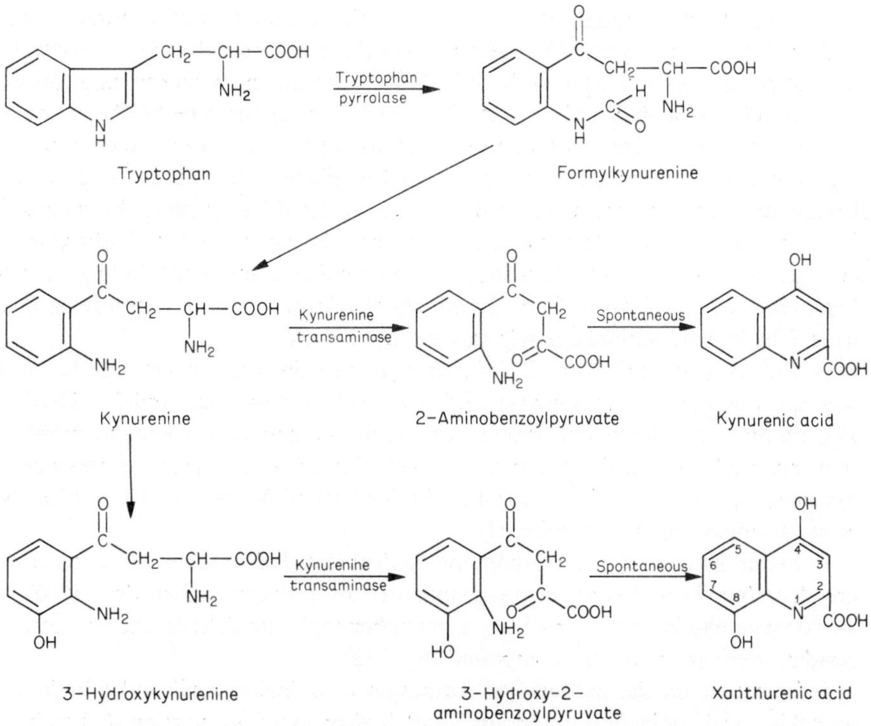

Fig. 251. Formation of quinoline carboxylic acids from tryptophan.

to form formylkynurenine (fig. 251). The degradation of tryptophan is initiated in the case of animals and micro-organisms by this reaction. Tryptophan pyrrolase has not yet been found in higher plants, but its occurrence is very likely.

Hydrolysis of formylkynurenine yields kynurenine, which in turn may be converted to 3-hydroxykynurenine by a mixed function oxygenase (cf. C.2.6). Metabolic pathways which lead to secondary natural products originating from kynurenine and 3-hydroxykynurenine are discussed in the following sections.

References for further reading

See D.10.

19.5.1 Formation and degradation of kynurenic acid and kynurenic acid derivatives

Kynurenic acid and compounds biogenetically related to it have been found in mammals, birds and insects as well as in various bacteria and higher plants.

By feeding isotopically labelled precursors and isolation of the corresponding enzymes it has been shown that animals as well as bacteria are capable of converting kynurenine into quinoline derivatives.

Two reactions compete with each other in the degradation of kynurenine and hydroxylated kynurenines. On the one hand these compounds may be degraded to alanine and derivatives of anthranilic acid by the enzyme kynureninase. On the other hand 2-aminobenzoylpyruvate is formed from kynurenine by the action of the enzyme kynurenine transaminase which as yet has only been found in micro-organisms and mammals, but not in higher plants. Derivatives of 2-aminobenzoylpyruvate are produced from the hydroxylated kynurenines, by means of the same or a similar enzyme. 2-Aminobenzoylpyruvate and its hydroxylated derivatives then cyclize spontaneously to the quinoline compounds. In this manner kynurenic acid is formed from 2-aminobenzoylpyruvate and xanthurenic acid from 3-hydroxy-2-aminobenzoylpyruvate (fig. 251).

5-Hydroxykynurenine and 3,4-dihydroxykynurenine are converted in the same way by kynurenine transaminase to 6-hydroxykynurenic acid and 7,8-dihydroxykynurenic acid. Transamination of kynurenine compounds by enzyme preparations proceeds only in the presence of α-ketoglutaric acid (also in the presence of pyruvate in the case of an enzyme obtained from *Neurospora*) to which the cleaved amino group is transferred.

If kynurenine or the corresponding hydroxylated derivatives of kynurenine are decarboxylated before being deaminated by transamination, derivatives of 4-hydroxyquinoline are formed via 2-aminobenzoylacetaldehyde and the corresponding derivatives of this compound (fig. 252).

Information on the pathway of formation of quinaldic acid and 8-hydroxyquinaldic acid, quinoline compounds not hydroxylated in position 4, has been provided in recent years. It has been shown in human beings and various

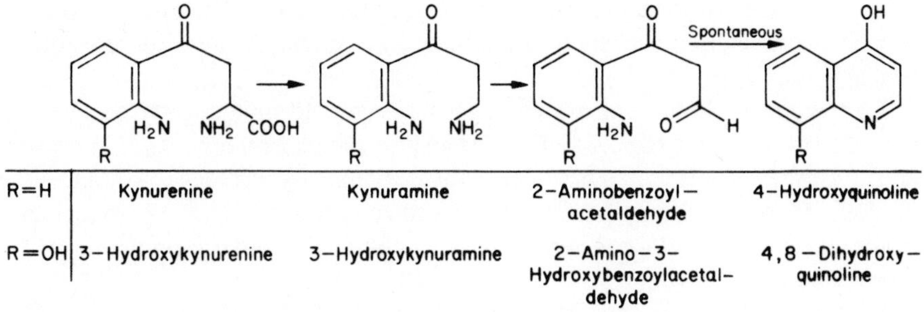

R=H	Kynurenine	Kynuramine	2-Aminobenzoyl-acetaldehyde	4-Hydroxyquinoline
R=OH	3-Hydroxykynurenine	3-Hydroxykynuramine	2-Amino-3-Hydroxybenzoylacetaldehyde	4,8-Dihydroxy-quinoline

Fig. 252. Formation of 4-hydroxyquinoline and 4,8-dihydroxyquinoline.

mammals that administered kynurenic acid and xanthurenic acid are converted to quinaldic acid and 8-hydroxyquinaldic acid respectively. Reduction proceeds mainly in the intestinal tract and is probably caused by the micro-organisms living there (fig. 253).

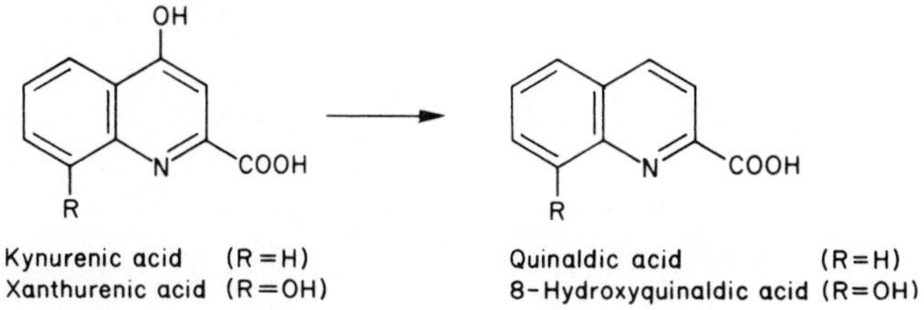

Kynurenic acid   (R=H)
Xanthurenic acid   (R=OH)

Quinaldic acid   (R=H)
8-Hydroxyquinaldic acid (R=OH)

Fig. 253. Formation of quinaldic acid and 8-hydroxyquinaldic acid.

In the case of mammals, birds and insects, compounds of the kynurenic acid group may be regarded as end products of metabolism, and are either excreted in the urine or stored in the organism. Micro-organisms (e.g. certain strains of *Pseudomonas fluorescens* as well as an unidentified *Micrococcus* species) on the contrary are capable of degrading kynurenic acid to aliphatic compounds.

This metabolic pathway has been very intensively investigated in *Pseudomonas fluorescens*. It was shown that the amino acid tryptophan can be metabolized in two different ways in the individual strains. In most of them degradation takes place by the so-called 'aromatic pathway' via formylkynurenine, kynurenine and anthranilic acid. In the case of others, however, it proceeds by the so-called 'quinolinic pathway' and tryptophan is first converted to kynurenic acid which is then further degraded to aliphatic compounds.

D- as well as L-tryptophan can be metabolized via the quinolinic pathway by these organisms. In some strains, D-tryptophan is degraded to 2-aminobenzoyl-pyruvate via D-formylkynurenine and D-kynurenine, while in others it is first converted to L-tryptophan by a tryptophan racemase (cf. D.7.1) and then changed to 2-aminobenzoylpyruvate via L-formylkynurenine in the usual way.

The possibilities for the complete degradation of the D-form of tryptophan present peculiarities which most micro-organisms and mammals are not capable of dealing with.

7,8-Dihydro-7,8-dihydroxykynurenic acid is first formed during the degradation of kynurenic acid (fig. 254). Cell-free extracts from *Pseudomonas fluorescens*, or from the *Micrococcus* mentioned above, require the presence of NADH (or in case of *Pseudomonas* also NADPH) and oxygen, so that the existence of a mixed function oxygenase is certain, and 7,8-epoxy-7,8-dihydrokynurenic acid is a probable intermediate.

7,8-Dihydrodihydroxykynurenic acid is changed in the next step to 7,8-dihydroxykynurenic acid by an NAD+-specific hydrogenase (cf. C.2.1). Xan-

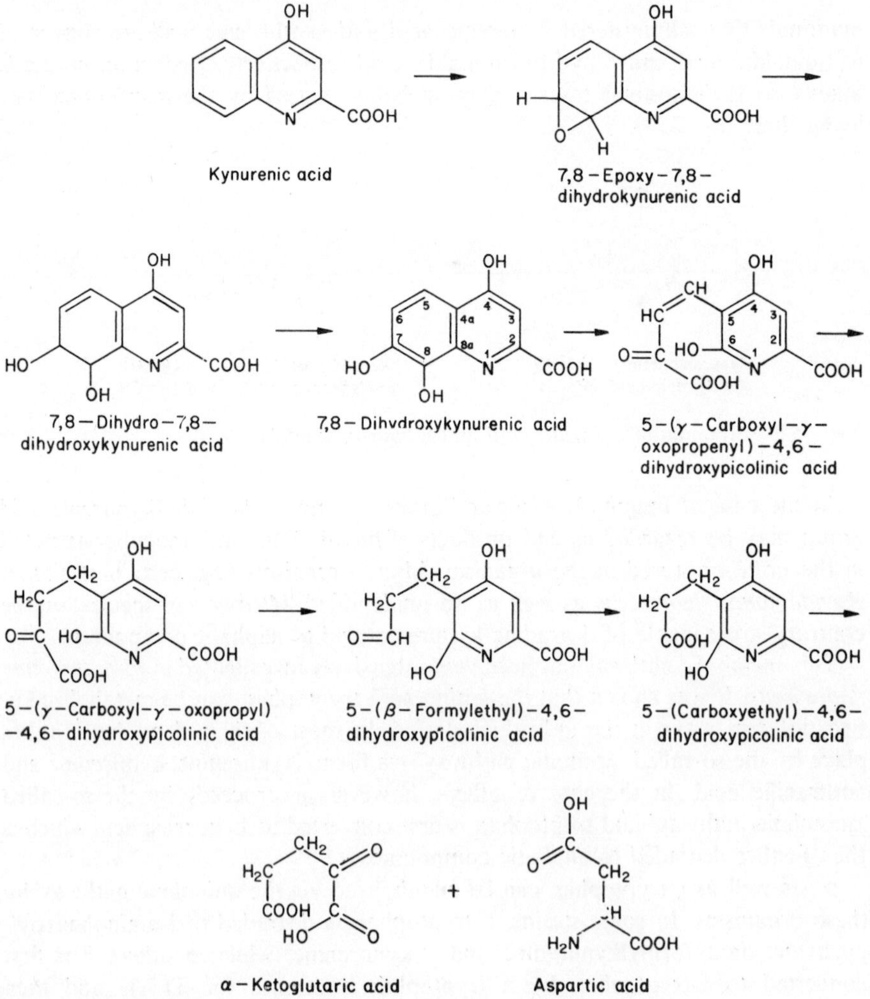

Fig. 254. Degradation of kynurenic acid by micro-organisms.

thurenic acid and 7-hydroxykynurenic acid are thus not intermediates of this reaction chain. Quinaldic acid is also not metabolized by extracts from *Pseudomonas fluorescens*.

In the further course of the degradation the quinoline ring system of 7,8-dihydroxykynurenic acid is broken between positions 8 and 8a by a dioxygenase (cf. C.2.5). The reaction is analogous to the opening of the aromatic ring of catechol, protocatechuic acid and 3,4-dihydroxyphenylacetic acid alongside both hydroxy groups.

5-(γ-Carboxyl-γ-oxopropenyl)-4,6-dihydroxypicolinic acid which in the presence of NADPH is reduced to 5-(γ-carboxyl-γ-oxopropyl)-4,6-dihydroxy-picolinic acid is formed as the initial pyridine derivative. The latter compound is then converted by decarboxylation to 5-(β-formylethyl)-4,6-dihydroxypicolinic acid which is further oxidized by 5-(carboxyethyl)-4,6-dihydroxypicolinic acid. This substance is decomposed to α-ketoglutaric acid and aspartic acid by crude extracts from the *Micrococcus* whose enzyme activity is partially inhibited by the addition of semi-carbazide, thus establishing a connection with primary metabolism. A large number of other products (e.g. glutamate, acetate, alanine, pyruvate, oxalacetate and ammonia) which have been described earlier as end products of degradation are formed from both the above compounds in crude extracts whose activity is not inhibited.

References for further reading

Behrmann, E. Tryptophan Metabolism in Pseudomonas, *Nature (London)* **196** (1962), pp. 150–2.

Luckner, M. Über neue Arbeiten zur Biosynthese der Alkaloide, 3. Teil: Die Bildung von Verbindungen mit Chinolinringsystem, *Pharmazie* **18** (1963), pp. 93–107.

Luckner, M. 'Chinoline', *Biosynthese der Alkaloide*, ed. K. Mothes and H. R. Schütte (VEB Deutscher Verlag der Wissenschaften, Berlin, 1969), pp. 510–50.

19.5.2 Biosynthesis of ommochromes

Ommochromes are a large group of pigments derived from 3-hydroxykynurenine. They possess a phenoxazine ring system and, as quinoid compounds, they are yellow brown in the oxidized form and red violet in the reduced form. Ommochromes occur most frequently in Arthropods, e.g. worms, crabs and insects. They are located in the eyes of the imagoes and as pigments on the wings.

Biosynthetic investigations with mutants of *Drosophila melanogaster* and *Ephestia kuehniella* showed that the formation of ommochromes is closely connected with tryptophan metabolism and that 3-hydroxykynurenine is a direct precursor of these compounds.

3-Hydroxykynurenine is converted to xanthommatin by tyrosinase (cf. C.2.3) if small amounts of DOPA are present. Dopaquinone formed as an intermediate probably dehydrogenates the 3-hydroxykynurenine. The DOPA thus resynthesized is in turn oxidized by the enzyme.

The quinonimine of 3-hydroxyanthranilic acid formed by dehydrogenation adds a second molecule of 3-hydroxyanthranilic acid as described in D.7.7 to form the phenoxazine ring system.

That portion of the molecule which is structurally similar to kynurenic acid is probably formed after transamination of one of the 3-hydroxykynurenine molecules via the corresponding substituted pyruvate derivative. The latter compound cyclizes spontaneously as described, with the formation of an N-hetero-cyclic ring (cf. D.19.5.1).

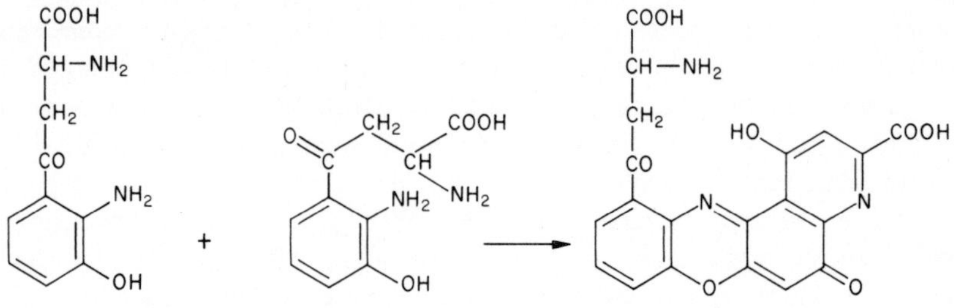

3-Hydroxykynurenine   Xanthommatin

Fig. 255. Formation of xanthommatin.

References for further reading

Luckner, M. Über neue Arbeiten zur Biosynthese der Alkaloide, 3. Teil: Die Bildung von Verbindungen mit Chinolinringsystem, *Pharmazie* **18** (1963), pp. 93–107.

Schütte, H. R. 'Verschiedenes', *Biosynthese der Alkaloide,* ed. K. Mothes and H. R. Schütte (VEB Deutscher Verlag der Wissenschaften, Berlin, 1969), pp. 645–77.

19.5.3 Biosynthesis of pyridine carboxylic acids

Kynurenine and 3-hydroxykynurenine are degraded to alanine and anthranilic acid and 3-hydroxyanthranilic acid by the enzyme, kynureninase, which is found in animals as well as in micro-organisms.

3-Hydroxyanthranilic acid, which possibly originates in a few organisms by direct hydroxylation of anthranilic acid, may then be cleaved between positions 3 and 4 by a dioxygenase (cf. C.2.5) to form α-amino-β-carboxymuconic acid-ϵ-semialdehyde. After a *cis-trans* rearrangement of the Δ^2-double bond of this aldehyde, spontaneous ring closure occurs and quinolinic acid is formed (fig. 256). Quinolinic acid is thus formed in a different way than in higher plants and in most bacteria (cf. D.14.2). It may be converted to nicotinic acid nucleotide as well as nicotinamide and nicotinic acid as shown in fig. 200.

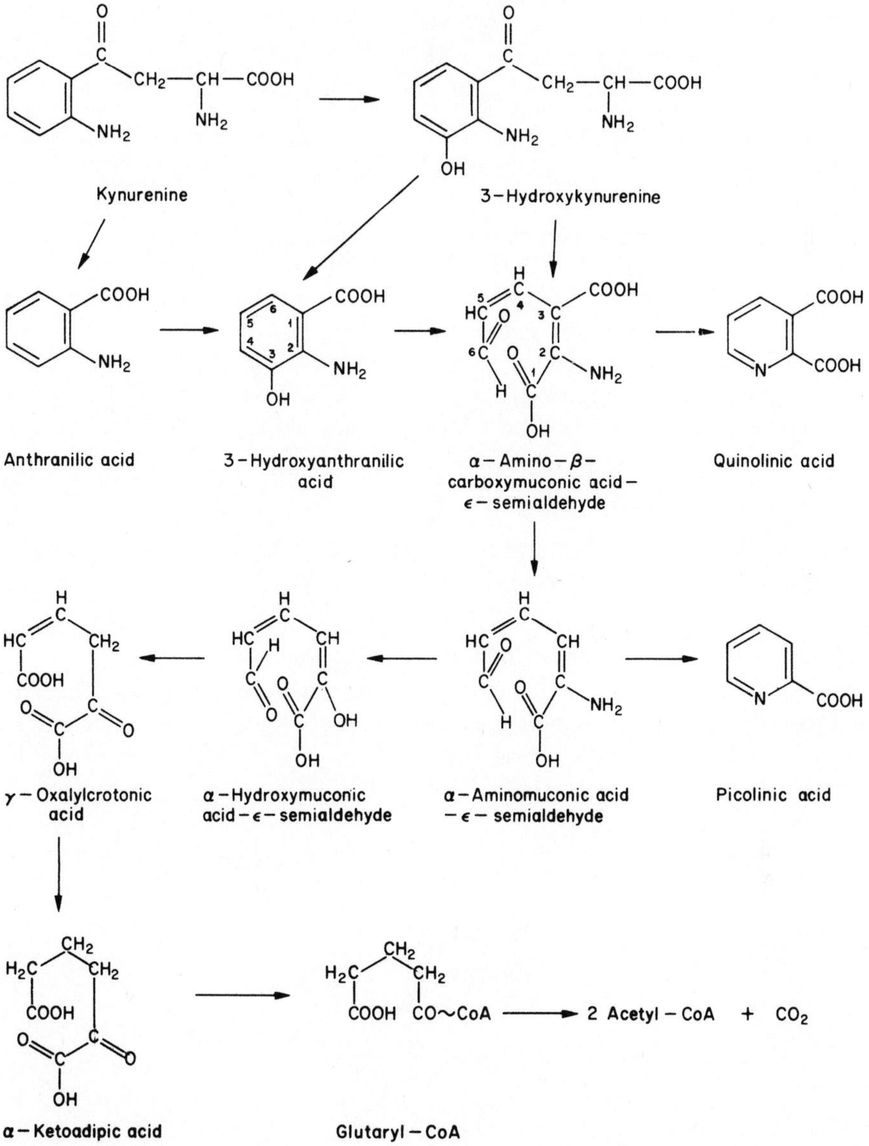

Fig. 256. Formation of pyridine carboxylic acids from tryptophan, and their degradation.

Nicotinamide, an end product of NAD^+- and $NADP^+$-degradation (cf. D.14.2), is converted in animals and humans to N^1-methylnicotinamide and N^1-methyl-5-carboxamido-2-pyridone (fig. 257). Both these compounds are execreted in the urine.

α-Amino-β-carboxymuconic acid-ϵ-semialdehyde can be decarboxylated to α-aminomuconic acid-ϵ-semialdehyde by a second metabolic pathway. Picolinic

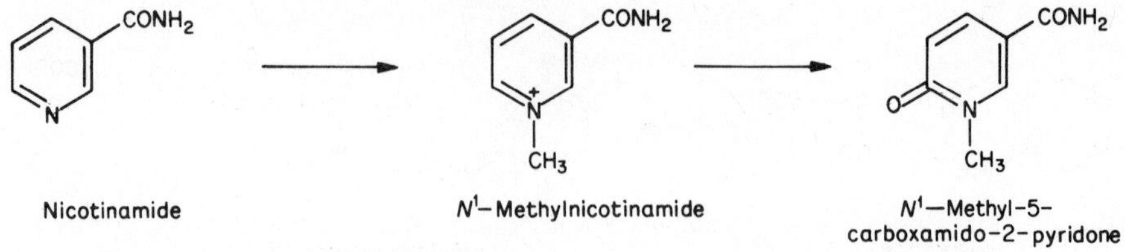

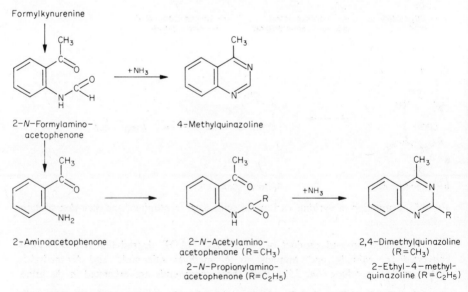

Nicotinamide — N^1–Methylnicotinamide — N^1–Methyl–5–carboxamido–2–pyridone

Fig. 257. Formation of excretory products from nicotinamide in animals.

acid is formed from this compound after a *cis-trans* rearrangement (fig. 256). Most of the α-aminomuconic acid semialdehyde is, however, decomposed hydrolytically with the formation of ammonia and α-hydroxymuconic acid semialdehyde, and then oxidized to γ-oxalylcrotonic acid. This compound is degraded to acetyl CoA and CO_2 via α-ketoadipic acid and glutaryl CoA (fig. 256).

References for further reading

Gross, D. Naturstoffe mit Pyridinstruktur und ihre Biosynthese, *Fortschritte Chem. org. Naturstoffe* **23** (1970), pp. 109–61.

19.5.4 Formation of quinazolines in *Pseudomonas aeruginosa*

Quinazoline compounds (e.g. 4-methylquinazoline and derivatives substituted in position 2) are formed in *Pseudomonas aeruginosa* during the degradation of

Formylkynurenine

2–*N*–Formylamino–acetophenone $+NH_3$ 4–Methylquinazoline

2–Aminoacetophenone

2–*N*–Acetylamino–acetophenone (R=CH₃)
2–*N*–Propionylamino–acetophenone (R=C₂H₅)

$+NH_3$

2,4–Dimethylquinazoline (R=CH₃)
2–Ethyl–4–methyl–quinazoline (R=C₂H₅)

Fig. 258. Possible pathway for the biosynthesis of quinazolines in *Pseudomonas aeruginosa.*

tryptophan. 2-*N*-Formylaminoacetophenone is probably formed first, via *N*-formylkynurenine, a compound which has been detected in cultures of *Pseudomonas*. In the presence of ammonia, 2-*N*-formylaminoacetophenone cyclizes to 4-methylquinazoline.

Quinazolines substituted in position 2 are derived from 2-aminoacetophenone, which is formed from the formyl compound by hydrolysis. The acyl amino group (cf. C.1.2) is probably formed by reaction of the aminoacetophenone with activated acids, such as acetic or propionic acid. The acylated amino-acetophenone then reacts with ammonia to form the quinazoline (fig. 258).

References for further reading

Gröger, D. 'Chinazolin-Alkaloide', *Biosynthese der Alkaloide*, ed. K. Mothes and H. R. Schütte (VEB Deutscher Verlag der Wissenschaften, Berlin, 1969), pp. 551–61.
Gröger, D. Anthranilic Acid as Precursor of Alkaloids, *Lloydia* **32** (1969), pp. 221–46.

19.6 Secondary products which originate from tryptophan by shortening or elimination of the side chain

The alanyl side chain of the amino acid tryptophan, and of tryptophan derivatives substituted in the indole ring, may be degraded to an acetyl group both in plants and animals. 3-Indolepyruvic acid is formed by transamination or deamination (cf. C.2.1.2) and is converted to 3-indoleacetaldehyde by decarboxylation (cf. C.3.4). Tryptamine (cf. D.7.3) is formed by decarboxylation and may be converted to 3-indoleacetaldehyde by oxidative deamination. 3-Indoleacetaldehyde is easily oxidized to 3-indoleacetic acid (auxin) (fig. 259), an important plant growth hormone.

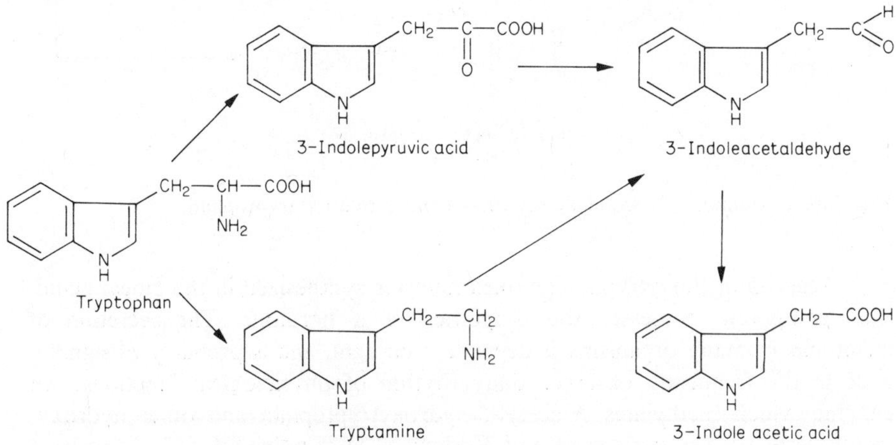

Fig. 259. Formation of indoleacetic acid.

5-Hydroxytryptophan (cf. D.19.1) is converted in a similar manner to 5-hydroxyindoleacetic acid (fig. 260). Other important substances originate from 5-hydroxytryptophan by methylation and acetylation (fig. 260).

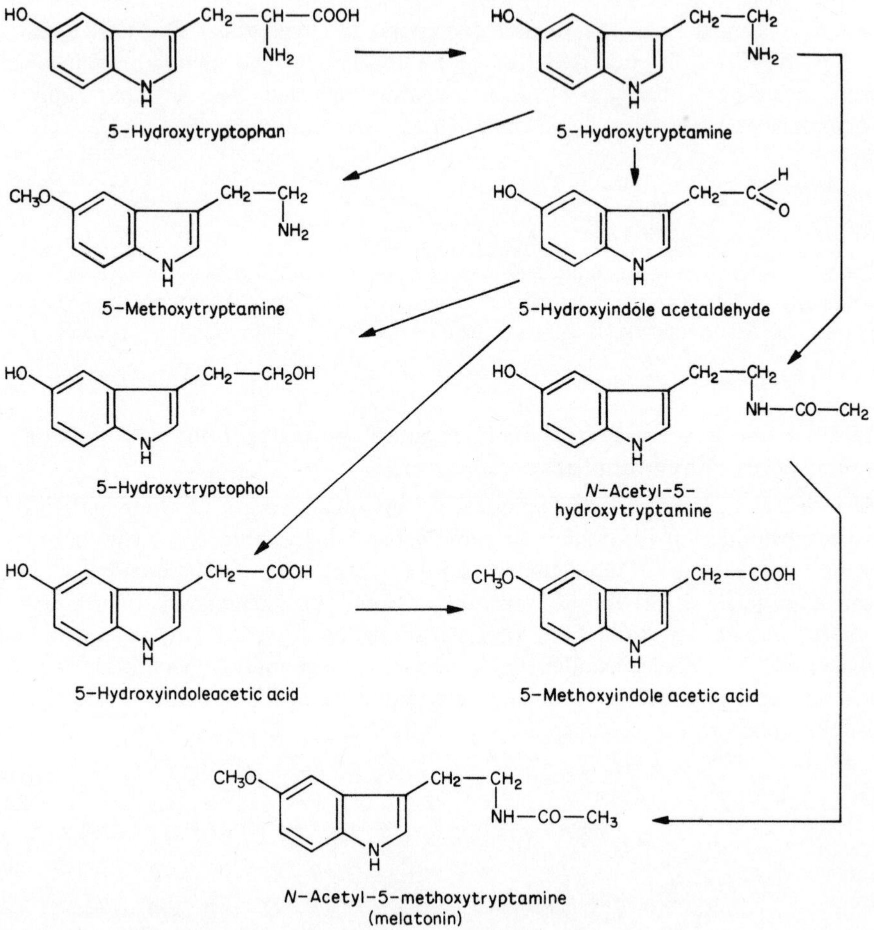

Fig. 260. Formation of secondary products from 5-hydroxytryptophan.

N-Acetyl-5-methoxytryptamine (melatonin) is synthesized in the pineal gland. This compound possesses the properties of a hormone. The secretion of melatonin in many organisms is dependent on light, and is probably of significance in the frequently observed daily rhythm of physiological functions. An enzyme which methylates *N*-acetyl-5-hydroxytryptophan and other hydroxyindole derivatives in the presence of *S*-adenosyl methionine (cf. C.3.3) has been isolated from the pineal gland.

A number of bacteria are capable of catalysing the degradation of tryptophan to indole, pyruvate and ammonia by the enzyme tryptophanase (fig. 261). The enzyme contains pyridoxal phosphate as coenzyme.

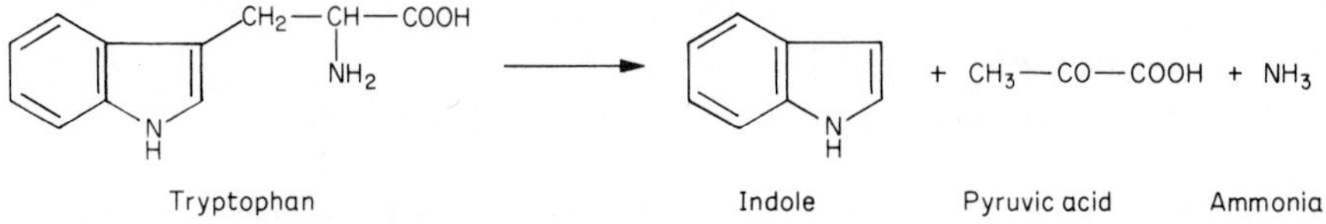

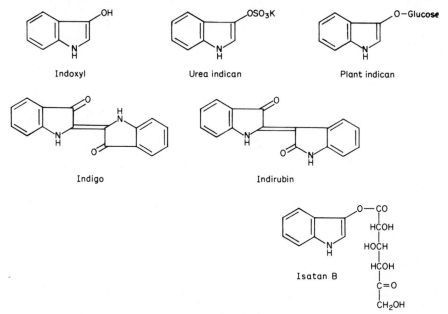

Fig. 261. Tryptophanase reaction

Though the degradative reaction catalysed by tryptophanase is irreversible, the enzyme isolated from *Escherichia coli* can form tryptophan from indole and serine. Thus tryptophanase, which is different from tryptophan synthetase (cf. D.19), can replace this enzyme in certain organisms.

The indole derivatives found in human urine, such as indoxyl, urinary indican, indigo and indirubin (fig. 262), originate from indole formed from tryptophan by intestinal bacteria through the reaction discussed above. Indole is converted to indoxyl by a mixed function oxygenase from liver mitochondria. The formation of urinary indican from indoxyl and 3'-phosphoadenosine-5'-phosphosulphate (cf. D.9) also takes place in the liver.

The indican (fig. 262) found in plants (e.g. in *Indigofera* species) is a glucoside, unlike the animal indican. Isatan B from *Isatis tinctoria* has been identified as

Fig. 262. The structural formulae of some indole derivatives.

indoxyl-5-ketogluconate. Indoxyl liberated during the degradation of both these compounds is oxidized to indigo by atmospheric oxygen. Natural indigo was of great importance historically as a blue dye.

References for further reading

Meister, A. *Biochemistry of the Amino Acids* (Academic Press, New York, 1965), Vol. II.

Schraudolf, H. Untersuchungen zur Biosynthese von Isatan B, der Indigovorstufe aus der Färberwaid (*Isatis tinctoria* L.), *Z. Naturforsch.* **23 b** (1968), pp. 572–3.

Libbert, E., Erdmann, N. and Schiewer, U. Auxinbiosynthese, *Biol. Rundschau* **8** (1970), pp. 369–90.

20. Formation of secondary natural products derived from phenylalanine and tyrosine

A large number of natural products possess a carbon skeleton in which a phenyl group is linked to an n-propyl side chain. These substances termed 'phenylpropane compounds' are derived from phenylalanine and tyrosine. Both these amino acids are synthesized by micro-organisms and higher plants. They are essential for animals and must be ingested in their food. Certain insects, however, have bacterial colonies living in their fatty body (an organ comparable to the liver of higher animals) which synthesize phenylpropane bodies.

The biosynthesis of the $C_6–C_3$-unit proceeds in the same way in all organisms investigated up until now. By means of the enzyme chorismic acid mutase, chorismic acid (cf. D.6) is first converted to prephenic acid. This may then be converted to phenylpyruvic acid by elimination of carbon dioxide and dehydration, or to p-hydroxyphenylpyruvic acid by the loss of carbon dioxide and dehydrogenation. The phenylpropane unit consisting of an aromatic ring and a

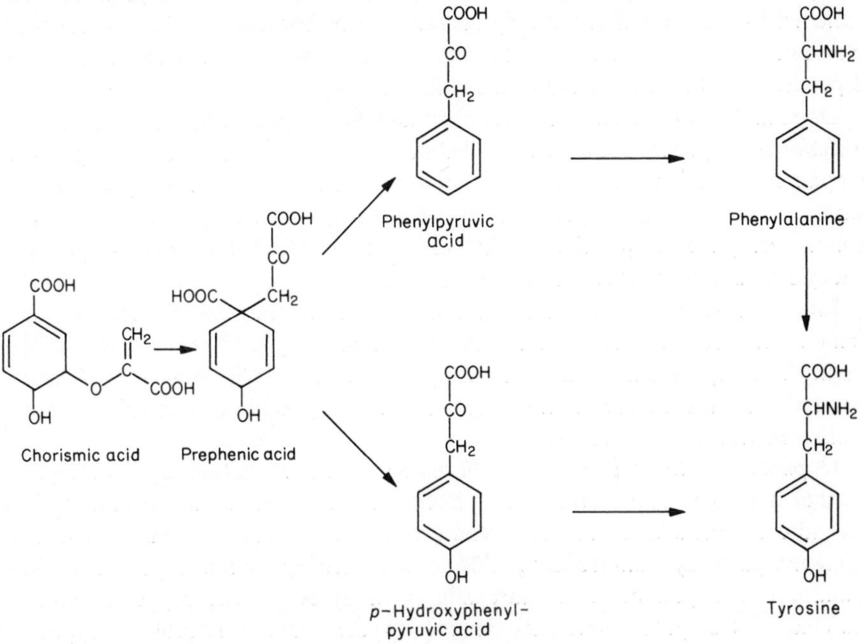

Fig. 263. Biosynthesis of phenylalanine and tyrosine.

side chain possessing three carbon atoms are recognizable for the first time in these compounds. Phenylalanine and tyrosine are formed by transamination from phenylpyruvic acid and *p*-hydroxyphenylpyruvic acid respectively (fig. 263). Tyrosine may also be formed from phenylalanine by the enzyme phenylalanine hydroxylase (cf. C.2.6.5).

References for further reading

See D.10.

20.1 Formation of secondary products with the retention of amino nitrogen

20.1.1 Biosynthesis of phenylalkylamines

The phenylethylamines occurring in nature originate by decarboxylation of the amino acids phenylalanine, tyrosine and DOPA (cf. D.7.3). They have an ethyl side chain and are frequently methylated. The methyl groups originate from methionine (cf. C.3.3). In the roots of barley the pathway tyrosine → tyramine → *N*-methyltyramine → hordenine (fig. 264) has been detected. Methylation is catalysed by the enzyme tyramine-methyl-transferase. The methyl groups are removed by transamination during degradation of hordenine, which proceeds at the same time as synthesis. The amount of amine present at any one time is thus dependent on the rates of both metabolic pathways.

Dopamine is the actual precursor for mescaline, a protoalkaloid isolated from *Anhalonium* species which causes hallucinations in human beings (fig. 264). Dopamine is probably first methylated to 3-methoxy-4-hydroxyphenylethylamine, which is then hydroxylated to 3-methoxy-4,5-dihydroxyphenylethylamine, methylated at position 5 (with the formation of 3,5-dimethoxy-4-hydroxyphenylethylamine) and converted to mescaline.

Investigations on *Penicillium notatum* have shown that the antibiotic xanthocillin is derived from two molecules of tyramine (fig. 264). The origin of the isonitrile group is unknown. The nitrogen of tyramine is more poorly incorporated than the carbon atoms. Acetate and compounds of one-carbon metabolism do not act as precursors.

Dopamine is the precursor for the tissue hormones noradrenalin (norepinephrine) and adrenalin (epinephrine). Both these compounds are synthesized in the adrenal medulla in animals. Noradrenalin also occurs in plants. The biosynthetic pathway outlined in fig. 264 is supported by feeding experiments. The oxidation of dopamine to noradrenalin is catalysed by the copper-containing enzyme, 3,4-dihydroxyphenylethylamine-β-hydroxylase. Molecular oxygen is necessary for this reaction.

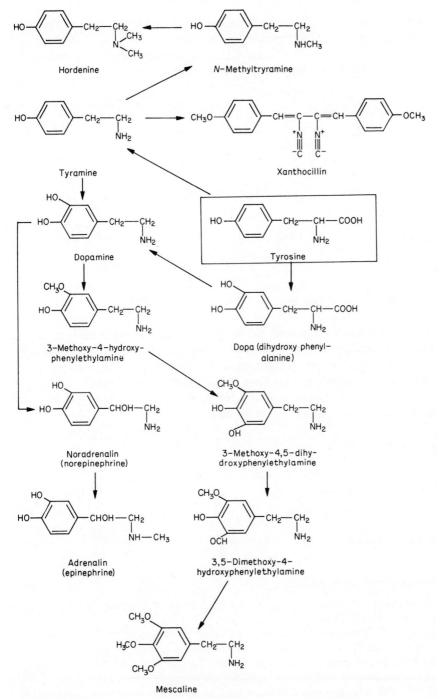

Hordenine

N–Methyltryramine

Tyramine

Xanthocillin

Dopamine

Tyrosine

3–Methoxy–4–hydroxy-
phenylethylamine

Dopa (dihydroxy phenyl-
alanine)

Noradrenalin
(norepinephrine)

3–Methoxy–4,5–dihy-
droxyphenylethylamine

Adrenalin
(epinephrine)

3,5–Dimethoxy–4–
hydroxyphenylethylamine

Mescaline

Fig. 264. Formation of hordenine, xanthocillin, mescaline and adrenalin.

Oxygen may be introduced into ephedrine by a similar mechanism. The pathway outlined in fig. 265 is suggested for this alkaloid. Since the *C*-methyl group does not originate from the carboxyl group of phenylalanine, a one-carbon fragment (cf. C.3.2) is probably involved in its formation.

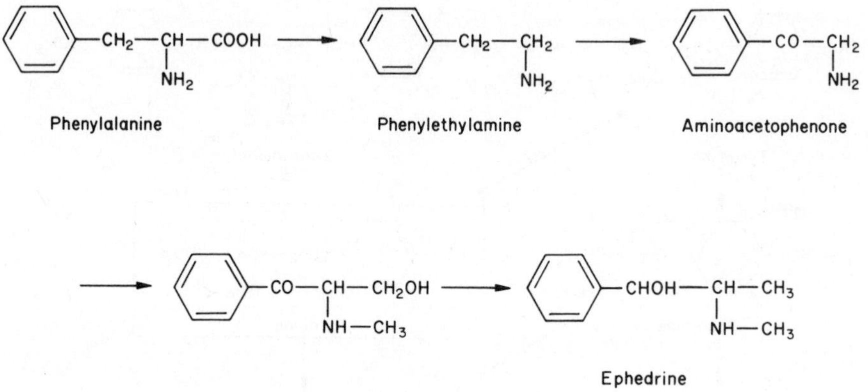

Fig. 265. Formation of ephedrine.

The biosynthesis of capsaicin, the pungent-tasting principle of fruits of various *Capsicum* species, presents a special case. The benzylamine component of this compound originates from phenylalanine in such a way that the nitrogen of the amino acid is retained. It must have shifted from the α- to the β-carbon atom similar to the shift involved in the formation of gramine (cf. D.19.1). The mechanism of these shifts is not understood.

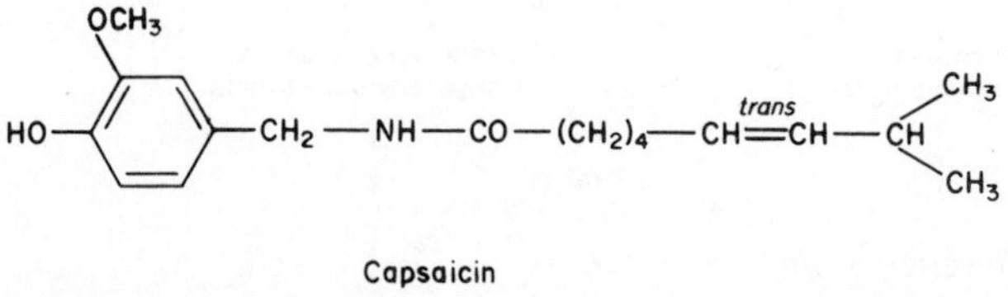

Capsaicin

Fig. 266. Capsaicin.

References for further reading

Achenbach, H. 'Xanthocillin', *Antibiotics*, Vol. II, ed. D. Gottlieb and P. D. Shaw (Springer Verlag, Berlin, 1967), pp. 26–8.

Schütte, H. R. 'Phenylalkylamine', *Biosynthese der Alkaloide*, ed. K. Mothes and H. R. Schütte (VEB Deutscher Verlag der Wissenschaften, Berlin, 1969), pp. 344–58.

Rosenberg, H., Khanna, K. L., Takido, M. and Paul, A. G. The Biosynthesis of Mescaline in *Lophophora williamsii*, *Lloydia* **32** (1969), pp. 334–8.

Molinoff, P.B. and Axelrod, J. Biochemistry of Catecholamines, *Ann. Rev. Biochem.* **40** (1971), pp. 465–500.

20.1.2 Formation of isoquinoline alkaloids by Mannich condensation

Isoquinoline alkaloids are found with varying frequency, in the orders Magnoliales, Ranales, Aristolochiales, Myrtales, Geraniales, and in the families Leguminosae and Tubiflorae as well as in Papaverceae, Rutaceae, Cactaceae and Chenopodiaceae.

The isoquinoline ring system probably originates in almost all cases by a Mannich condensation, involving a more or less strongly substituted phenyl-ethylamine derivative and an aldehyde (cf. D.7.7). On the basis of the structure of the aldehyde, the following alkaloid types may be distinguished.

(a) Anhalonium bases

In the formation of the anhalonium base anhalonidine, in the Cactaceae, acetalde-hyde reacts with 3,4-dimethoxy-5-hydroxyphenylethylamine which is formed from tyramine via 3-methoxy-4,5-dihydroxyphenylethylamine (cf. D.20.1.1) (fig. 267).

The fact that an acid and not an aldehyde is the actual precursor, and that an N-acyl derivative is formed first, cannot be ruled out, however (cf. D.19.3).

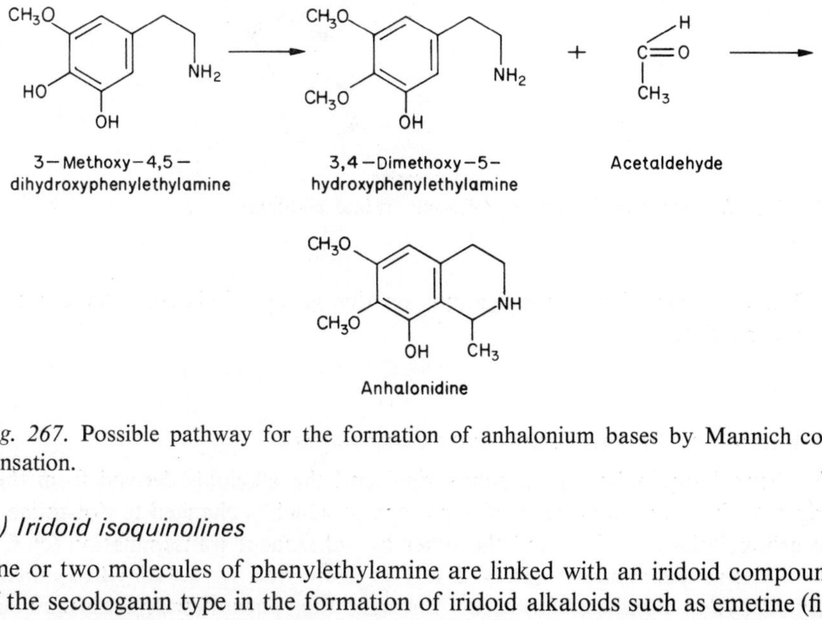

3−Methoxy−4,5− dihydroxyphenylethylamine	3,4−Dimethoxy−5− hydroxyphenylethylamine	Acetaldehyde

Anhalonidine

Fig. 267. Possible pathway for the formation of anhalonium bases by Mannich con-densation.

(b) Iridoid isoquinolines

One or two molecules of phenylethylamine are linked with an iridoid compound of the secologanin type in the formation of iridoid alkaloids such as emetine (fig. 268). Ipecoside is an isoquinoline glucoside which structurally resembles vinco-side (cf. D.19.3) and is formed in an analogous way to this compound from dopamine and secologanin. Ipecoside occurs in *Ipecacuanha* species as a minor alkaloid. It may be regarded as an acylated intermediate of the biosynthetic pathway of cephaëline and emetine. In tubolosine an isoquinoline ring is linked to a carboline ring through an iridoid bridge.

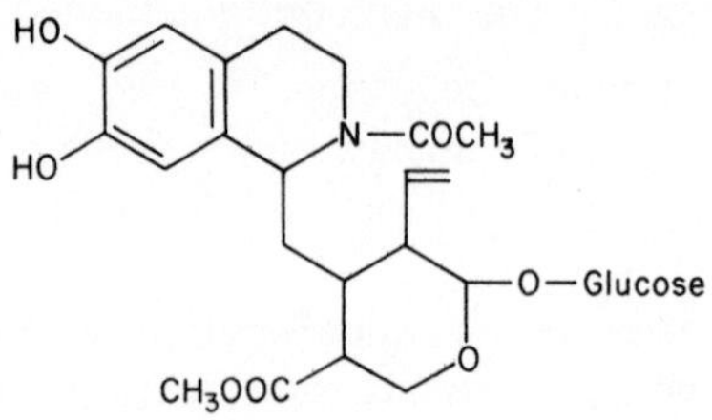

Ipecoside

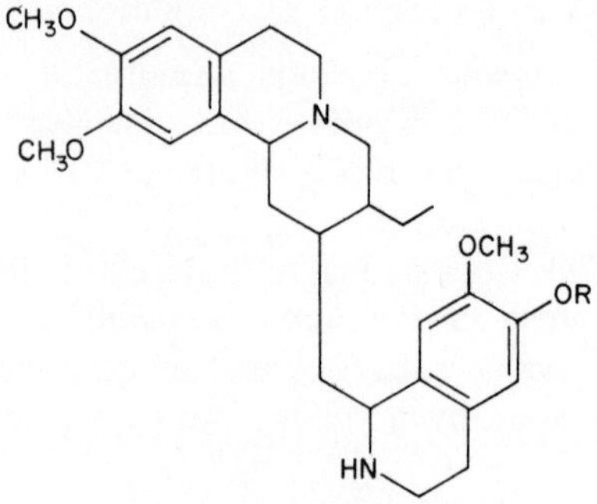

Cephaeline (R=H)
Emetine   (R=CH)

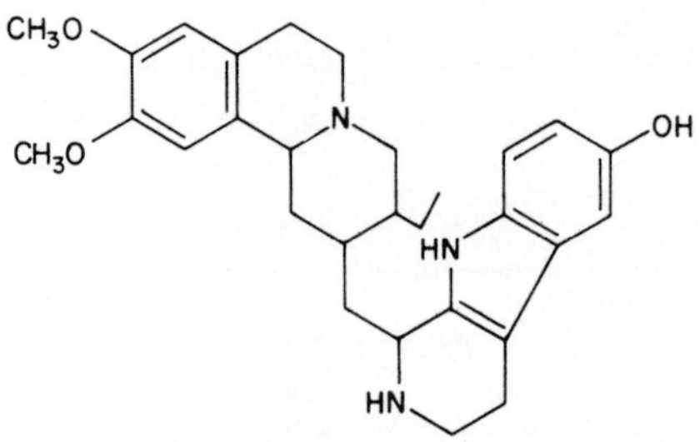

Tubolosine

Fig. 268. The structural formulae of some iridoid alkaloids.

Detailed biosynthetic investigations on this group of alkaloids have not yet been carried out.

(c) Benzylisoquinolines and related compounds

The benzylisoquinolines (e.g. papaverine) and the alkaloids derived from them originate from two molecules of dopa, one of which is changed to dopamine by decarboxylation (cf. C.4) and the other by subsequent transamination (cf. C.4) or oxidative deamination (cf. C.2.1.2) is converted to 3,4-dihydroxyphenyl-acetaldehyde. Synthesis proceeds through the phenyltetrahydroisoquinoline derivative, norlaudanosine. This tetrahydroisoquinoline may be dehydrogenated to aromatic compounds in subsequent reactions (fig. 269).

It is not known whether ring closure must always occur with the participation of an amine or whether the corresponding amino acids can also react. Several isoquinoline alkaloids possessing a carboxyl group in position 3 and probably originating directly from amino acids have recently been found.

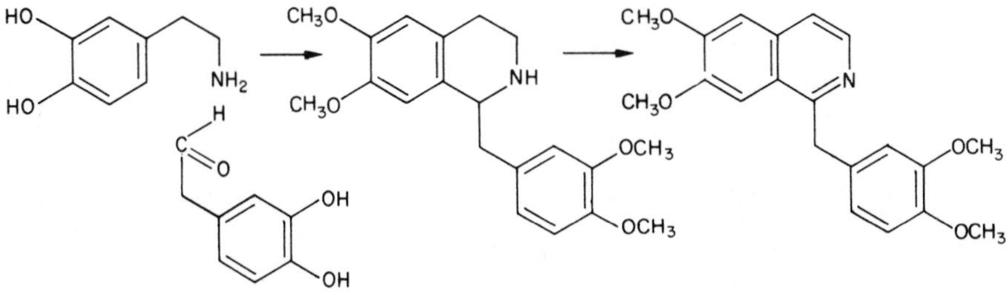

Dopamine   3,4–Dihydroxyphenyl   Norlaudanosine   Papaverine
acetaldehyde

Fig. 269. Biosynthesis of papaverine.

At what step in the biosynthetic pathway the methylations take place is still unknown.

Other alkaloid types are derived from the benzylisoquinolines by renewed ring closures and rearrangements. If norlaudanosoline is written as shown in fig. 270 so that one part of the molecule is rotated around the dotted line, the relationship of this alkaloid to morphine and similar alkaloids becomes clear.

The feeding of labelled precursors has shown that norlaudanosoline is a precursor of codeine and morphine. However, the corresponding dimethylated alkaloid (−)-reticuline appears to be the actual precursor. Reticuline has been isolated from *Anona reticulata* and is also found in opium. By attack of a phenol oxidase (cf. C.2.3) on the free hydroxyl groups, a biradical is formed from (−)-reticuline which stabilizes to form (+)-salutaridine, a dienone. Salutaridine is an alkaloid occurring in small quantities in *Papaver somniferum*.

Salutaridine changes to the dienol I by reduction. This yields thebaine by formation of an ether bridge and with the elimination of water. Thebaine is comparatively widespread and is present in all *Papaver* species investigated.

The conversion of thebaine to codeine and morphine is initiated by demethylation of one of the methoxy groups. Neopinone is formed at first. This alkaloid is converted to codeinone by a shift of the double bond. Codeine originates from codeinone by reduction, and by further demethylation morphine is formed. Both these alkaloids have until now been found only in two *Papaver* species.

The formation of the aporphine alkaloids of the bulbocapnine and glaucine types also appears to proceed via norlaudanosoline as the intermediate (fig. 271). Detailed investigations, however, are lacking.

If norlaudanosoline is methylated to orientaline (orientaline differs from reticuline due to the pattern of substitution in ring C) aporphine alkaloids of the isothebaine type are formed in the course of further rearrangement reactions (fig. 272) instead of morphine bases.

Orientaline is converted to orientalinone via the biradical formed by the action of a phenol oxidase and is then reduced to orientalinol. Isothebaine, which occurs

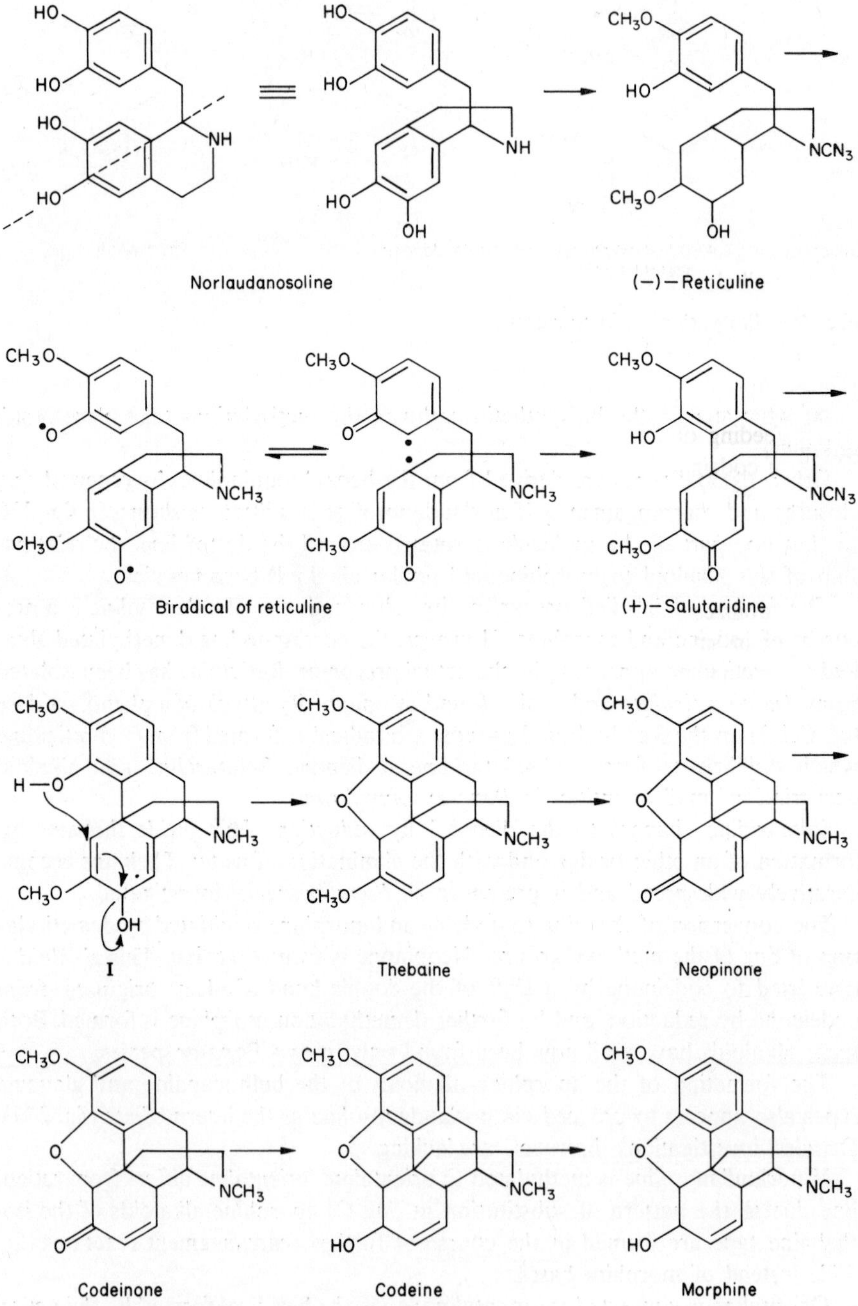

Fig. 270. Biosynthesis of codeine and morphine.

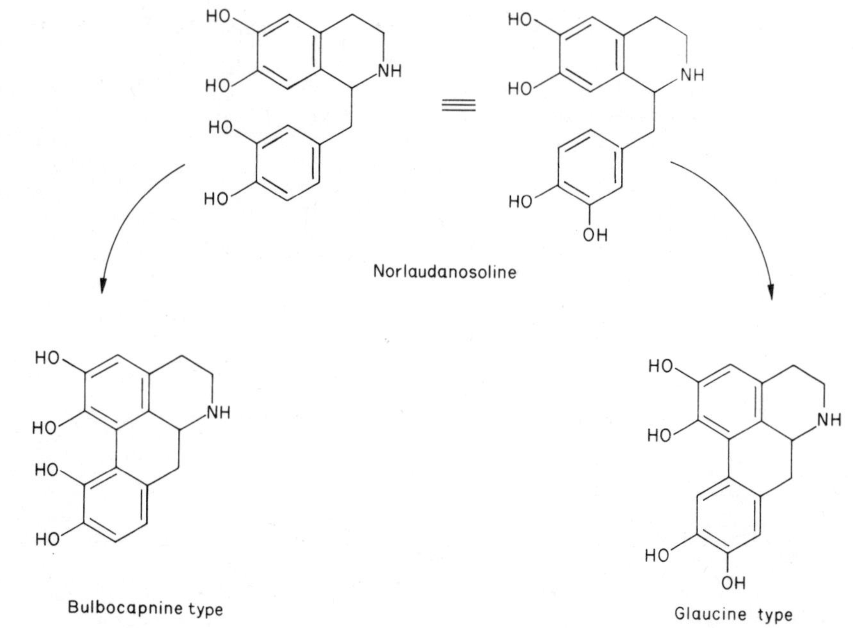

Fig. 271. Formation of aporphine alkaloids of the bulbocapnine and glaucine types.

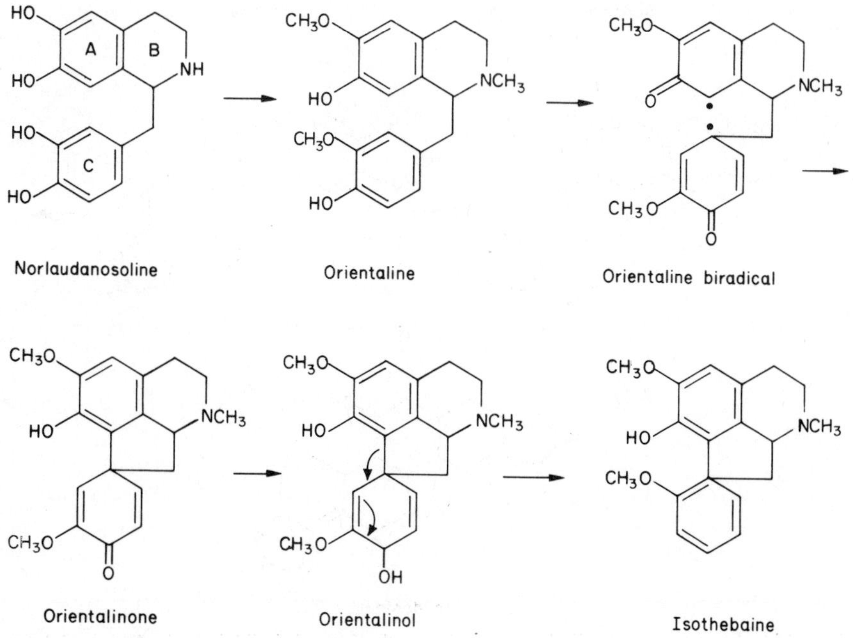

Fig. 272. Biosynthesis of isothebaine.

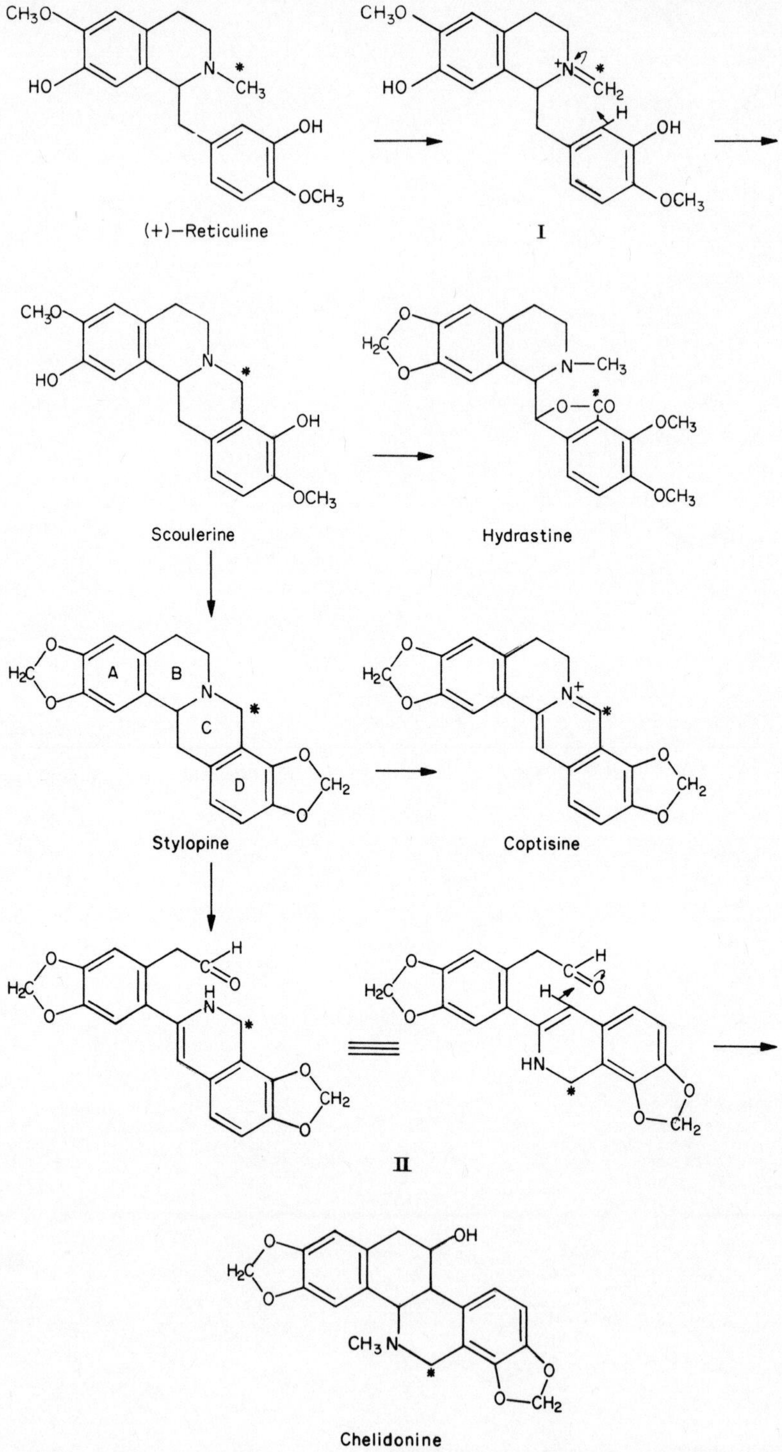

Fig. 273. Formation of alkaloids of the berberine, phthalideisoquinoline and benzo-phenanthridine groups.

in large amounts in *Papaver orientale*, is formed from orientalinol by the elimination of water, opening of the five-membered ring and renewed ring closure (cf. D.20.1.5 for the rearrangement to stephanine).

Reticuline is not incorporated into isothebaine.

Berberine and similar alkaloids, such as coptisine, originate from (+)-reticuline. The carbon atom of the so-called 'berberine bridge' marked '*' in fig. 273 originates from the *N*-methyl group. The quarternary imine I is possibly the actual precursor. However, other formulations may also be possible.

An alkaloid of the tetrahydroberberine type, scoulerine is formed first. This can then be dehydrogenated to the berberine derivative, coptisine.

The phthalideisoquinoline alkaloid, hydrastine, is formed by oxidation from the tetrahydroberberine alkaloid, scoulerine (fig. 273). The actual mechanism of this reaction is still unknown.

The alkaloids of the benzophenanthridine group, such as chelidonine, are also very closely related to the tetrahydroberberine alkaloids. The tetrahydroberberine alkaloid, stylopine, might be converted into the alkaloid, chelidonine, by cleavage of ring B, formation of a double bond in the heterocyclic ring (compound II), rotation of the heterocyclic part of the molecule about the single bond linking rings A and C, and renewed ring closure.

The minor alkaloid, berberastine, isolated from *Hydrastis canadensis* and differing from berberine by an additional hydroxyl group, is not formed from the latter. However, a good incorporation of noradrenalin has been observed (cf. D.20.1.5 also). It is not yet clear whether this compound is a physiological precursor, since noradrenalin has not been found in *Hydrastis*.

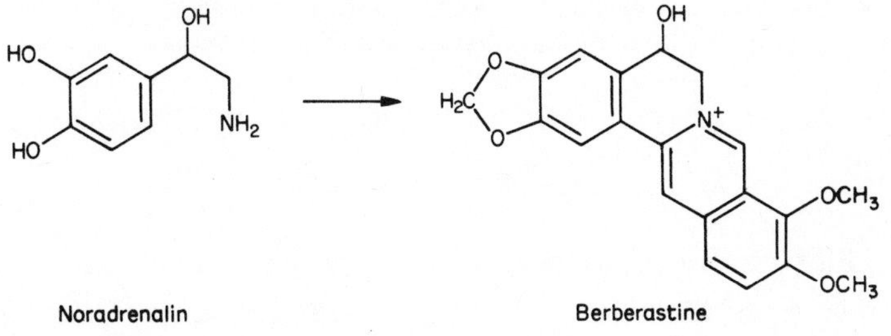

Noradrenalin   Berberastine

Fig. 274. Formation of berberastine.

The bisbenzylisoquinoline derivatives to which tubocurarine belongs originate by oxidative coupling of two benzylisoquinoline molecules. However 'mixed' dimers, such as thalicarpine, are also known which consist of an aporphine portion and a benzylisoquinoline portion (fig. 275). Formation of radical precursors by phenol oxidases (cf. C.2.3) is a prerequisite for dimerization in all cases.

The methylenedioxy groups frequently observed in isoquinoline alkaloids originate by oxidative condensation from a methyl and a hydroxyl group. The

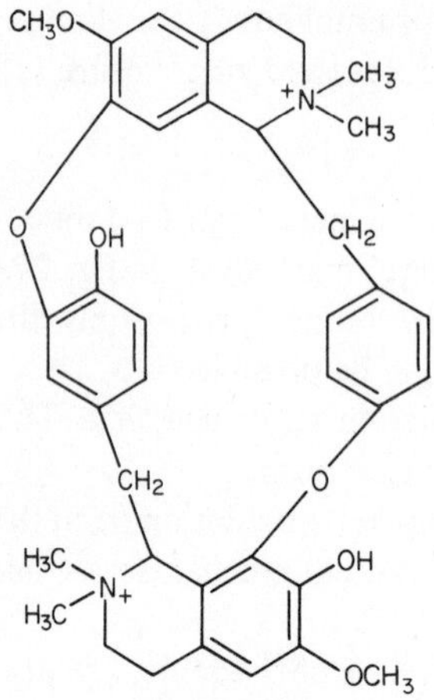

Tubocurarine

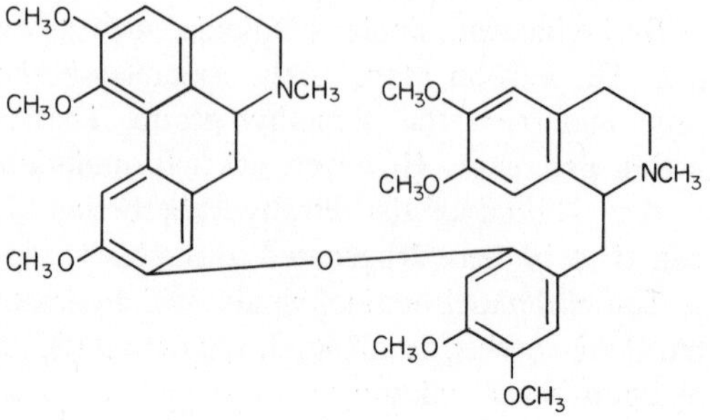

Thalicarpine

Fig. 275. The structural formulae of tubocurarine and thalicarpine.

same mechanism is effective in the formation of this grouping within other classes of natural products.

References for further reading

Battersby, A. R. 'Phenol Oxidations in the Alkaloid Field', *Oxidative Coupling of Phenols*, ed. W. I. Taylor and A. R. Battersby (Marcel Dekker Inc., New York, 1967), pp. 119–65.

Schutte. H. R. 'Isochinolinalkaloide', *Biosynthese der Alkaloide*, ed. K. Mothes and H. R. Schütte (VEB Deutscher Verlag der Wissenschaften, Berlin, 1969), pp. 367–419.

Gross, D. Die Biosynthese iridoider Naturstoffe, *Fortschritte der Botanik* **32** (1970), pp. 93–108.

20.1.3. Biosynthesis of Erythrina and Amaryllidaceae alkaloids

In addition to a Mannich condensation, phenylethylamine derivatives and aldehydes can react with each other to form Schiff bases. The latter compounds are precursors for a few interesting alkaloid groups.

(a) Erythrina alkaloids

The formation of the alkaloids found in the genus *Erythrina* of Leguminosae appears to proceed via the Schiff base I which is formed from dopamine and dihydroxyphenylacetaldehyde. By reduction of this compound the amine II is formed, which is first converted to the alkaloid erysodienone by the action of phenol oxidases via the biradical III, coupling (compound IV), renewed oxidation

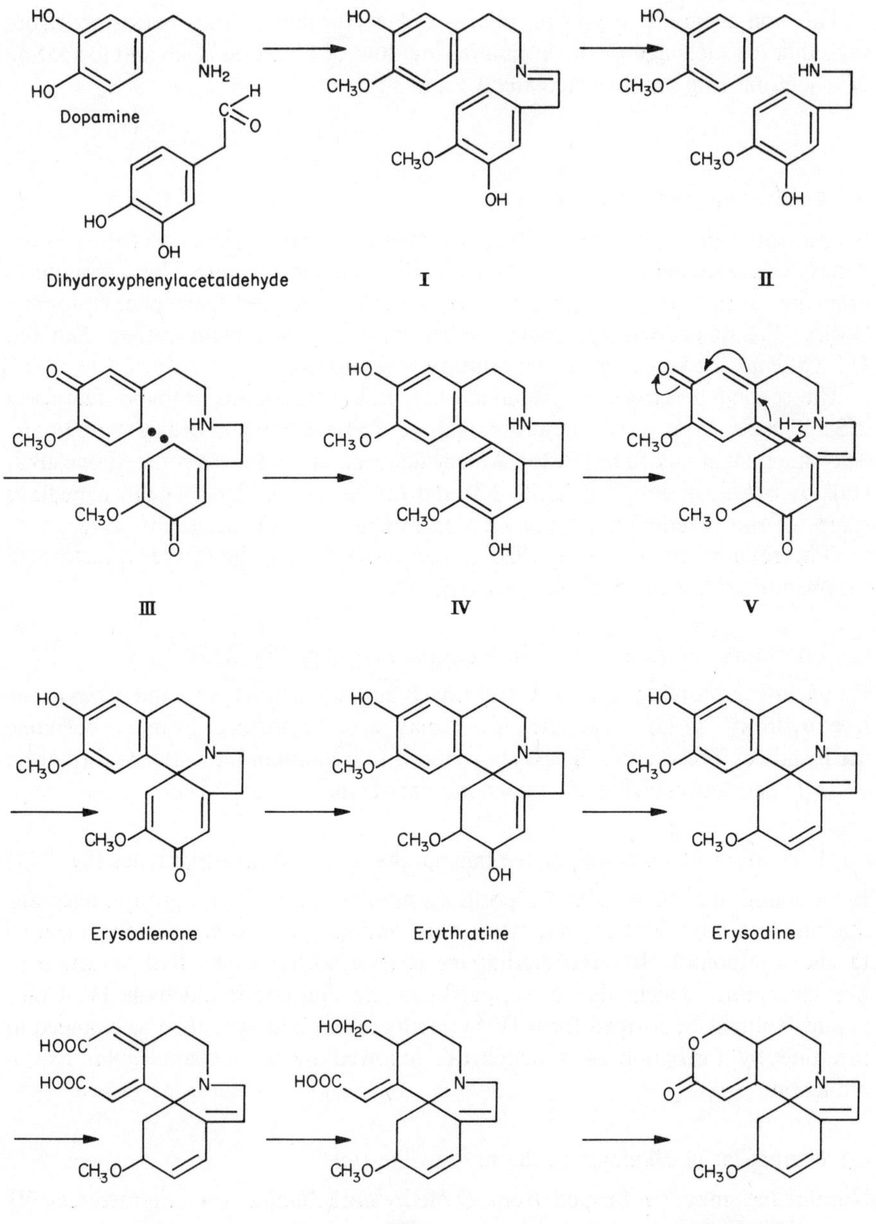

Fig. 276. Possible pathway for the formation of Erythrina alkaloids.

(compound V) and addition of an amino group to the quinoid system. Erythratine is formed from it by reduction, and elimination of water with the shift of a double bond gives erysodine.

The non-aromatic Erythrina alkaloid, α-erythroidine, may be formed from erysodine by cleavage of the aromatic ring, loss of a carbon atom and formation of a lactone ring via intermediates VI and VII.

(b) Amaryllidaceae alkaloids

In spite of their very different chemical structure the alkaloids occurring in the Amaryllidaceae represent a biogenetically uniform group. The compounds originate from two precursors, both of which are derived from phenylpropane bodies. Protocatechuic aldehyde, which may originate from caffeic acid (cf. D.20.6) may be regarded as the primary building stone.

The second precursor is tyramine. Both these compounds probably first form a Schiff base, from which the protoalkaloids norbelladine and belladine originate. The other alkaloids found in the Amaryllidaceae originate from the above alkaloids by oxidative coupling (cf. C.2.3) and further conversions. The intermediary stages of the reactions have not been established with any certainty so far.

O-Methylnorbelladine (norbelladine methylether) may be further transformed by phenol oxidases in the following ways:

(1) Formation of alkaloids of the galanthamine type (fig. 277)

By oxidative coupling of ring A and ring B at the positions *para* and *ortho* to the free hydroxyl groups, the dienone I may first be formed from norbelladine methylether. The dienone I is then reduced to galanthamine with the formation of an O-heterocyclic ring via the enone narwedine.

(2) Formation of alkaloids of the haemanthamine and tazettine types (fig. 277)

By coupling of both rings at the positions *para* to the hydroxyl groups, haemanthamine may be formed first from norbelladine methylether via the dienone II and compound III. Haemanthamine is then converted by hydroxylation to haemanthidine, which also exists partly as the tautomeric aldehyde IV. Compound V might be formed from IV by methylation, and may then be changed to tazettine by formation of a hemiketal, followed by an intramolecular oxido-reduction.

(3) Formation of alkaloids of the norpluviine type

Norpluviine may be formed from O-methylnorbelladine via intermediates VI and VII by coupling of ring A and ring B at the positions *ortho* and *para* to the free hydroxyl groups. Lycorine and galanthine may originate from this alkaloid by further minor changes (fig. 278).

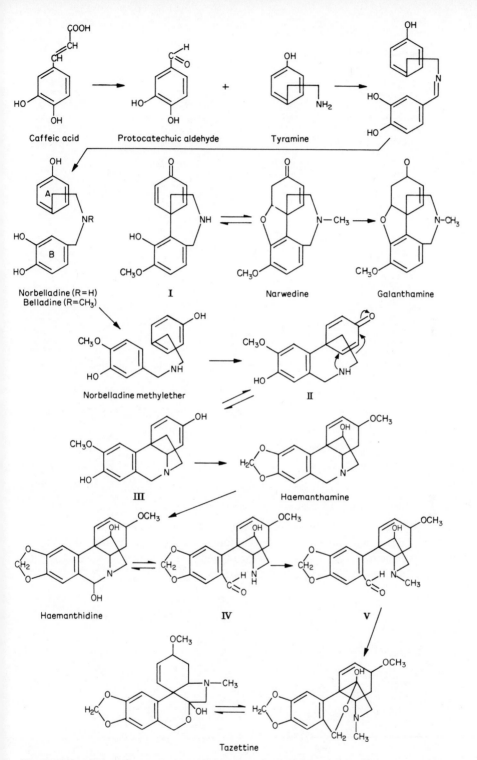

Fig. 277. Possible pathways for the formation of the alkaloids galanthamine, haemanthamine and tazettine.

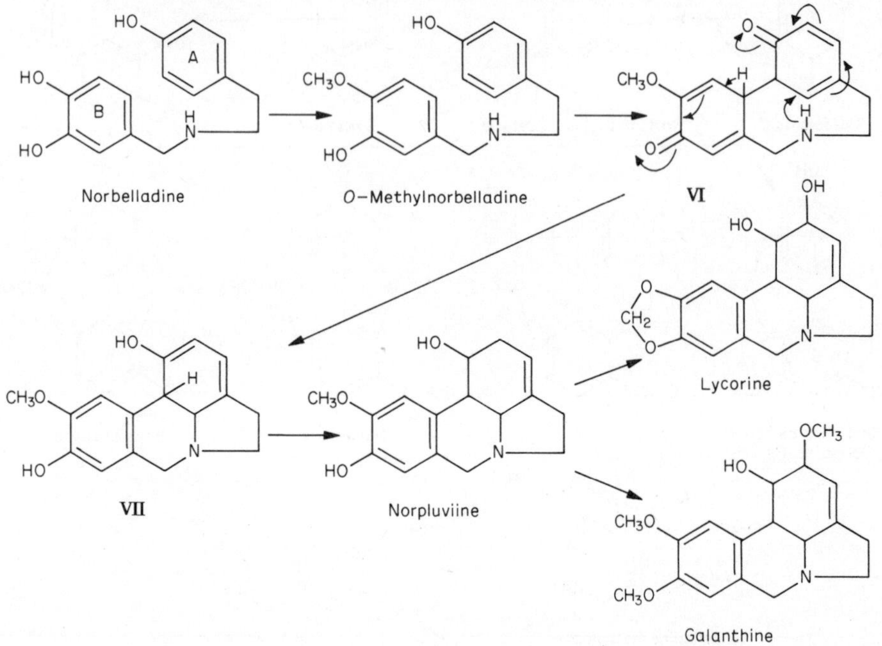

Fig. 278. Probable pathway for the formation of norpluviine, lycorine and galanthine.

References for further reading

See D.20.1.2.

20.1.4 Biosynthesis of colchicine

Colchicine is a protoalkaloid which occurs in Liliaceae species such as *Colchicum autumnale*, the autumn crocus. The compound consists of an aromatic ring A, with three methoxy groups, a seven-membered ring B, which is substituted with an acetylated amino group, and the tropolone ring C (cf. fig. 280).

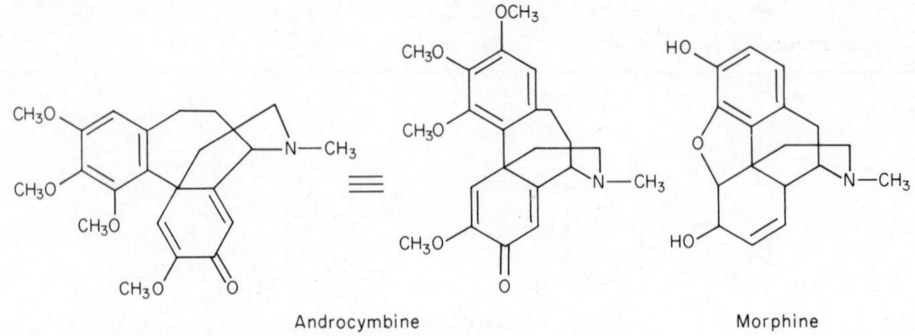

Fig. 279. Structural formulae of androcymbine and morphine.

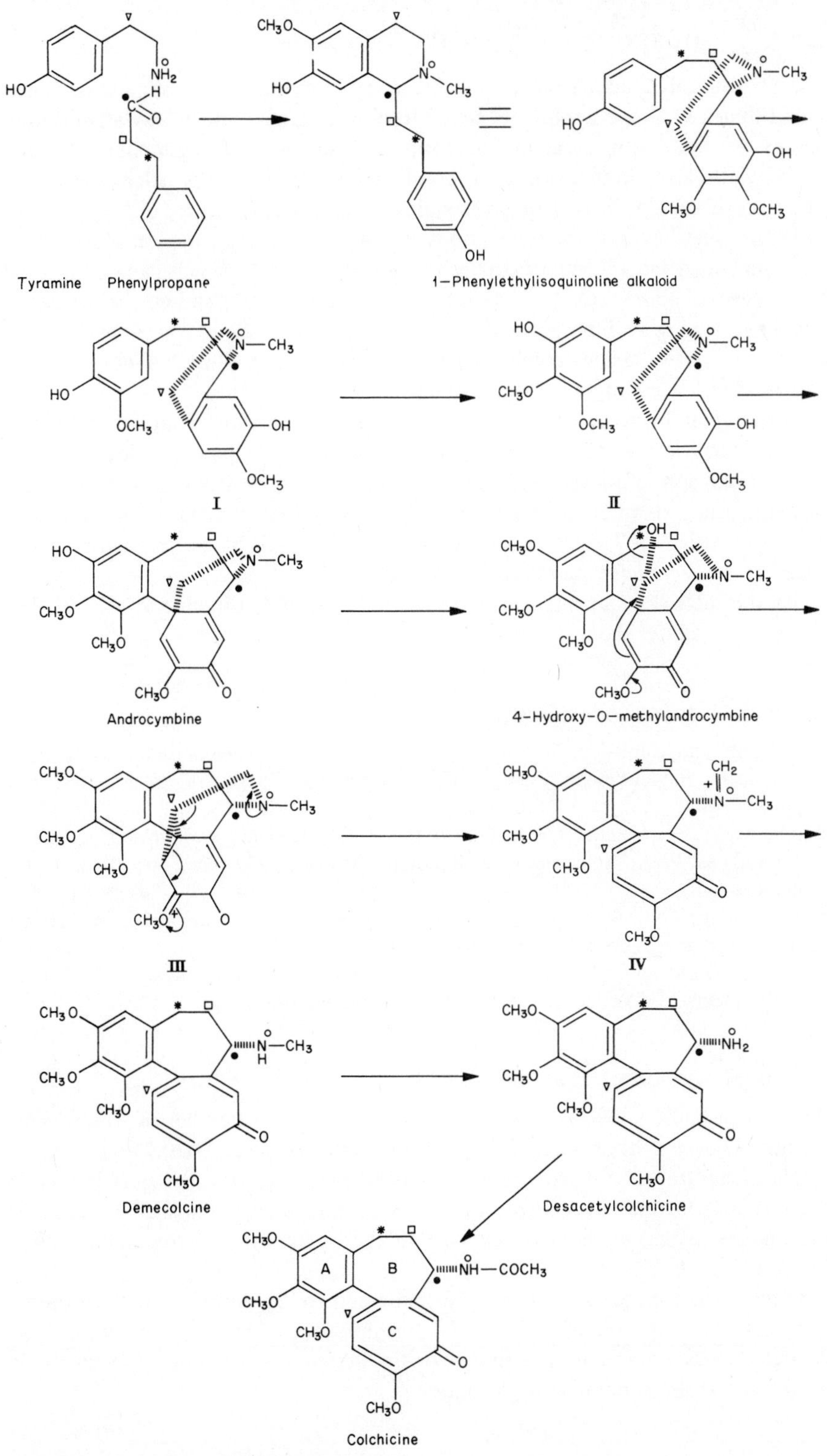

Tyramine Phenylpropane

1—Phenylethylisoquinoline alkaloid

I

II

Androcymbine

4-Hydroxy-O-methylandrocymbine

III

IV

Demecolcine

Desacetylcolchicine

Colchicine

Fig. 280. Biosynthesis of colchicine.

This unusual structure has motivated many biosynthetic investigations. It was found that a C_6–C_3-component related to cinnamic acid and phenylalanine condenses at first with tyramine to form a 1-phenylethylisoquinoline alkaloid. 1-Phenylethylisoquinolines occur as co-alkaloids in a number of colchicine plants. They are homologues of 1-benzylisoquinolines (cf. D.20.1.2).

Hydroxylations and methoxylations at ring A, the sequence of which is still not clear, lead via the intermediate I to the diphenol II which can be attacked by phenol oxidases (cf. C.2.3) and converted to androcymbine by oxidative coupling. Androcymbine is structurally similar to morphine. This is particularly clear in fig. 279. Besides colchicine, androcymbine, and O-methylandrocymbine formed from it by methylation, also occur in various plants.

O-Methylandrocymbine is then hydroxylated at position 4 and converted to demecolcine via the labile intermediates III and IV with the expansion of ring C by that carbon atom corresponding to the β-carbon atom of tyramine. By demethylation, desacetylcolchicine is formed, which in turn can be acetylated to colchicine (fig. 280). Demecolcine and desacetylcolchicine regularly occur together in *Colchicum* plants as co-alkaloids.

All the methyl groups of colchicine originate from methionine (cf. C.3.3). The nitrogen originates from tyramine.

References for further reading

Fell, K. R. and Ramsden, D. Colchicum, A Review of Colchicums and the Sources, Chemistry, Biogenesis and Assay of Colchicine and its Congeners, *Lloydia* **30** (1967), pp. 123–40.

Battersby, A. R. 'Phenol Oxidations in the Alkaloid Field', *Oxidative Coupling of Phenols*, ed. W. I. Taylor and A. R. Battersby (Marcel Dekker Inc., New York, 1967), pp. 119–65.

Leete, E. 'Alkaloid Biogenesis', *Biogenesis of Natural Compounds*, ed. P. Bernfeld (Pergamon Press, Oxford, 1967), pp. 953–1023.

Gross, D. 'Colchicin', *Biosynthese der Alkaloide*, ed. K. Mothes and H. R. Schütte (VEB Deutscher Verlag der Wissenschaften, Berlin, 1969), pp. 359–66.

20.1.5 Formation of aristolochic acids

Aristolochic acids possess a nitro group and occur in a number of *Aristolochia* species. They are accompanied by aporphine alkaloids (cf. D.20.1.2).

The structural relationship between both these groups of substances has led to the supposition that aristolochic acids originate from isoquinoline alkaloids. In fact feeding experiments have shown that the biosynthetic pathway outlined in fig. 281 is probable.

Orientalinol (cf. D.20.1.2) may be converted to alkaloid I with the elimination of water, rotation of ring D, and formation of a new bond between rings A and D. This compound is a precursor for stephanine. Aristolochic acid I may be formed from stephanine by oxidation.

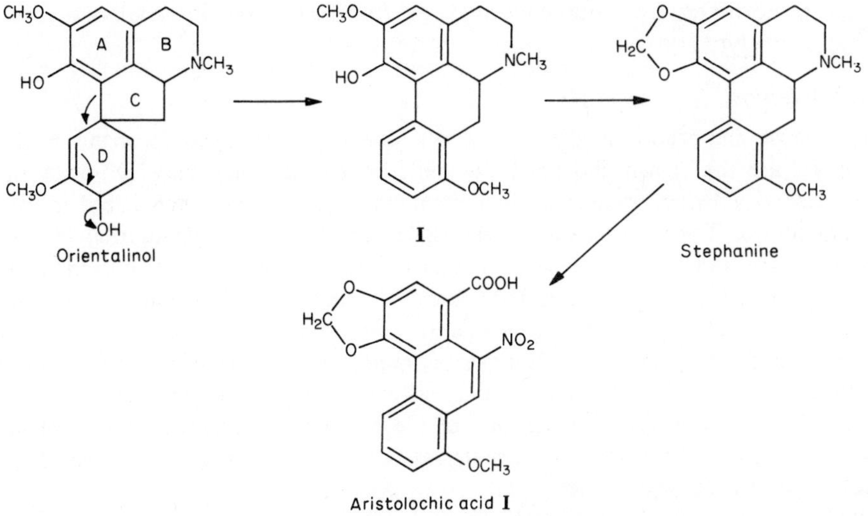

Orientalinol

I

Stephanine

Aristolochic acid **I**

Fig. 281. Possible pathway for the formation of aristolochic acid.

It is interesting to note that noradrenalin is incorporated into aristolochic acid with good specific incorporation rates (cf. D.20.1.2 also). It is concluded from this result that 4-hydroxynorlaudanosoline, which is structurally derived from noradrenalin, is an important precursor and that the presence of a hydroxyl group in position 4 is a prerequisite for oxidative degradation of the heterocyclic ring.

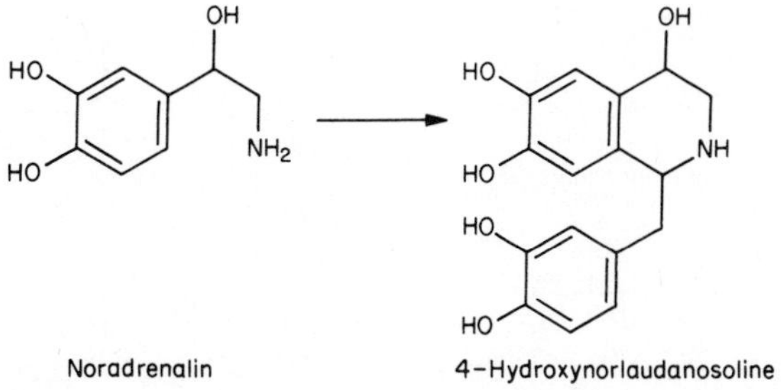

Noradrenalin

4-Hydroxynorlaudanosoline

Fig. 282. Conversion of noradrenalin to 4-hydroxynorlaudanosoline.

References for further reading

Schütte, H. R. 'Isochinolinalkaloide', *Biosynthese der Alkaloide,* ed. K. Mothes and H. R. Schütte (VEB Deutscher Verlag der Wissenschaften, Berlin, 1969), pp. 367–419.

20.1.6 Formation of indole derivatives from dihydroxyphenylalanine and m-tyrosine

(a) Melanins

The most important indole derivatives derived from phenylpropane components are the brown and black pigments known as melanins, which occur in all classes of living organisms. In animals the melanins are synthesized in the melanocytes. They are mainly responsible for the dark pigmentation of the human skin and hair, and help to protect the body against excessive irradiation by UV-light (cf. D.5.8), since they form free radicals which interact with the reactive radicals formed during irradiation of the tissue of the skin.

Melanins are found in the feathers of many birds, in the skin of reptiles and fishes, and in the skeleton of insects as well as in plants and micro-organisms. As the colouring constituent of the 'ink' of Cephalopods (cuttlefishes) the pigment serves in the defence of these organisms. Cuttlefish melanin is the colouring constituent of the dye 'sepia'.

Chemically, melanins are insoluble amorphous polymers, which are frequently linked with proteins. Their molecular weight is not determinable. They are extraordinarily stable and have been isolated in unchanged form from fossils which are 150 million years old.

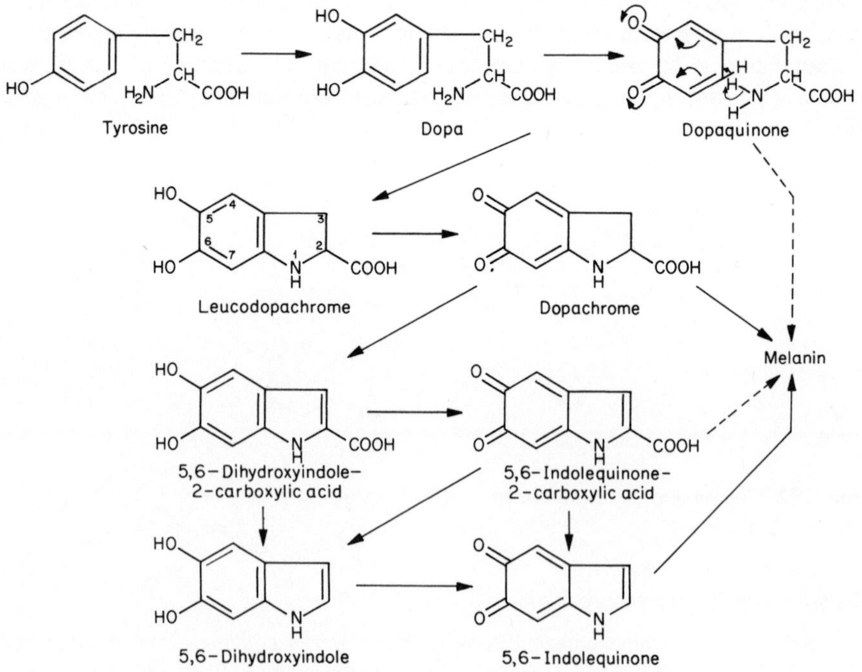

Fig. 283. Formation of melanin.

It has been shown by feeding experiments that the melanins usually originate from tyrosine.* All reactions which do not proceed spontaneously are catalysed by the enzyme tyrosinase.

On the one hand tyrosinase hydroxylates tyrosine in the manner of a mixed function oxygenase (cf. C.2.6) with the formation of dihydroxyphenylalanine. Dihydroxyphenylalanine may act as a hydrogen donor and is oxidized to the corresponding o-quinone. On the other hand the enzyme, like the laccases and peroxidases (cf. C.2.3 and C.2.4), catalyses the oxidation and coupling of poly-phenols with hydroxyl groups at the *ortho* position.

Dopaquinone changes to leucodopachrome by addition of the amino group to the quinoid system (cf. D.7.7). This compound is then oxidized to the quinone form by phenol oxidase to form dopachrome. By a shift of the double bond,

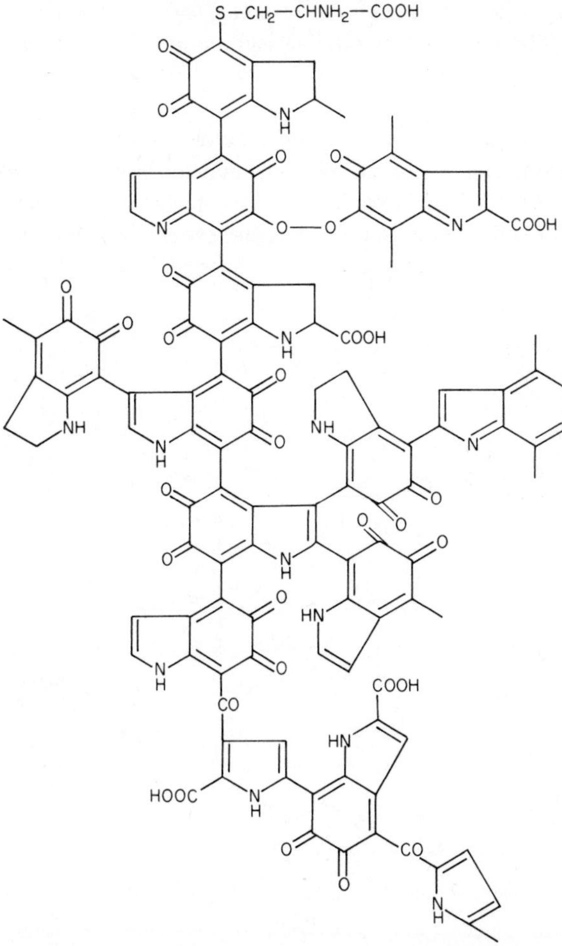

Fig. 284. Part of a melanin molecule from the 'ink' of *Sepia officinalis*.

* In certain cases formation from other precursors, such as tryptophan, is also possible.

5,6-dihydroxyindole-2-carboxylic acid is formed which may be either dehydro-genated to the corresponding quinone, or converted to 5,6-dihydroxyindole by decarboxylation. 5,6-Indolequinone is ultimately formed from this compound by dehydrogenation, or by elimination of carbon dioxide from 5,6-indole-quinone-2-carboxylic acid (fig. 283).

All the quinones shown in fig. 283 are capable of undergoing addition reactions and tend to polymerize (cf. D.7.7). Polymerization products with low molecular weights are red and those with higher ones are black. Positions 4 and 7 are especi-ally active in the formation of new bonds. They may, however, originate from the carbon atoms in positions 2 and 3 so that highly interlaced three-dimensional macromolecules are formed. Amino acids also react with the quinones in a manner similar to the synthesis of humic acids (cf. D.2.4.4), so that melanins are very complex substances. Part of the melanin of *Sepia officinalis* is shown in fig. 284. The pyrrol rings possibly originate by oxidative degradation of indolequinones.

(b) Gliotoxin

Gliotoxin, synthesized by various fungi, is another indole compound which is derived from the phenylpropane amino acids. Phenylalanine or *m*-tyrosine, as well as serine and the methyl group of methionine (cf. C.3.3), make up the skeleton of this compound. A dethio compound is perhaps synthesized first and is then changed to gliotoxin (fig. 285) analogous to the formation of biotin from dethio biotin (cf. D.9.4).

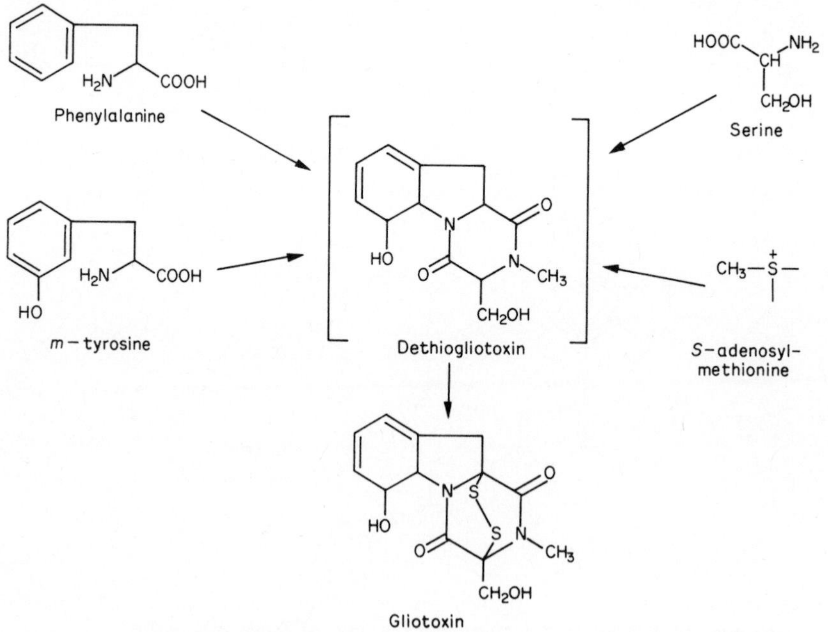

Fig. 285. Formation of gliotoxin.

(c) Betalains

The coloured compounds synthesized by the Centrospermae, which, according to their structure may be red (betacyanins) or yellow (betaxanthins), are termed betalains. They contain a substituted tetrahydropyridine ring, which is linked via a two-carbon fragment with leucodopachrome in the case of the betacyanins, and with proline in the case of the betaxanthins.

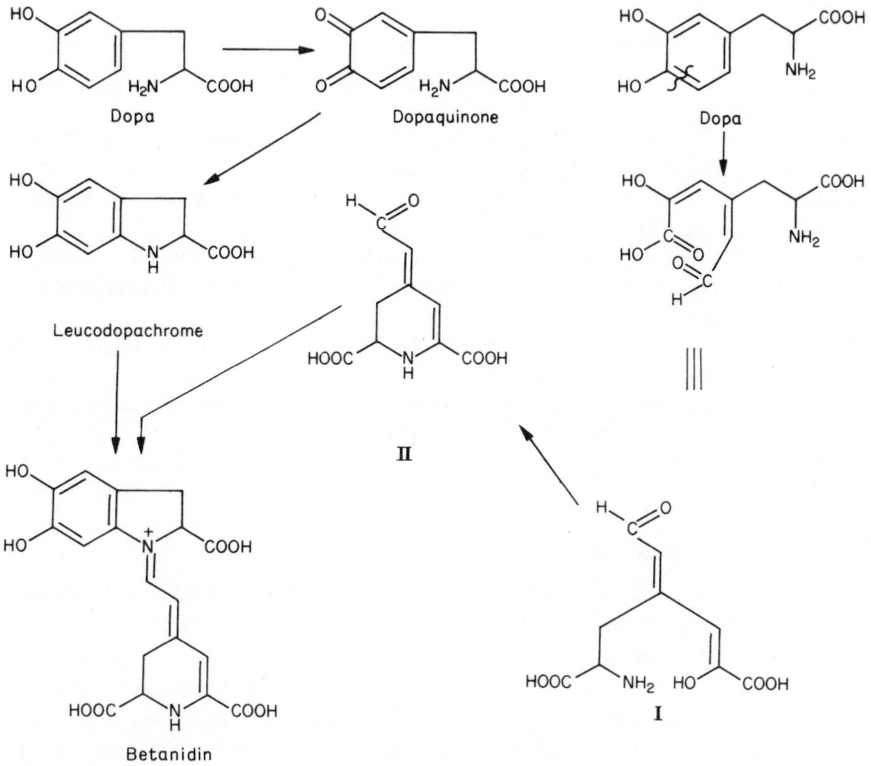

Fig. 286. Possible pathway for the formation of betanidin.

It is of physiological interest that betacyanins and anthocyanins (cf. D.20.5.2) never occur together. In contrast, flavonols and betacyanins are found side by side in certain plants.

Experiments with isotopically labelled precursors have shown that two molecules of DOPA are incorporated into the betacyanins. The biosynthetic mechanism outlined in fig. 286 is postulated for betanidin. Leucodopachrome is formed from one molecule of DOPA via dopaquinone. The ring of the second molecule of DOPA is opened immediately adjacent to the hydroxyl groups by a dioxygenase (cf. C.2.5), thus forming an aldehyde and a carboxylic acid grouping (compound I). The N-heterocyclic compound II may be formed from it by elimination of water. Leucodopachrome and compound II may then condense to form betanidin, again with the elimination of water.

In *Opuntia ficus-indica*, proline is incorporated into the indicaxanthin molecule in place of leucodopachrome.

References for further reading

Nicolaus, R. A. Biogenesis of Melanins, *Rassenga di Midicina Sperimentale* (1962), Supplement No. 1, pp. 1–32.

Nicolaus, R.A., Piatelli, M. and Fattorusso, E. The Structure of Melanins and Melanogenesis IV, *Tetrahedron* **20** (1964), pp. 1163–72.

Suhadolnik, R. J. 'Gliotoxin' *Antibiotics*, Vol. II, ed. D. Gottlieb and P. D. Shaw (Springer Verlag, Berlin, 1967), pp. 29–31.

Brown, B. R. 'Biochemical Aspects of Oxidative Coupling of Phenols', *Oxidative Coupling of Phenols*, ed. W. I. Taylor and A. R. Battersby (Marcel Dekker Inc., New York, 1967), pp. 167–201.

Danneel, R. Die Entstehung der Farbmuster bei Säugetieren, *Naturwiss. Rundschau* **21** (1968), pp. 420–4.

Gross, D. Vorkommen, Struktur und Biosynthese natürlicher Piperidinverbindungen, *Fortschritte Chem. org. Naturstoffe* **29** (1971), pp. 1–59.

20.1.7 Formation of novobiocin and anisomycin

The antibiotic novobiocin, from *Streptomyces niveus*, contains besides noviose, a sugar with a branched carbon chain (cf. D.1.2.3), a coumarin ring substituted by an amino group originating from tyrosine. Experiments with tyrosine-$U^{14}C,C^{18}OOH$ showed that the oxygen present in the heterocyclic ring, B, originates from the carboxyl group of this amino acid. The C-methyl group attached to the coumarin as well as one of the two C-methyl groups and the O-methyl group of noviose originate from methionine (cf. C.3.3). The formation of the coumarin ring takes place here in a fundamentally different way to that described in sections D.20.4.2, D.20.5.3 and D.20.5.4.

Part C of novobiocin originates from p-hydroxybenzoic acid. This is probably first substituted by isopentenyl pyrophosphate, and then undergoes an ATP dependent activation of the carboxyl group to form a peptide with the glycosylated coumarin moiety.

Anisomycin is an antibiotic which is found in certain *Streptomyces* species. According to the results obtained from feeding experiments, tyrosine forms the benzyl side chain and carbon atom 2 of the pyrrolidine ring; carbon atoms 4 and 5 of the pyrrolidine ring originate from glycine or acetate. The origin of the nitrogen atom and that of carbon atom 3 is unknown.

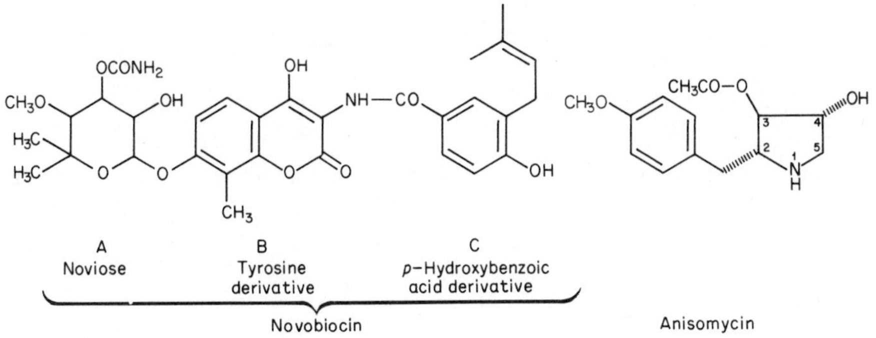

Fig. 287. The structural formulae of novobiocin and anisomycin.

References for further reading

Schütte, H. R. 'Verschiedenes', *Biosynthese der Alkaloide*, ed. K. Mothes and H. R. Schütte (VEB Deutscher Verlag der Wissenschaften, Berlin, 1969), pp. 645–77.

Kominek, L. A. 'Novobiocin', *Antibiotics*, Vol. II, ed. D. Gottlieb and P. D. Shaw (Springer Verlag, Berlin, 1967), pp. 231–9.

20.2 Formation of plastoquinones, tocopherol quinones and tocopherols from *p*-hydroxyphenylpyruvic acid

p-Benzoquinones, and hydroquinone derivatives substituted by an isoprenoid side chain, are widespread in nature. Besides the ubiquinones (cf. D.6.3.2), plastoquinones, α-tocopheryl quinones and α-tocopherols which are formed in higher plants belong to this group. All these compounds form groups whose individual representatives differ from each other with respect to the length of the side chain (fig. 288). They are essential for human beings and a number of animals.

Experiments on *Zea mays* have shown that the isoprenoid side chain of these compounds is formed from isopentenyl pyrophosphate as described in D.5.2, and when first formed, exists in the free state as the pyrophosphate.

The aromatic nucleus and one of the *C*-methyl groups originate from tyrosine. This amino acid is first converted to *p*-hydroxyphenylpyruvic acid by transamination (cf. C.4), and then changes to homoarbutin via homogentisic acid (toluquinol, the aglycone of homoarbutin, is not an intermediate of the reaction chain) (fig. 289).

The conversion of *p*-hydroxyphenylpyruvic acid to homogentisic acid by the enzyme *p*-hydroxyphenylpyruvate oxygenase resembles hydroxylation with an NIH-shift (cf. C.2.6.5). However, it is not usually the hydrogen atom that shifts, but the side chain. Intermediates of the conversion have not yet been detected.

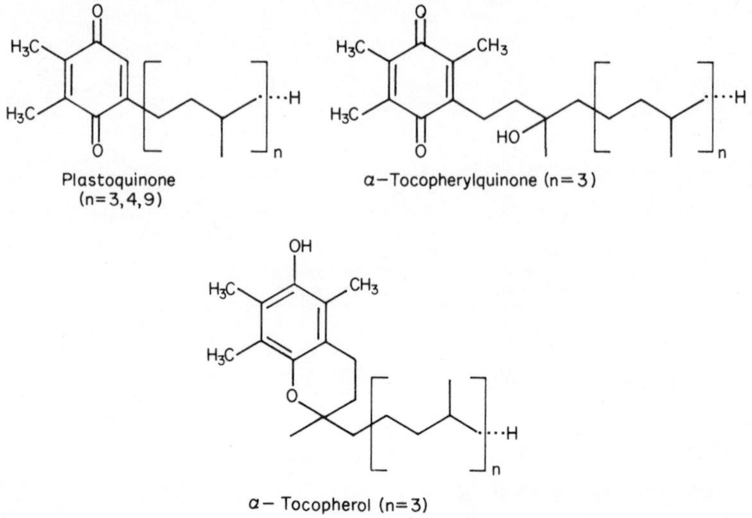

Fig. 288. The structural formulae of plastoquinone, α-tocopherylquinone and α-tocopherol.

Homoarbutin is probably the compound which is substituted by the prenyl-pyrophosphate grouping serving as the precursor of the side chain. Pyrophosphate is eliminated in this reaction. Methylprenylhydroquinone is hydroxylated, oxidized and methylated in further reaction steps. The sequence of these reactions is unknown. The methyl groups added originate from methionine (cf. C.3.3).

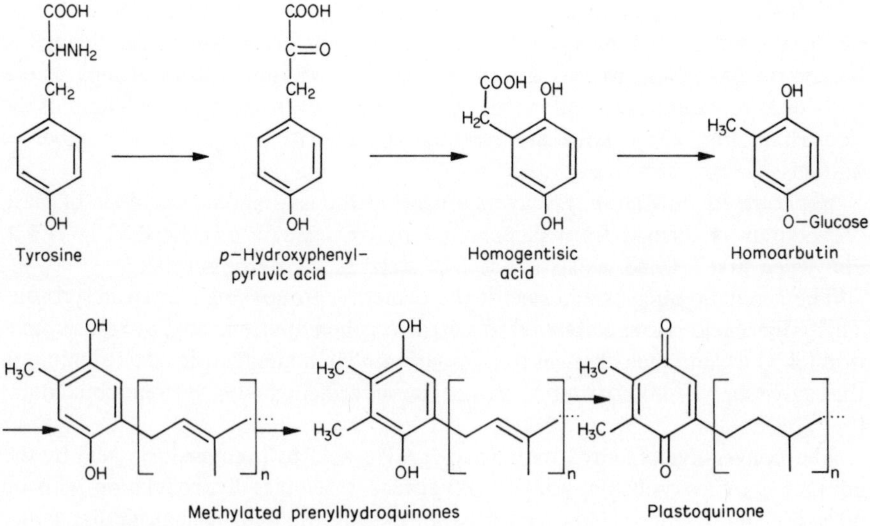

Fig. 289. Biosynthesis of plastoquinones.

References for further reading

Threlfall, D. R. 'Biosynthesis of Terpenoid Quinones', *Terpenoids in Plants*, ed. J. B Pridham (Academic Press, London, 1967), pp. 191–222.

Zenk, M. H. and Leistner, E. Biosynthesis of Quinones, *Lloydia* **31** (1968), pp. 275–92.

Threlfall, D. R. and Whistance, G. R. 'Biosynthesis of Isoprenoid Quinones and Chromanols', *Aspects of Terpenoid Chemistry and Biochemistry*, ed. T. W. Goodwin (Academic Press, London, 1971), pp. 357–404.

20.3 Formation of tropic acid, atropic acid and phenylglyceric acid

Phenylpropane compounds with branched carbon chains, e.g. L-tropic acid, atropic acid and L-α-phenylglyceric acid, are constituents of a number of tropane alkaloids (cf. D.15.1). They originate from phenylalanine by rearrangement of the side chain.

It was shown by feeding specifically labelled phenylalanine that the carboxyl group of the amino acid forms the carboxyl group of tropic acid. It must therefore shift from the carbon atom marked '+' to the one marked 'o' (fig. 290). The mechanism by which the reaction proceeds is unknown. Radioactivity from phenylalanine-α-^{3}H is not incorporated into tropic acid. It may be concluded that there is an equilibrium with phenylpyruvate or phenylpyruvate may be an intermediate.

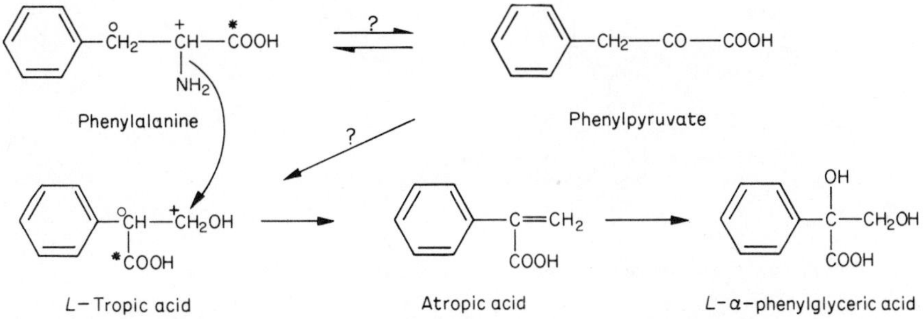

Fig. 290. Formation of L-tropic acid, atropic acid and L-α-phenylglyceric acid.

Atropic acid and L-α-phenylglyceric acid are structurally related to tropic acid and may originate from it. It is still unclear, however, whether the conversions take place before or after esterification with tropane alcohols (cf. D.15.1)

References for further reading

Luckner, M. Über neue Arbeiten zur Biosynthese N-heterocyclischer Verbindungen, 4. Teil: Die Bildung von Verbindungen mit Pyridin-, Piperidin-, Chinolizidin-, Pyrrolizidin- und Tropanringsystem, *Pharmazie* **19** (1964), pp. 1–14.

Liebisch, H. W. 'Tropanalkaloide und Pyrrolidinbasen', *Biosynthese der Alkaloide*, ed. K. Mothes and H. R. Schütte (VEB Deutscher Verlag der Wissenschaften, Berlin, 1969), pp. 183–214.

20.4 Cinnamic acid and cinnamic acid derivatives as secondary natural products

20.4.1 Formation of cinnamic acid and cinnamic acid derivatives

Secondary natural products derived from cinnamic acid are synthesized mainly by higher plants. Cinnamic acid and its derivatives occur partly free and partly bound as esters and are the starting materials for the biosynthesis of further groups of secondary natural products.

The unsubstituted cinnamic acid which originates from phenylalanine by

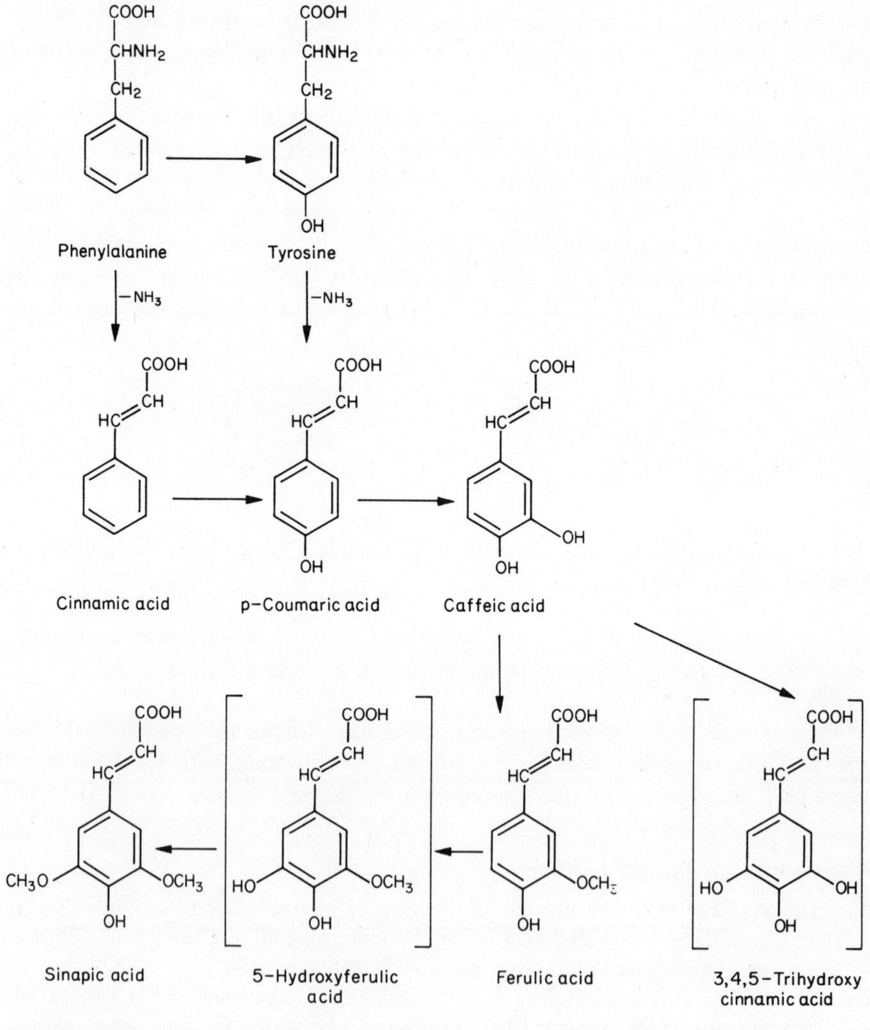

Fig. 291. Formation of cinnamic acid and cinnamic acid derivatives.

elimination of ammonia is the precursor for the various cinnamic acid derivatives found in nature. The conversion of tyrosine to *p*-coumaric acid is possible only in the case of the Graminea. The elimination of ammonia from phenylalanine and tyrosine is catalysed by ammonia lyases which possess the trivial names, phenylalanine deaminase and tyrase.

The other cinnamic acid derivatives shown in fig. 291 originate from cinnamic acid and *p*-coumaric acid by hydroxylation and methoxylation reactions. During the formation of *p*-coumaric acid from cinnamic acid, catalysed by a mixed function oxygenase the hydrogen atom at the *p*-position is shifted to the *m*-position in an NIH-shift (cf. C.2.6.5).

Caffeic acid, ferulic acid and sinapic acid are widespread. 3,4,5-Trihydroxycinnamic acid and 5-hydroxyferulic acid, which have not yet been found in nature, are, however, important as precursors in the formation of other natural products, e.g. gallic acid (cf. D.20.6) and sinapic acid (see above).

References for further reading

Higuchi, T. and Kawamura, I. 'Enzymes of Aromatic Biosynthesis', *Moderne Methoden der Pflanzenanalyse*, Vol. VII, ed. K. Paech and M. V. Tracy (Springer Verlag, Berlin, 1964), pp. 260–89.

Neish, A. C. 'Coumarins, Phenylpropanes, and Lignin', *Plant Biochemistry*, ed. J. Bonner and J. E. Varner (Academic Press, New York, 1965), pp. 581–617.

Pridham, J. B. Low Molecular Weight Phenols in Higher Plants, *Ann. Rev. Plant Physiol.* **16** (1965), pp. 13–36.

Brown, S. A. Lignins, *Ann. Rev. Plant Physiol.* **17** (1966), pp. 223–44.

20.4.2 Biosynthesis of coumarins

The coumarins are typical metabolic products of higher plants. Their formation by micro-organisms has been detected in only a few cases. A few simple coumarins are shown in table 17.

Table 17. Naturally occurring coumarins

Name	R_6	R_7	R_8
Coumarin	H	H	H
Umbelliferone	H	OH	H
Herniarin	H	OCH_3	H
Aesculetin	OH	OH	H
Aesculin	*O*-Glucose	OH	H
Scopoletin	OCH_3	OH	H
Scopolin	OCH_3	*O*-Glucose	H
Fraxetin	OCH_3	OH	OH

Other more complex compounds occur in which the coumarin ring is substituted by an isopentenyl group or a furan ring, e.g. in suberosin, braylin and pimpinellin (fig. 292). The additional carbon atoms of the furocoumarins correspond to carbon atoms 1 and 2 of isopentenyl pyrophosphate (cf. D.5.1). The isopentenyl and furocoumarin compounds are thus very closely related.

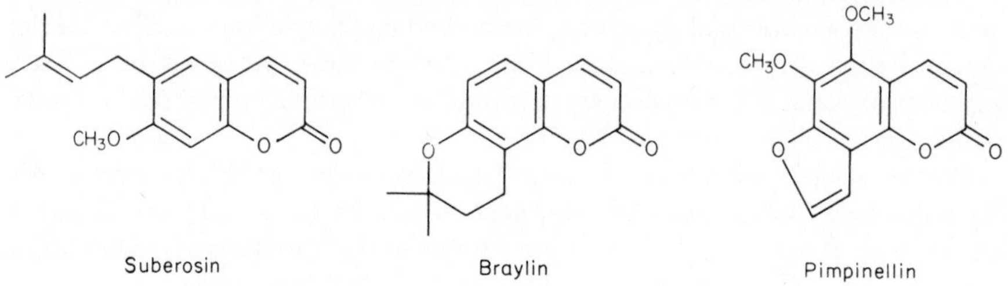

Suberosin   Braylin   Pimpinellin

Fig. 292. The structural formulae of suberosin, braylin, and pimpinellin.

See D.20.5.3 and D.20.5.4 for the coumarins of the coumoestrol group and the 3- or 4-phenyl coumarins.

Most coumarins are formed from the corresponding substituted *trans* cinnamic acid derivatives as outlined in fig. 293 for unsubstituted coumarin. Hydroxylation at the *o*-position of the particular cinnamic acid in question takes place first. The *o*-coumaric acid derivative that is formed is subsequently glucosylated. The *o*-coumaric acid glucoside is then rearranged in a spontaneous light-dependent reaction to the corresponding coumarinic acid glucoside, which is structurally derived from *cis*-cinnamic acid. By enzymatic elimination of glucose, free coumarinic acid is formed which spontaneously cyclizes to coumarin (fig. 293).

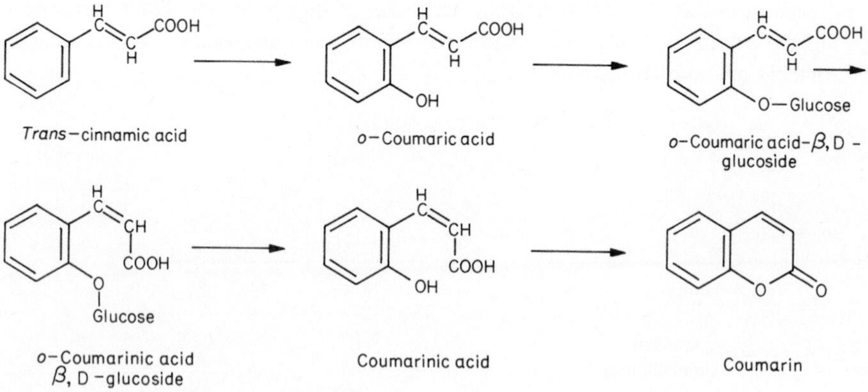

Trans—cinnamic acid   *o*—Coumaric acid   *o*—Coumaric acid–β, D – glucoside

o—Coumarinic acid β, D –glucoside   Coumarinic acid   Coumarin

Fig. 293. Formation of coumarin.

Direct synthesis of coumarinic acid appears to be possible to a small extent from *cis*-cinnamic acid. The latter probably originates spontaneously from *trans*-cinnamic acid by the action of UV-light.

The biosynthesis of herniarin and umbelliferone occurs in a somewhat modified form via glucosidized intermediates (fig. 294). The glucosides aesculin and scopolin on the contrary are not intermediates in the biosynthesis of aesculetin and scopoletin, but originate from these compounds by subsequent glycosylation.

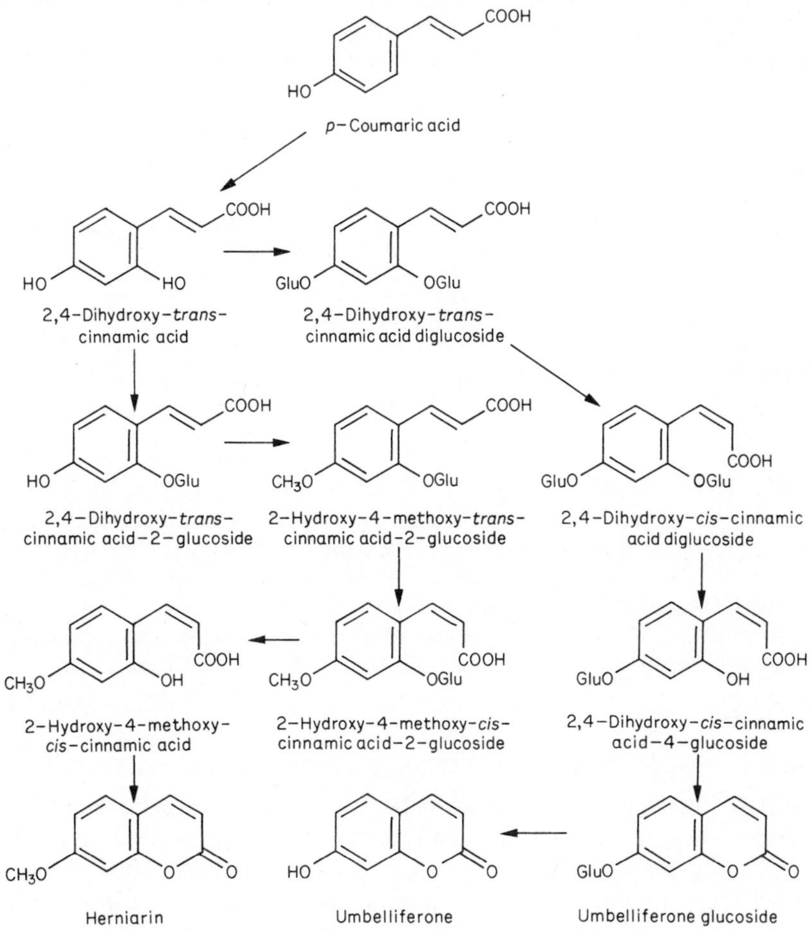

Fig. 294. Formation of umbelliferone and herniarin.

Coumarinic acid glucosides are exclusively present in fresh plants. Cleavage of glucose and formation of coumarins occurs on injury to the cells or on drying. Thus the plants woodruff (*Asperula odorata*), sweet clover (*Melilotus officinalis*) and *Anthoxanthum odoratum* develop the characteristic smell of coumarin only during drying.

The so-called dicoumarol which is the toxic factor of decaying sweet clover (*Melilotus officinalis*) may be formed by bacteria in dead plants containing coumarins. This compound inhibits the coagulation of blood.

The direct precursor of dicoumarol is 4-hydroxycoumarin, which is formed from *o*-coumaric acid via β-hydroxy melilotic acid and β-oxo melilotic acid (fig. 295). The carbon of the methylene bridge appears to originate from formaldehyde formed during decomposition of the plants. Biological one-carbon donors (e.g. methionine, serine and choline) are not effective precursors.

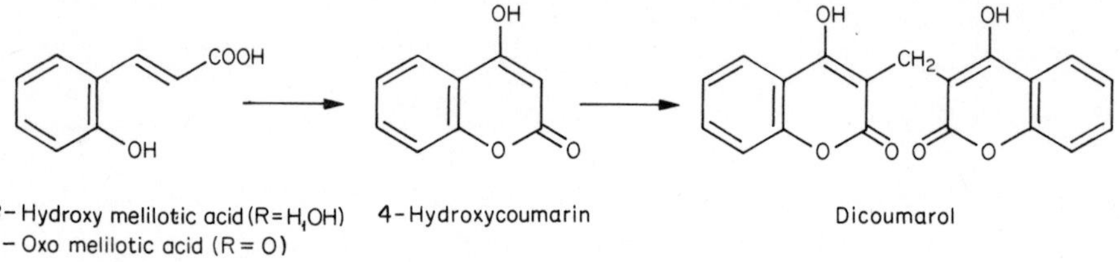

β-Hydroxy melilotic acid (R=H,OH) 4-Hydroxycoumarin   Dicoumarol
β-Oxo melilotic acid (R=O)

Fig. 295. Formation of dicoumarol.

References for further reading

Neish, A. C. 'Coumarins, Phenylpropanes, and Lignin', *Plant Biochemistry*, ed. J. Bonner and J. E. Varner (Academic Press, New York, 1965), pp. 581–617.

Brown, S. A. 'Biosynthesis of Coumarins', Biosynthesis of Aromatic Compounds, ed. G. Billek (Pergamon Press, Oxford, 1966), pp. 15–24.

Bye, A. and King, H. K. The Biosynthesis of β-Hydroxycoumarin and Dicoumarol by *Aspergillus fumigatus* Fresenius, *Biochem J.* **117** (1970), pp. 237–45.

20.4.3 Formation of cinnamic alcohol glucosides from cinnamic acids

Substituted cinnamic alcohols occur in large quantities in higher plants and are involved in the synthesis of lignin and lignans. They are all synthesized from the corresponding substituted cinnamic acids, probably in the form of the CoA-esters of these acids (cf. the reduction of benzoic acid derivatives to the corresponding benzyl alcohols, D.20.6). In metabolism the alcohols are glucosylated at the position *para* to a side chain and are thus converted to stable transport and storage forms. The most important naturally occurring cinnamic alcohols with their glucosides and the cinnamic acids from which they are derived are given in table 18.

References for further reading

See D.20.4.4.

Table 18. Biosynthesis of cinnamic alcohol glucosides important in the formation of lignin

R_1	R_2			
H	H	*trans-p*-Coumaric acid	*trans-p*-Coumaryl alcohol	*trans-p*-Glucocoumaryl alcohol
H	OCH_3	*trans*-Ferulic acid	*trans*-Coniferyl alcohol	D-Coniferin
OCH_3	OCH_3	*trans*-Sinapic acid	*trans*-Sinapyl alcohol	Syringin

20.4.4 Formation of lignin

Lignin, which is found in the walls of various cell types of Pteridophytes (fern-like plants) and Spermatophytes (seed-bearing plants), is a polymeric substance built up of phenylpropane units. Cell walls containing lignin are termed lignified, and the xylem in which lignified cells (e.g. vessels, tracheids and wood fibres) occur in large numbers is the woody portion of the plant. Lignin constitutes 20–30% of the total mass of raw wood.

In all the plants examined so far lignin is formed by polymerization of the cinnamic alcohol derivatives *trans-p*-coumaryl alcohol, *trans*-coniferyl alcohol and *trans*-sinapyl alcohol. These compounds are incorporated in different amounts into the lignin of individual plant species (cf. table 19). It seems likely that sinapyl alcohol and *p*-coumaryl alcohol became important as precursors of lignin at a later stage in the development of the plant kingdom, since coniferyl alcohol is the major precursor of lignin in those species which are considered to be more primitive.

Table 19. Precursors of lignin in various plants

Systematic unit	Precursors
Gymnosperms	Mostly *trans*-coniferyl alcohol but some *trans-p*-coumaryl alcohol and *trans*-sinapyl alcohol
Angiosperms	
Dicotyledons	Mostly *trans*-coniferyl alcohol and *trans*-sinapyl alcohol and less *trans-p*-coumaryl alcohol
Monocotyledons	*trans*-Coniferyl alcohol, *trans*-sinapyl alcohol and *trans-p*-coumaryl alcohol

The formation of lignin from the glucosides of these alcohols (serving as storage and transport forms) may be divided into five stages:

(a) The free alcohols are formed from the glucosides by the action of β-glucosidases.

(b) These are then attacked at the unsubstituted p-hydroxyl group by phenol oxidases (e.g. laccase, cf. C.2.3), thus giving rise to a single electron on the oxygen atom. The radical (a) thus formed remains in equilibrium with other radicals (b, c and d) which are formed by shift of electrons, and possess a quinoid structure. In these cases the single electron is located at a carbon atom which is either o- or p- to the oxidized hydroxyl group, or on the side chain in the β-position (fig. 296).

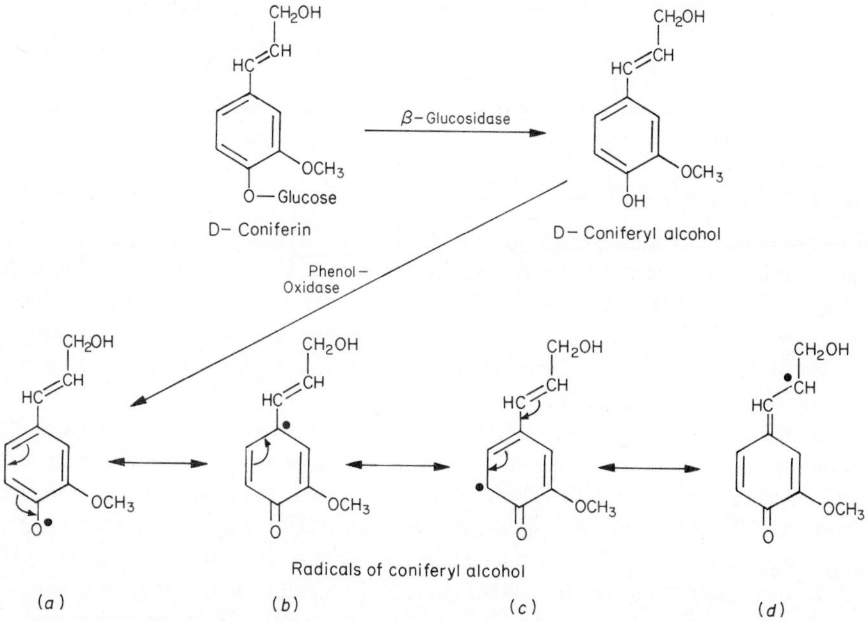

Fig. 296. Formation of the radicals of coniferyl alcohol.

(c) The radicals are capable of addition to unsaturated compounds such as cinnamic alcohols. Since their radical nature is retained (fig. 297), further unsaturated compounds may be added and a reaction chain is initiated which ultimately leads to the formation of high molecular weight substances.

The reactions taking place during polymerization have been investigated in vitro by experiments with laccase and coniferyl alcohol. By stopping the reaction after a short time, dimeric and trimeric derivatives of coniferyl alcohol could be isolated, while longer experimental times give

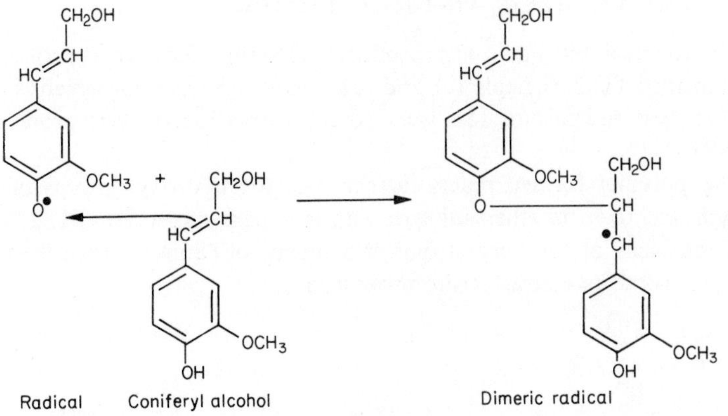

Fig. 297. Addition of an unsaturated compound to the radical of coniferyl alcohol.

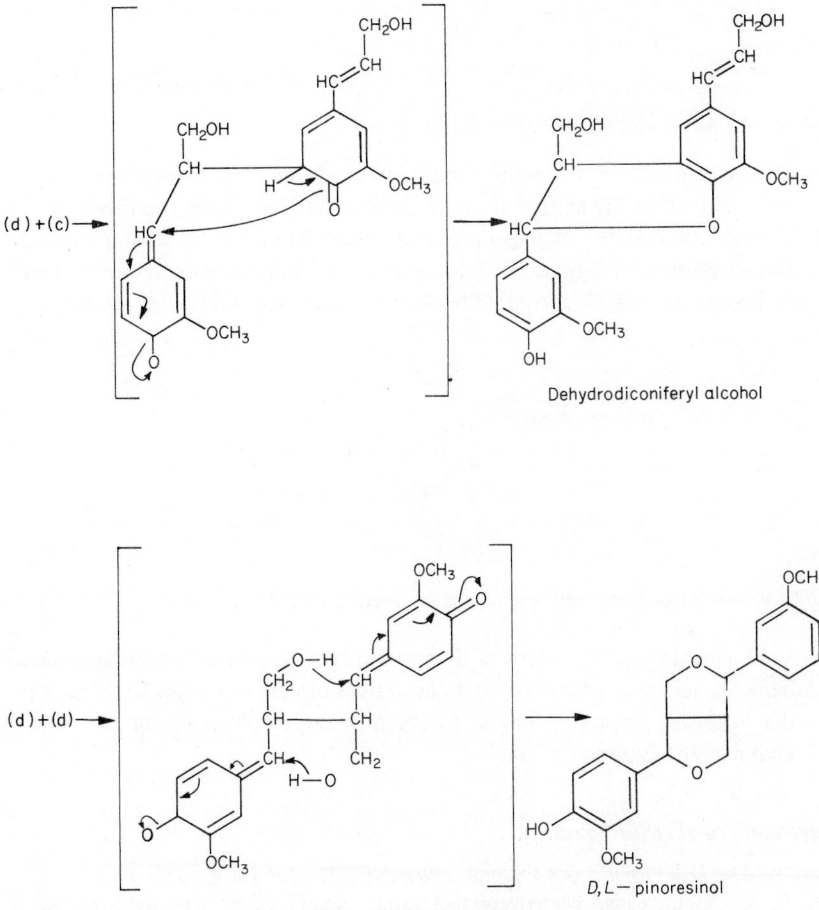

Dehydrodiconiferyl alcohol

D,L — pinoresinol

Fig. 298. Formation of dimeric derivatives of coniferyl alcohol *in vitro*.

rise to insoluble lignin-like products. Dehydrodiconiferyl alcohol which is formed from radicals (c) and (d), and D,L-pinoresinol which is formed from two radicals of (d), were found within the dimeric products (fig. 298).

(*d*) The polymer formed reacts further with the hydroxyl groups of carbohydrates, to form ether linkages with the sugar molecules (fig. 299). This is the basis of the very strong attachment of lignin to the cellulose, and gives wood its characteristic properties.

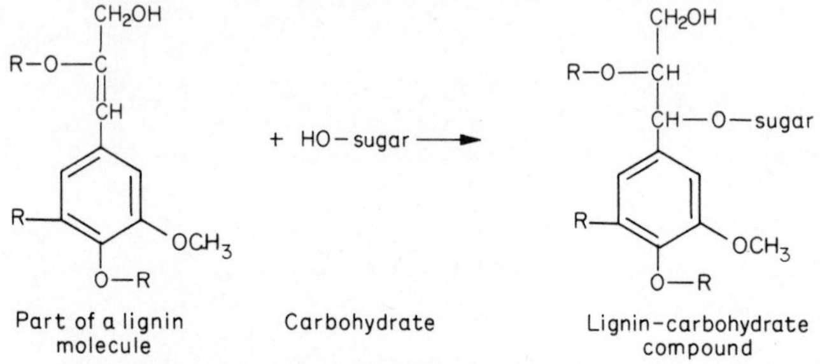

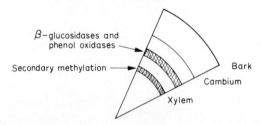

| Part of a lignin molecule | Carbohydrate | Lignin–carbohydrate compound |

Fig. 299. Linkage of lignin to carbohydrates.

(*e*) The last step is the secondary methylation of the lignin, which occurs after polymerization. That this is so in conifers is clear from the fact that the location of activity of β-glucosidases and phenol oxidases which initiate the formation of lignin, lies nearer the cambium than the zone in which radioactivity was found after feeding methionine-$^{14}CH_3$ (fig. 300).

Fig. 300. Zones of enzyme activity in a cross section of a branch of fir.

In certain plants, esters of lignin with hydroxylated and methoxylated benzoic acid and cinnamic acid derivatives are also formed. In these cases the acids are probably linked to the primary alcoholic group of the side chain of the lignin skeleton.

References for further reading

Schenk, W. Die Biosynthese des Lignins, *Pharmazie* **16** (1961), pp. 585–94.

Neish, A. C. 'Coumarins, Phenylpropanes, and Lignin', *Plant Biochemistry*, ed. J. Bonner and J. E. Varner (Academic Press, New York, 1965), pp. 581–617.

Isherwood, F. A. 'Biosynthesis of Lignin', *Biosynthetic Pathways in Higher Plants*, ed. J. B. Pridham and T. Swain (Academic Press, New York, 1965), pp. 133–46.

Harkin, J. M. Recent Developments in Lignin Chemistry, *Fortschr. chem. Forsch.* **6** (1966), pp. 101–58.

Brown, S. A. Lignins, *Ann. Rev. Plant Physiol.* **17** (1966), pp. 223–44.

Kratzl, K and Okabe, J. 'The Incorporation of *p*-Hydroxybenzoic Acid into Lignins', *Biosynthesis of Aromatic Compounds*, ed. G. Billek (Pergamon Press, Oxford, 1966), pp. 67–73.

Brown, B. R. 'Biochemical Aspects of Oxidative Coupling of Phenols', *Oxidative Coupling of Phenols*, ed. W. I. Taylor and A. R. Battersby (Marcel Dekker Inc., New York, 1967), pp. 167–201.

Harkin, J. M. 'Lignin—a Natural Polymeric Product of Phenol Oxidation', *Oxidative Coupling of Phenols*, ed. W. I. Taylor and A. R. Battersby (Marcel Dekker Inc., New York, 1967), pp. 243–321.

Nord, F. F. and Schubert, W. J. 'The Biogenesis of Lignins', *Biogenesis of Natural Compounds*, ed. P. Bernfeld (Pergamon Press, Oxford, 1967), pp. 903–40.

20.4.5 Formation of lignans

Lignans consist of two phenylpropane units which are linked to each other by a C–C bond between carbon atoms 2 and 2' of the side chain. Lignans are widespread in higher plants. Cubebin is found in large quantities in the unripe fruits

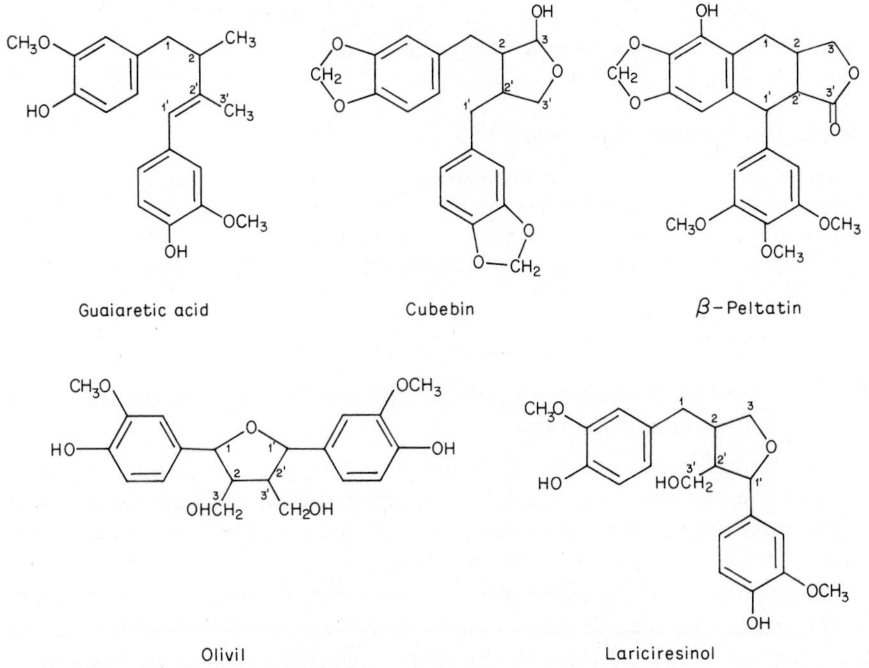

Guaiaretic acid   Cubebin   β-Peltatin

Olivil   Lariciresinol

Fig. 301. Naturally occurring lignans.

of *Piper cubeba*, olivil constitutes 50% of the resinous excretions of the olive tree (*Olea europaea*) and guaiaretic acid is an important constituent of guaiacum resin (obtained from *Guajacum officinale*). The formulae of a few important lignan derivatives are given in fig. 301.

Little is known about the biosynthesis of these and other compounds of similar structure. However, from their structure it may be concluded that they originate from two molecules of a phenylpropane alcohol or one molecule of a phenylpropane alcohol and a molecule of a phenyl proprionic acid. Since the lignans usually possess L-configuration at carbon atom 2 and D-configuration at carbon atom 2′, an enzymatic coupling the two phenylpropane units must probably have taken place. The coupling most probably takes place reductively, since higher polymers do not occur here as is the case during the formation of lignin, where radical polymerization takes place after oxidation (fig. 302).

Lignans which have additional rings in the molecule (e.g. cubebin, peltatin, olivil and lariciresinol) may originate by further reactions from products formed in the above manner.

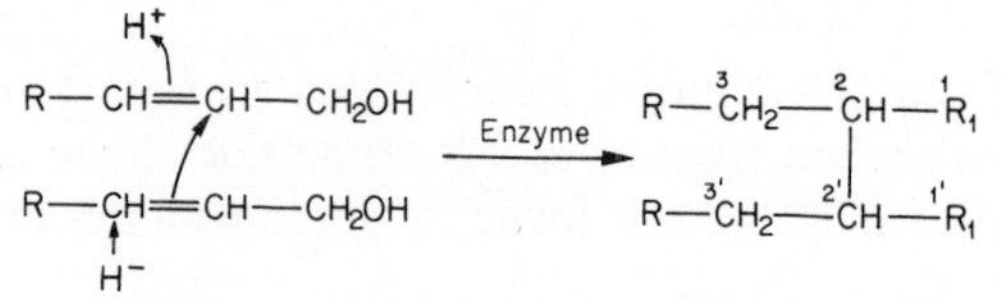

Basic unit of lignans

Fig. 302. Reductive dimerization of phenylpropane units.

References for further reading

Neish, A. C. 'Coumarins, Phenylpropanes, and Lignin', *Plant Biochemistry*, ed. J. Bonner and J. E. Varner (Academic Press, New York, 1965), pp. 581–617.

Weinges, K. and Spänig, R. 'Lignans and Cyclolignans', *Oxidative Couling of Phenols*, ed. W. I. Taylor and A. R. Battersby (Marcel Dekker Inc., New York, 1967), pp. 323–55.

20.5 Formation of secondary natural products from cinnamic acid derivatives and acetate

A large number of natural products originate from one molecule of cinnamic acid and one, two or three molecules of acetate. Particularly widespread are those compounds formed with the participation of three acetate units, such as the stilbene derivatives, flavonoids, aurones, isoflavones and 3- or 4-phenyl coumarins.

The naturally occurring derivatives of monocyclic α-pyrones such as para-cotoin and kawain (fig. 303) may be considered to be condensation products of one cinnamic acid molecule with one or two molecules of acetate. No experiments, however, have been carried out as yet to prove this assumption.

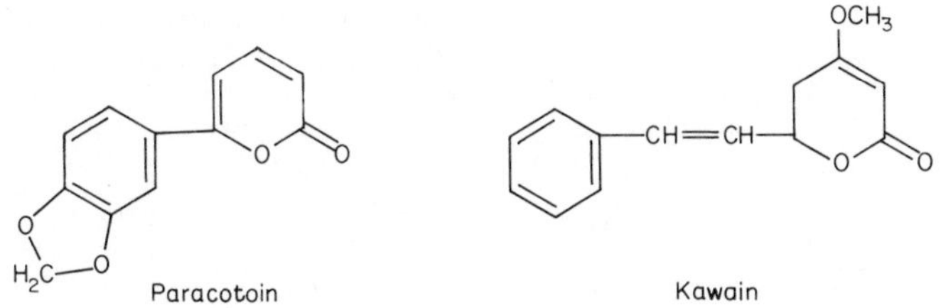

Paracotoin   Kawain

Fig. 303. The structural formulae of paracotoin and kawain.

20.5.1 Biosynthesis of stilbene derivatives

Stilbene derivatives are a small group of secondary plant products which occur in various families. They are characterized by the fact that ring A is almost always substituted by two *m*-hydroxyl groups while ring B shows the presence of a substitution pattern characteristic for cinnamic acids.

It has been shown by feeding experiments in various plants that the stilbenes originate from three molecules of acetate and one molecule of a cinnamic acid. The substitution pattern of ring B is thus determined by the particular cinnamic acid involved. A few important stilbene derivatives and the cinnamic acids serving in their biosynthesis are shown in table 20.

Table 20. Formation of stilbenes from cinnamic acid derivatives

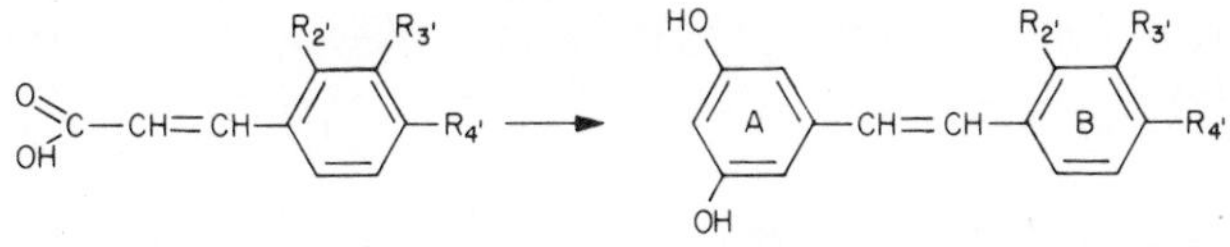

$R_{2'}$	$R_{3'}$	$R_{4'}$	Cinnamic acid	Stilbene
H	H	H	Cinnamic acid	Pinosylvin
H	H	OH	*p*-Coumaric acid	Resveratrol
OH	H	OH	2,4-Dihydroxy cinnamic acid	Hydroxyresveratrol
H	OH	OH	Caffeic acid	Piceatannol
H	OH	OCH_3	Isoferulic acid	Rhapontigenin

The actual mechanism of synthesis is, however, still unknown. It is supposed that the CoA derivatives of the acids react with each other by head to tail condensation. Acetyl CoA is probably first converted to malonyl CoA as in the case of biosynthesis of fatty acids. During the condensation an intermediate made up of fifteen carbon atoms may be formed. This, like intermediates formed during the synthesis of anthracene and the tetracycline ring systems (cf. D.2.4.1 and D.2.4.2), is unstable as a polyketo compound and possibly only exists bound to

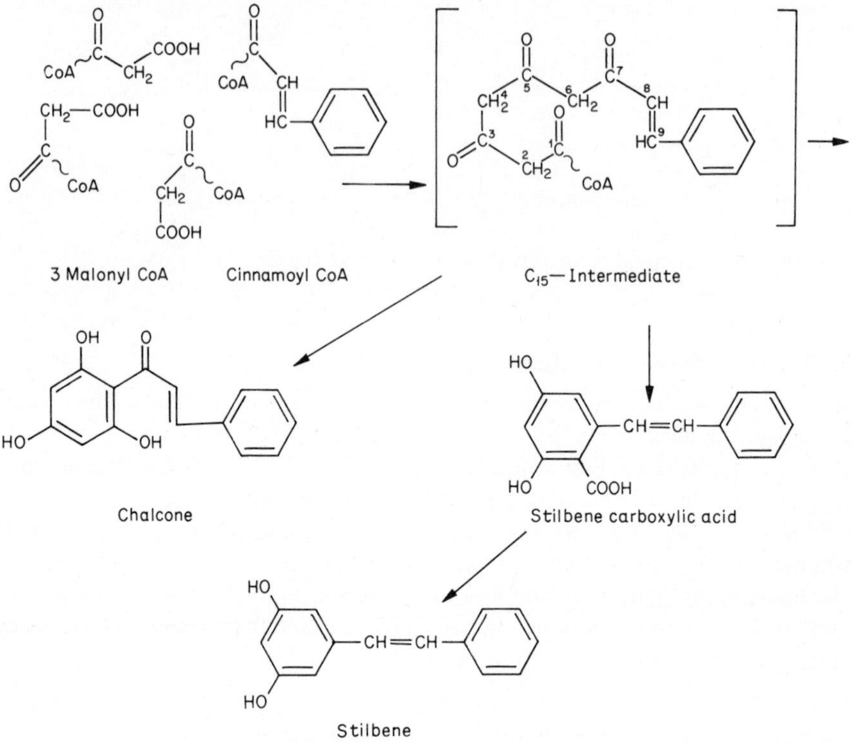

Fig. 304. Formation of stilbene derivatives and chalcones from acetyl CoA and cinnamoyl CoA.

an enzyme. It may be capable of cyclizing to a stilbene carboxylic acid by form-ming a bond between carbon atoms 2 and 7 or to a chalcone (cf. D.20.5.2) by the formation of a bond between carbon atoms 1 and 6 (fig. 304).

The naturally occurring stilbenes may be formed from stilbene carboxylic acid by decarboxylation, but the carboxyl group is retained as in hydrangic acid. This compound is a precursor of hydrangenol in *Hydrangea macrophylla* (fig. 305).

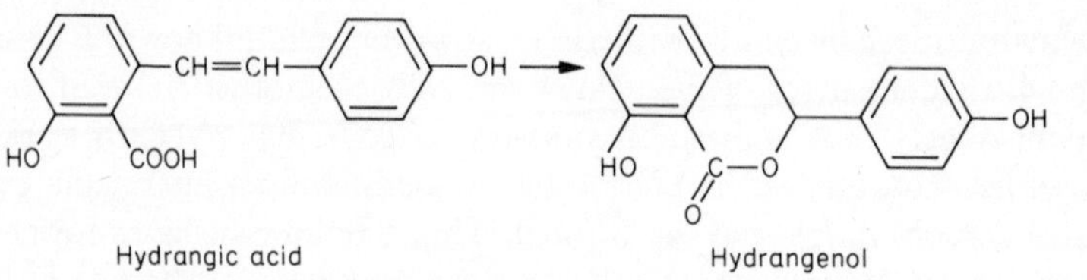

Fig. 305. Biosynthesis of hydrangenol from hydrangic acid.

References for further reading

Billek, G. and Schimpl, A. 'Biosynthesis of Plant Stilbenes', *Biosynthesis of Aromatic Compounds*, ed. G. Billek (Pergamon Press, Oxford, 1966), pp. 37–44.

Billek, G. Biosynthese aromatischer Systeme, *Angew. Chem.* **79** (1967), p. 586.

20.5.2 Biosynthesis of compounds with flavan structure

Substances derived from 2-phenylchroman are termed flavans, or flavonoid compounds. A large number of secondary products synthesized by higher plants belong to this group. Flavans make their first appearance in the plant kingdom in ferns and mosses. The flavonoids found in the wings of the butterflies are derived from compounds ingested by these insects. The most important types of naturally occurring flavonoid derivatives are given in fig. 306 (the substituents in rings A and C are not considered here).

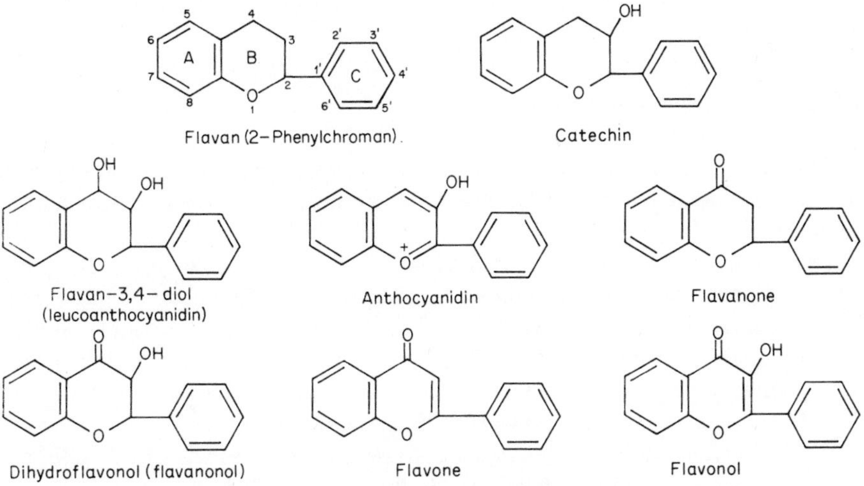

Fig. 306. Various types of naturally occurring flavan derivatives.

It was established during structural elucidation of the individual representatives of this group of substances that certain regularities in the substitution pattern exist. With few exceptions the oxygen-containing substituents are arranged at the *m*-position of ring A, while the substitution pattern in ring C resembles that of the cinnamic acid derivatives. As is known today this is due to the fact that ring A originates from three molecules of acetate by head to tail condensations, while the rest of the carbon atoms originate from a cinnamic acid. It is thought that the synthesis proceeds via an intermediate consisting of fifteen carbon atoms as shown in fig. 304 (where it is shown cyclizing to a chalcone) (cf. D.20.5.1).

In aqueous solution, chalcones and flavanones remain in an equilibrium which is shifted in favour of the flavanone if ring A of the chalcone carries a hydroxyl group at positions 2 or 6. Strong hydrogen bonding to the carbonyl group is then possible so that the side chain is so fixed that addition of the hydroxyl group to the double bond can proceed easily. The adjustment of this equilibrium between a chalcone and a flavanone is catalysed by an enzyme. Naturally occurring flavanones are optically active at the carbon atom marked '•' and have the 2S configuration (fig. 308).

Conversion to a flavanone as described above is impossible after reduction of the double bond of the side chain, which would result in the formation of a dihydrochalcone. Thus phloridzin is formed by reduction of the corresponding substituted chalcone in *Malus* species (fig. 307).

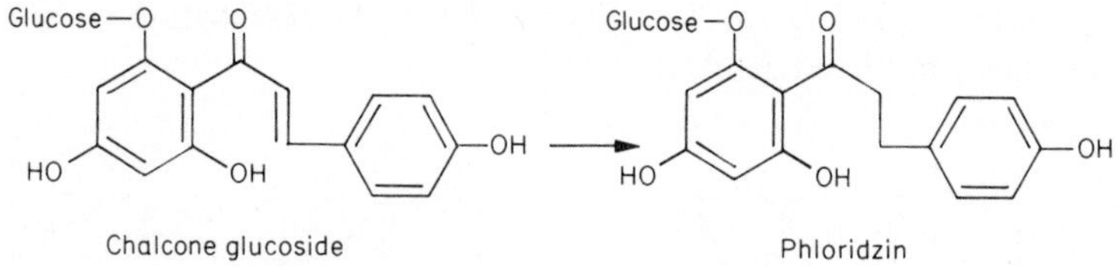

Fig. 307. Formation of phloridzin.

The flavanones may probably be regarded as precursors for the other flavan derivatives outlined in fig. 308. They are converted to flavones by dehydrogenation.

Two pathways are possible for the formation of dihydroflavonols. One pathway is thought to involve the direct hydroxylation of flavanone at position 3, while the second pathway predicts the addition of oxygen to the double bond of the side chain of a chalcone with the formation of an epoxide ring from which dihydroflavonol can then originate. Flavonols may be formed from this by dehydrogenation while reduction yields flavan-3,4-diols.

Anthocyanidins and catechins appear to be the end products of two different metabolic pathways starting from dihydroflavonol (fig. 308).

As already mentioned, the substitution pattern in ring C of the flavan derivatives is usually determined by the cinnamic acid involved in the biosynthesis (cinnamic acid initiator theory), and there are specific enzymes for the incorporation of the particular cinnamic acid. In a few cases, however, subsequent hydroxylation of the exogenously supplied chalcone is possible. According to recent investigations, however, it is questionable whether the latter findings have any importance under normal conditions and whether they can be generalized (substitution theory).

The last step in flavonoid biosynthesis is usually a glycosylation. It has been shown by genetic experiments and by isolation of the corresponding enzymes that in those compounds which contain several sugar molecules, these sugars are added one at a time. In the case of *Streptocarpus hybrida*, five genes are

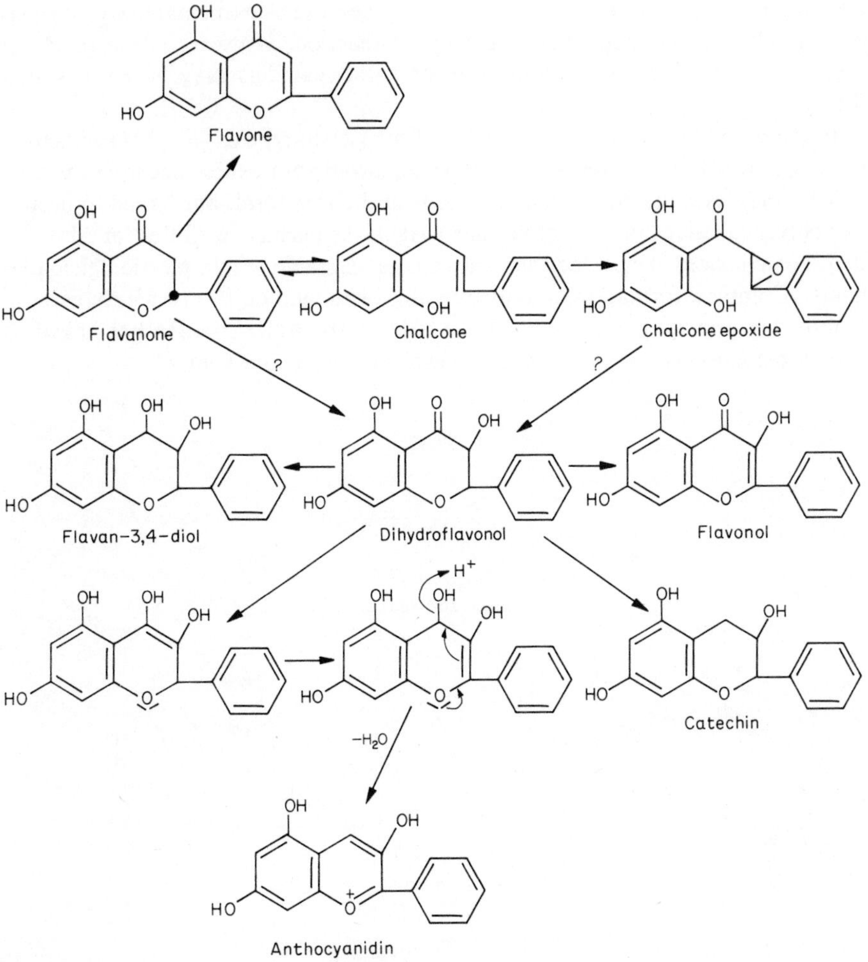

Fig. 308. Possible pathway of formation of various flavan derivatives from chalcones.

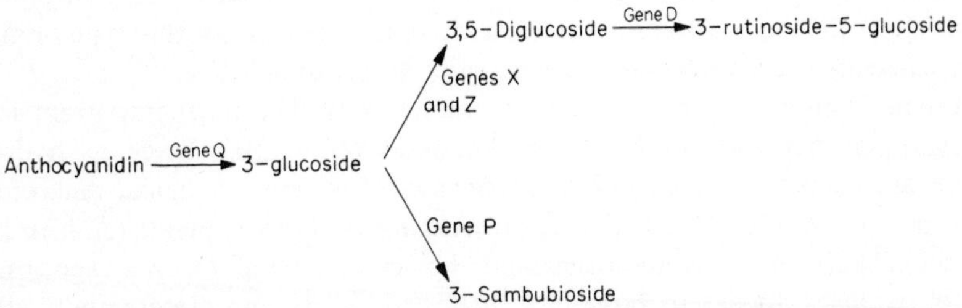

Fig. 309. Important genes for the formation of anthocyanidin glycosides in *Streptocarpus hybrida.*

known which code for the formation of enzymes which participate in glycosylation. Each enzyme catalyses a particular reaction. However, their substrate specificity is not very great and different anthocyanidins may be glycosylated (fig. 309).

In the case of C-glycosyl compounds of the vitexin type (cf. fig. 311) an activated sugar (cf. D.1.1.1) reacts with a carbon atom with excess negative charge.

The anthocyanins do not usually occur in the free form, but linked to pectin-like polysaccharides of high molecular weight. As chelates with Fe^{3+} or Al^{3+} ions they are coloured deep blue, and this colour is stable within physiological pH-limits. Complex formation is only possible with anthocyanins which carry a hydroxyl group at positions 3' and 4' (fig. 310). Anthocyanins not linked as complexes exist as red oxonium salts at physiological pH-values.

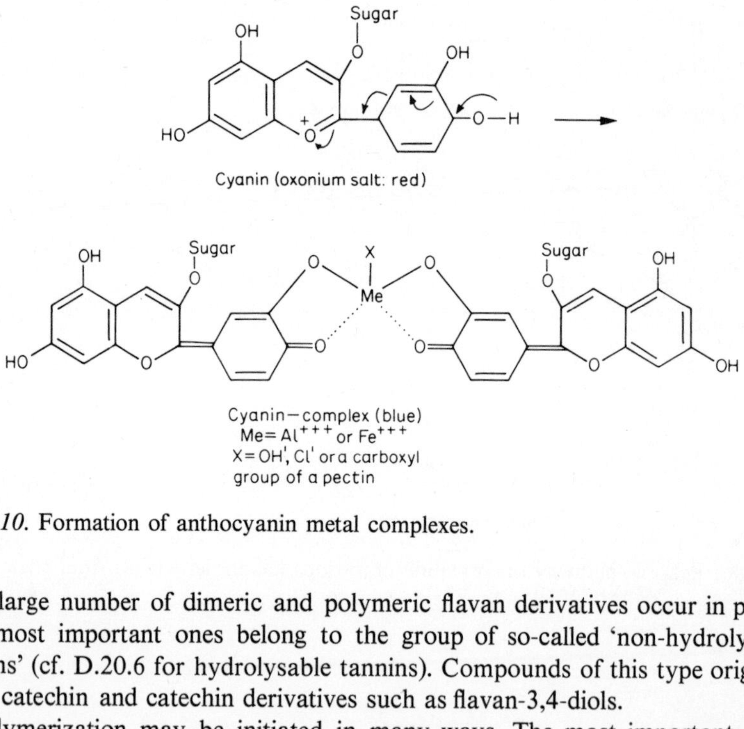

Cyanin (oxonium salt: red)

Cyanin—complex (blue)
Me= Al^{+++} or Fe^{+++}
X= OH', Cl' or a carboxyl
group of a pectin

Fig. 310. Formation of anthocyanin metal complexes.

A large number of dimeric and polymeric flavan derivatives occur in plants. The most important ones belong to the group of so-called 'non-hydrolysable tannins' (cf. D.20.6 for hydrolysable tannins). Compounds of this type originate from catechin and catechin derivatives such as flavan-3,4-diols.

Polymerization may be initiated in many ways. The most important is the formation of free radicals by phenol oxidases (cf. C.2.3). Since all hydroxyl groups are usually attacked with about the same frequency, irregular macromolecules are formed (fig. 311, I) during polymerization in most plants (cf. formation of lignin, D.20.4.4). In some organisms, however, such as *Quercus* species, the phenol oxidases appear to possess a higher specificity and preferentially attack the hydroxyl groups of ring C of catechin derivatives. Less highly branched molecules are formed in these cases (fig. 311, II). Only water-soluble polymerization products can function as tanning substances. The high molecular weight, insoluble substances (phlobaphenes) are of no use in this process.

Besides tannins, dimeric or polymeric flavone derivatives occur in some plants. Amentoflavone (fig. 311) is one such compound. These compounds also probably originate through phenol oxidase action.

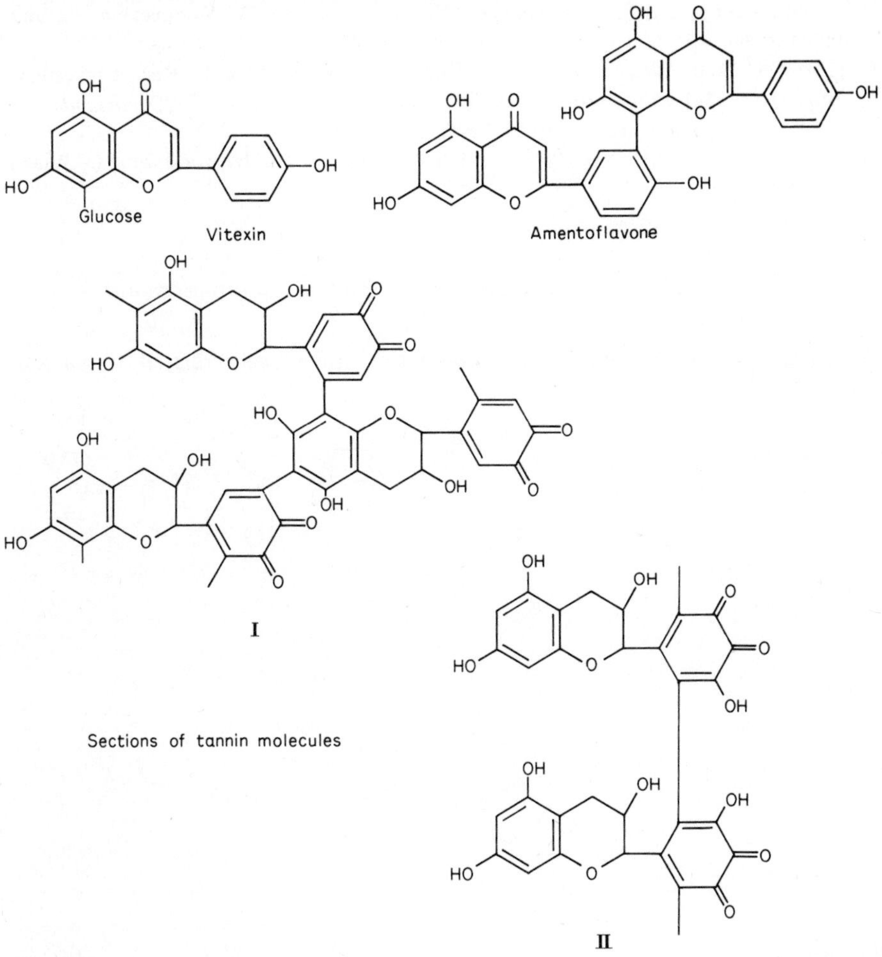

Vitexin

Amentoflavone

I

Sections of tannin molecules

II

Fig. 311.

References for further reading

Swain, T. 'The Tannins', *Plant Biochemistry*, ed. J. Bonner and J. E. Varner (Academic Press, New York, 1965), pp. 552–80.

Grisebach, H. 'Anthocyanidins and Flavonoids, A Summary', *Biosynthetic Pathways in Higher Plants*, ed. J. B. Pridham and T. Swain (Academic Press, London, 1965), pp. 159–61.

Williams, A. H. 'Dihydrochalcones', *Comparative Phytochemistry*, ed. T. Swain (Academic Press, London, 1966), pp. 297–307.

Wagner, H. 'Flavonoid *C*-Glycosides', *Comparative Phytochemistry*, ed. T. Swain (Academic Press, London, 1966), pp. 309–20.

Humphries, S. G. 'The Biosynthesis of Tannins', *Biogenesis of Natural Compounds*, ed. P. Bernfeld (Pergamon Press, Oxford, 1967), pp. 801–27.

Bayer, E., Egerer, H., Fink, A., Nether, K. and Wegmann, K. Komplexbildung und Blütenfarben, *Angew. Chem.* **78** (1966), pp. 834–41.

Grisebach, H., Barz, W., Hahlbrock, K., Kellner, S. and Patschke, L. 'Recent Investigations on the Biosynthesis of Flavonoids', *Biosynthesis of Aromatic Compounds*, ed. G. Billek (Pergamon Press, Oxford, 1966), pp. 25–36.

Grisebach, H. *Biosynthetic Patterns in Microorganisms and Higher Plants* (J. Wiley, New York, 1967).

20.5.3 Formation of aurones, isoflavones, 3-phenylcoumarins and rotenones

Aurones, isoflavones and 3-phenylcoumarins are closely related to flavans (fig. 312).

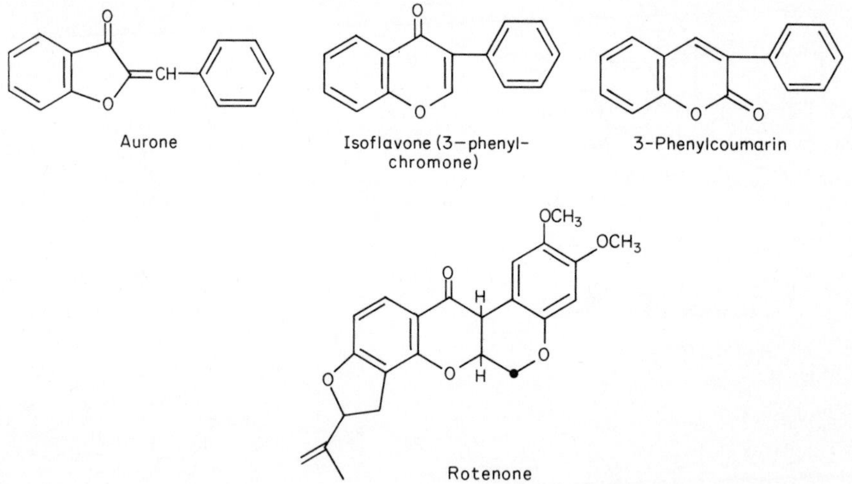

Aurone Isoflavone (3–phenyl-chromone) 3-Phenylcoumarin

Rotenone

Fig. 312.

The aurones are usually found together with the corresponding substituted chalcone derivatives and are easily derived from them *in vitro*. Enzymes which catalyse the conversion of chalcones to aurones have also been detected. Thus butein may be converted to sulphuretin as shown in fig. 313.

The isoflavones have until the present time only been detected in the Leguminosae. Their formation takes place by the shift of a phenyl group from carbon atom 2 to carbon atom 3 of the flavan ring system (fig. 314). Since flavanones

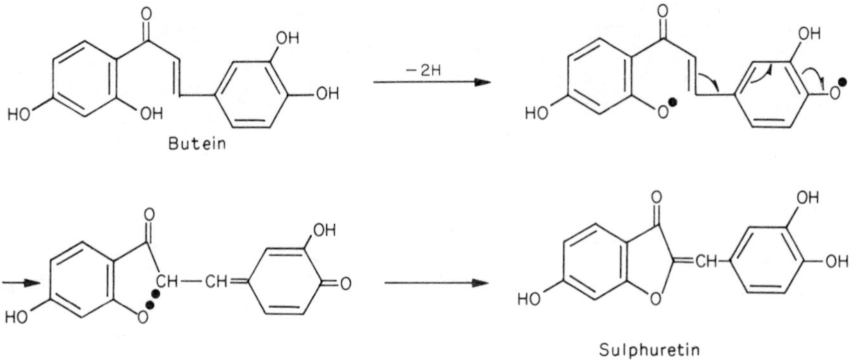

Butein

−2H

Sulphuretin

Fig. 313. Possible intermediates in the conversion of butein to sulphuretin.

are good precursors of isoflavones, the shift probably takes place at this stage of the biosynthetic pathway. Its exact mechanism is, however, unknown.

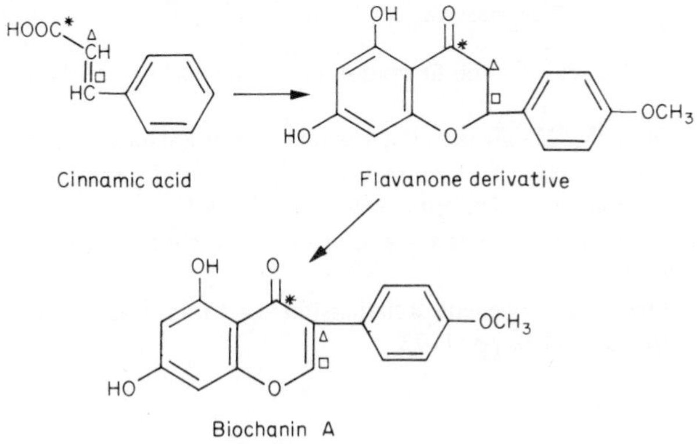

Cinnamic acid

Flavanone derivative

Biochanin A

Fig. 314. Formation of isoflavonoids.

3-Phenylcoumarins may be synthesized from isoflavones. Thus daidzein, which occurs together with coumoestrol in alfalfa, is a precursor for the latter compound. The pathway outlined in fig. 315 is proposed for this conversion.

Rotenones also possibly originate from isoflavones. Both these types of compounds occur together in *Derris elliptica*. The carbon atom marked '•' in the rotenone molecule (fig. 312) probably originates from a methyl group in a manner similar to that described for the formation of berberine (cf. D.20.1.2).

References for further reading

Grisebach, H. 'Biosynthesis of Flavonoids', *Chemistry and Biochemistry of Plant Pigments*, ed. T. W. Goodwin (Academic Press, London, 1965), pp. 279–308.

Swain, T. 'Nature and Properties of Flavonoids', *Chemistry and Biochemistry of Plant Pigments*, ed. T. W. Goodwin (Academic Press, London, 1965), pp. 211–45.

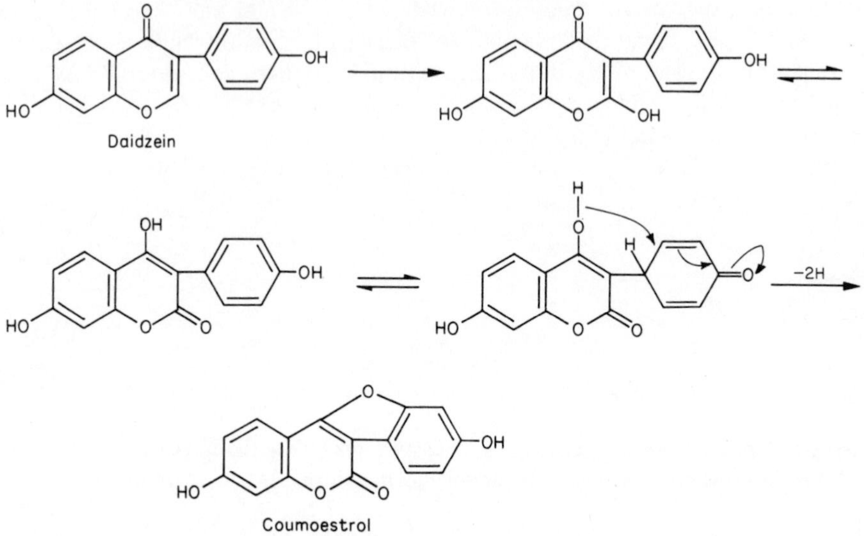

Fig. 315. Possible pathway for the formation of coumoestrol from daidzein.

Grisebach, H., Barz, W., Hahlbrock, K., Kellner, S. and Patschke, L. 'Recent Investigations on the Biosynthesis of Flavonoids', *Biosynthesis of Aromatic Compounds*, ed. G. Billek (Pergamon Press, Oxford, 1966), pp. 25–36.

Grisebach, H. 'Biosynthetic Patterns', in *Microorganisms and Higher Plants* (J. Wiley, New York, 1967).

Wong, E. Structural and Biogenetic Relationships of Isoflavonoids, *Fortschritte Chem. org. Naturstoffe* **28** (1970), pp. 1–73.

20.5.4 Biosynthesis of neoflavonoids (4-phenylcoumarins)

As well as the 3-phenylcoumarins described in the previous section other coumarins are found which have a phenyl group at position 4. Calophyllolid, which

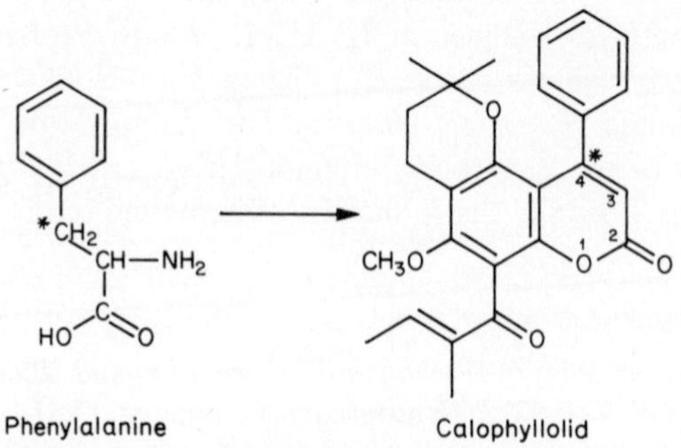

Fig. 316. Incorporation of phenylalanine-3-^{14}C into calophyllolid.

occurs in *Calophyllum inophyllum*, belongs to this group. The question of whether these substances originate by means of a double shift of the phenyl group from flavans or by direct condensation has been settled in favour of the latter possibility. The incorporation of phenylalanine-3-^{14}C gave calophyllolid which was specifically labelled at position 4 (fig. 316). Thus in the case of the 4-phenylcoumarins a new type of condensation between phenylpropane units and acetate is seen.

References for further reading

Kunesch, G. and Polonsky, J. On the Biosynthesis of Neoflavonoids: Calophyllolid (4-Phenylcoumarin), *Chem. Commun.* (1967), pp. 317–18.

20.6 Shortening of the side chain of cinnamic acid and phenylethylamine derivatives with the formation of C_6C_2-, C_6C_1- and C_6-bodies

Nitrogen-free derivatives of phenylethane occur quite rarely in nature (in contrast to phenylethylamines, cf. D.20.1.1). The most important are the benzophenone derivatives, picein and pungenin isolated from needles of conifers, and phenylacetic acid which is formed in animal organisms during the degradation of phenylalanine either by decarboxylation of phenylpyruvic acid (cf. C.3.4) or by the transamination of phenylethylamine (cf. C.4) via phenylacetaldehyde. Alcohols, aldehydes and acids with a phenyl ring and a one-carbon side chain have, however, been shown to occur frequently in micro-organisms, higher plants and animals. In most cases they are found as esters or glycosides, but in the urine of mammals they are present in the free form. C_6-bodies which are derived from phenylpropane derivatives, e.g., hydroquinone and substances derived from it like the glycosides arbutin and methyl arbutin, are synthesized in large amounts by Ericaceae species among others.

(a) β-Oxidation of cinnamic acid derivatives

The most important biosynthetic pathway for these compounds starts from cinnamic acid or its derivatives. The pathway has been detected in micro-organisms as well as in higher plants and animals. The shortening of the side chain takes place by a mechanism resembling that of the oxidation of fatty acids (cf. D.2.3.3) (fig. 317). After activation of the particular cinnamic acid to form the CoA-ester, water may be added to the double bond and the β-hydroxy acid that is formed may be oxidized to the β-keto acid. The benzophenone derivatives may be formed therefrom by the elimination of the carboxyl group. This mechanism has been made probable in the case of pungenin by feeding the corresponding labelled compounds.

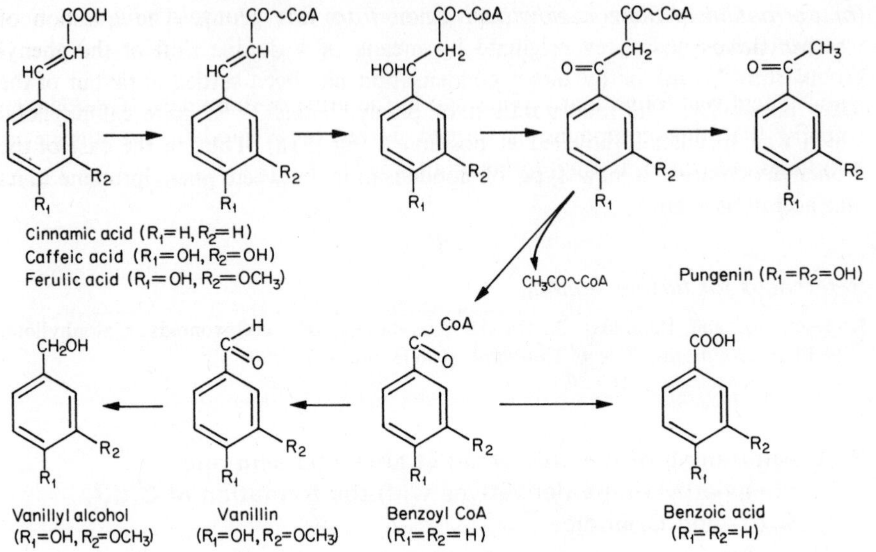

Cinnamic acid (R₁=H, R₂=H)
Caffeic acid (R₁=OH, R₂=OH)
Ferulic acid (R₁=OH, R₂=OCH₃)

Pungenin (R₁=R₂=OH)

Vanillyl alcohol
(R₁=OH, R₂=OCH₃)

Vanillin
(R₁=OH, R₂=OCH₃)

Benzoyl CoA
(R₁=R₂=H)

Benzoic acid
(R₁=R₂=H)

Fig. 317. Shortening of the side chain of cinnamic acid derivatives.

The β-keto acid which is formed may be converted by thiolytic elimination of acetyl CoA to a CoA-ester of benzoic acid or the substituted benzoic acids. The free acids may be formed by hydrolytic degradation. It is thought that the benzoic acid derivatives shown in table 21 are formed in this way from the corresponding cinnamic acid derivatives. Hydroxylation and subsequent methylation of benzoic acid derivatives appear to be of greater importance in microorganisms only.

Table 21. Conversion of cinnamic acid to benzoic acid derivatives

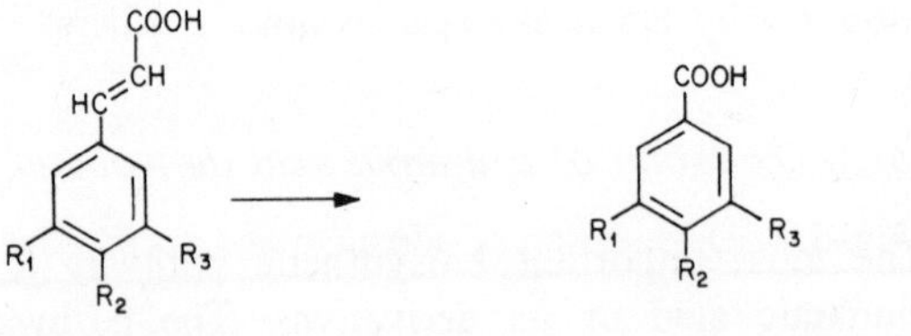

R_1	R_2	R_3		
H	H	H	Cinnamic acid	Benzoic acid
H	OH	H	p-Coumaric acid	p-Hydroxybenzoic acid
OH	OH	H	Caffeic acid	Protocatechuic acid
CH₃	OH	H	Ferulic acid	Vanillic acid
CH₃	OH	CH₃	Sinapic acid	Syringic acid
OH	OH	OH	3,4,5-Trihydroxycinnamic acid	Gallic acid

(b) Formation of benzoic acid derivatives from phenylethylamine derivatives

Vanillic acid was found many years ago in the urine of mammals. It was shown recently that this compound is formed during the degradation of the tissue hormones, adrenalin and noradrenalin (fig. 318).

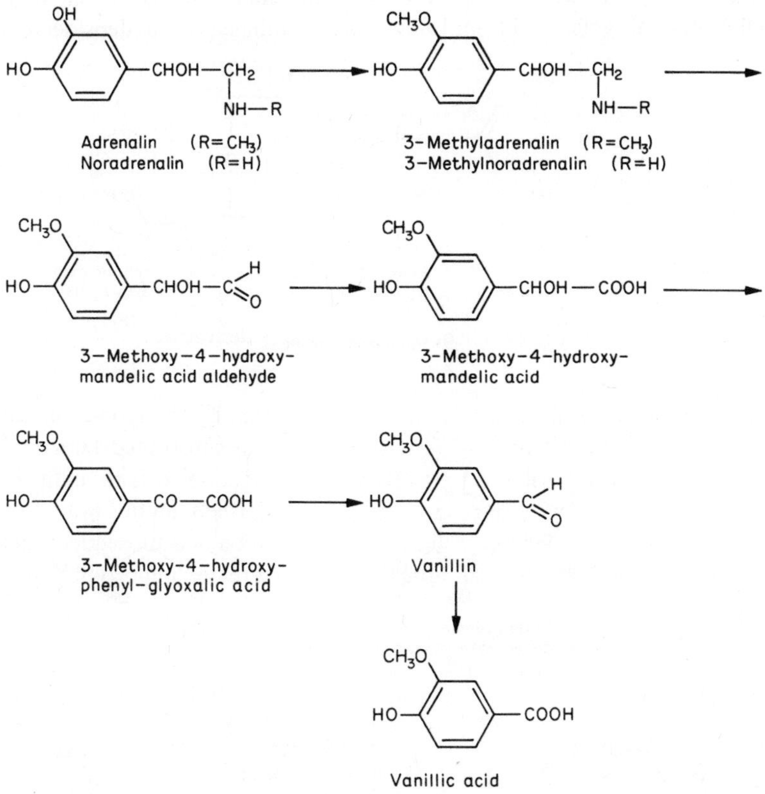

Adrenalin (R=CH$_3$)
Noradrenalin (R=H)

3-Methyladrenalin (R=CH$_3$)
3-Methylnoradrenalin (R=H)

3−Methoxy−4−hydroxy−mandelic acid aldehyde

3-Methoxy-4-hydroxy-mandelic acid

3−Methoxy-4−hydroxy−phenyl−glyoxalic acid

Vanillin

Vanillic acid

Fig. 318. Degradation of adrenalin and noradrenalin to vanillic acid.

The conversion is initiated by methylation at the *m*-hydroxyl group. Then the methylamino group or the amino group is removed by an amine oxidase (cf. C.2.1.1 and D.7.3). Oxidation of the aldehyde formed leads to 3-methoxy-5-hydroxymandelic acid (vanillylmandelic acid). This compound may be oxidized to the corresponding α-keto acid by a dehydrogenase, and is converted to vanillin by decarboxylation (cf. C.3.4). Vanillic acid, which is excreted as the end product of the degradation pathway, is formed by oxidation of the latter compound.

(c) Conversion of the synthesized benzoic acid derivatives into other secondary natural products

Formation of hydrolysable tannins.

Hydrolysable tannins are formed from gallic acid and a sugar, usually glucose. Unlike the non-hydrolysable tannins (cf. D.20.5.2) they may be degraded to their constituents by heating with acids. Groups of these compounds, differing only in the number of gallic acid groups linked to the sugar, usually occur together. Normally several gallic acid molecules are combined as a depside to form

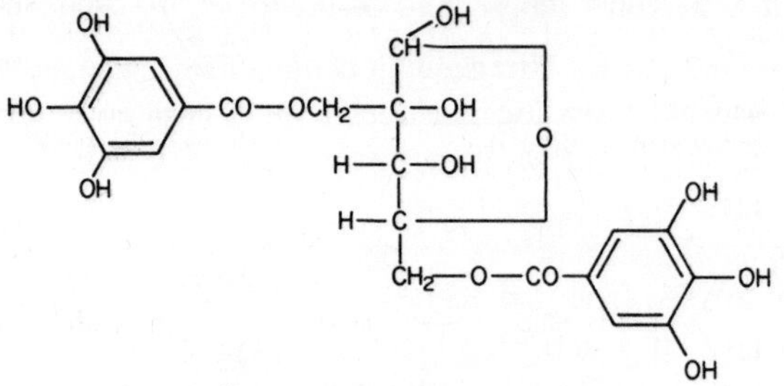

β-Hamamelistannin

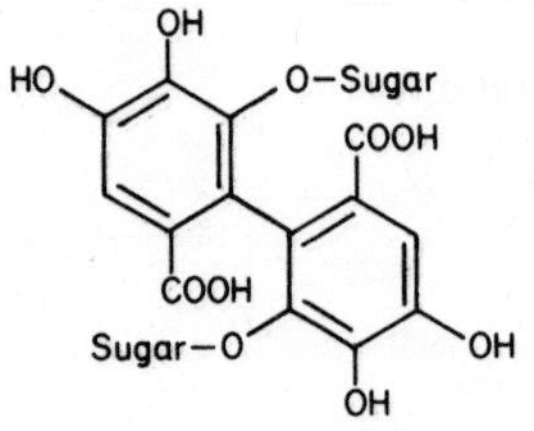

Ellagictannin
(hexahydroxy diphenic acid
diglucotide)

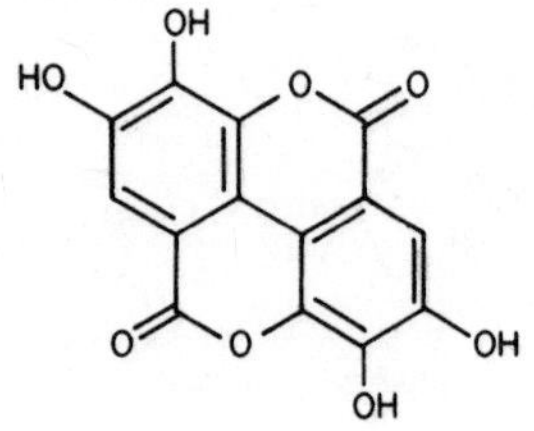

Ellagic acid

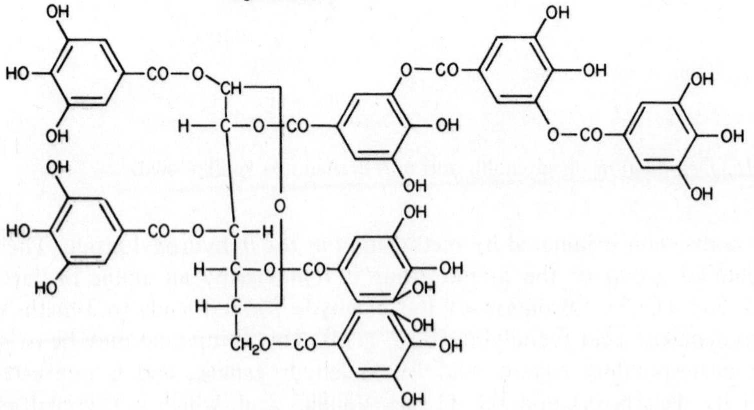

1,3,4,6-Tetra-O-galloyl-2-O-trigalloyl-glucose

Fig. 319. Formulae of some hydrolysable tannins.

digallic acid, trigallic acid and possibly higher molecular weight compounds. Carbon–carbon bonds, formed through phenol oxidase action (cf. C.2.3) between the gallic acid groups, are also found as, for example, in the case of ellagic tannins derived from hexahydroxydiphenic acid. Some of the hydrolysable tannins found in higher plants are shown in fig. 319. Ellagic acid probably does not occur free in living organisms, or else occurs in very minute quantities. It is formed from ellagic tannins by spontaneous ring closures after elimination of the sugar groupings.

Reduction of activated benzoic acid derivatives.

The starting material for the reduction of benzoic acid derivatives are usually the CoA-esters. They are formed during the β-oxidation of cinnamic acids (see above). The activation of the free acid is also possible, at least in the case of micro-organisms.

Reduction to the aldehyde is coupled with the elimination of the CoA group. Alcohols are formed from the corresponding aldehydes by the action of alcohol dehydrogenases. Thus salicin, which occurs in *Salix* species, is formed from *o*-coumaric acid via salicylaldehyde. After glucosylation of the hydroxyl group (to form helicin), reduction to salicin takes place (fig. 320).

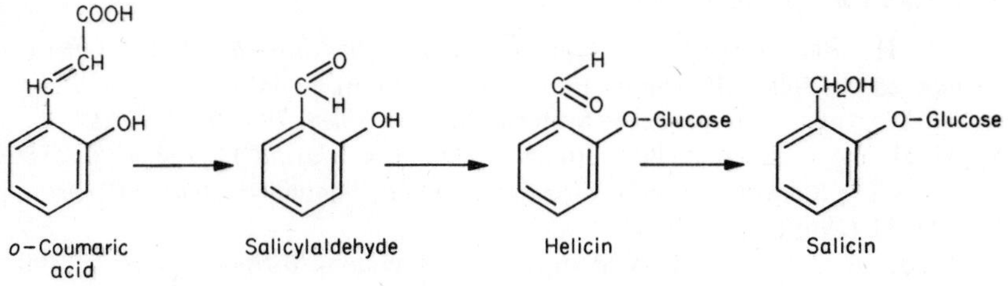

Fig. 320. Conversion of *o*-coumaric acid to salicin.

Oxidative decarboxylation of benzoic acid derivatives.

Arbutin and methyl arbutin are synthesized by higher plants from *p*-hydroxybenzoic acid. It is thought that this compound is oxidatively decarboxylated and that the hydroquinone thus formed is converted to arbutin (fig. 321).

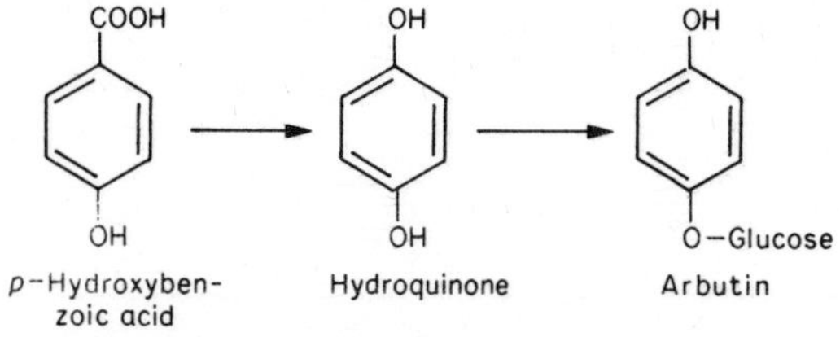

Fig. 321. Formation of arbutin from *p*-hydroxybenzoic acid.

The hydroquinone found in the defensive secretions of cockroaches is probably formed in this manner.

The secretion containing the hydroquinone may be mixed with one containing hydrogen peroxide in a special chamber of the secreting gland, thus forming p-quinone and oxygen. During this reaction the mixture becomes heated to 100°C, when the oxygen blows the boiling liquid out of the gland. The secretion is very corrosive, so it terrifies the enemy and injures the aggressor.

2-Methoxyhydroquinone is formed in higher plants by oxidative decarboxylation of vanillic acid. 2,6-Dimethoxyhydroquinone is synthesized from syringic acid. Both these compounds are easily oxidized to the corresponding benzoquinone derivatives.

Hydroquinonemethyl ether and methyl arbutin are not, however, synthesized from anisic acid by *Arctostaphylos* (see above).

Benzoquinone derivatives are probably precursors for some naphthoquinone derivatives methylated at position 7 (e.g. chimaphilin in *Chimaphila umbellata*). These compounds appear to be formed via prenylated derivatives as described in detail in D.6.2.1 for purpurin.

References for further reading

Zenk, M. H. 'Biosynthesis of C_6C_1-Compounds', *Biosynthesis of Aromatic Compounds*, ed. G. Billek (Pergamon Press, Oxford, 1966), pp. 45–60.

Billek, G. Biosynthese aromatischer Systeme, *Angew. Chem.* **79** (1967), p. 587.

Zenk, M. H. and Leistner, E. Biosynthesis of Quinones, *Lloydia* **31** (1968), pp. 275–92.

Luckner, M. Die Biosynthese von Hydrochinon- und p-Chinonderivaten, *Fortschritte der Botanik* **31** (1969), pp. 110–22.

Aneshansley, D. J., Eisner, T., Widom, J. M. and Widom, B. Biochemistry at 100°C: Explosive Secretory Discharge of Bombardier Beetles (Brachinus). *Science (Washington)* **165** (1969), pp. 61–3.

Dirscherl, W. Biochemie der Vanillinsäure, *Arzneimittelforschung* **20** (1970), pp. 405–9.

21. Peptides, peptide derivatives and proteins possessing the character of secondary products

Peptides are formed from two or several amino acids by the elimination of water. They possess the structure of acid amides. When two amino acids are coupled together the compound is known as a dipeptide, in the case of three amino acids, as a tripeptide, and so on. In the case of more than ten amino acids they are termed oligopeptides. Those composed of a large number of amino acids are known as polypeptides (proteins).

The name of the peptide is formed by using the ending 'yl' for those amino acids whose carboxyl group participates in the peptide bond (fig. 322).

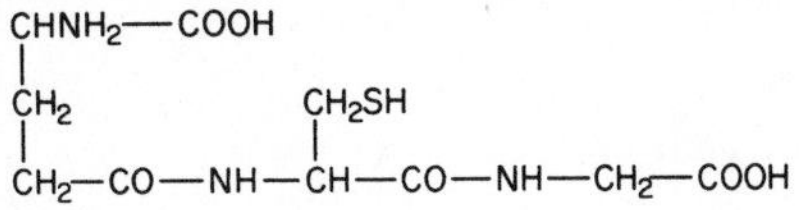

$$CHNH_2-COOH$$
$$|$$
$$CH_2 \qquad\qquad CH_2SH$$
$$| \qquad\qquad\quad |$$
$$CH_2-CO-NH-CH-CO-NH-CH_2-COOH$$

Fig. 322. Structural formula of γ-glutamyl-cysteinyl-glycine (glutathione).

While low molecular weight peptides are synthesized by the action of soluble enzymes on activated amino acids (cf. C.1.1), a complex mechanism partly linked to the cell structures is active in the biosynthesis of proteins. This permits the use of the genetic information stored in the nucleic acids, which determine the sequence of amino acids in the protein being synthesized.

21.1 Formation of diketopiperazines and compounds derived from them

Cyclic peptides formed from two amino acids possess a diketopiperazine structure. Compounds of this type are synthesized by a number of micro-organisms. If one of the two amino acids is replaced by anthranilic acid, benzo-diazepines are formed (cf. D.6.4.2).

The formation of pulcherrimic acid in *Candida pulcherrima* and of the aspergillic acids has been experimentally investigated. Pulcherrimic acid originates from two molecules of L-leucine via cyclo-L-leucyl-L-leucine, while neohydroxyaspergillic acid is formed from leucine via flavacol and neoaspergillic acid (fig. 323).

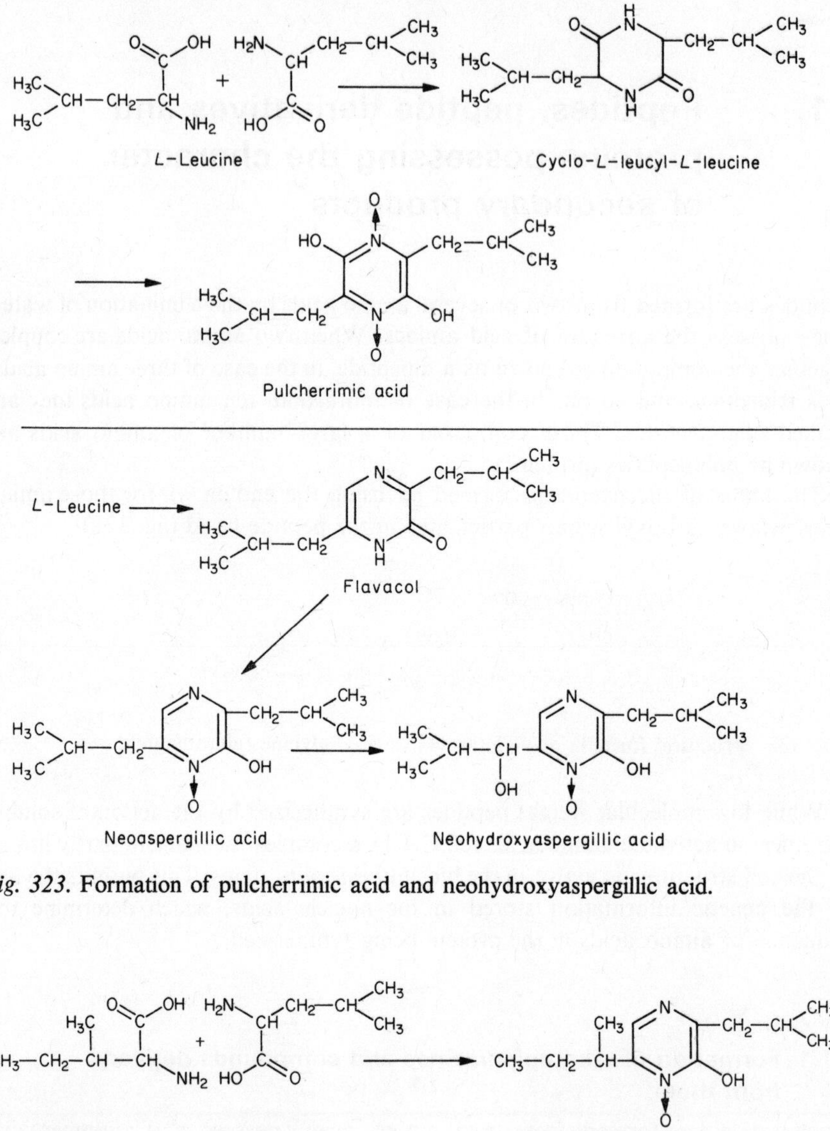

Fig. 323. Formation of pulcherrimic acid and neohydroxyaspergillic acid.

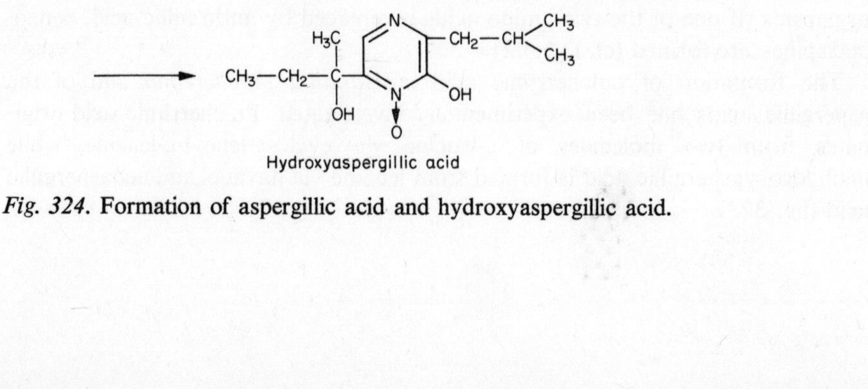

Fig. 324. Formation of aspergillic acid and hydroxyaspergillic acid.

Aspergillic acid is formed from one molecule of leucine and isoleucine. The compound can in turn be converted to hydroxyaspergillic acid (fig. 324).

See C.2.6.2 and D.21.2 for the mechanism of oxidation at the nitrogen atom leading to the formation of hydroxamic acids.

The diketopiperazine echinulin (fig. 325) which is synthesized by various *Aspergillus* strains originates from tryptophan, alanine and mevalonic acid.

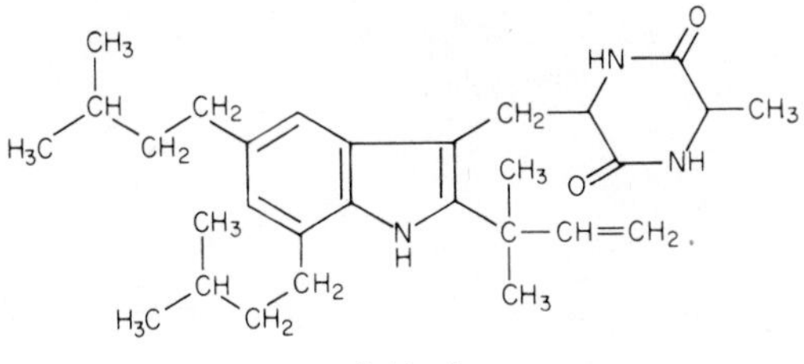

Echinulin

Fig. 325. Echinulin.

References for further reading

MacDonald, J. C. 'Aspergillic Acid and Related Compounds', *Antibiotics*, Vol. II, ed. D. Gottlieb and P. D. Shaw (Springer Verlag, Berlin, 1967), pp. 43–51.

Schütte, H. R. 'Verschiedenes', *Biosynthese der Alkaloide*, ed. K. Mothes and H. R. Schütte (VEB Deutscher Verlag der Wissenschaften, Berlin, 1969), pp. 645–77.

21.2 Formation of hydroxamic acids

Compounds which have one or several oxidized peptide bonds are called hydroxamic acids (fig. 326).

Substances of this class are synthesized especially by micro-organisms with the exception of 2,4-dihydroxy-7-methoxy-2H-1,4-benzoxazin-3-one (DIMBOA) which occurs in higher plants. Hadacidin occurs in *Penicillium aurantioviolaceum*, aspergillic acid (cf. D.21.1) in *Aspergillus flavus*.

Hydroxamic acids, after dissociation of a proton, form stable five-membered rings with metals and therefore are important in the iron metabolism of many organisms. Compounds with one iron-trihydroxamate centre are termed siderochromes. If they possess growth-promoting properties, for example like ferrichrome, they are called sideramines; if they are antibiotically active like ferrimycin A they are called sideromycins. If the Fe^{3+} iron is reduced to Fe^{2+}, the stability of the complex is greatly reduced and iron may be transfered from ferrichrome, for example, to compounds with a porphyrin ring system, and may be incorporated into haemoproteins (cf. D.8.2). The activity of sideramines is competitively inhibited by sideromycins.

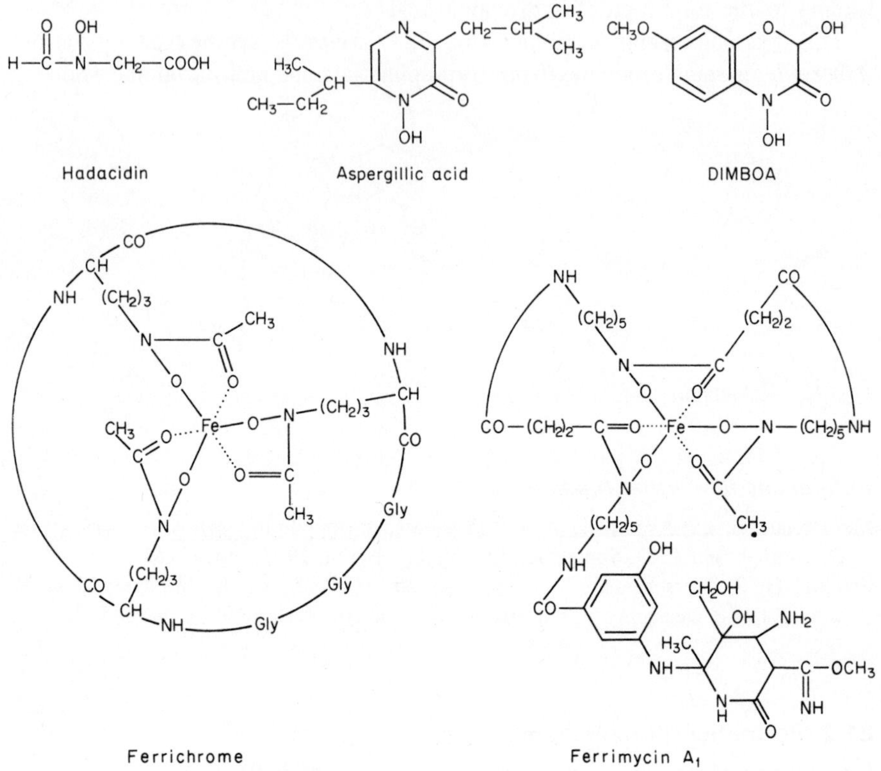

Hadacidin   Aspergillic acid   DIMBOA

Ferrichrome   Ferrimycin A₁

Fig. 326. Hydroxamic acid derivatives.

The $\begin{matrix} O & OH \\ \| & | \\ -C & -N- \end{matrix}$ group originates from a $\begin{matrix} O & H \\ \| & | \\ -C & -N- \end{matrix}$ group by mixed function oxygenation (cf. C.2.6.2). On the one hand the acid amide itself serves as a substrate for oxygenation (cf. formation of aspergillic acid). On the other, the amino group of the free amino acid is oxygenized, as in the case of the biosynthesis of hadacidin, and the formation of an acid amide bond takes place in a subsequent reaction (fig. 327). N-Hydroxylated amino acids, such as δ-N-hydroxyornithine, serve as a precursor in the synthesis of ferrichrome. They are converted to hydroxamic acids by acetylation. Three molecules of hydroxamic acid are coupled together and with glycine to form the cyclic peptide ferrichrome.

N-Hydroxy amino acids are probably precursors of cyanogenic glycosides and glucosinolates in higher plants (cf. D.7.5 and D.7.6).

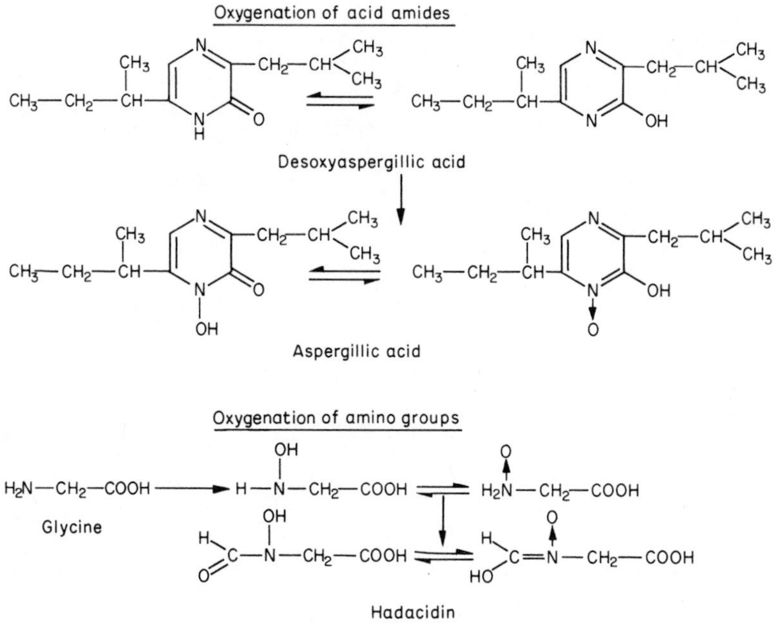

Oxygenation of acid amides

Desoxyaspergillic acid

Aspergillic acid

Oxygenation of amino groups

Glycine

Hadacidin

Fig. 327. Formation of hydroxamic acids by mixed function oxygenases.

References for further reading

Keller-Schierlein, W., Prelog, V., and Zähner, H. Siderochrome, *Forstschritte Chem. org. Naturstoffe* 22 (1964), pp. 279–322.

Neilands, J. B. Hydroxamic Acids in Nature, *Science (Washington)* 156 (1967), pp. 1443–7.

Emery, T. F. 'Hadacidin', *Antibiotics,* Vol. II, ed. D. Gottlieb and P. D. Shaw (Springer Verlag, Berlin, 1967), pp. 17–25 and 439.

21.3 Biosynthesis of penicillins and cephalosporins

A number of fungi (e.g. of the genera *Penicillium, Aspergillus, Trichophyton* and *Epidermophyton*) produce sulphur containing antibiotics termed penicillins. The most important naturally occurring compounds of this group are given in table 22. Substances with a similar structure but possessing a thiazine ring in place of the thiazolidine ring occur in *Cephalosporium* and *Emericello* species. They are termed cephalosporins.

Both penicillins and cephalosporins may be regarded as cyclic peptides in whose formation the amino acids α-amino adipic acid, cysteine and valine take part. Though some details of their biosynthesis until now have not been elucidated, the biosynthetic pathway outlined in fig. 328 is assumed to be correct.

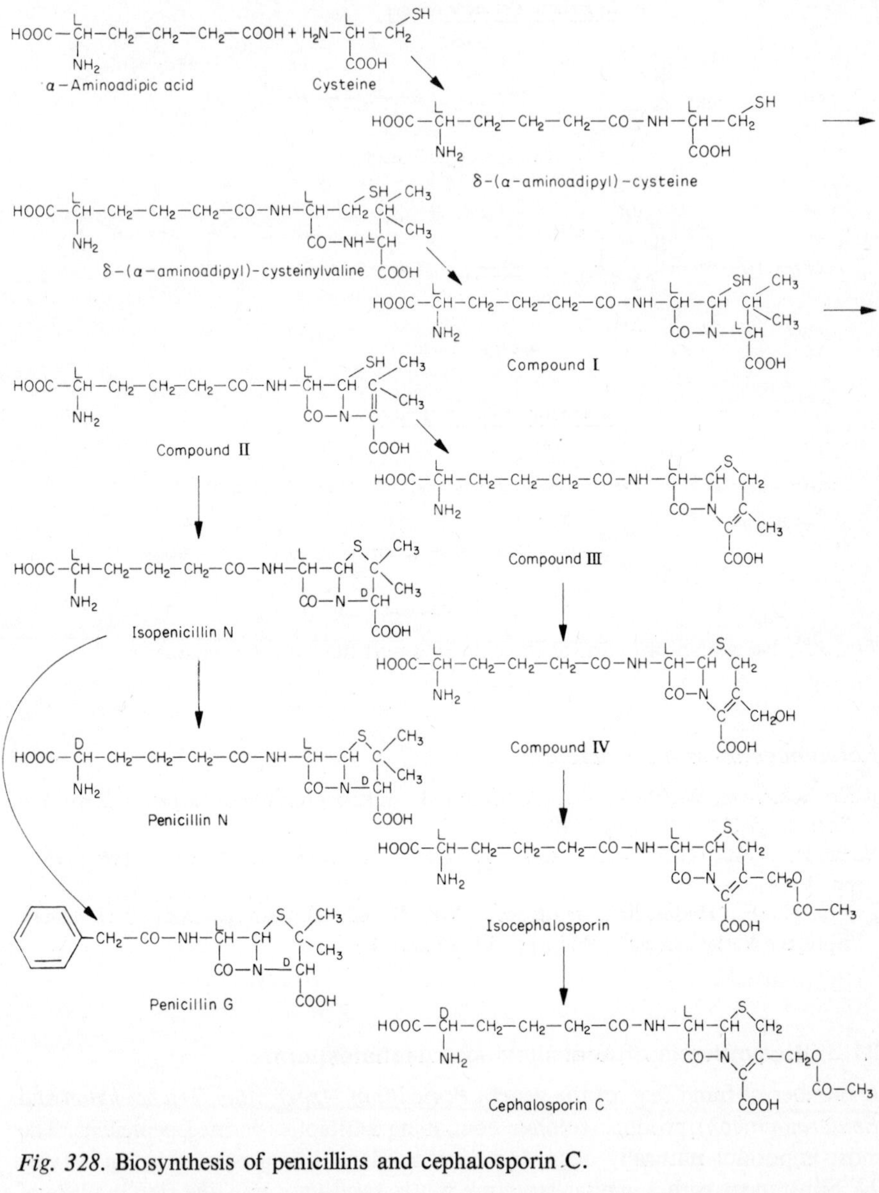

Fig. 328. Biosynthesis of penicillins and cephalosporin .C.

L-α-Amino adipic acid (cf. D.16) reaets at first with L-cysteine with the formation of δ-(α-aminoadipyl)-cysteine, which in turn is converted to δ-(α-aminoadipyl)-cysteinylvaline by reaction with valine. A β-lactam ring (compound I)
is then formed by dehydrogenation, and after that a double bond to give compound II. Compound II may then be converted to penicillins or through several
intermediate steps to cephalosporins. The thiazolidine ring characteristic of

penicillins, in which the carbon atom of valine possesses the D-configuration, is formed by addition of the sulphydryl group to the double bond.

Inversion of configuration at the α-carbon atom of the α-aminoadipic acid in isopenicillin N yields penicillin N. Replacement of the α-aminoadipic group of this compound by other acids gives the other penicillins outlined in table 22. The corresponding acylases can form penicillinic acid from isopenicillin N. Penicillin production may be increased by the addition to the culture medium of certain acids or acid derivatives which can be enzymically activated to form the corresponding CoA-esters. In certain cases new penicillins are formed (e.g. formation of the acid stable, and therefore also orally active, phenoxymethyl-penicillin, after addition of synthetically prepared phenoxyacetic acid.

Penicillinic acid obtained by enzymatic hydrolysis of penicillin G is nowadays usually converted to the so-called semisynthetic penicillins, by reaction with acid chlorides. Representatives of this group are characterized by stability to acids and resistance to β-lactamases (see below), and possess a broad spectrum of activity in contrast to the naturally occurring penicillins.

It was found by feeding experiments with radioactively labelled precursors that the cephalosporins are synthesized from the same amino acids as the penicillins. It is thought that compound II is first converted to compound III by dehydrogenation, forming the dihydrothiazine ring which is characteristic of cephalosporins. By hydroxylation, compound IV may be formed which can then

Table 22. Naturally occurring and semisynthetic penicillins

$$R-NH-CH-CH \overset{S}{\underset{O=C-N-CH-COOH}{\bigvee}} C\overset{CH_3}{\underset{CH_3}{\diagup}}$$

Name	R: Name	R: Structure
p-Hydroxybenzylpenicillin (Penicillin X)	p-Hydroxyphenylacetyl-	$HO-\langle\bigcirc\rangle-CH_2-CO-$
Benzylpenicillin (Penicillin G)	Phenylacetyl-	$\langle\bigcirc\rangle-CH_2-CO-$
n-Propylpenicillin	n-Butyryl-	$CH_3-(CH_2)_2-CO-$
n-Butylpenicillin	n-Valeryl-	$CH_3-(CH_2)_3-CO-$
n-Amylpenicillin	n-Capronyl-	$CH_3-(CH_2)_4-CO-$
Δ^2-Pentenylpenicillin (Penicillin F)	Δ^3-Dehydro-n-capronyl-	$CH_3-CH_2-CH=CH-CH_2-CO-$
n-Heptylpenicillin (Penicillin K)	n-Caprylyl-	$CH_3-(CH_2)_6-CO-$
Penicillin N	δ-(α-Aminoadipyl)-	$HOOC-\underset{NH_2}{\underset{\vert}{CH}}-(CH_2)_3-CO-$
Penicillinic acid	Hydrogen	$H-$

be acetylated to isocephalosporin. By conversion of the L-configuration of α-aminoadipic acid to the D-configuration, cephalosporin C, the most important antibiotic of the cephalosporin group, is formed.

A few derivatives which are used in the treatment of patients sensitive to penicillin are prepared semisynthetically from cephalosporin C.

Penicillins may be broken down to penicillinic acid by cleavage of the side chain by acylases which are present in a number of micro-organisms. The β-lactam ring is broken by other enzymes (penicillases). The antibiotic activity is lost in both cases. In many bacteria similar enzymes cause inactivation of cephalosporins. The widespread use of penicillins, especially penicillin G, in recent years has led to the spreading of resistant micro-organisms which possess penicillin-degrading enzymes.

References for further reading

Abraham, E. P. and Newton, G. G. F. 'Penicillins and Cephalosporins', *Antibiotics*, Vol. II, ed. D. Gottlieb and P. D. Shaw (Springer Verlag, Berlin, 1967), pp. 1–16.

Schütte, H. R. 'Verschiedenes', *Biosynthese der Alkaloide*, ed. K. Mothes and H. R. Schütte (VEB Deutscher Verlag der Wissenschaften, Berlin, 1969), pp. 645–77.

Morin, R. B. and Jackson, B. G. Chemistry of Cephalosporin Antibiotics, *Fortschritte Chem. org. Naturstoffe* **28** (1970), pp. 343–403.

21.4 Formation of the framework of the bacterial cell wall

The cell wall of the bacteria usually has a very complicated constitution. The matrix is built up from a carbohydrate macromolecule and peptide chains linked with each other to form a network. This network is covered by a layer of globular proteins. In the case of gram-negative bacteria (e.g. *Escherichia coli*, species of *Salmonella* and *Shigella*) the cell wall also contains large quantities of lipid.

The carbohydrate portion of the peptidoglycan serving as the framework consists of up to twelve N-acetylglucosamine and N-acetylmuramic acid units alternating with each other and linked with each other through β-1,4-bonds. The muramic acid molecules are either wholly or partly substituted by peptide chains which, for example, may consist of L-alanine, D-glutamic acid, a dibasic amino acid (e.g. L-lysine or *meso*-, D,D- and L,L-diaminopimelic acid) and D-alanine. These peptide chains are linked with each other directly or via further amino acids, thus forming a two-dimensional (rarely three-dimensional) network. Coupling usually takes place via the ε-amino group of lysine or diaminopimelic acid (fig. 329).

While the structure of the carbohydrate chains and of the peptide linked to

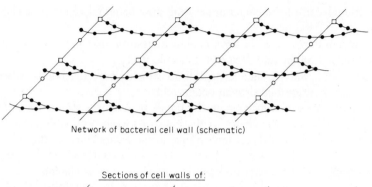

Network of bacterial cell wall (schematic)

Sections of cell walls of:

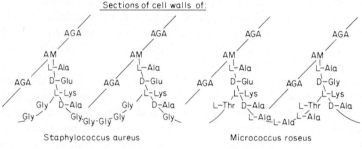

Staphylococcus aureus

Micrococcus roseus

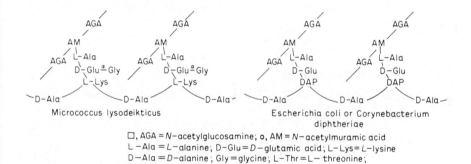

Micrococcus lysodeikticus

Escherichia coli or Corynebacterium
diphtheriae

□, AGA = N−acetylglucosamine; o, AM = N−acetylmuramic acid
L−Ala = L−alanine; D−Glu = D−glutamic acid; L−Lys = L−lysine
D−Ala = D−alanine; Gly = glycine; L−Thr = L− threonine;
DAP = diaminopimelic acid; • amino acid

Fig. 329. Structure of cell wall of bacteria.

muramic acid is probably fundamentally the same in gram-positive as well as gram-negative bacteria, the mode of linkage of the peptide chains in the individual species and strains appears to be specific. In *Staphylococcus aureus* the cross linkages form a fine network (made up of five glycine units). In one strain of *Micrococcus roseus* the cross links are composed of a peptide made up of three molecules of L-alanine and one molecule of L-threonine, while in another strain three molecules of L-alanine are involved. The peptide chains are linked directly with each other in the case of *Micrococcus lysodeikticus, Escherichia coli* and *Corynebacterium diphtheriae* (fig. 329).

The peptidoglycan of the bacterial cell wall may be attacked by three classes of enzymes:

(a) by glycosidases (e.g. lysozyme) which destroy the bonds between *N*-acetylglucosamine and *N*-acetyl muramic acid;

(b) by acetylmuramyl-L-alanine peptidases which cleave the bonds between the carbohydrate and protein chains and

(c) by endopeptidases which break the bonds between the amino acids of the peptides that join the chains or the bonds between two peptide chains.

This bond breaking is of direct importance in the enlargement of the cell-wall molecule, since the precursors, e.g. UDP-*N*-acetyl glucosamine and the peptide UDP-*N*-acetylmuramyl-L-alanyl-D-glutamyl-L-lysyl-D-alanyl-D-alanine, can add to the points of breakage. In *Staphylococcus aureus*, for example, 7% of the peptides attached to the muramic acid always have a free end. A certain percentage of pentaglycine also is only linked at the carboxyl end to the ϵ-amino group of lysine while the amino end is free.

During reaction of new units with the carbohydrate chain, the UDP-group of the precursors is lost (cf. D.1.1.2). The coupling of the peptide grouping located on one *N*-acetyl muramine moiety with the peptide chain attached to another muramic acid moiety is a transpeptidization reaction. Here the bond between the last D-alanine group of the peptide and the one before it is broken, and a new bond is formed with a peptide or with another peptide chain linked to muramic acid. The last D-alanine unit of the above-mentioned precursor is thus lost in this reaction.

References for further reading

Strominger, J. L. and Ghuysen, J.-M. Mechanism of Enzymatic Bacteriolysis, *Science (Washington)* **156** (1967), pp. 213–21.

Raftery, M. A. and Dahlquist, F. W. The Chemistry of Lysozyme, *Fortschr. Chem. org. Naturstoffe* **27** (1969), pp. 340–81.

21.5 Biosynthesis of cyclic polypeptides

Peptide antibiotics with cyclic molecular structure, e.g. bacitracins, gramicidins and tyrocidines (tyrocidine A, fig. 330), are synthesized by various species of *Bacillus*. The cyclic peptides of the polymyxin and valinomycin groups occurring in species of *Streptomyces* are similar.

While proteins (the relatively low molecular weight insulin among others) are synthesized with the help of ribosomes on specific messenger ribonucleic acids, enzymes alone are involved in the biosynthesis of these peptides. The mechanism thus corresponds to the formation of the tripeptide glutathione which plays an important role in primary metabolism (cf. fig. 322). Enzyme complexes are involved into the biosynthesis of the cyclic polypeptides.

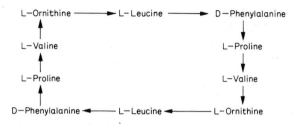

Fig. 330. Tyrocidine A.

The amino acids which are used as the precursors are activated by the transformation into the corresponding enzyme-bound AMP-derivatives and then transferred to SH-groupings of the enzyme complex forming thioester linkages. They are arranged in the sequence later present in the resulting polypeptide.

By a transfer reaction the thioester bonds are then cleaved by the simultaneous formation of peptide bonds between the amino acids forming a growing peptide chain. At the last step the formed peptide is split off from the enzyme complex by cyclization, i.e. formation of a peptide bond including the terminal amino group of the chain.

The D-amino acids which are often found in the molecule of cyclic polypeptides originate from the L-form at the level of the AMP-derivatives.

References for further reading

Weinberg, E. D. 'Bacitracin, Gramicidin and Tyrocidine', *Antibiotics*, Vol. II, ed. D. Gottlieb and P. D. Shaw (Springer Verlag, Berlin, 1967), pp. 240–53.

Paulus, H. 'Polymyxins', *Antibiotics*, Vol. II, ed. D. Gottlieb and P. D. Shaw (Springer Verlag, Berlin, 1967), pp. 254–67.

Perlman, D. and Bodanszky, M. Biosynthesis of Peptide Antibiotics, *Ann. Rev. Biochem.* **40** (1971), pp. 449–64.

Roskoski, R., Ryan, G., Kleinkauf, H., Gevers, W. and Lipmann, F. Polypeptide Biosynthesis from Thioesters of Amino Acids, *Arch. Biochem. Biophysics* **143** (1971), pp. 485–92.

21.6 Formation of sclerotins in insects

The proteins which make up the stiff outer covering, the so-called exoskeleton, of insects belong to protein which possess characteristics of secondary metabolites. The exoskeleton consists of a system of plates and tubes, known as sclerites, connected by flexible joints. Muscles attached to the sclerites enable the insect to execute precise movements. The rigidity of the exoskeleton is due to fibrous proteins called sclerotins. By filling the spaces between the peptide chains with polymeric aromatic substances, a tough structure is obtained, which resembles plastic strengthened by glass fibres. This solid layer is covered on the outside by chitin, a polymeric substance made up of *N*-acetylglucosamine (cf.

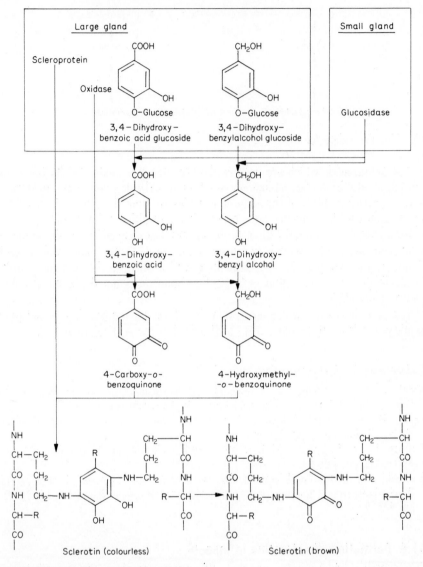

Fig. 331. Formation of scleroproteins.

D.1.1.2) and is thus watertight. Egg capsules and cocoons which are formed by a number of insects are made up of the same substances, but here the chitin layer is missing.

While the individual stages of formation of the insect skeleton have not been investigated, due to experimental difficulties, it has been possible to study the sclerotins which make up cocoons and egg capsules. The necessary material for their formation is synthesized in two glands in the case of cockroaches, of which the larger one synthesizes the scleroproteins as well as 3,4-dihydroxybenzoic acid glucoside, or 3,4-dihydroxybenzylalcohol glucoside and an oxidase. The sclero-proteins are contained in the gland as sizeable globules. Considerable cross link-ing of the peptide chains exists in these proteins. A glucosidase is produced in the smaller gland which can degrade the glucosides formed in the larger gland. During secretion the contents of the large and the small gland are mixed, thus forming 3,4-dihydroxybenzoic acid or 3,4-dihydroxybenzyl alcohol. Both these compounds are susceptible to oxidation by oxidases and are oxidized to 4-carboxy-*o*-benzoquinone or 4-hydroxymethyl-*o*-benzoquinone. Both these compounds are capable of polymerization and condensation with the peptide chains of the scleroproteins (cf. D.7.7). Subsequent oxidation to the quinoid form gives the brown quinone and quinhydrone pigments which give sclerotin its dark colour (fig. 331).

The quinones react with the free amino groups, e.g. the ε-L-lysyl groups of the scleroproteins. Thus masses formed by the polymerization of the quinones and resembling plastic are linked to the peptide chains. The number of free tyrosine groups in the protein is diminished during this process. The latter phenomenon is probably due to a dimerization effected by phenol oxidases (cf. C.2.3), and leads to additional interlacing of the peptide chains.

In the formation of the skeleton of most insects, acetyldopamine (cf. D.20.1.1) is used in place of the above-mentioned benzoic acid or benzylalcohol derivatives, whereas in the formation of the cocoons of the American silk-moth (*Samia cecropia*) 3-hydroxyanthranilic acid (cf. D.6.4.1) is used. These compounds can also be converted to quinoid compounds; solidification and interlacing proceed as mentioned above.

References for further reading

Brunet, P. C. J. Sclerotin, *Endeavour* **26** (1967), pp. 68–74.

Index

The sign § after the page number indicates that the structural formula is given on this page, with or without additional information.

The sign + after the page number indicates that the biosynthesis is treated on this page; additional information may also be given.

Abietadiene 133§+, 135
Abietic acid 133§+, 135+
Absolute incorporation rate 15
Accumulation of secondary products 5
Acetaldehyde 51§+, 188§+, 192, 321
 activated 51§+, 81, 82§, 246, 266
Acetic acid 7§
Acid amides 198+, 371
Activated acetic acid 81
Activation of acetic acid 82
Acetidine-2-carboxylic acid 234+, 235§+
Acetoacetic acid 262
Acetoacetyl CoA 123+
Acetobacter suboxydans 75
α-Acete-α-hydroxybutyric acid 246§+
α-Acetolactic acid 246§+
2-N-Acetylaminoacetophenone 312§+
Acetyl-AMP 82
Acetyl CoA 81§
 reactions of 30, 31
Acetyl CoA carboxylase 82
N-Acetyl-L-cysteine 231
Acetyldopamine 383
Acetylene derivatives 98+
N-Acetyl glucosamine 64, 378
N-Acetylglucosamine-6-phosphate 67§+
N-Acetylglutamic acid 259+, 260§
N(α)-Acetylglutamic acid-γ-semialdehyde 259+, 260§
O-Acetylhomoserine 283
N-Acetyl-5-hydroxytryptamine 314§+
N-Acetyl-5-hydroxtryptophan 314
Acetyllipoic acid 81+, 82§+
N-Acetyl-5-methoxytryptamine 314§+
N-Acetyl muramic acid 378
Acetylmuramyl-L-alanine peptidases 380
N-Acetyl neuraminic acid 96
N-α-Acetylornithine 259+, 260§
Acetyl phosphate 27, 82

N-Acetyl tryptamine 200, 296§+
Aconitic acid 119§+
Acridone alkaloids 178§+
Acrylic acid 236
Actinidine 137§
Actinocine 175, 176+
Actinomycin D 175§
Actinomycins 175, 176
Activated acetaldehyde 51
Activated acid derivatives 31
Activated compounds 28
Activated formic acid 46
Activated methanol 46
Activated pyruvate 51
Acyl carrier protein (= ACP) 83
Acyl phosphate 30
Addition reactions of acetylene derivatives 100
Adenosine amino hydrolase 211
Adenosine deaminase 211
Adenosine monophosphate 210+, 211§+, 283§, 284+
Adenosine-3-phosphate-5-phosphosulphate (= 3-Phosphoadenosine-5-phosphosulphate) 225§, 315
Adenosine-5-phosphosulphate 225§
Adenosine triphosphate 282, 283§+
S-Adenosylhomocysteine 48
S-Adenosylmethionine (= active methionine) 48, 234§, 235
S-Adenosyl-S-methylmercaptopropylamine 235§+
Adenylsuccinate lyase 210
Adenylsuccinate synthetase 210
Adonitol 75§+
ADP-sugars 62
Adrenaline 318+, 319§+, 367§
Adrenodoxin 40
Aescigenin 152§+

Aescigenin glycosides 152
Aesculetin 345§+
Aesculin 345§+
Agaricic acid 118+, 119§+
Aglycon 64
Agmatine 187§+, 259+, 260§+, 267
Agroclavine 294§+, 295+
Alanine 193§+, 194+, 226§+, 295, 306, 373, 379
β-Alanine 186§+, 251
Alcohol dehydrogenase 33
Aldehyde 192§+
Aldehyde transaminases 52
Aldebionic acids 73+
Aldol condensation, in sugars 57
Aldose-1-phosphate 27, 61
Aldoses 57
Aldosterone 156+, 157§+
Aldoxime 192§+
Alginic acid 66, 73
Alizarin 169+
Alkaloids 198
    protoalkaloids 198
    pseudoalkaloids 198
    real 198
n-Alkanes 97+
Alkyl citric acids 118+
Alkylcysteine derivatives 228§+
S-Akylcysteine sulphoxide 228§+
2-Alkyl-4-quinolones 178§+
Allantoins 220+
    methylated 220
Allene groupings, formation of 99
Allicin 229§+
Allinase 228, 229
Alliins 228§+
Alloxan 221+
Allyl alliin 228§+, 229§
Allyl cysteine 228§
Allyl mustard oil 194§
Allyl sulphenic acid 229§+
Amadori rearrangement 213
Amaryllidaceae alkaloids 330+
Amentoflavone' 361§+
Amines 186+
2-Aminoacetophenone 312§+
Aminoacetophenone 320§+
Amino acid decarboxylases 52
Amino acid oxidases 187
Amino acid oxygenases 41
Amino acids 184
    methylated 189§+
    non-protein 184

oxidative deamination of 33
    protein 184
D-Amino acids 184+
    incorporation in bacterial cell wall 379
    incorporation in cyclic polypeptides 378
L-Amino acids 34
α-Aminoacrylic acid 229§+
α-Aminoadipic acid 269+, 271§+, 273, 376§
α-Aminoadipic acid semialdehyde 269, 271§+ 272§, 273+
Aminoadipyl cysteine 376§+
Aminoadipyl cysteinyl valine 376§+
p-Aminobenzoate synthetase 171
p-Aminobenzoic acid 171§+, 215
2-Aminobenzoyl acetaldehyde 306+, 307§+
O-Aminobenzoyl acetyl CoA 177§+
2-Aminobenzoyl pyruvate 305§+, 306+
γ-Aminobutyric acid 187§+
γ-Aminobutyraldehyde 259+, 260§+, 266§+
α-Amino-γ-butyrolactone 234§+
4-Amino-5-carboxyimidazole 221§+
α-Amino-β-carboxymuconic acid
    semialdehyde 310+, 311§+
4-Amino-4-cyanobutyric acid 193§+
α-Aminodimethyl-γ-butyrothetin 234, 235§
2-Amino-3-hydroxybenzoylacetaldehyde 307§+
2-Amino-4-hydroxy-6-hydroxymethyl-7, 8-dihydropteridine 214§+
2-Amino-4-hydroxy-6-(3-triphosphoglyceryl)-7, 8-dihydropteridine 214§+
4-Aminoimidazole 221§+
β-Aminoisobutyric acid 251§+
α-Amino-β-ketoadipic acid 202+, 203§+
α-Amino-ε-ketopimelic acid 269+, 270§+
δ-Aminolevulinic acid 202+, 203§
3-Aminomethylindole 292+, 293§+
α-Aminomuconic acid-ε-semialdehyde 311§+, 312+
2-Amino-3-oxoisophenoxazine 175§+
o-Aminophenol 175§+
2-Amino-4-phenylbutyric acid 195§, 196+
2-Amino-4-phenyl-4-hydroxybutyric acid 196§+
4-Aminopipecolic acid 272§+, 273+
α-Aminopropionitrile 193§+
β-Aminopropionitrile 193§+
Amino purines, methylated 212+
    degradation of 212
6-Aminopyrrolopyrimidine 218+, 219§+
6-Aminovaleric acid 41
Amino sugars 67+
Ammodendrine 277§+

Ammonia lyases 345
Amygdalin 191§
i-Amylamine 186§+
Amylases 65
Amylopectin 64
Amylose 65
Amylo-(1, 4–1, 6)-transglucosidases 65
Amylpenicillin 377§
Amyrin 151§+
Anabasine 276+, 277§+
    piperidine ring of 276+
Androcymbine 322§
Anagyrine 275§+, 276+
Androstane 156
Androstenedione 157§+
Angelic acid 266+
Angustifoline 275§+, 276+
Anhalonidine 321§+
Anhydrocitroforum factor 46+, 48§+, 47§. 210
Anhydrorhodovibrin 143§+
Anisomycin 340+, 341§+
Anteisoalkanes 97§+
Anthocyan-metal complexes 360§+
Anthocyanidin 357, 359§+, 358
Anthocyanidin glycoside 359§+
Anthocyanidins 357, 358
Anthocyanins, see Anthocyanidin glycoside
Anthracene derivatives 104+, 105§+
Anthranilate synthetase 171
Anthranilic acid 171§+, 172+, 174§, 175§, 289§, 306, 311§
Anthrone derivative 104, 105§+
Apiin 69, 70§
Apiose 69, 70§
Arabans 64
Arabinose 58§
Arabinose-1-phosphate 61
Arabitol 75§+
Arachidonic acid 88+, 89§+
Arbutin 369§+
Arginine 79§, 259+, 260§+, 267, 279‡, 280§
Arginine succinic acid 243
Aristolochic acid 334+, 335§+
Aromatic pathway, in tryptophan degradation 307
Artemesia ketone 98§
Artemesia lactone 98§
Arthritis, chronic 220
Ascorbic acid 40, 74§+
Ascorbigen 197§+
Asparagine 193§+, 249+
Aspartase 249

Aspartate carbamyl transferase 250
Aspartate-β-decarboxylase 225
Aspartic acid 249§+, 279§+, 308§+, 309+
Aspartic acid-β-semialdehyde 269, 270§
Aspartyl ionosine monophosphate 211§+
Aspergillic acid 373, 374§, 375+
Aspidosperma skeleton, alkaloids with 296, 297§, 299‡, 300+
Atranorin 109+, 110§
Atropa belladonna, toxicity 8
Atropic acid 221, 343§+
Atropine 137§
Aurines 362§+
Auxin 313§+
Axerophthol, see Vitamin A
Azomethines 199§+
Azulenes 130

Bacitracins 380
Bacterial cell wall 378, 379
Bacterial polysaccharides 64
Baikiain 272§+, 273
Balata 126
Barbiturase 252
Barbituric acid 251§+
Barbituric acid amino hydrolase 252
Belladine 330+, 331§+
Benzodiazepine derivatives 179§+, 371
Benzoic acid 261, 276, 366§+
    glycine conjugates 222
Benzoic acid derivatives 267
Benzophenone derivatives 104+, 365+
Benzopteridines 216+
Benzoquinoline alkaloids, see Acridone alkaloids
p-Benzoquinone 7§
Benzoquinone derivatives 110+
    isoprenoidal 341
Benzoyl CoA 267
Benzyl alcohol derivatives 366+
Benzylisoquinolines 322+
Benzyl mustard oil 194§+
Benzylpenicillin 377§
Berberastine 327§+
Berberine 327§+
Berberine bridge 326, 327
Betacyanins 339+
Betain 189§+, 190§+
Betaines 188+
Betanidin 339§+, 340
Betaxanthins 339+
Betonicin 189+
Betulic acid 150§+

Betulin 150$^{§+}$, 151
Bile acids 154, 155$^+$
Bile pigment protein complexes 207
Bile pigments 205$^+$, 206$^{§+}$
Bilirubin 205, 206$^{§+}$
Bilirubin glucuronide 205
Bilirubin protein complex 205
Biliverdin 205, 206$^{§+}$
Biochanin A 363$^{§+}$
Biotin 45, 230$^{§+}$
Bisabolene 130$^+$, 132$^{§+}$
Bisabolol 130$^+$, 132$^{§+}$
Bisbenzylisoquinolines 327$^+$
Bishomo-γ-linolenic acid 88$^+$, 89$^{§+}$
Boivinose 59§
Braylin 346§
Bromtetracyclin 106§
Bufadienolide 158§
Bufotenine 290$^{§+}$
Bufotenine-N-oxide 290$^{§+}$
Bulbocapnine type alkaloids 323$^+$, 325$^{§+}$
Butein 362, 363§
Butenyl mustard oil 194$^{§+}$
Butylamine 186$^+$, 188
Butylcitric acid 118$^+$, 120$^{§+}$
Butylpenicillin 377§
Byssochlamic acid 118, 120$^{§+}$

Cadaverine 187$^{§+}$, 273, 274§, 275§, 276
Caffeic acid 331, 344$^{§+}$, 345, 355, 366§
Caffeine 212$^{§+}$
     degradation of 220
Calciferols, see Vitamin D
Calcium oxalate 120$^+$
Callose 66
Callus cultures 17
Calophyllolid 364$^{§+}$, 365$^+$
Calotropin 5
Calycanthidine 301$^{§+}$
Calycanthine 300, 301$^{§+}$
Cantharidine 6
Caprylic acid 7§
Capsaicin 320$^{§+}$
Carbamic acid 251$^{§+}$
Carbamyl-β-alanine 251§
Carbamyl-β-aminoisobutyric acid 251§
Carbamyl aspartic acid 250$^{§+}$
Carbamyl glutamic acid 286$^{§+}$
Carbamyl phosphate 230§, 237§, 250§, 279§
Carbamyl putrescine 259$^+$, 260$^{§+}$, 267$^+$
Carboline alkaloids 295$^+$
Carboxyethyl cysteine 228$^{§+}$

5-Carboxyethyl-4, 6-dihydroxy-picolinic
     acid 308$^{§+}$, 309$^+$
4-Carboxy-o-benzoquinone 382$^{§+}$
Carboxybiotin 45
β-Carboxy-β-hydroxyisocaproic acid
     243$^{§+}$
Carboxyisopropyl cysteine 228$^{§+}$
Carboxylases 45
5-Carboxyoxopropenyl-4, 6-dihydroxypico-
     linic acid 308$^{§+}$, 309$^+$
5-Carboxyoxopropyl-4, 6-dihydroxypicolinic
     acid 308$^{§+}$, 309$^+$
Cardenolide type 158$^{§+}$
Carene 129$^{§+}$, 130$^+$
Carlina oxide 98§, 101$^{§+}$, 102
Carnauba wax 97
Carnosolic acid 133§
α-Carotene 141$^{§+}$, 142$^+$
β-Carotene 141$^{§+}$, 142$^+$
γ-Carotene 141$^{§+}$, 142$^+$
δ-Carotene 141$^{§+}$, 142$^+$
ϵ-Carotene 141$^{§+}$, 142$^+$
Carotenes 144
Carotenoid compounds 140
Carpaine 112$^+$, 113§
Caryophyllene 130$^+$, 132$^{§+}$
Cassiine 112$^+$, 113§
Castoramine 5
Catechin, see Catechol
Catechol 38, 175$^{§+}$, 309, 357, 359$^{§+}$, 358
Catharanthine 299$^{§+}$
CDP-choline 93$^{§+}$
CDP-ethanolamine 93
Cellobiose 73
Cellulose 64, 66
Cephaeline 321, 322§
Cephalosporin C 376$^{§+}$, 378
Cephalosporins 375$^+$
Ceramide 95$^{§+}$, 96$^+$
Cerebrosides 95$^{§+}$, 96$^+$
Chalcone epoxide 358$^+$, 359§
Chalcones 356$^{§+}$, 358, 359§
Chanoclavine 294$^{§+}$, 295$^+$
Chelidonine 326$^{§+}$, 327$^+$
Chimonanthine 300$^+$, 301$^{§+}$
Chitin 67, 381
7-Chlor-6-demethyltetracyclin 106§
Chlorogenic acid 167$^{§+}$
Chlorophyll 203, 204
     a 203$^+$, 204$^{§+}$
     other 205
Chlorophyllide a 204$^{§+}$
Chlororaphin 182$^+$

Chlortetracyclin 106§, 107§+
Choleglobin 205, 206§+
Cholestane 146§
Cholesterol 153+, 154§+, 156§, 157§, 159, 161
Cholic acid 156§+
Choline 93§
Choline plasmologen 96§+
Chondroitin sulphate 66
Chorismic acid 166§+, 317§
Chorismic acid mutase 317
Cinchona alkaloids 301+
Cinchonamine 302§
Cinchonidine 302§
Cinchonine 302§
Cinnabaric acid 175§+
Cinnamic acid 276, 344§+, 355,366
    initiator theory in flavonoid biosynthesis 358
    shortening of side chain of 365
Cinnamic alcohol glucosides 348, 349§+
Cinnamic alcohols 348, 349§+
Cinnamoyl formic acid 196§+
Cinnamoyl quinic acid 167§+
Citric acid 119§+
Citrulline 260§+, 279§+
Claviceps purpurea, alkaloid production in 8
Clavine alkaloids 293
Cobalamines, see Vitamin B₁₂
Cocaine 261
Codeine 323+, 324§+
Codeinone 323+, 324§+
Coenzyme A 81§, 226+, 227§+
Coenzyme Q₁₀ 173
Coffee 212
Colamine 186§+
Colchicine 332+, 335§+
Collagen 40
Compartmentation 9
Competitive inhibition 215
Coniceine 112§+
Coniferin 349§+
Coniferyl alcohol 349+, 351+
Coniine 112§+
Conium alkaloids 112+
Conquinamine 302
Copper 36
Coproporphyrin I 203§+, 205+
Coproporphyrinogen I 203§+, 205+
Coproporphyrinogen III-oxidase 205
Coprostane 146§
Coptisine 326§+, 327+
Corrin 208+

Cortisol 156+, 157§+
Cortisone 156+
Corynanthe-Strychnos skeleton 297§+, 298§+, 299§+, 300§
Corynomycolic acid 87§+
Cotinine 256§+
o-Coumaric acid 346§+, 369
    glucoside 346§+
p-Coumaric acid 344§+, 345§, 366§
Coumarin ring 340
Coumarinic acid 346§+
    glucoside 346§+, 347
Coumarins 345§+, 346§+
p-Coumaroyl quinic acid 167§+
p-Coumaryl alcohol 349§+
Coumoestrol 363+, 364§+
Creatinine 189§+
Cresols 7§
Cubebin 353§, 354+
Cucurbitacins 147§+, 148§+
β-Cyanoalanine 193§+
Cyanogenic glycosides 190+, 191, 192§+
    degradation of 192
Cyanohydrin 192§+
Cyclitols 76+
Cycloalliin 228
Cycloartanol 148§+, 153
Cycloheximide 112+, 113§+
Cycloleucanol 153+, 155§
Cycloleucylleucine 371+, 372§+
Cyclopenase 180
Cyclopenine 179§+, 180§+
Cyclopenol 179§+, 180§+
Cyclopentanoperhydrophenanthrene 145
Cymarose 59§
Cystanthionine 121§+, 120+, 233§+
Cysteamine 225+, 226§+
Cysteic acid 225+, 226§
Cysteine 225+, 226§
Cysteine sulphenic acid 225+, 226§+
Cysteine sulphinic acid 225+, 226§+
Cysteine sulphoxide 196§, 225+
Cytidine triphosphate 251+
Cytisine 275§+, 276+
Cytosine 251§

Daidzein 363, 364§
Dammaradienol 149§+
Damascenine 174§+
2-Decaprenylphenol 147§+
Decarboxylation of amino acids, enzymic 186
    of sugars 58

Defective mutants 20
Dehydrobufotenine 290[§+]
7-Dehydrocholesterol 163, 164[+]
Dehydrocrepis acid 99[§+']
Dehydrodiconiferyl alcohol 351[§+], 352
Dehydrogenases 32
   containing pyridine nucleotides 32
Dehydrogriseofulvin 108[§+], 109[+]
Dehydroluciferin 229[+], 230[+]
Dehydromatricaria acid 99[§+]
Dehydroquinic acid 166[§+], 167[+]
Dehydrorhodopin 143[§+]
Dehydroschikimic acid 166[§+], 167[+]
Demecolcine 333[§+], 334
Demethylocotinine 256[§+]
Demethyltetracycline 106[§]
Dendrobine 135[§], 136[+]
de nove synthesis 25
Deoxyallose 59[§]
Deoxyamine sugar, *see* Amino sugars
Deoxyarabinoheptulosonic acid phosphate
   166[§+]
Deoxyaspergillic acid 375[§]
2-Deoxyglucose 59[§]
6-Deoxyglucose 59[§]
Deoxygulose 59[§]
Deoxylapachol 169
Deoxymethyl allose 59[§]
Deoxypeganine 181
Deoxyribonucleic acid, methylated bases 212
Deoxyribose 57, 58[§]. 68[+]
Deoxyriboside diphosphate 68[§+]
Deoxysugars 57, 68[+]
Deoxytalose 59[§]
Deoxyuridine monophosphate 251
Dephosphocoenzyme A 226[+], 227[§+]
Depside 109, 368
Desacetylcolchicine 333[§+], 334[+]
Desmosine 278[§+]
Desoxynupharidin 5
Desthiobiotin 230, 231[§]
Dethiogliotoxin 338[§+]
Detoxication 4
Dextran 66[+]
Dhurrin 191[+]
Diallyldisulphide 229[§+]
Diaminobutyric acid 193[§+]
Diaminopimelic acid 257, 269[+], 270[§+], 378
Diatrene 98[§]
N,N'-Dibenzoylornithine, *see* Ornithuric acid
Dicarboxylic acid, from alkanes 92[+]
Dicatenarin 105[§+]
Dicoumarol 348[§+]

Digitalose 59[§]
Digitoxigenin glycoside 158[§+]
Dihydrobiopterin 40
Dihydrochalcon 358[+]
7, 8-Dihydro-7, 8-dihydroxykynurenic acid
   308[§+]
Dihydrodiol 41
Dihydrodipicolinic acid 257, 258[§], 269[+],
   270[§+]
Dihydroflavanol 357, 358[+], 359[§+]
Dihydrofolic acid 215[§+]
Dihydrofolic acid reductase 33
Dihydroisopropylfuroquinolines 177[§+]
Dihydrolipoic acid amide dehydrogenase 33
Dihydroorotase 250
Dihydroorotic acid 250[§+]
Dihydroorotic aminohydrolase, *see*
   Dihydroorotase
Dihydroorotic dehydrogenase 250[§]
Dihydropteroic acid 215[§+]
Dihydropteroic acid sythetase complex 215
Dihydropyrimidine hydrase 251
Dihydrosphingosine 95[§+], 96[+]
Dihydrostreptomycin 78
Dihydrothymine 251[§+]
Dihydrouracil 251[§+]
Dihydroxy derivatives 41
Dihydroxyacetone 58[§]
3, 4-Dihydroxybenzoic acid 382[§+], 383[+]
2, 5-Dihydroxybenzoic acid 42
3, 4-Dihydroxybenzoic acid glucoside
   382[§+]
3, 4-Dihydroxybenzyl alcohol 382[§+], 383[+]
3, 4-Dihydroxybenzyl alcohol glucoside
   382[§+]
Dihydroxycholesterol 156[+], 157[§+]
2, 4-Dihydroxycinnamic acid 347[§+], 355
2, 4-Dihydroxycinnamic acid glucoside
   347[§+]
5, 6-Dihydroxyindole 336[§+], 338[+]
5, 6-Dihydroxyindole-2-carboxylic acid
   336[§+], 338[+]
Dihydroxyisovaleric acid 246[§+]
Dihydroxykynurenine 306
Dihydroxykynurenic acid 306[+], 308[§+], 309
2,4-Dihydroxy-7-methoxy-2,H-1,4-
   benzoxazine-3-one, *see* DIMBOA
α,β-Dihydroxy-β-methylvaleric acid 246[§]
Dihydroxyphenylacetaldehyde 323[§], 329[§]
Dihydroxyphenylacetic acid 309
Dihydroxyphenylalanine 318, 320[§], 337[+]
Dihydroxyphenylethylamine-β-hydroxylase
   318

DIMBOA 373, 374§
Dimethoxyhydroquinone 370+
3,5-Dimethoxy-4-hydroxyphenylethylamine 318+, 319§+
Dimethyladenine 212, 213§
Dimethylallylpyrophosphate 124+, 125§+
Dimethylallyltryptamine 295
Dimethylallyltryptophan 294§+, 293
Dimethylamino sugar 68
Dimethyldisulphide 237+
Dimethylglucose 59§
Dimethylglycine 189§+
Dimethylguanine 213§+
Dimethylpropiothetin 234, 235§+
Dimethylquinazoline 312§+
6,7-Dimethyl-8-ribityllumazin 216+, 217§+
Dimethylsulphide 236+
Dimethyltaurine 226§+
Dimethyltryptamine 291§+
Dimethyltryptamine-N-oxide 290§+
Diosgenin 159§+
Dioxygenases, in degradation of tryptophan 38, 366
  in hypotaurine formation 226
Diphosphothiamine disulphide 239§+
2,6-Dipicolinic acid 257+, 258§+
Disaccharides 62
Disulphides 229
Diterpenes 122, 133+
Dodecenal 7§
DOPA, see Dihydroxyphenylalanine
Dopamine 318, 319§+, 323§, 329§
Dopachrome 336§+, 337+
dUDP-sugar 62

Eburicoic acid 154, 155§
Ecgonine 261§, 262§+
Echimidinic acid 265§+, 266
Echinorine 176, 177§+
Echinulin 373§+
Eicosapentaenoic acid 88+, 89§+
Eicosatetraenoic acid 88+, 89§+
Eicosatrienoic acid 88+, 89§+
Elastin 278
Ellagic acid 368§
Ellagic tannin 368§
Elymoclavine 294§+, 295+
Emetine 322§
Emodin 105§+
Emulsin 65, 192
Endocrocin 105§+
Endopeptidases, splitting of peptidoglucans 380

Enzyme systems 22
Enzymes 22
  activating 27
  of primary metabolism 25
Enzymology 25
Ephedrine 320§+
Epicinchonidine 302§
Epicinchonine 302§
Epimerases, action on UDP-sugars 62
Epimerization of sugars 57
Epinephrine 318+, 319§+
Epiquinamine 302§
Epiquinidine 302§
Epiquinine 302§
Epoxy groups, formation in acetylene compounds 102
  in mixed function oxygenation 89, 147, 232, 308
7,8-Epoxy-7,8-dihydrokynurenic acid 308§+
Epoxypipecolic acid 272§+, 273§
Ergochromes 104
Ergoline alkaloids 293+
Ergosterol 50, 154+, 155§, 163
Ergot, alkaloids in 293
Ergothioneine 188, 285§+
Erysodienone 328+, 329§+
Erysodine 329§+, 330+
Erythratine 329§+, 330
Erythrina alkaloids 328+
Erythritol 75§+
Erythroidine 329§+, 330+
Erthyromycins 115+, 116§+
Erythronolides 115+, 116§+
Erythrose 58§
Erythroskyrin 241§+
Erythrulose 58§
Esterase, formation of depsides 109
Ethanolamine 186§+
Ethylamine 188+
2-Ethyl-4-methylquinazolone 312§+
Euphol 149§+
Excretions 3
Excretory metabolism 4

Falcarinone 98§
Farnesene 130+
Farnesol 130+
Farnesyl pyrophosphate 125§+
Fats 92+
Fatty acid elongation complex 85, 94
Fatty acid ester 92+
Fatty acid peroxidase 91
Fatty acid synthesis 83

Fatty acid synthetase 83
 model for the structure of 84 §
Fatty acids, degradation of 90
 activation of 90
 hydroxylated 87⁺
 long chain 97
 methylation of 86
 saturated 83⁺
 thiolytic cleavage of 82, 90
Ferrichrome 373, 374 §⁺
Ferrimycins 373, 374 §
Ferriprotoporphyrin IX 36
Ferroprotoporphyrin IX 38
Ferulic acid 344 §⁺, 345, 349, 366 §
Fervenulin 218 §
Flavacol 371⁺, 372 §
Flavan-3,4-diols 358⁺, 359 §
Flavanones 358
Flavans 358⁺, 359 §
Flavin adenine dinucleotide 34, 216
Flavin enzymes 34, 36
Flavin mononucleotide 34, 216
Flavone 359 §⁺
Flavonoid 358 §⁺
Flavonol 359 §⁺
Floral pigments, ecological significance 8
Foetor hepaticus 236
Folicanthine 301 §⁺
Folic acid derivatives 213
Formaldehyde, active 46
Formamide 287⁺
Formic acid 120⁺, 185
 activated 46, 47 §
Formimino glutamic acid 46, 286 §⁺
Formimino glycine 46, 221 §⁺
Formiminotetrahydrofolic acid 46⁺, 48 §
2-N-Formylaminoacetophenone 312 §, 313⁺
Formyl aspartic acid 287 §⁺
5-(Formylethyl)-4,6-dihydroxypicolinic acid
 308 §, 309⁺
Formyl glutamic acid 286 §⁺
Formyloglycinamide ribonucleotide 209 §,
 210⁺
Formylglycinamidine ribonucleotide 209 §,
 210⁺
Formylkynurenine 305 §⁺
Formyl orsellinic acid 109⁺, 110 §
Formyl tetrahydrofolic acids 46⁺, 47 §
Formyl tetrahydrofolic acid synthetase 46
Fraxetin 345 §
Friedelin 151 §⁺
Fructose 58 §
Fructose-6-phosphate 67 §

D-Fucose 59 §
L-Fucose 58 §, 68⁺
Fumaric acid 119⁺, 210, 249 §, 279 §
Fumigatin 110⁺
Funtumafrine 161 §⁺
Funtumine 161 §⁺
Furan derivatives, formation from
 acetylene derivatives 101
Furocoumarins 346
Furoquinoline alkaloids 177 §⁺
Fusaric acid 257⁺, 258 §

Galactaric acid 73 §⁺
Galactinol 65⁺, 76
Galactonic acid 73⁺
Galactose 58 §
Galactose-1 phosphate 61⁺, 72 §
Galacturonic acids, methylated, as precursors
 of pectins 64
Galanthamine 330⁺, 331 §
Galanthine 330⁺, 332 §⁺
Galegine 280 §⁺
Gallic acid 110⁺, 111 §⁺, 170 §⁺, 366
Gangliosides 95, 96
GDP-fucose 69 §⁺
GDP-mannose 69 §
GDP-mannuronic acid 72⁺
GDP-sugar 62
Genins 64
Gentianin 138⁺, 139 §⁺
Gentianose 62, 63 §
Gentiopicroside 137⁺, 139 §⁺
Geraniol 130⁺, 304
Geranylgeraniol 133⁺
Geranylgeranyl pyrophosphate 125⁺
Geranyl pyrophosphate 125 §⁺
Germanicol 151 §⁺
Giberellic acid 134 §⁺, 135
Giberellins 135⁺
Gitogenin 159
Glaucanic acid 118⁺, 120 §⁺
Glaucine type 323⁺, 325 §⁺
Glauconic acid 118⁺, 120 §⁺
Glochicidine 282 §
Glochidine 282 §
Glomerin 7 §
Glow worm 229
Glucaric acid 73 §⁺
Glucobarabarin 196 §⁺
Glucobrassicanapin 194 §
Glucobrassicin 195 §⁺, 196⁺, 197 §
 degradation of 197 §
Glucoconringiin 195 §⁺, 196⁺

Glucocoumaryl alcohol 349 §+
Glucoibervirin 194 §, 195 §+, 196
Glucomannans 64
Gluconapin 194 §
Gluconasturtiin 194 §, 195 §+
Gluconic acid 73+
Glucosamine 67+
Glucosamine-6-phosphate 67 §+
Glucosamine-6-phosphate-N-acylase 67
Glucose 58 §
Glucose-1,6-diphosphate 61
Glucose oxidase 34
Glucose-1-phosphate 61+, 65, 72
Glucose-6-phosphate 61, 72 §+
Glucose-6-phosphate dehydrogenase 33
β-Glucosidases 65
    formation of sclerotin 382
    in degradation of cyanogenetic
        glycosides 192
Glucose inolates 194+
Glucosides, see Glycosides
Glucotropaeolin 194 §, 195 §+, 196+
Glucuronic acid 77 §+
Glucuronic acid lactone 74 §+
Glucuronic acid-1-phosphate 74 §+
Glucuronides 65+
Glutamineh-S-aryl transferase 232
Glutamic acid 33, 193 §+, 194+, 259+, 260 §+,
    266, 286 §+, 287+, 378
L-Glutamic acid dehydrogenase 34
Glutamic acid-γ-semialdehyde 259+, 260 §+
γ-Glutamyl-β-aminopropionitrile 193 §+
γ-Glutamyl-β-cyanoalanine 193 §+
Glutaryl CoA 273, 311 §+, 312+
Glutathione 371+, 380
Glycerine, see Glycerol
Glyceraldehyde 58 §
Glyceraldehyde phosphate 252, 253 §, 289 §+,
    292+
Glycerol 75 §, 76+, 252
Glycerol-1-phosphate 92, 93 §
Glycine 202+, 223 §, 375 §
Glycinamide ribonucleotide 209 §+, 210+
Glycinamide ribonucleotide synthetase 210
Glycinamide ribonucleotide transformylase
    210
Glycine betaine, see Betaine
Glycine conjugates 222+
Glycine oxidase 187
Glycogen 64, 65+, 66+
Glycolaldehyde 202+
Glycolate oxidase 120
Glycolic acid 121 §, 202+

Glycosidases, in degradation of
    peptidoglycans 380
Glycosides 59+
    cyanogenic 190+, 191 §, 192 §+
    degradation of 190, 191
N-Glycosides 59, 60 §, 200+
O-Glycosides 59, 60 §
S-Glycosides 59, 60 §
Glycosidization of flavonoids 358, 360
Glycosyl compounds 60 §
Glyoxylic acid 119 §+, 121+, 187+, 202+, 219+
Glyoxylic acid cycle 118
Glycyrrhetic acid 152 §
Gout 220
Gramicidins 380
Gramine 50, 291 §+, 292+, 293 §+
Griseofulvin 108 §+
Griseophenone 108 §+, 109+
Guaiacum resin 354
Guaianolides 130+
Guaiaretic acid 353 §, 354
Guanidine acetic acid 189 §
γ-Guanidino butyramide 41
Guanidine compounds, secondary 280+
Guanosine monophosphate 211 §+
Gulonic acid 74 §+
Gulonic acid lactone 74 §+
Gutta-percha 122

Hadacidin 373, 374 §, 375+
Haem 38, 204 §+, 205+
Haem a 205+
Haematoporphyrin IX 205
Haemoglobin 206 §
    degradation of 205
Haemoproteins 374
Halogenation 37
Halogen peroxidases 37
Harmalan 296 §+
Harman 296 §
Helicin 369 §+
Heliosupin 264 §
Helleborin 158+
Hemicellulose 72+, 73+
Heptoses 57, 58 §
Heptylpenicillin 377 §
Herbipoline 208
Hercynine 189 §+, 285 §+
Herniarin 345 §, 347 §+
Heterosides 62+
Hexoses 57, 58 §
Hexylamine 186 §, 188 §+
Himalachene 130+, 132 §+

Hippuric acid 222[+], 223[§]
Histamine 187[§+], 282[+], 287[§+]
  degradation 287
Histidase 286
Histidinal 284[§+]
Histidine 189[§], 284[§+], 285[§], 286[§], 287[§]
Histidinaemia 287
Histidine deaminase 286
Histidine transaminase 287
Histidinol 284[§+]
Histidinol dehydrogenase 284
Histidinol phosphate 284[§+]
Histidinol phosphate phosphatase 284
Holaphyllamine 161[§+]
Holaphylline 161[§+]
Holochrome 205
Holosides 62[+]
  degradation of 65
Homoaconitic acid 269[+], 271[§+]
Homoarbutin 342[§+]
Homoarginine 281[§+]
Homocitric acid 269[+], 271[§+]
Homocysteine 233[§+]
Homocysteine methyltransferase 47
Homogenates 22
Homogentisic acid 342[§+]
Homoisocitric acid 269[+], 271[§+]
Homothionine 195[§], 196[+]
Homoserine 233[§+]
Homostachydrine 189[§+]
Hordenine 318[+], 319[§+]
Humic acids 110[+]
Humulene 130[+], 132[§+]
Hyaluronic acid 66[+], 73
Hydantoin propionic acid 286[§+]
Hydrangenol 356[§+]
Hydrangic acid 356[§]
Hydrastine 326[§+], 327[+]
Hydrogen cyanide 7[§], 193[§+]
Hydrogen peroxide 34
Hydroquinone 369[§+], 370
Hydroquinone derivatives, isoprenoids 341
Hydroxamic acid 373[+], 375[§]
ω-Hydroxy acids 92[+]
3-Hydroxy-2-aminobenzoyl pyruvate 305[§+]
N-Hydroxyamino acids 191, 192[§+], 196, 374
3-Hydroxy anthranilic acid 174[§+], 175[§+],
  310[+], 311[§+], 383
Hydroxy aspergillic acid 372[§+], 373[+]
o-Hydroxybenzoic acid, see Salicylic acid
p-Hydroxybenzoic acid 172[§+], 340, 366[+],
  369[§]
Hydroxybenzylpenicillin 377[§]

Hydroxybenzyl mustard oil 194[§+]
2-Hydroxy-3-carboxyisocaproic acid 243[§+]
Hydroxy corticosterone 156[+], 157[§+]
Hydroxycotinine 256[§]
Hydroxycoumarin 348[§+]
Hydroxy dimethyltryptamine 290[§+], 291[§+]
Hydroxy dimethyltryptamine-N-oxide 290[§+],
  291[§+]
α-Hydroxyethyl-2-thiamine pyrophosphate
  51[§+]
5-Hydroxyferulic acid 344[§+], 345
Hydroxyglutinen 151[§+]
Hydroxyhomoarginine 281[§+]
6-Hydroxyhyoscyamine 263[§+]
4-Hydroxyimidazole 221[§+]
5-Hydroxyindole acetaldehyde 314[§+]
5-Hydroxyindole acetic acid 314[§+]
3-Hydroxykynuramine 307[§+]
6-Hydroxykynurenic acid 306[+]
3-Hydroxykynurenine 174, 306[+], 307, 309
5-Hydroxykynurenine 306
Hydroxyl radicals 37
Hydroxylation, NIH shift 41
Hydroxylupanine 275[§+], 276
Hydroxylysine 278
2-Hydroxy-4-methoxycinnamic acid 347[§+]
2-Hydroxy-4-methoxycinnamic acid glucoside
  347[§+]
4-Hydroxy-0-methylandrocymbine 333[§+]
4-Hydroxymethyl-o-benzoquinone 382[§+],
  383[+]
5-Hydroxymethylcytosine 251[+]
3-Hydroxy-3-methylglutaryl CoA 123[§+],
  244[§+], 245
Hydroxymethylindole 197[§+]
Hydroxymethylpyrimidine 253[§+]
Hydroxymethyl transferase 46
Hydroxymethyltryptamine 290[§+]
α-Hydroxymuconic acid-ε-semialdehyde 311[§+]
Hydroxynicotine 256[§+], 257[§+]
6-Hydroxynicotinic acid 254[§+]
α-Hydroxynitriles 193[§+]
4-Hydroxynorlaudanosoline 335[§+]
Hydroxyornithine 374
m-Hydroxyphenyl alanine 43
Hydroxyphenyl pyruvate oxygenase, NIH
  shift 41, 341
Hydroxyphenyl pyruvic acid 317[§+], 318[+],
  341, 342[§+]
Hydroxypipecolic acid 272[§+], 273[+]
Hydroxyprogesterone 157[§+]
Hydroxyproline 189[§]
Hydroxypyridyloxobutyric acid 256[§+]

8-Hydroxyquinaldic acid 306+, 307§+
4-Hydroxyquinoline 306+, 307§+
Hydroxytryptamine 43, 291§+, 314§+
5-Hydroxytryptophan 41, 290§+, 292+, 314§
5-Hydroxytryptophol 314§+
Hygric acid 189§+
Hygrine 262§+
Hyoscyamine 261, 262+, 263§
Hypaphorine 189§+
Hypotaurine 225+, 226§+
Hypotaurocyamine 226§+
Hypoxanthine 219+, 220§

Iboga type alkaloids 296+, 297+, 299+
Iditol 75§+
Imidazole acetaldehyde 287§+, 288+
Imidazole acetic acid 287§+, 288+
Imidazole acetol phosphate 284§+
Imidazole ethanol 285§+
Imidazole glycerol phosphate 283§+, 284§
Imidazole glycerol phosphate dehydrase
     284
Imidazole hydroxyacetone phosphate 284
Imidazole lactic acid 285§+
Imidazole propionic acid 285§+
Imidazole pyruvic acid 285§+
Imidazolone acetic acid 287§+, 288+
Imidazolone propionic acid 286§+, 287+
Indican, plant 315§
Indicaxanthin 340+
Indigo 315§+
Indirubin 315§+
Indole 174+, 289§+, 292+, 315§+
Indoleacetaldehyde 313§+
Indole-3-acetamide 41
Indoleacetic acid 313§+
Indole alkylamines 292+
Indole glycerol phosphate 289§+, 292+
Indole derivatives, formation from
     m-tyrosine 336
Indolenine derivatives 300+, 301§+, 304§+,
     305+
5,6-Indolequinone-2-carboxylic acid 336§+
Indolequinones 336§+, 338+
Indolylacetonitrile 197§+
Indoxyl 315§+
Indoxyl-5-ketogluconate 316
Inosine monophosphate 208+, 209§+, 210+,
     283§+
D-Inositol 77§+
L-Inositol 77§+
meso-Inositol 77§+, 79§
scyllo-Inositol 77§+

meso-Inositol hexaphosphate 76
meso-Inositol-1-phosphate 77§+
Inositols 76+
Inososes 77
Insulin 380
Internal mixed function oxidation 40
Ionone ring system 142§
Ionones 142+
Ipecoside 322§
Iridodial 136, 137§
Iridoids 136+, 138, 296, 302
     in indole alkaloids 296
     in quinuclidine nucleus 302
Isatan B 315
Isethionic acid 226§+
Isoalkanes 97§+
Isoamylase 65
Isobutyryl CoA 85
Isocaproic acid 157§+
Isocaproyl CoA 85
Isocephalosporin 376§+
Isocitrate lyase 121
Isocitric acid 121§+
Isoferulic acid 355+
Isoflavones 362§+
Isoquinoline alkaloids 321+
Isoleucine 246§+, 265§, 373
Isomerizations
     in acetylene compounds 99, 100
     in mustard oils and rhodanid 197
     in sugars 57
Isopelletierine 273+, 274§+
Isopenicillin N 376§+, 377
Isopentenyl pyrophosphate 123§+, 124, 125,
     126, 244§+ 293, 340
     polymerization of 124
Isopentenyl pyrophosphate isomerase 124
Isoprene 122
     activated 123
Isoprene rule 122
2-Isopropylmalic acid 243§+
Isopyrrolnitrin 304§+
Isoretronecanol 264+, 265§+
Isothebaine 323+, 325§+
Isothiocyanates, see Mustard oils
Isotopes
     atom % excess 15
     radioactive 15
     stable 15
Isotopically labelled compounds, feeding of
     15
Isovaleryl CoA 85, 244§+
Isoxanthopterin, degradation of 222§

Javanicin 168⁺
Juglon 168, 169§⁺

Kanamycin 78
Kaurenal 134§⁺, 135⁺
Kaurene 134§⁺, 135⁺
Kaurenic acid 134§⁺, 135⁺
Kaurenol 134§⁺, 135⁺
Kawain 354⁺, 355§
β-Ketoacyl thiolase 90
α-Keto acid decarboxylase 50, 237
α-Keto acid oxime 191⁺, 192§⁺, 196
α-Ketoadipic acid 269⁺, 271§⁺, 311§⁺, 312⁺
α-Keto-ε-aminocaproic acid 272§⁺, 277§⁺
α-Keto-ε-aminovaleric acid 259⁺, 260§⁺, 261⁺
α-Ketobutyric acid 246§
2-Keto-3-carboxyisocaproic acid 243§⁺
α-Ketoglutaric acid 119§⁺, 269, 271§, 306, 308§⁺, 309
2-Keto-L-gulonic acid lactone 74§⁺
Ketoinositols 77§⁺
2-Ketoisocaproic acid 243§⁺, 244§⁺
α-Ketoisovaleric acid 240§⁺, 243§⁺, 244, 246
A-Keto-β-methylvaleric acid 246§⁺, 247§⁺
Ketones 192⁺
α-Ketopantoic acid 240§⁺
α-Ketopropionic acid 246§
Ketosequoyitol 77§⁺
Ketoses 57
Kinases 61
Kinins 209
Krebs-Henseleit cycle 279
Kryptogenin 159§⁺
Kynuramine 307§⁺
Kynurenic acid 305§⁺, 306⁺, 307§, 308§
  degradation of 307
Kynureninase 174, 306
Kynurenine 305§⁺, 306⁺, 307§
Kynurenine transaminase 305, 306

Laburnine 264, 265§⁺
Laccases 36, 337, 350
Lactose 62, 63⁺, 73
α-Lactyl-2-thiamine diphosphate 51§⁺
Lanosterol 148§⁺, 153, 154§
Lariciresinol 353§, 354⁺
Latex 23, 126
Lathyrine 281§⁺
Lathyrism factor 194
Laurencine 98§
Lawson 168, 169§⁺
Lecanoric acid 109⁺, 110§⁺
Lecithin 93§⁺

Leucine 243§⁺, 372§
Leucodopachrome 336§⁺, 337⁺
Leucopterin 214§⁺
Lichens, phenol carboxylic acid derivatives of 109
Light emitting by micro-organisms 229
Lignans 353
Lignin 50, 349⁺
  attachment to cellulose 352
  mechanical function of 7
  secondary methylation 352
Limonene 129§, 130⁺
Linamarin 191§⁺
Lindefolidine 264§⁺, 265
Lindefoline 264§
Linoleic acid 89§
  degradation of 91
Linolenic acid 89§⁺
Linolyl CoA 88§⁺, 91§
Lipases 93
Lipoic acid 81, 82§
Lipopolysaccharides 64⁺
Lipoproteins 96⁺
Lobelia alkaloids 273⁺
Lobeline 274⁺
Lobinaline 274§⁺
Loganin 137§, 138§⁺, 139§
Longifolene 132§⁺
Lotaustralin 191§⁺
Luciferase 229, 230
Luciferin 229⁺, 230
Lupanine 199⁺, 275§⁺, 276⁺
Lupeol 150§⁺, 151⁺
Lupinine 275§⁺, 276⁺
Lupulon 103§
Lycoctonine 135§, 136⁺
Lycopene 140§⁺, 141⁺, 143§⁺
Lycorine 330⁺, 332§⁺
Lysergic acid 293, 294§⁺, 295⁺
Lysergic acid ethanolamide 293, 294§⁺
Lysine 269⁺, 270§⁺, 271§⁺, 272§, 378
Lysozyme 380

Macdougallin 155§
Macrolide antibiotics 115⁺
Magnamycin 116§, 117⁺, 241
Magnesium protoporphyrin IX 204§⁺, 205⁺
Malic acid 252⁺, 254§⁺
Malonic acid 251§⁺, 252⁺
Malonamoyl CoA, as starter in tetracyclin formation 107§
Malonyl CoA 82⁺
  as starter in tetracyclin formation 107§

Maltose 73
Mannans 64
Mannich condensation 199
Mannitol 75 §+, 76
Mannoheptulose 58 §
Mannose 58 §
Mannose-1-phosphate 61+
Manool 133 §+
Matricaria ester 100 §
Matricaria lactone 98+
Matricin 130+, 132 §+
Melanins 336+, 337 §
Melanocytes 336
Melatonin 314 §+
Melilotic acid, β-hydroxy 348+
Menadione 169 §+
Menthol 130+
Mercapturic acid 231 §+, 232+
Mescaline 318+, 319 §+
Mesobilin 206 §+, 207+
Mesobilirubin 206 §+, 207+
Mesobilirubinogen 206 §+, 207+
Metal flavoproteins 35
Metapyrocatechase 38
Methacrylic acid 7 §
Methional 236 §+
Methionine 47, 233 §+
Methoxydimethyltryptamine 290 §+
Methoxydimethyltryptamine-N-oxide 290 §+
Methoxyhydroquinone 370+
5-Methoxyindole acetic acid 314 §+
5-Methoxytryptamine 314 §+
Methylacetoacetyl CoA 247 §+
Methyl adenine 50, 213 §+
5-Methyl adenosine 234 §+, 235
4-Methyl-5-alanylthiazole 237+, 238 §+
Methylalliin 228 §+
Methylamine 187+, 188 §+
N-Methylated amino acids 188
N-Methylaminoethanol 187+, 188 §+
γ-Methylaminobutyraldehyde 266 §+
N-Methylaminobutyric acid 257 §
2-Methylamino-L-glucose 68+
3-Methylaminomethylindole 292+, 293 §+
O-Methylandrocymbine 334+
Methylarbutin 365, 369+
Methylaspartic acid 237 §+
Methylated fatty acids 115+
Methylbarbituric acid 251 §+
Methylbutadiene 122
Methylbutyric acid 247+, 261
2-Methylbutyryl CoA 85, 247 §+

N'-Methyl-5-carboxamido-3-pyridone 311, 312 §+
Methyl cobalamine 47
3-Methylcrotonyl CoA 244 §+
Methylcrotonyl CoA-carboxylase 45, 244
N-Methyl-3-cyanopyridine 253+, 254 §+
N-Methyl-3-cyanopyridone 253+, 254 §+
Methylcysteine 50, 228 §+
5-Methylcytosine 46, 249+, 250, 251
Methylecgonine 261
Methylenedihydrofolic acid 46, 47 §+
Methylenedioxy groups 327+
γ-Methyleneglutaric acid 254 §+
Methylenestearic acid 86 §+
Methylenetetrahydrofolic acid 46, 47 §+
Methylfucose 64
3-Methylglutaconyl CoA 244 §+
Methylguanine 212+, 213 §
Methylhistamine 284+
Methylhistidine 284+
Methylhydroxybutyryl CoA 247 §+
4-Methyl-5-hydroxyethylthiazole 238+
4-Methyl-5-hydroxyethylthiazole phosphate 238 §+
2-Methyl-5-hydroxymethyl-6-amino-pyrimidine diphosphate 238 §+
6-Methyl-7-hydroxy-8-ribityllumazine 217 §+
N'-Methylhypoxanthine 213 §+
Methylimidazole acetaldehyde 284+
Methylimidazole acetic acid 284+
3-Methyl-2-ketobutyric acid 265 §+
Methylmalonic acid 251 §+
Methylmalonyl CoA 114 §+, 240 §+
Methylmercaptopropyl amine 186 §+
S-Methylmethionine 234+, 235 §+
N¹-Methylnicotinamide 311+, 312 §+
O-Methylnorbelladine 330, 331 §+
Methylpentoses 57
Methylpipecolic acid 189 §+
Methylprenylhydroquinone 342 §+
6-Methylpretetramide 106, 107 §+
N-Methylputrescine 262 §, 266 §+
Methylpyrrolinium cation 254, 266 §+
4-Methylquinazoline 312 §+
Methyltetrahydrofolic acid 47 §+
5'-Methylthioadenosine 234 §+, 235 §+
Methylthiopropyl mustard oil 194 §+
Methyl transferases 48
N-Methyltyramine 291 §+, 300+, 301 §+, 318+, 319 §+
5-Methyluracil, see Thymine
Methyl urea 220+
7-Methyluric acid, degradation of 220

Methylvaleric acid 115 $^{\S+}$

Methylxanthine 212 $^{\S+}$, 220
  degradation of 220

Methylxylose 64

Methymycin 116 $^{\S+}$, 117 $^{+}$

Mevaldic acid 123 $^{\S+}$

Mevaldic acid reductase 33, 124

Mevalonic acid 123 $^{+}$, 124 $^{\S+}$, 373

Mevalonic acid monophosphate 123 $^{+}$, 124 $^{\S+}$

Mevalonic acid pyrophosphate 123 $^{+}$, 124 $^{\S+}$

Mimosine 278 $^{\S+}$

Molybdenum 35

Monomethylspirilloxanthine 143 $^{\S+}$

Monosaccharides 57

Monoterpenes 122, 129 $^{+}$

Morphine 323 $^{+}$, 324 $^{\S+}$

Morphine alkaloids 323 $^{+}$, 324 $^{\S+}$, 325 §

Mucolipids 96

Multiflorenol 151 $^{\S+}$

Multiflorine 275 $^{\S+}$, 276 $^{+}$

Muramic acid 67 $^{\S+}$, 68 $^{+}$

Mustard oils 194 $^{+}$, 197 $^{+}$

Mutants 20

Mutases 61

Mutations 20

Mycaminose 68 $^{+}$

Mycarose 50, 70 $^{\S+}$

Mycocerosic acid 115 $^{\S+}$

Mycophenolic acid 103 §

Myoinositol 76

Myrcene 129 $^{\S+}$

Myrosinase 197

NAD, A side of 32
  B side of 32
  H$_A$ of 32
  H$_B$ of 32

Naphthalene 231 §
  epoxide 231 $^{\S+}$

Naphthalene premercapturic acid 231 $^{\S+}$

Naphthohydroquinone 169 $^{\S+}$, 168 $^{+}$

Naphthol 168 $^{+}$, 169 $^{\S+}$

Naphthoquinone derivatives 168 $^{+}$, 169 $^{\S+}$

Naphthyl mercapturic acid 231 $^{\S+}$

Narwedine 330 $^{+}$, 331 $^{\S+}$

Natural products, secondary
    as defence secretions 8
    diminution of vitality 8
    elimination from metabolism 4
    general 3
    in hair 6
    in haemolymph 6

  in skin 6
  location of accumulation 5
  location of synthesis 5, 9
  mechanical function of 7
  nutrition by animals 5
  sexual life and reproduction 8
  significance of 7
  storage of 5, 7
  toleration of 6

Necic acids 264 $^{+}$

Necine bases 264

Neoaspergillic acids 371 $^{+}$, 372 $^{\S+}$

Neoflavenoids 364 $^{+}$

Neohydroxyaspergillic acid 371 $^{+}$, 372 $^{\S+}$

Neomycin 78

Neopinone 323 $^{+}$, 324 $^{\S+}$

Neryl pyrophosphate 129 $^{\S+}$

Neurosporene 140 $^{\S+}$, 141 $^{\S+}$, 142

Nicotinamide 252 $^{+}$, 253 $^{\S+}$, 254, 311, 312 §

Nicotinamide adenine dinucleotide 252 $^{+}$, 253 $^{\S+}$

Nicotine 50, 255 §, 256 §, 257 §, 266 $^{\S+}$
  formation of pyrrol ring of 266

Nicotine-N-oxide 256 $^{\S+}$

Nicotinic acid 252 $^{+}$, 253 $^{\S+}$, 255 §, 266 $^{\S+}$, 277
  conjugation of 222, 267

Nicotinic acid adenine dinucleotide 252 $^{+}$, 253 $^{\S+}$

Nicotinic acid mononucleotide 252, 253 $^{\S+}$

NIH-shift 42, 292, 341, 345

Nitriles 190 $^{+}$, 192 $^{\S+}$, 197 $^{+}$
  Strecker synthesis of 194

Nitro compounds 249 $^{+}$, 304 $^{+}$, 305 $^{+}$, 334 $^{+}$

β-Nitropropionic acid 249 $^{+}$

Nonacosan 97 §

Nonacosanol-(15) 97 §

Nonacosanone-(15) 97 §

Nonadrides 118, 120 $^{+}$

Noradrenaline 318 $^{+}$, 319 $^{\S+}$, 327 §, 335 §

Norbelladine 331 $^{\S+}$, 332

Norbelladine methyl ether 330 $^{+}$, 331 $^{\S+}$, 332 $^{\S+}$

Norepinephrine 318 $^{+}$, 319 $^{\S+}$

Norlaudanosine 322 $^{+}$, 323 $^{\S+}$

Norlaudanosoline 323 $^{+}$, 325 §

Norlupinan 275

Nornicotine 255 $^{\S+}$, 256 $^{\S+}$

Norpluviine 330 $^{+}$, 332 $^{\S+}$

Noviose 50, 70 §, 71 $^{+}$, 340

Novobiocin 340 $^{+}$, 341 §

Nucleoside diphosphate sugars 61 $^{+}$

Nucleoside phosphate 61 $^{+}$

Nucleosides 65 $^{+}$

Ocimene 129 §+
Octopine 280 §+
Oestrone 156+, 157 §+
Oils, fatty 92+
Oleandrose 59+
Oleanolic acid 151 §+
Oleic acid 86 §
Oleyl CoA 88 §+
Oligopeptides 371
Oligosaccharides 62, 64
Olivil 353 §, 354
Omnochromes 309+
One-carbon metabolism 46, 213
Onic acids 73+
Onocerin 152+, 153 §+
Onocol, see Onocerin
Orcinol 111 §+
Orientaline 323, 325 §+
Orientalinol 323+, 325 §+, 334, 335 §
Orientalinone 323+, 325 §+
Ornithine 79 §, 259+, 260 §+, 261, 266, 267,
    279 §+
Ornithine conjugates 267+
Ornithuric acid 267+
Orotic acid 249+, 250 §+
Orotidine monophosphate 250 §+
Orotidyl decarboxylase 250
Orsellinic acid 109+ 110 §+
Oxalacetic acid 119 §+
Oxalic acid 120+, 121 §
Oxaloglutaric acid 269+, 271 §+
Oxalosuccinic acid 119 §+
γ-Oxalylcrotonic acid 311 §+, 312+
O-Oxalylhomoserine 120+, 121 §, 233
Oxidases 32
    in formation of sclerotin 382, 383
α-Oxidation 91
β-Oxidation 90
γ-Oxidation 92
Oxidative demethylation 41
Oxidoreductases 32
Oxodecanoylhistamine 282 §
2-Oxo-4-phenylbutyric acid 196 §+
2-Oxo-4-phenyl-4-hydroxybutyric acid 196 §+
Oxopipecolic acid 272 §+, 273+
Oxygenases 32
    mixed function 39
        in clavine alkaloids formation 295
        in p-coumaric acid formation 345
        in giberelline formation 135
        in histamine degradation 288
        in homogentisic acid formation 341
        in 3-hydroxyanthranilic acid formation 174

        in indole alkyl amines formation 292
        in kynurenic acid degradation 308
        in premercapturic acid formation 232
        in prostaglandin formation 88
        in sterol synthesis 153
        in tetracycline biosynthesis 106
        in tryptophan hydroxylation 42
        in xanthophyll formation 142
        NIH shift 42
        reaction mechanism of 41, 42, 43
        tyrosinase 337
Oxygenation of amines and thio compounds
    40
Oxynitrilases 192
Oxypyrrolnitrin 304
Oxytetracycline 106 §

Palmitic acid 87
Palmiton 87 §+
Panthetheine phosphate, see 4'-Phospho-
    pantetheine
Pantoic acid 240 §+
Panthothenic acid phosphate, see
    4'-Phosphopantothenic acid
Pantothenyl cysteine phosphate, see
    4'-Phosphopantothenylcysteine
Papaverine 323 §+
Paracotoin 354+, 355 §
Pectin 50, 64+, 66
Peganine 181 §+
Peltatin 353 §, 354
Penicillic acid 377 §, 378+
Penicillinase 378
Penicillins 375+, 376 §+, 377 §, 378+
    semisynthetic 377, 378
Pentenyl mustard oil 194 §+
Pentenylpenicillin 377 §
Pentoses 57
Peptide alkaloids from lysergic acid 295
Peptide bonds, see Acid amides
Peptides 371+
    cyclic 375+, 380+
Peptidoglucans 380
Peroxidases 36
    degradation of histidine 287
    dehydrogenation by 36
    hydroxylation by 36
    oxidation of polyphenols 337
Petroleum bacteria 92
Phenazine-1-carboxylic acid 182 §+
Phenazines 182 §+
Phenol carboxylic acids 109+

Phenol oxidases 36
  action on tyrosine 383
  in biosynthesis of colchicine 334
  in biosynthesis of *Erythrina* alkaloids 327
  in dimerization of anthracene derivatives 104
  in flavonoid polymerisation 361
  in isoquinoline alkaloid formation 323
  in lignin synthesis 350
  in melanin formation 337
  in radical formation 350
  in xanthommatin formation, tyrosinase 309
Phenol radical, addition to amines 201
Phenoxazine derivatives 175+, 309
Phenoxymethylpenicillin 377§
Phenylacetic acid 365
  conjugation 222, 267, 268
Phenylacetylglutamine 267§
Phenylalanine 274§, 317§+, 318+
Phenylalanine deaminase 345
Phenylalanine hydroxylase 40
Phenylalkylamines 318+
Phenyl bodies, C_6-bodies 365
2-Phenylchroman 357§
3-Phenylcoumarins 362§, 363+
4-Phenylcoumarins 364+
Phenylethane derivatives 365+
Phenylethylamine 187§+, 320§
Phenylethylisoquinoline alkaloid 333§, 334+
Phenylethyl mustard oil 194§+
Phenylglyceric acid 343§+
5-Phenyl-2-oxazolinethione 195§, 196+
Phenylpropane unit 317+
Phenyl pyruvate 343§
Phenyl pyruvic acid 317§+
2-Phenyl-4-quinolones 179+
Phenyl rings, formation from acetylene derivatives 100+
Phlobaphenes 361
Phloridzin 358§+
Phoroacetophenone derivatives 109+, 111§+
Phosphate cycle 29
Phosphate ester 27
Phosphates
  high-energy 27, 29
  low-energy 27, 29
Phosphatidic acid 92+, 93§+
Phosphatidylcholine 190§+
Phosphatidyldimethylethamolamine 190§+
Phosphatidylethanolamine 93+, 190§+
Phosphatidylserine 190§+
Phosphoenolpyruvate 27
Phosphohistidine 285

4'-Phosphopantetheine 83, 84§, 226+, 227§+
4'-Phosphopantothenic acid 226+, 227§+
4'-Phosphopantothenylcysteine 226+, 227§+
5-Phosphoribosyl-1-amine 209§+, 210+
1-Phosphoribosyl-5-aminoimidazole 209§+, 210+
N-(5'-Phosphoribosyl)-anthranilic acid 289§+
Phosphoribosyl–ATP 282+, 283§+
1-Phosphoribosyl-4-carboxamido-5-amino-imidazole 209§+, 210+, 283§+, 284+
5-Phosphoribosyl pyrophosphate 209§, 283§+, 289§
1-Phosphoribosyl-4-(*N*-succinyl)-carboxamido-5-aminoimidazole 209§+, 210+
(5'-Phosphoribulosyl)-anthranilic acid 289§+
Phosphorylated compounds 27
Phosphorylation of sugars 61
Phthienoic acid 115§+
Phycobiliproteins 207, 208+
Phycocyanins 207
Phycoerythrins 207
Phylloquinones 168
Phytane 122
Phytic acid 77+
Phytoene 127+, 128+, 140§, 141
Phytofluene 140§+, 141
Phytol 122, 133+, 205
PICE 205
Piceatannol 355§+
Picein 365
Picolinic acid 311§+
Pilocarpine 282§
Pilosine 282§
Pimaradiene 133§+
Pimelic acid 230§, 231
Pimpinellin 346§
Pinene 7§, 129§, 130+
Pinitols 77§+
Pinoresinol 351§+, 352+
Pinosylvin 355§+
Pipecolic acid 189§, 272§+, 273
Piperideine 254, 255§, 277§+
Piperideine carboxylic acids 255§, 272§+, 277§+
Piperideine-2,6-dicarboxylic acid 269+, 270§+
Piperidine derivatives, from lysine 112+
  polyketides 112+
Plasmologens 95+, 96
Plastoquinones 341+, 342§+
Polyketides 103+
Polymyxin 380
Polypeptides, cyclic 380

Polysaccharides 62, 66⁺

Polysaccharides 62, 66[+]
Polyterpenes 122
Polyuronides 73
Porphobilinogen 202, 203[§+]
Porphyria 202
Porphyrinogen decarboxylase 204
Porphyrin ring system, degradation of 205
Porphyrins 202[+]
Precursor 16, 18
Pregnane derivatives, sugar 57
Pregnenolone 156[+], 157[§+], 161[§]
Premercapturic acid 231[§+], 232
Prenylnaphthoquinone 169[§+]
Prenylpyrophosphate 342
Prephenic acid 317[§+]
Pretetramide derivatives 107[§+]
Previtamins D 163[+], 164[§+]
Primary metabolism 3
Primer, polysaccharide biosynthesis with 66
Pristane 122
Progesterone 156[+], 157[§+], 158[§], 160, 161[§+]
Proline 189[§], 259[+], 260[§+], 266, 339, 340
Propenylalliin 228[§+]
Propenylcysteine 228[§+]
Propenylsulphenic acid 228[§+]
Propionic acid 114[§+], 254[+]
2-N-Propionylaminoacetophenone 312[+]
Propionyl CoA 85, 114[§+], 247
    reactions of 115
Propylalliin 228[§+]
Propylamine 186[§+]
Propylcysteine 228[§+]
Propylpenicillin 377[§]
Prosopine 112[+], 113[§]
Prostaglandins 38, 88[+], 89[§+]
Protoalkaloids 198
Protocatechuic acid 170[§+], 309, 366[§+]
Protocatechuic aldehyde 330[+], 331[§+]
Protochlorophyll a 205
Protochlorophyllide a 204[§+], 205[+]
Protoporphyrin IX 204[§+], 205[+]
'Protoporphyrin-iron-chelating-enzyme' 205
Protoporphyrinogen IX 203[§+], 205[+]
Prunasine 191[§+]
Pseudans 178[§+]
Pseudoalkaloids 198
Pseudo indicans 136
Pseudopelletierine 273[+], 274[§+]
Pseudouridine 251[+]
Psilocine 291[§+]
Psilocybine 291[§+]
Pteridines 213[+], 214[+]
Pterins 213, 214

Punica alkaloids 273[+]
Purine alkaloids 212[+]
Purine derivatives 208[+]
    degradation of 219, 221
    degradation of methylated derivatives 220
Purpurin 169[§+]
Purpurin-3-carboxylic acid 169[§+]
Putrescine 189[§+], 234, 235[§], 259[+], 260[§+], 261, 264, 267, 266[§]
Pyocyanine 182[§+]
Pyridine carboxylic acids 310[+]
Pyridine nucleotide coenzymes 32
Pyridoxal-5-phosphate 52
Pyridoxamine-5-phosphate 52
Pyridylacetic acid 256[§+]
γ-(3-Pyridyl)-γ-oxobutyric acid 256[§+]
Pyrogallol 111[§+]
α-Pyrones, monocyclic 354
Pyrophosphate 27
Pyrroline 259[+], 260[§+], 267
Pyrroline carboxylic acids 259[+], 260[§+]
Pyrrolizidine alkaloids 264[+]
Pyrrolnitrins 304[§+]
Pyrrolopyrimidines 218[§+]
Pyruvate 306
    activated 51
    oxidative decarboxylation of 81
Pyruvate decarboxylase 50
Pyruvate-α-lactyl-2'-thiamine diphosphate 50
Pyruvic acid 269, 270, 280

Quebrachitol 77[§+]
Quercetin-D-glucopyranosido uronic acid 64
Quercitols 76
Quillaic acid 152[§]
Quinaldic acid 306[+], 307[§+]
Quinamine 302[§]
Quinazoline alkaloids 181[§+]
Quinazolines 312[+]
Quinic acid 167[§+]
Quinidine 302[§]
Quinine 303[§+], 304[+]
Quinoline alkaloids 302[+]
Quinolinic acid 252[+], 253[§+], 310[+], 311[§+]
Quinolinic acid mononucleotide 252[+]
Quinolinic pathway, in tryptophan degradation 307
Quinolizidine alkaloids 275[§+]
Quinones, addition of amines to 201
Quinonimines, in actinomycin synthesis 176
Quinuclidine nucleus 301, 302[+], 303[§+], 304

Radical formation by phenol oxidases 350
Radioactivity, specific 15
Raffinose 62, 63§
Randomization 18
Reserve starch 64, 65+
Resorcinol 111§+
Resveratrol 355§+
(−)-Reticuline 323, 324§+
(+)-Reticuline 326§, 327
Retinal 144§+
Retinol 144§+
Retinolic acid 144§+
Retrosine 264§
Rhamnose 58§, 69§+
Rhapontigenin 355§+
Rhodanid 197§+
Rhodopin 143§+, 144
Rhodovibrin 143§+
Ribitol 75§+, 216
4-Ribitylamino-5-aminouracil 216+, 217§+
Riboflavin 216+, 217§+
Riboflavin synthetase 216
Ribonucleic acid, methylated bases 212
Ribose 58§
Ribose-1-phosphate 61+
Ribose-5-phosphate 210
Ribulose 58§
Ricinine 253+, 254§+
Ricinolyl-CoA 88§+
Rotenone 362§, 363+
Rubredoxin 39
Rutaecarpine 181§+

Saccharopine 269+, 271§+
Salamander alkaloids 5, 161
Salicin 369§+
Salicylaldehyde 7§, 369§+
Salicylic acid 171+, 172§+
Salutaridine 323+, 324§+
Samandarine 161+, 162§
Sapogenins, steroid 159+
    acidic 152+
Sarcosine 188§, 189+
Schiff bases, see Azomethines 199+
Scleroproteins 382§+, 383
Sclerotin 381, 382§+
Scopine 261§
Scopolamine 261, 263§+
Scopoletin 345§+
Scopolin 345§+
Scoulerine 326§+, 327+
Secoanthraquinones 104+

Secondary metabolism
    general 3
    methods of investigation of 13
    regulation of 9
    significance of 7, 8
Sedamine 274§+
Sedoheptulose 58§
Sedum alkaloids 273+
Senecioyl CoA 244§+
Seneciphyllic acid 264+, 265§+
Sepia 336
Sequoyitol 77§+
Serine 233§+, 278§, 289, 292, 315
    as C_1-donor 46
Serine hydroxymethylase 46
Serotonin 187§+, 290§+
Sesquiterpenes 122, 129+
Shikimic acid 166§+
Sialic acid 96
Sideramine 373, 374
Siderochrome 373
Sideromycins 373, 374
Sinalbin 194§
Sinapic acid 50, 344§+, 345+, 349§, 366
Sinapyl alcohol 349§+
Sinigrin 194§, 195§+, 196+
β-Sitosterol 154+, 155§
Smilagenin 159
Solanidine 160§+
Sorbitol 75§+
Sorbose 58§, 75+
Sparteine 275§+, 276+
Spermidine 234, 235§+
Sphaerophysine 280§+
Sphingolipidases 96
Sphingolipids 95+
Sphingomyelin 94, 95§+
Sphingosine 95§+
Spinacin 282§
Spirilloxanthin 143§+
Spiroketal grouping 159
Spontaneous reactions 25
Squalene 127+, 128§+
Squalene diepoxide 153§+
Squalene epoxide 147§+, 149
Stachydrin 189§+
Stachyose 63§, 65+
Starch 64, 66+
Stearyl CoA 88§
Stephanine 334+, 335§+
Stercobilin 206§+, 207+
Stercobilinogen 206§+, 207+
Steroid alkaloids 159+

Steroid hormones 155[+]
Steroid hydroxylases 39
Steroid ring system 145[§]
Steroid sapogenins 159[+]
Sterols 153[+]
Stigmasterol 154[+], 155[§]
Stilbene carboxylic acid 356[§+]
Stilbene derivatives 355[+], 356[§+]
Streptamine 79[§+]
Streptidine 78[+], 79[§+]
Streptomycin 78[§], 80[+]
Streptose 70[§]
Stylopine 326[§+], 327
Suberosin 346[§]
Suberylarginine 158
Substitution theory of flavonoid biosynthesis 358
Succinic acid 119[§+], 121[§+]
Succinyl AMP 283[§+]
Succinyl CoA 119[+], 202, 203[§]
Succinyldiaminopimelic acid 269[+], 270[§+]
O-Succinylhomoserine 233
Sucrose 62, 63[§], 65[+]
Sucrose-6-phosphate 65[+]
Sugar 57
  abnormal 67
  methylation of carbon atoms in 69
  with branched carbon chain 69[+]
Sugar alcohols, aliphatic 76[+]
Sugar dicarboxylic acids 73[+]
Sulphate, see Sulphuric acid
Sulphatides 95[§+], 96[+]
Sulphide 224
Sulphite, see Sulphurous acid
Sulphocysteine 224[+], 225[§+]
Sulphonamides 215
Sulphonium compounds 234[+]
Sulphuretin 362[+], 363[§+]
Sulphuric acid 225[+], 226[§+]
Sulphurous acid 225[+], 226[§+]
Sweet clover 348
Syringin 349[§+]
Syringic acid 366[§+], 370

Tannin molecule, section of 361[§]
Taraxerol 151[§+]
Taurine 225[+], 226[§+]
Taurocyamine 226[§+]
Tazzetine 330[+], 331[§+]
TDP-sugar 62
Tea 212
Tenuazonic acid 247[+], 248[§+]
C$_{20}$-terpene 133[+]

Terpenes 124[+]
Testosterone 156[+], 157[§+]
Tetracyclines 106[+], 107[§+]
Tetrahydroanabasine 277[§+]
Tetrahydroberberine type 327[+]
Tetrahydrobiopterine 40
Tetrahydrofolic acid 46
Tetrahydroharman 296[§+]
Tetrahydroisoquinolines 322[+]
Tetrehydro-6-oxynicotinic acid 252[+], 254[§+]
Tetrahydroquinoline 300
Tetramethyldecanoic acid 115[§+]
Tetramethylundecanoic acid 115[§+]
Tetraoxooctahydropteridine 221[+], 222[§+]
Tetraoxopteridine isomerase 221
Tetrasaccharides 62
Tetraterpenes 122
Tetroses 57
Thalicarpine 327[+], 328[§]
Thebaine 323, 324[§+]
Theobromine 212[§+]
  degradation of 220
Theophylline 212
  degradation of 220
Thevetose 59[+]
Thiamine 237[+]
Thiamine diphosphate, see Thiamine pyrophosphate
Thiamine monophosphate 238[§+]
Thiamine monophosphate synthetase 238
Thiamine pyrophosphate 51[§+], 81, 82, 238, 239[§]
Thiamine pyrophosphokinase 238
Thiazine ring 375
Thiazolidine ring 375
Thiobinupharidine 135[§], 136[+]
Thiocyanates, see Rhodanid
Thioester 30
  bending 82
Thioether 41
  formation from acetylene derivatives 100
Thiokinases 30
Thiolases 123
Thiolurocanic acid 188[+], 285[§+]
Thiooxazolidone 195[§+]
Thiophorases 30
Thiosulphate 224[+], 225[§+]
Thiotaurine 226[§+]
Threonine 379
Thymidine monophosphate 251[+]
Thymine 46, 249[+], 250[§]
Tiglic acid 7[§], 247[+], 261, 266, 276
Tigloyl CoA 247[§+]

Tocopherols 341+, 342§+
Tocopheryl quinones 341+, 342§+
p-Toluoquinone 7§
Tomatidine 160§+
Toxoflavin 218§
Toyocamycin 218§+
Trachelanthamidine 264, 265§+
Transaminases 52
Transamination 52
Transcarboxylases 45
Transformylases 46
Transglycosidase, in streptomycin formation 80
Transglycosylases 65
Transglycosylation reactions 61, 64
Transketolases 237
  in sugars 58
Transmethylation 48
Transpeptidization, in formation of bacterial cell wall 380
Transphosphorylases 27, 28
Transphosphorylation 28
Tricarboxylic acid cycle 118
Trigylcerides 72+, 73+
Trigonelline 253
Trihydroxycinnamic acid 344§+, 345+, 346§+
Trihydroxycoprostane 155+, 156§+
Trihydroxycoprostanic acid 155+, 156§+
Trimethylamine 188§+, 285+
Trimethylarsine 50
Trioses 57
Trioxohexahydropteridine 221+, 222§+
Trisaccharides 62
Triterpenes 122
  acid sapogenins 152
  cyclic 147+
  pentacyclic ring system 145§
Tropane 261§
Tropane alkaloids 261+, 343
Tropic acid 18, 261, 263§, 343§+
Tropine 261§, 262§+, 263§
Tropine esterase 268
Tropinone 262§+
Tryptamine 187§+, 291§+, 296§, 298§, 313§+
Tryptophan 174+, 187§, 189§, 289§+, 291§, 293§, 294§, 313, 373
  degradation of 307
  shortening and elimination of side chain of 313
Tryptophanase 315
Tryptophan-5-hydroxylase 42
Typtophan pyrrolase 305
Typtophan synthetase 292

Tubercidin 218§+
Tuberculostearic acid 86§+
Tubocurarine 327+, 328§
Tubolosine 322§
Turicine 189§+
Turmerones 130
Tyramine 187§+, 318+, 319§+, 321, 330, 331§, 333§, 334
Tyramine methyl transferase 318
Tyrase 345
Tyrocidine A 381§
Tyrosinase (phenyl alanine hydroxylase), NIH shift 36, 40
  action of 337
  in xanthommatin synthesis 309
Tyrosine 41, 187§, 317§+, 318+
m-Tyrosine 338§

Ubiquinones 173§+
UDP-N-acetylglucosamine 68
UDP-N-acetylmuramyl-L-alynyl-D-glutamyl-L-lysyl-D-alanyl-D-alanine 380
UDP-amino sugar in higher plants 62
UDP-apiose 71§+
UDP-arabinose 72§+
UDP-galactose 72§+
UDP-galacturonic acid 72§+
UDP-glucose 70, 72§+
UDP-glucose dehydrogenase 33, 72
UDP-glucuronic acid 72§+
UDP-mycarose 70§+
UDP-rhamnose 69§+
UDP-streptose 71§+
UDP-sugar 61+
  in higher plants 62
UDP-uronic acids 62+
  in higher plants 62
UDP-xylose 72§+
Umbelliferone 345§, 347§+
Umbelliferone glucoside 347§+
Unedoside 137§+
Uracil 249, 250§, 251§+
Urea 219+, 220+, 251§+, 279§+
  methylated 220
Urea cycle 249
Urea indican 315§+
4-Ureido-5-carboxyimidazole 221§+
Uric acid 219+, 220§+
Uricase 219, 220, 221
Uridine diphosphate glucose 61§+
Uridine monophosphate 250§+
Uridine triphosphate 61§
Urocanic acid 286§+

Uronic acids 72+, 77
  conjugation with 64
Uroporphyrinogen-III-cosynthetase 203
Uroporphyrinogen-I-synthetase 203
Uroporphyrinogens 203§+
Ursolic acid 151§+
Usnic acid 109+, 111§+

Valeriana alkaloids 137§, 138+
Valine 240§, 246§+, 265§
Valinomycin 380
Vanillic acid 366§+, 367§+
Vanillin 366§+
Vanillyl alcohol 366§+
Veracevine 162§
Veratramine 162§
Veratric acid 276
Verbacose 65+
Verdoglobin, see Choleglobin
Viomycin 218§
Viridicatine 179§+
Viridicatol 179§+
Vitamin A 144§+
Vitamin B₁, see Thiamine
Vitamin B₂, see Riboflavin
Vitamin B₁₂ 50, 207§, 208+
Vitamin C, see Ascorbic acid 40
Vitamin D 163+, 164§+, 165
Vitamin E, see Tocopherols
Vitamin H, see Biotin

Vitamin K, see Menadione and Phylloquinones
Vitexin 360+, 361§

Wax, epicuticular 94
Wood, raw 349

Xanthine 212§+, 219, 220§+, 221§+
Xanthine-8-carboxylic acid 221+, 222+
Xanthine-8-carboxylic acid 221+, 222§+
Xanthine dehydrogenases 35, 219
Xanthine oxidases 35, 219, 221
  in degradation of histidine 287
  in degradation of histamine 288
Xanthocillin 318+, 319§+
Xanthommatin 309+, 310§+
Xanthone derivatives 104+, 105§+
Xanthophylls 141, 142+
Xanthopterin 214§+, 222§
  degradation of 222
Xanthosine monophosphate 210+, 211§+,
  305§+, 307§
Xanthurenic acid 305§+, 306, 307
Xylans 64
Xylitol 75§+
Xylodextrin 66+
Xylose 58§
Xylulose 58§

Zapotidin 282§
Zeacarotenes 141§+, 142
Zeatin 209
Zymosterol 153+, 154§+